AF531505

INSTRUMENTATIONS
AND
NANOSTRUCTURES

INSTRUMENTATIONS AND NANOSTRUCTURES

A.S. BHATIA

Foreword by

DR. S.M. ISHTIAQUE
Professor
Department of Textile Technology
Indian Institute of Technology, Delhi

from

DEEP & DEEP PUBLICATIONS PVT. LTD.
F-159, Rajouri Garden, New Delhi - 110 027

INSTRUMENTATIONS AND NANOSTRUCTURES

ISBN 978-81-8450-147-6

Typeset by THE COMPOSERS,
260 C.A. Apt., Paschim Vihar, New Delhi - 110 063

Printed in India at MAYUR ENTERPRISES
WZ Plot No. 3, Gujjar Market, Tihar Village, New Delhi - 110 018

NuTech Books are published by DEEP & DEEP PUBLICATIONS PVT. LTD.
F-159, Rajouri Garden, New Delhi - 110 027 • Phone : 25435369, 25440916
E-mail : ddpubs@gmail.com • ddpbooks@yahoo.co.in
Showroom :
2/13, Ansari Road, Daryaganj, New Delhi - 110 002 • Telefax : 23245122

Contents

Foreword

A good book is essential for an understanding of the fundamentals of a subject and development of ability to apply the knowledge in real life situations. This book covers the elements of instrumentations and nanostructures. The treatment is at a level commensurate with the understanding of research scholars, educationists and even by students who are known to learn about nano science and technology.

The text is very lucid and the illustrations add immeasurably to the value of the book. The book is well written using simple language, consistent with prevision and brevity.

I am sure that this book will prove quite useful for increasing knowledge on nano.

Dr. S.M. ISHTIAQUE
Professor,
Department of Textile Technology
Indian Institute of Technology, Delhi
Hauz Khas, New Delhi-110016

Preface

The essence of nanotechnology is the possibility to work at the atomic and molecular level, in the length scale between 1 and 100 nm, in order to produce and use materials and devices characterized by new properties and functions because of the small scale of their structure. Thus, the term "nanotechnology" relates to the sizes of structural elements.

In the past few years, intense scientific research has been done by researches at various leading laboratories. A profound understanding of the physics of nanostructure formation and the development of tools for controlling and turning geometrical parameters is a prerequisite for any significant success in nanotechnology. To achieve this goal it requires complementary studies including specially designed and growth experiments, theoretical modeling of the growth and structural and optical characterization on a nanoscale to allow significant progress in fabricating nanostructures suitable for devices.

The quality and the resolution of the devices have increased in parallel with the developments in scientific research. It is possible to fabricate and study these materials at very small (atomic) scales. Also the physical properties of the materials at such a small scale are observed to change dramatically in nanoscale as compared to their microscopic parts.

A number of methods based on physical principles have been developed for manufacturing and characterization of nanostructured materials.

In this book, I have tried to give in detail various techniques such as electron microscopy techniques, including Transmission Electron Microscopy (TEM) and Scanning Electron Microscopy (SEM), and scanning probe microscopy techniques, including Scanning Tunneling Microscopy (STM) and Atomic Force Microscopy (AFM) which are necessary to probe the physical characteristics of Quantum Dot (QD) structures.

The book is organized into seven chapters. In Chapter 1 a detailed study on electron microscopy, TEM, SEM, STEM, HRTEM is made, whereas in Chapter 2 SPM, AFM, Laser-microscopy, X-ray techniques, magnetic force microscopy etc. are given. Spectroscopy techniques are in Chapter 3. The details on this field are given in this chapter. Chapter 4 is about Lithography. This chapter provides an overview of lithography and its characterization.

Chapter 5 is on nanostructures. It encounters a large number of techniques and results in the engineering of nanostructures. Nanomaterials, nano particles, their fabrication, characteristics etc. are all detailed in this chapter. Chapter 6 tells about nanolayers while Chapter 7 is on nanopolymers.

New Delhi

A.S. BHATIA

Microscopy-I

ELECTRON MICROSCOPE

An **electron microscope** is a type of microscope that uses electrons instead of photon (due to shorter wavelength of electrons) to illuminate and create an image of a specimen. It has much higher magnification and resolving power than a light microscope, with magnification up to about million times. Unlike a light microscope, which used glass lenses to focus light, the electron microscope uses electrostatic and electromagnetic lenses to control the illumination and imaging of the specimen. Electron microscope by using electrons instead of photons, allows to observe matter with atomic resolution.

There are two types of electron microscope: *the Scanning Electron Microscope (SEM)* that scans an electron beam over the surface of an object and measures how many electrons are scattered back, and *the Transmission Electron Microscope (TEM)* that shoots electrons through the sample and measures how the electron beam changes because it is scattered in the sample.

Using electron beams however requires working in a vacuum environment. All electron microscope work under low pressures and in high vacuum chambers to avoid scattering the electrons in the gas. In environmental electron microscopes, differential pumping systems are used to actually have gases present by the sample together with the electron beam.

ELECTRON MICROSCOPY

Electron microscopy is an extremely important technique for the examination and analysis of both the surface and subsurface of nanostructured systems (SEM) as well as the bulk structure of thin samples, usually averaged through thickness (TEM and STEM). The state-of-the-art electron microscopy technique is the single, most powerful method for studying the full range of advanced materials.

GENERAL CONSIDERATION FOR IMAGING

The general purpose of microscopy is to resolve finer details than are discernible with the naked eye. which is 0.1-0.2 nm at the near point of vision, 250 nm. For nanostructured materials resolution of details down to a level of 100nm and below is required. In transforming an object to an image, three methods are available, as shown in Figure 1.1.

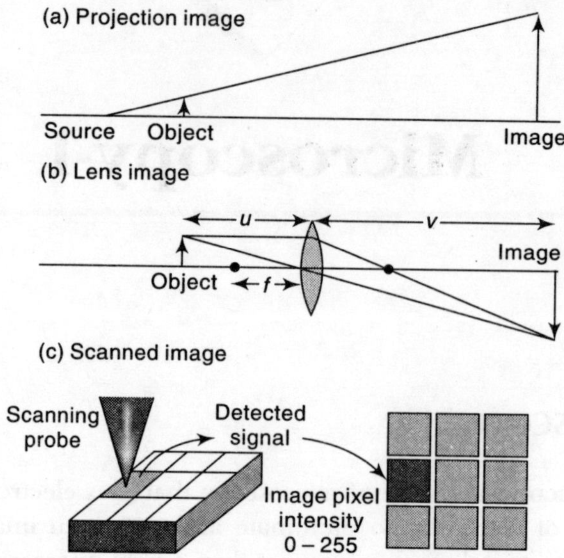

Fig. 1.1 *Formation of (a) a projection image, (b) a magnified image by a lens and (c) a scanned image.*

Analytical and Imaging Techniques

Before using a particular characterization technique it is essential to consider what information is required about a sample and at what resolution. This information can be as:

- morphology (the microstructural or nanostructural architecture):
- crystal structure (the detailed atomic arrangement in the chemical phases contained within the microstructure):
- chemistry (the elements and possibly molecular groupings present);
- electronic structure (the nature of the bonding between atoms).

In addition, *information of the mechanical, thermal or electronic properties of the sample may also be required.* However, many of these properties will arise as a direct consequence of the structural factors outlined above and, ultimately, we need to establish a correlation between nanostructure and properties and determine how both these factors depend on the processing conditions employed as input to computer-based modelling procedures, which may then allow the prediction of nanostructure property-processing relationships.

Having decided on the type of information required, the next question is, from what sample volume or area should this information be obtained? The resolution of a particular technique may be conveniently divided into lateral or spatial resolution (i.e., from what area on a sample surface or from what volume within a sample a particular signal originates?) and depth resolution (i.e., from how far techniques may be classified as either surface or bulk (volume) techniques. Different types of nanostructure (e.g., films, dots, particles) require different combinations of lateral and depth resolutions.

- A projection image formed in parallel; e.g., field ion microscopy (FIM).
- A lens image formed in parallel; e.g., TEM and light microscopy. Here a lens of focal length f will form an image of an object at a magnification v/u, where u and v are the object and image distances, respectively. The relationship between these parameters is given approximately by the thin lens formula

$$\frac{1}{f} = \frac{1}{v} + \frac{1}{u}$$

  from which the magnification v/u can be obtained.
- A scanned image formed in serial; e.g., scanning electron microscopy (SEM), confocal laser scanning microscopy (CLSM), scanning tunnelling microscopy (STM) and scanning force microscopy (SFM).

Signals used to form an image can be *reflected* (e.g., reflected light microscopy of backscattered electrons in SEM), *transmitted* (e.g., transmitted or polarized light microscopy, or electrons in TEM and scanning TEM) or *stimulated* (e.g., secondary electrons, cathodoluminesence, X-ray emission or electron beam induced current (EBIC) in SEM, tunneling current in STM or force in SFM). The concept of picture elements (pixels) is applicable to all images collected in either serial or parallel. Each pixel in the image is defined by x and y coordinates and may be assigned an intensity value; e.g., in terms of grey levels a number between 0 and 225, allowing images to be stored digitally on a computer. Imaging techniques capable of a 3D representation of the object (e.g., tomographic techniques) involve 3D volume elements (voxels). Digital images may be processed to reduce noise (e.g., by use of a framestore), alter contrast (e.g., by use of a non-linear intensity scalling) or analysed and quantified mathematically.

It is possible in both SEM and TEM to make radiation more monochromatic, thereby minimizing chromatic aberrations. Recently, computer-controlled correction of spherical aberration in electromagnetic lenses in SEM, TEM and STEM; aberration-corrected STEM has produced sub-angstrom probes for imaging.

In many respects, electron optics may be regarded as analogous to light optics; the main differences may be summarized as follows. Firstly, the wavelength of accelerated electrons is very much smaller than the wavelength of visible or UV photons, which implies a much greater resolution for electron imaging. Secondly, electrons interact much more strongly with matter than photons, thus all optical paths in an electron microscope must usually be under a vacuum of at least 10^{-4} Pa. However a major benefit for nanostructure is that even though the signal is derived from a small amount of material, it will be relatively strong compared to that derived using other incident radiations. Thirdly, electrons are charged particles and may thus be focused by magnetic or electric fields. Furthermore, this focused electron beam may be scanned relatively easily using electrostatic fields.

The electron beam may be produced by thermionic emission, where thermal energy is used to overcome the surface potential barrier (work function) of a solid source thus allowing extraction of electrons from the conduction band of an emitter, or by field emission, in which an extremely high electric field is employed to reduce the surface potential barrier of an emitter. The finite width of the barrier allows quantum tunnelling of electrons out of the source at room temperature (cold field emission), although higher temperatures have been used more recently, in Schottky or thermally assisted field emission, (which requires a lower field strength). A typical thermionic electron gun employs a tungsten cathode filament which is resistively heated to 2800 K in a vacuum of 10^{-4} Pa to give the electrons are collimated and focused to a beam of typically 50 μm in diameter. The anode potential is typically 1-20kV in the SEM and 100-200kV in the TEM.

Increased brittleness of the source β allows either the use of higher currents (the current density per unit solid angle) which improve sensitivity and image contrast, or smaller probe areas, which yield higher spatial resolution. The brightness may be increased by using a lanthanum hexaboride (LaB_6) filament that possesses a lower work function than tungsten and a smaller source size (~ 1μm), giving an order or magnitude increase in β. Alternatively, a field emission gun (FEG), which employs a high field strength surrounding a pointed tungsten cathode of radius 100 Å, gives a source size as low as 5nm an a brightness increase of 10^4 over a thermionic tungsten filament. Both sources possess a longer lifetime than tungsten filaments but require a higher vacuum in the gun region. These source also give a much reduced energy spread of the emitted, leading to improved resolution as a result of reduced chromatic aberration in spectroscopic analysis.

Figure 1.2(a) summarises the various signals produced as a result of electron-specimen interactions in SEM and shows the calculated volumes and penetration depths which give rise to the different signal from thick specimens. The *interaction volume* is defined as the volume within which 95 % of the electrons are brought to rest by scattering, and has a characteristic teardrop shape (as shown in the figure). The depth and lateral width of electron penetration in the specimen are roughly proportional to V^2 and $V^{3/2}$, respectively, where V is the accelerating voltage. Materials of high (average) atomic number will exhibit reduced electron penetration depths and increased lateral spread relative to those of lower atomic number. For comparison.

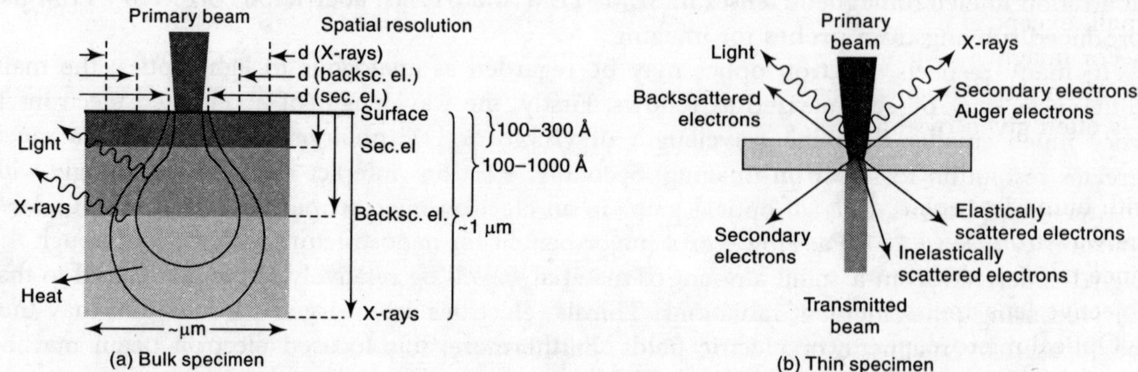

Fig. 1.2 *The beam-specimen interaction in (a) a thick specimen and (b) a thin specimen.*

Figure 1.2 (b) shows the signals produced in thin specimens appropriate for TEM imaging. Besides the primary electrons, there are various types of emitted electrons which leave the surface of a bulk sample and may used for imaging or analysis, as summarized in the following sections.

1. Secondary Electrons

The term 'secondary electrons' is used in general to describe any electrons that escape from the specimen with kinetic energies below about 50 eV. They are most likely to arise from ionized electrons previously associated with atoms close to the surface of the solid which have gained a small amount of kinetic energy and escaped. Alternatively they may be primary electrons which have lost nearly all their energy though scattering and have reached the surface. Secondary electrons are extremely abundant and the secondary electron yield, δ (the number emitted per primary electron), is dependent on the accelerating voltage and can even exceed 1, hence they are extensively used for imaging in SEM.

2. Backscattered Electrons

Some primary electrons undergo large deflections and leave the surface with little change in their kinetic energy; these are termed backscattered electrons (BSE). The backscattered electron yield, η, is almost independent of accelerating voltage and is small compared with δ, but backscattered electrons are also used for imaging in SEM as η is strongly dependent on Z (unlike δ) owing to the fact that they arise from Rutherford backscattering from the nucleus. Owing to this nuclear interaction, backscattered electron imaging can be used to differentiate phases of differing average atomic number.

3. Auger electrons and emitted X-rays

Atoms which have undergone inner shell ionization and have been promoted into an excited state by the primary electron beam relax when electrons from higher energy levels drop into the vacant inner shells. This process results in the release of the excess energy between the electron energy levels involved in the transition (ΔE), producing low-energy (100-1000 eV) Auger electrons, or X-rays, or visible photons of wavelength $\lambda = /hc/\lambda E$. The Auger yield is small, except in the case of light elements, and is extremely useful for surface analysis

For imaging of nanoscale objects, optical microscopy has limited resolution since the objects are often much smaller that the wavelength of light. The achievable resolution for a wavelength λ is often given by diffraction limit d as

$$d = 0.61 \quad \textit{(Diffraction limit)}$$

with numerical aperture NA, which can be approximated by the largest angle of incidence α of the wavefront toward the sample, $NA \approx \alpha$.

Since $\alpha << 1$, therefore $\sin(\alpha) \approx \tan(\alpha) \approx (\alpha)$ and hence $NA \approx (a) \approx r_d / r_{wd}$ where r_a is radius of the objective lens aperture and r_{wd} the working distance.

Optical microscopes can often a resolution of about $d = 0.5$ nm For nanoscale resolution this is not sufficient to distinguish; for instance a single nanotube from two adhering to each other, since they have diameters of less than 100 nm.

The figure 1.3 below gives an overview typical magnification achievable by the different electron microscopers compared to an light microscope.

Electron optical system use electrical and magnetic fields to control the electron beam. Although the law of refraction in optics is exchanged with the Lorentz force in electrodynamics, the electron optical system has similar diffraction limits as optical systems, since they depend

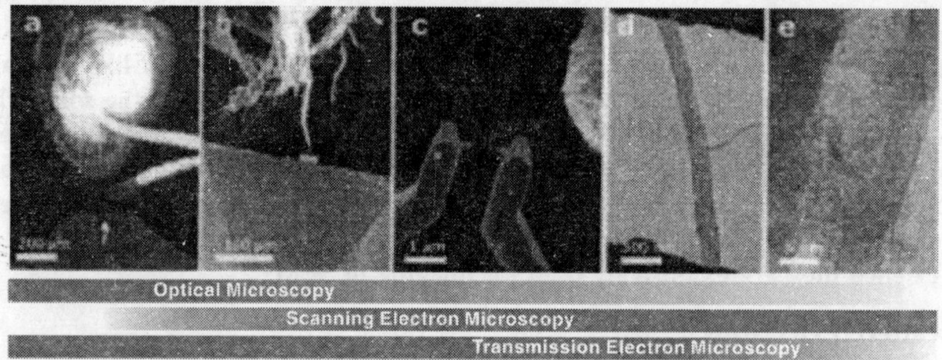

Fig. 1.3 *The different methods for microscopy cover a range of magnification roughly indicated by the in the figure. The resolution of optical microscopy is limited to about 200 nm.(a) SEM image of the head of an ant facing a microfabricated chip with a pair of microfabricated grippers. The grippers are barely vsibly at the tip of the arrow. (b) SEM image of a gripper approaching a large bundle of carbon nanotubes. (c) Closeup in SEM of the gripper and nanotubes. (d) TEM image of a carbon nanotube suspended between two grippers.(e) TEM closeup of the shells of carbon atoms in a carbon nantoube.*

on the wave nature of the electron beam.

One can achieve a considerable improvement in resolution with instruments such as the transmission electron microscope and the scanning electron microscope that use electrons with De Broglie wavelength much smaller than that of visible light. The De Broglie wavelength λ of an electron with momentum p is

$$\lambda = \frac{h}{p} = \frac{h}{\sqrt{2m_e E_b}}, \quad \textit{(De Broglie wavelength)}$$

where h is Plancks constant. The electron has rest mass m_e and energy $E_e = m_e c^2 = 511$ keV.

If an electron with charge qe is accelerated from rest by an electrical potential U, to the electron beam energy $E_b = q_e U$, it will have a wavelength of 1 nm 1 eV decreasing to 1 pm at 100 keV where it will be travelling with 50 % the speed of light.

Types of Electron Microscopes

Transmission Electron Microscope (TEM)
Scanning Electron Microscope (SEM)
Reflection Electron Microscope (REM)
Scanning Transmission Electron Microscope (STM)

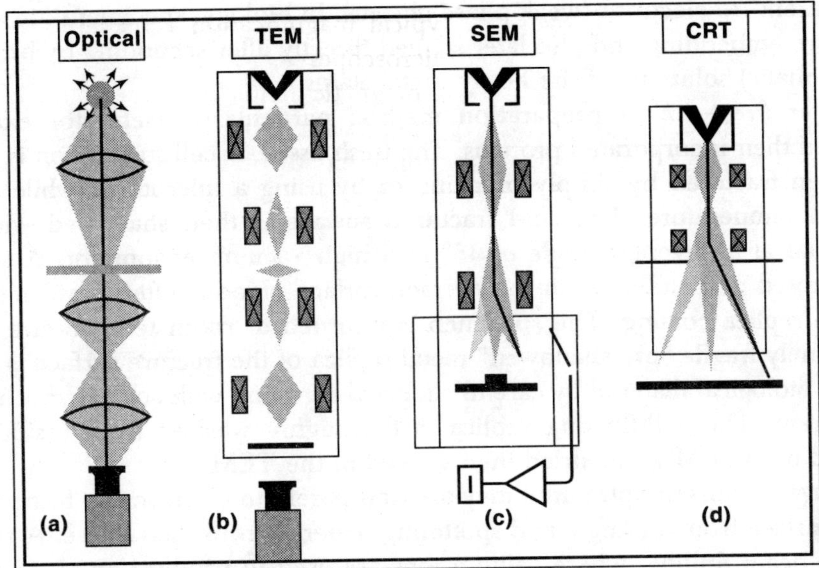

Fig. 1.4 *A very simple sketch of a Transmission Electron Microscope (TEM) and Scanning Electron Microscope (SEM) compared to an optical transmission microscope and a cathode ray tube (CRT) TV screen- both system have many things on common with the electron microscope. The optical microscope uses lenses to control the lights pathway through the system and is in many ways built up like a TEM- only the TEM uses electromagnetic lenses to direct the beam of electrons. The CRT uses electromagnetic lenses as the TEM and SEM to control the eletron beam, and generates an image for the viewer by scanning the beam over a fluorescent screen in the same way the a SEM generates an image by scanning the electron beam over a small sample.*

Sample Preparation

Materials to be viewed under an electron microscope may require processing to produced a suitable sample. The technique required to prepare a sample for electron microscopy, varies depending on the specimen and the analysis required. These are as;

Cryofixation—freezing a specimen so rapidly, to liquid helium temperatures, that the water forms vitreous (non-crystalline) ice. This preserves the specimen in a solution state. With the development of cryo-electron microscopy of vitreous sections (CEMOVIS), it is now possible to observe virtually any biological specimen close to its native state.

Dehydration—replacing water with organic solvents such as ethanol or acetone.

Embedding—infiltration of the tissue with a resin (such as araldite) or an epoxy for sectioning. After this embedding process begins, the specimen must be polished to a mirror-like finish using ultra-fine abrasives. The polishing process must be done accordingly, or it may lead to scratches imposing on the image quality.

Sectioning—produces thin slices of specimen, semitransparent to electrons. These can be cut on an ultramicrotome with a diamond knife to produce very thin slices. Glass knives are also used.

Staining—uses heavy metals such a lead, uranium or tungsten to scatter imaging electrons and thus give contrast between different structures, since many (especially biological) materials are nearly "transparent" to electrons (weak phase objects). In biology, specimens are usually stained "en bloc" before embedding and also later stained directly after sectioning by brief exposure to aquous (or alcohalic) solutions of the heavy metal stains.

Freeze-fracture or freeze-etch—a preparation method particularly useful for examining lipid membranes and their incorporated proteins. The fresh tissue or cell suspension is frozen rapidly (cryofixed), then fractured by simply breaking or by using a microtome while maintained at liquid nitrogen temperature. The cold fractured surface is then shadowed with evaporated platinum of gold at an average angle of 45° in a high vacuum evaporator. A second coat of carbon, evaporated perpendicular to the average surface plane is often performed to improve stability of the replica coating. The specimen is returned to room temperature and pressure, then the extremely fragile "pre-shadowed" metal replica of the fracture surface is released from the underlying biological material by careful chemical digestion with acids, hypochlorite solution of SDS detergent. The still-floating replica is thoroughly washed from residual chemicals, carefully fished up on EM grids, dried then viewed in the TEM.

Ion Beam Milling—It thins samples until they are transparent to electrons by firing ions (typically argon) at the surface from an angle and sputtering material from the surface. A subclass of this is Focused ion beam milling, where gallium ions are used to produce an electron transparent membrane in a specific region of the sample, for example through a device within a microprocessor.

Conductive Coating—An ultrathin coating of electrically conducting material, deposited either by high vacuum evaporation or by low vacuum sputter coating of the sample. This is done to prevent the accumulation of static electric fields at the specimen due to the electron irradiation required during imaging. Such coating include gold, gold/palladium, platinum, tungsten, graphite etc. and are specially important for the study of specimens with the scanning electron microscope. When an osmium coater is used, a layer far thinner is possible.

Disadvantages of Electron Microscopes

Electron microscopes are dynamic rather than static in their operation; requiring extremely stable high-voltage supplies, extremely stable currents to each electromagnetic soil/lens, continuously-pumped high-/ultra-high vacuum systems, and a cooling water supply circulation through the lenses and pumps. As they are very sensitive to vibration and external magnetic fields. Microscopes aimed at achieving high resolutions must be housed in building (sometimes underground). Newer generations of TEM operating at lower voltages (around 5kV) do not have stringent voltage supply, lens coil current, cooling water or vibration isolation requirements.

The samples have to be viewed in vacuum, as the molecules that make up air scatter the electrons. Recently hydrated samples are imaged while using a scanning electron microscope.

Scanning electron microscopes are usually image conductive or semiconductive materials best. Non-conductive materials can be imaged by an environmental scanning electron microscope. A common preparation technique is to coat the sample with a several-nanometer layer of conductive material, such as gold, from a spouting machine.

The samples have to be prepared in many ways to give proper detail, which may result in *artifacts* purely the result of treatment. This gives the problem of distinguishing artifacts from material, particularly in biological samples. By comparing various preparation techniques it is found that all of them produce similar artifacts, So electron microscopy features correlate with living cells. In addition, higher-resolution work on comparing with X-ray crystallography confirms, the validity of this technique.

TRANSMISSION ELECTRON MICROSCOPE

Transmission Electron Microscope (TEM) is an imaging technique whereby a beam of electrons is transmitted through a specimen, then an image is formed, magnified and directed to appear either on a fluorescent screen or layer of photographic film (see electron microscope), or to be detected by a sensor such as a CCD camera. The original form of energy microscopy, Transmission electron microscopy (TEM) involves a high voltage electron beam emitted by a cathode, usually a tungsten filament and focused by electrostatic and electromagnetic lenses. The electron beam that has been transmitted through a specimen that is in part transparent to electrons carries information about the inner structure of the specimen in the electron beam that reaches the imaging system of the microscope. The spatial variation in this information (the "image") is then magnified by a series of electromagnetic lenses until it is recorded by hitting a fluorescent screen, photographic plate, or light sensitive sensor such as a CCD (charge-coupled device) camera. The image detected by the OCD may be displayed in real time on a monitor or computer.

The basic TEM instrumentation is summarized in Figure 1.5. The electron gun is usually thermionic tungsten of LaB_6, however FEGs are becoming increasingly common. The accelerating voltage is considerably higher than in an SEM and is typically 100–400 kV, and although a number of specialist high-voltage instruments are designed to operate at 1 MV and above. The benefits of high voltage include increased imaging resolution, due to the decreased electron wavelength, and also increased penetration and thus the ability to study thicker samples. Two or more condenser lenses demagnify the probe to typically 1 μm in diameter, although this can be reduced via use of a condenser–objective nanoprobe system. The condenser excitation controls the beam diameter and convergence. The first condenser (Cl) controls the demagnification of the crossover (the spot size), whilst the second condenser (C2) controls the size and convergence of the probe at the specimen, and hence the area of the sample that is illuminated. The specimen must be no more than a few hundred nanometres in thickness, and is usually in the form of a 3 mm diameter disc. The specimen is located between the pole pieces of the objective lens. The combination of the objective lens and the projector lens system provides an overall magnification of around 10^6. The *selected are diffraction* (SAD) *aperture* allows selection of a minimum sample area of ~0.1 μm diameter for electron diffraction. Smaller areas (down to a few nanometres in diameter) can be selected using a focused probe rather than an aperture this is convergent beam electron diffraction (CBED).

TEM specimens are normally in the form of an ultrathin disc, and prepared by cutting mechanical polishing or chemical dissolution, followed by electropolishing (for conducting materials), chemical polishing (for semiconductors and ceramics) or ion beam milling, includes focused ion beam (FIB) machining. Powders may be ground and dispersed onto an amorphous support film on a circular metal grid.

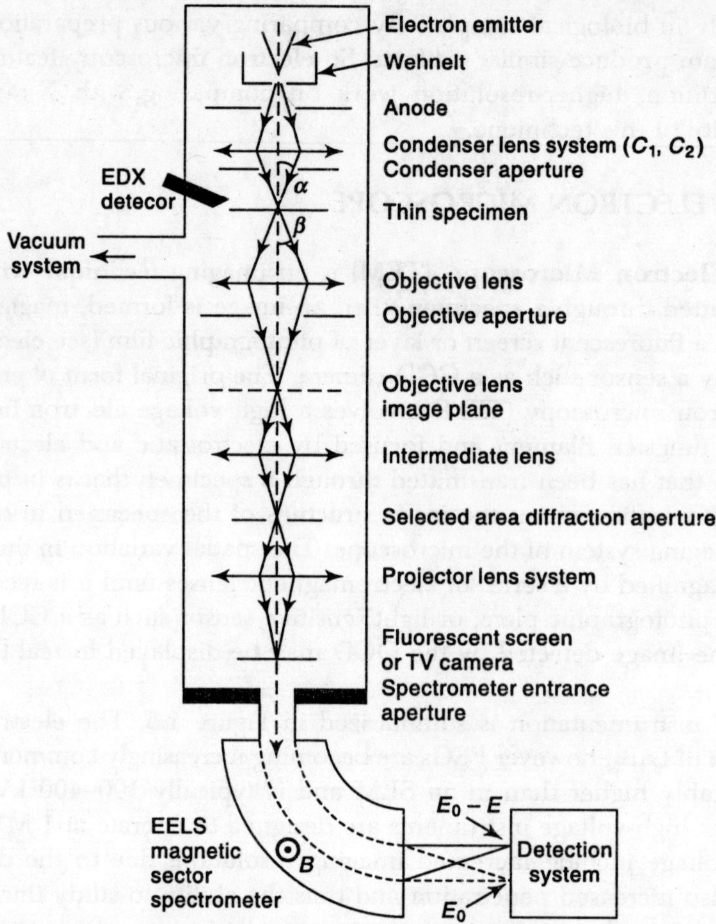

Fig. 1.5 *Layout of an analytical transmission electron microscope*

Imaging

A TEM image is a two dimensional projection of the internal structure of a material. There are three basic contrasts mechanisms which contribute to all TEM images.

Mass thickness contrast

During transmission through a thin TEM sample, electrons undergo a range of scattering processes which change their energy and angular distribution. The intrinsic bore of the microscope column prevents electrons that have been scattered through an angle greater than ~10^{-2} mrad for contributing to the image. In the presence of a sample, regions that are thicker or of higher density will scatter the electrons more strongly and hence more electrons well be scattered through an angle greater than 10^{-2} mrad making these areas appear darker in the image. This simple mass–thickness contrast is exhibited by all specimens, whether amorphous or crystalline.

Biological samples are often deliberately stained with a heavy metal such as osmium or uranium in order to increase the image contrast from the structural features into which the stain becomes incorporated.

Contrast may be increased by placing aperture in the back focal plane of the objective lens, which will further limit the angular range of electrons allowed to contribute to the image. If this objective aperture is centred on the optic axis, then in the absence of a specimen the image is bright, hence this is termed bright-field imaging (Figure 1.6).

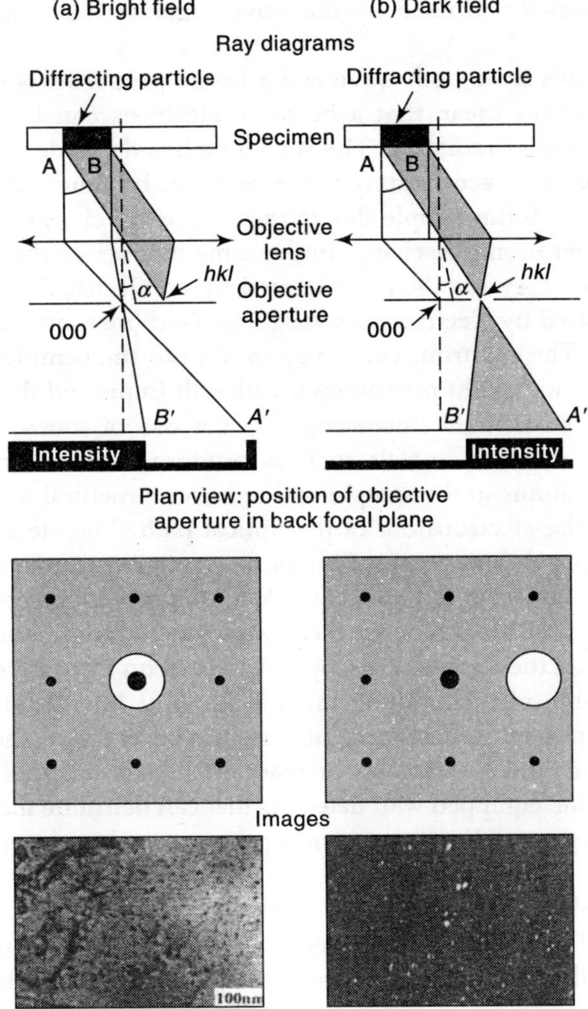

Fig. 1.6 *Tem ray diagrams for the formation of (a) bright-field and (b) dark-field images together with some typical images of carbide precipitates in a vanadium steel.*

Resolution of the Tem is limited primarily by spherical aberration, but a new generation of aberration correctors have been able to partially overcome spherical aberration to increase resolution. Software correction of spherical aberration for the High Resolution TEM (HRTEM) has allowed the production of images with sufficient resolution to show carbon atoms in diamond separated by only 0.89 angström and atoms in silicon at 0.78 angstöm at magnifications of 50 million times. The ability to determine the positions of atoms within materials has made the HRTEM an important tool for nano-technologies.

Background

In the past, light microscopes were of the used for imaging. However, the maximum resolution of the image obtained was determined by the wavelength of the photons used to probe the sample.

Like all matter, electrons have both wave and particle properties (Louis-Victor de Brogile), and their wavelike properties mean that a beam of electrons can in some circumstances be made to behave like a beam of radiation. The wavelength is dependent on their energy, and so can be tuned by adjustment of accelerating fields, and can be much smaller than that of light, yet they can still interact with the sample due to their electrical charge. Electrons are generated by a process known as thermionic discharge in the same manner as the cathode in a thiode ray tube, or by field emission; they are then accelerated by an electric field (measured in V/m, or volts per meter) and focused by electrical and magnetic field (measured in A/m, or amperes per meter) onto the sample. The electrons can be focused onto the sample providing a resolution far better than is possible with light microscope, and with improved depth of vision. Details of a sample can be enhanced in light microscopy by the use of stains; similarly with electron microscopy, compounds of heavy metals such as osmium, lead or uranium can be used to selectively deposit heavy atoms in the sample and enhance structural detail, the dense nuclei of the heavy atoms scatter the electrons out of the optical path. The electrons that remain in the beam can be detected using a photographic film, or fluorescent screen among other technologies. So areas where electrons are scattered appear dark on the screen, or on a positive image.

The capabilities of the TEM can be further extended by additional stages and detectors, sometimes incorporated on the same microscope. An **electron cryomicroscope** is a TEM with a specimen holder capable of maintaining the specimen at liquid nitrogen or liquid helium temperatures. This allows imaging specimens prepared in vitreous ice, the preferred preparation technique for imaging individual molecules or macromolecular assemblies.

An analytical TEM is one equipped with detectors that can detrmine the elemental composition of the specimen by analysing itx X-ray spectrum or the energy-loss spectrum, of the transmitted electron.

Modern research TEMs include aberration correctors, to reduce the amount of distortion in the image, allowing information on features on the scale of 0.1 nm to be obtained at magnifications of 50 million times. Monochromators are also used which reduce the energy spread of the incident electron beam to less than 0.15 eV.

The transmission electron microscope (TEM) has been used for many years to investigate the morphology and atomic structure of microelectronic device structures, while their analytical properties–electronic structure and chemical composition and bonding–have been the subject of surface-sensitive, medium-resolution technique: ellipsometry for oxide thickness and dielectric determinations, deep-level transient spectroscopy and capacitance-voltage measurements for defect electronic structure, X-ray diffraction for crystal strain, and secondary (SIMS) for composition. A device structure imaged using a Tem could be identified by inference with such analytical properties.

The transmission electron microscope (TEM) has been used for many years to investigate the morphology and atomic structure of microelectronic device structures, while their analytical properties–electronic structure and chemical composition and bonding–have been the subject of surface-sensitive, medium-resolution technique: ellipsometry for oxide thickness and dielectric determinations, deep-level transient spectroscopy and capacitance-voltage measurements for defect electronic structure, X-ray diffraction for crystal strain, and secondary (SIMS) for composition. A device structure imaged using a Tem could be identified by inference with such analytical properties.

Application of the TEM

The TEM is used in both material science/metallurgy, the biological science and semiconductor processing. The specimens are very thin and able to withstand the high vacuum present inside the instrument.

For biological specimens, the maximum specimen thickness is roughly 1 miocrometre. To withstand the instrument vacuum, biological specimens are typically held at liquid nitrogen temperatures after embedding in vitreous ice, or fixated using a negative staining material such as uranyl acetate or by plastic embedding. Typical biological applications include tomographic reconstructions of small cells or thin sections of larger cells and 3-D reconstructions of individual molecules via Single Particle Reconstruction.

In material science/metallurgy the specimens tend to be naturally resistant to vacuum, but must be prepared as a thin foil, or etched so some portion of the specimen is then enough for the beam to penetrate. Preparation techniques to obtain as electron transparent region include ion beam milling Fig. 1.7 and wedge polishing. The focused ion beam (FIB) is a relatively new technique to prepare thin samples for TEM examination from larger specimens. Because the FIB can be used to micro-machine samples very precisely, it is possible to mill very thin membranes from a specific area of a sample, such a semiconductor or metal. Materials that have dimensions small enough to be electron transparent, such as powders or nanotubes, can be quickly produced by the deposition of a dilute sample containing the specimen onto support grids. The suspension is normally a volatile solvent, such as ethanol, ensuring that the solvent rapidly evaporates allowing a sample that can be rapidly analysed.

Fig. 1.7 *SEM image of a thin TEM sample milled by FIB. The thin membrane is suitable for Tem examination; however, at ~300-nm thick, it would not be suitable for high-resolution Tem without further milling.*

The imaging techniques explained below are particularly important in materials science. Faults in crystal affect both the mechanical and the electronic properties of materials, so understanding how they behave gives a powerful insight. By careful selecting the orientation of the sample, it is possible not just to determine the position of defects but alto to determine the type of defect present. If the sample is oriented so that one particular plane is only slightly tilted away from the strongest diffracting angle (known as the Bragg Angle), any distortion of the crystal plane that locally tilts the plane to the Bragg angle will produce particularly strong contrast variations. However, defects that produce only displacement of atoms that do not tilt the crystal to the Bragg angle (i.e. displacements parallel to the crystal plane) will not produce strong contrast.

Furthermore, the HRTEM technique allows the direct observation of crystal structure and therefore has an advantage over other methods in that there is no displacement between the location of a defect and the contrast variation caused in the image. However, it is not always possible to interpret the lattice images directly in terms of sample structure or composition. This is because the image is sensitive to a number of factors (specimen thickness and orientation, objective lens defocus, spherical and chromatic aberration), and although quantitative interpretation of the contrast shown in lattice images is possible, it is inherently complicated and may require extensive simulation of the images. Computer modeling of these images has added a new layer of understanding to the study of the crystalline materials.

Imaging in the TEM

The contrast in a TEM image is not like the contrast in a light microscope image. A crystalline material interacts with the electron beam mostly by diffraction rather than absorption, although the intensity of the transmitted beam is still affected by the volume and density of the material through which it passes. The intensity of the diffraction depends on the orientation of the planes of atoms in a crystal relative to the electron beam; at certain angles the electron beam is diffracted strongly from the axis of the incoming beam, while at other angles the beam is largely transmitted. Modern TEMs are often equipped with specimen holders that allow the user to tilt the specimen to a range of angles in order to obtain specific diffraction conditions, and apertures placed below the specimen allow the use to select electrons diffracted in a particular direction.

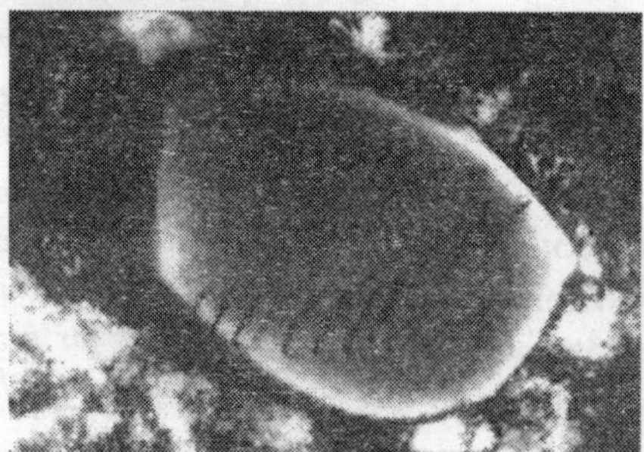

Transmission Electron Micrograph of Dislocations

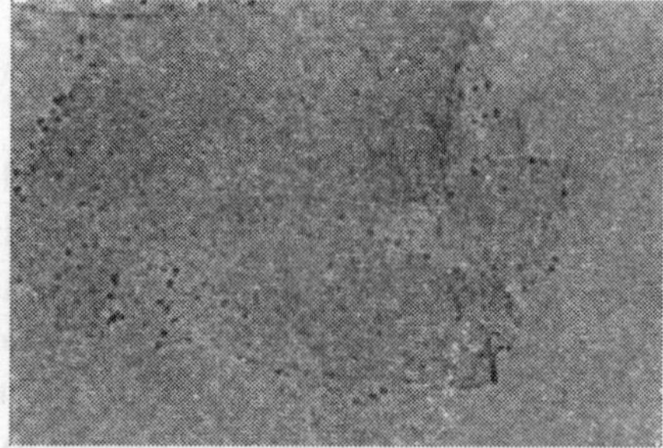

Transmission Electron Micrograph of a cobalt catalyst (darker spots) supported on coal carbonized at 850 °C and added by ion-exchange technique

Fig. 1.8

A high-contrast image can therefore be formed by blocking electrons deflected away from the optical axis of the microscope by placing the aperture to allow only unscattered electrons through. This produces a variation in the electron intensity that reveals information on the crystal structure, and can be viewed on a fluorescent screen, or recorded on photographic film or captured electronically.

This technique (known as *Bright Field* or *Light Field*) is particularly sensitive to extended crystal lattice defects in an otherwise ordered crystal, such as dislocations. As the local distortion of the crystal around the defect changes the angle of the crystal plane, the intensity of the scattering will vary strong the defect As the image is formed by the distortion of the crystal planes around the defect, the contracts in these images does not normally coincided exactly with the defect, but is slightly to one side.

It is also possible to produce an image from electrons deflected by a particular crystal plane. By either moving the aperture to the position of the deflected electrons, or tilting the electron beam so that the deflected electrons pass through the centred aperture, an image can be formed of only deflected electrons, known as *Dark Field* image.

In the most powerful diffraction contrast TEM instrument, crystal structure can also be investigated by High Resolution Transmission Electron Microscopy (HRTEM), also known as phase contact imaging as the images are formed by due to difference in phase of electron waves scattered through a thin specimen.

Resolution of the HRTEM is limited by spherical and chromatic aberration, but a new generation of aberration correctors has been able to overcome spherical aberration. Software correction of spherical aberration has allowed the production of images with sufficient resolution to show carbon atoms in diamond separated by only 0.89 ångströms (89 pm) and atoms in silicon at 0.78 ångström (78 pm) at magnifications of 50 million times. Improved resolution has also allowed the imaging of lighter atoms that scatter electrons less efficiently–lithium atoms have been imaged in lithium battery materials. The ability to determine the positions atoms within materials has made the HRTEM an indispensable tool for nanotechnology including heterogeneous catalysis and the development of semiconductor devices for electronics and photonics.

HIGH RESOLUTION TRANSMISSION ELECTRON MICROSCOPY (HRTEM)

The electron microscope was initially conceived as a response to the limitations of light microscopes for imaging very small objects shortly after de Brogile's discovery that electrons could be treated theoretically as waves. The resolution of a microscope is the minimum distance at which two objects may be distinguished and its theoretical maximum is proportional to the wavelength of the radiation used. Hence, the maximum theoretical resolution of a light microscope is of the order of a few hundred nanometers (nm). The wavelength of an electron is related to its energy such that an electron accelerated to 100 kilo-electron Volts (keV) will have a wavelength, and a maximum theoretical resolution, of the order of a few picometres (pm), about 100 times small than the diameter of an atom. The potential of the electron microscope to attain such sub-atomic resolution has not yet been fulfilled because of the considerable practical difficulties involved in microscope construction. However, atomic-scale resolution is attainable using modern instruments.

As well as this imaging capability other important advantages of using electron beams were soon recognised. Electrons interact strongly with atoms and can be employed in crystallographic diffraction experiments in the Transmission Electron Microscope (TEM). Unlike X-rays, electrons are charged and can be directed and focused easily using magnetic and electric fields. This allows electron diffraction to be performed down to the nano-scale to extract crystallographic information from tiny areas of a sample, such as a single crystallite or part of a crystallite. The resulting Selected Area Electron Diffraction (SAED) patterns consists of a 2-D array of "spots" for a single crystal which can be related to a slice through the reciprocal lattice of the diffracting crystal. Inter-planar distances and the angles between planes are thus obtained. If any crystals are selected, the angular information is lost and the pattern resembles a series of concentric rings from which inter-planar distances only can be calculated.

The Transition Electron Microscope consists of a vertical column down with the electron beam passes from the electron source at the top, through the specimen and down to the bottom of the column where the image is formed. Figure 1.9 contains a simplified representation of the TEM. The column is half under vacuum by system of a high performance pumps to reduce scattering of the electron beam by gas atoms. Electromagnetic coils, known as lenses, are positioned around the column along its length and are used to modify the motion of the electrons in the beam, either concentrating or spreading the beam as required, an analogues way to the optical lenses in a light microscope. Apertures of different diameters can be inserted into the electron beam at several positions along the column. This is done to select part of the beam, for image formation, for example, and exclude the contribution of the rest.

The electron beam is generated in the electron gut at the top of the microscope column. Free electrons are obtained from a filament in one of two ways. Still the most common arrangement involves the thermionic emission of the electrons from a filament of tungsten or LaB_6. These filaments are heated by an electric current to about 2773K and 1773K, respectively, at which temperature a significant number of electrons obtain energy equal to or greater than the work function of the material and are able to leave the filament material and pass into the vacuum. These emergent electrons are focused by a cylinder to which a small negative bias is applied. At the same time, they are accelerated by a very large electrostatic field applied between the filament and the anode plate and they pass through an aperture in the anode plate, then into the TEM column. The strength of this field determines the kinetic energy of the electron beam and therefore the wavelength of the electrons and their maximum theoretical resolving power (see above). In practice, an applied potential of at least 100 kilovolts (kV) is advisable for HRTEM. The electron beam may also be generated in a Field Emission (EFG). These work at much lower temperatures, about 573K, and contain two successive electrostatic fields, between the filament and two anode plates positioned one after the after. The tungsten filament in a EFG has an extremely fine tip. The first electrostatic potential extracts electrons from the filament through this tip. The second potential is used, as above, to accelerate the electrons and form the high energy beam. Because the filament tip is so fine, the beam form a FEG is much more intense and much more spatially coherent than the beam from a thermionic filament.

Furthermore, the variation in energy within the beam is very small. These factors make FEGs of great interest for HRTEM where a narrow, coherent and intense beam is required.

The electron gun generates a thin electron beam by focusing and accelerating electrons from the filament. Electrons with energies or paths different from those required are excluded from the beam at the apertures. The Condenser Lens system is used to further adjust the beam so that it has the desired properties when it arrives at the specimen. Usually, a parallel beam is required. Different beam diameters may be desired and these can be set using the Condenser Lenses (there are usually two or three). This beam then passes through the specimen which is mounted on the specimen holder. The incident beam (see Figure 1.9), interacts with the internal crystal structure of the sample and emerges as a set of diffracted and non-diffracted beams. These beams are focused by the Objective Lens. In Figure 1.9, a ray diagram is drawn containing the non-diffracted beam and two diffracted beams. These are focused to points, or spots, at Back Focal Plane (BFP). There is no direct relationship between the position of these spots and the point of origin of the beam in the sample. The beams diffracted at the same angle are focused to the same spot in the BFP. Considering the rule for Bragg diffraction, and extending to three dimensions, each spot is expected to be related to one particular value of inter-planar spacing at one particular orientation-that is, to one family of crystal planes at one particular orientation. Following at three rays further, they are again focused at the image plane. Here, all the rays from each individual spot can be traced back to the same point in the sample, confirming that the image of the sample is formed at this plane. In fact, the image and the diffraction pattern formed by a particular crystal in a particular orientation are related mathematically, the latter being the Fourier Transform of the former. This is illustrated in

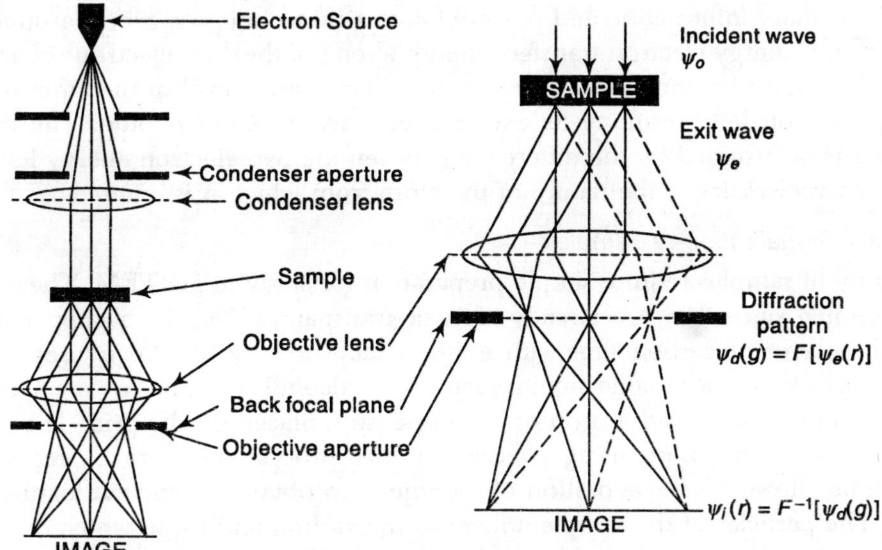

Fig. 1.9 *Simplified schematic ray diagram of the HRTEM in imaging mode representing the functions of electron source and condenser and objective lenses. On the right, the relationship between the exit wave, the diffraction pattern formed at the back focal plane and the image are represented.*

Figure 1.9 where the incident wave ψ_o, exit wave, ψ_e, and the wave functions of the diffraction pattern ψ_d, and the image, ψ_i are labelled. The incident wave, ψ_o, interacts with the internal potential of the same and emerges as ψ_e, which therefore contains information from the sample. The Objective Lens effectively performs a Fourier Transform on ψ_e to form the diffraction pattern which is a Fourier Transform, F (in reciprocal space, denoted by the vector, **g**), of the exit wave (in real space, denoted by the vector, **r**). The image is an inverse Fourier Transform, F^{-1} (in real space), of the diffraction pattern and is therefore closely related to the internal structure of the sample. A very widely used technique in the analysis of HRTEM images is to perform a Fourier Transform mathematically on the area of interest of an image in order to calculate a Digital Diffraction Pattern, or DDP. Information about the crystallography of the lattice images and the amplitude and phase of the Fourier components of the image (lattice planes), can be easily extracted. This in turn allows the inter-planar spacings and inter-planar angles to be obtained form the selected area of the image. The use of DDPs is often more convenient and more precise than collecting a SAED pattern experimentally, especially when examining very small regions.

Images and diffraction patterns are visualised on a fluorescent screen and may be recorded either or photographic film, or, increasingly, digitally, using a camera positioned below the fluorescent screen. To record a diffraction pattern, the strength of the Objective Lens is changed in order to throw the diffraction pattern onto the image plane. Further electromagnetic lenses are incorporated between the BFG and the image plane in commercial TEM instruments to allow the operator to focus and adjust the magnification of the final image.

The elastic interactions of electrons with the specimen which give rise to the image and the diffraction pattern, their inelastic interactions, in which energy is transferred, also provide useful complementary information and powerful analytical techniques, grown around the TEM. An incoming high energy electron transfers energy to one of the core electrons of an atom such that it is ejected from the atom. An electron at a higher energy level in the atom then fall into the resulting electron hole resulting its excess energy as an X-ray photon. The frequency of such a photon is determined by the difference between the two electron energy levels and will therefore be characteristics of the nature of the atom from which it is emitted.

Preparation and Visualisation of Samples

The majority of samples require simple preparation for study in HRTEM. There is, no need to thin the sample since they are already electron-transparent. Usually a copper of gold grid covered with a layer of carbon film with either a lacy or holy texture is used. The sample powder is suspended in a volatile liquid such as an alcohol, acetone, or an alkane such as hexane, by ultrasonication, and a drop of the suspension placed on the grid. The grid may be passed through the suspension using tweezers. Either process needs to be repeated several times, with time allowed for evaporation of the liquid, to obtain a sufficient loading of sample on the grid. The particles of the sample adhere to the carbon and those which overhang holes in the structure may be imaged in transmission. Sample loading of the grid should be verified using an optical microscope before use in the TEM.

In the technique of microtomy the sample is sliced into layers a few tens of nanometers thick using a quartz of diamond blade mounted on a mechanical arm, a microtome. This preparation technique is used extensively for the investigation of biological samples held in solidified resin. The overlap of particles is reduced since the thickness of the sample is limited by the thickness of the section made. It may then be easier to visualise the crystal structure of the particles, especially if they tend to form agglomerates. In addition, since particles or agglomerates may be broken open, differences between the surface and the bulk structure and composition may be imaged. This may be of interest to determine the distribution of supported metal particles or the nature of an encapsulating surface layer.

For, dense objects, semiconductor wafers or ceramics, ion beam milling is employed to reduce the thickness of the sample in the area of interest. A hole is milled from one side of the sample and the thin region around the hole is examined in the TEM since the edge of the hole will be thin enough to transmit electrons. This method provides more information about the disposition of the active phases relative to the support material.

In order to visualise the crystal structure of a particle in HRTEM, the particle must be oriented in such a way that its atom columns are aligned with the electron beam. That is, a zone axis of the crystal structure must be aligned with the electron beam. A zones axis describes a direction in the crystal which is parallel with at least one family of crystal planes. More electron microscope sample holders have the facility for tilting the sample in at least one direction to allow the operator to align particles with the electron beam. The maximum angle through which the sample can be tilted is limited by the design of the sample holder and by the internal geometry of the TEM instrument.

Imaging Simulation

Because of the complexity of HRTEM image interpretation, for the detailed analysis of the contrasts in HRTEM images and to model accurately the diffraction and image formation processes, Image Simulation (IS) techniques are necessary. The procedure for Image Simulation can be broken down into three steps. Firstly, a mathematical representation of the structure observed in the experimental image must be constructed on Cartesian co-ordinates. Sophisticated software packages are available for the generation of these structural computer models. Many variables must be considered in this step. An example relevant to heterogeneous catalysis is that of a metal particle supported on an oxide support. The crystallographic phases of the particle and the support must first be decided. There may be several possible alternatives. For example, tetragonal and cubic structures of the oxide may be known. The relative orientation of the crystal structures of the particle and the support, as well as the nature of the interface where the two meet, must be decided. For example, is the particle epitaxial with the support, and if so which planes are parallel in the two structures? The thickness of the support crystal and the size of the supported particle are set by inputting the number of crystal planes in each direction. The morphology of the supported particle is set by adding the facetting of the particle, i.e. by identifying the crystallogrpahic planes which make up the particle's surface. Close inspection of the experimental images and a through knowledge of the crystallographic of the phases which may be present in the sample are essential in setting these variables. Even so, some uncertainly will remain in some of them. To accommodate this, and to follow good practices, several alternative structural models are generally prepared for any particular experimentally-observed feature. These are then passed onto the second step of the IS process in which the interaction of these candidate structures with the electron beam of the TEM is modelled. Each structural model, its alignment with respect to the electron beam and the operating parameters of the microscope used to obtain the experimental images are input into the computer. The effects of the interactions of the electron beam with the atoms of the model structure, and with the apertures and electromagnetic lenses of the TEM itself, are modelled mathematically and the images which would result are simulated. Important variables in this step are the exact geometry of the alignment of the model structure to the electron beam and the defocus value–a measure of the extent to which the specimen is out of focus. Both of these variables can cause dramatic changes in the contrasts seen in the simulated images. Because of this, it is normal practice to collect both experimental and simulated images over a range of defocus–a through-focus series. In the final step of the IS process, the series of images simulated for each model structure are compared with the experimental images and those proposed structures giving a poor match are excluded, leaving the more probable alternatives for further study.

Particle size Analysis

In catalyst materials containing dispersed over a support, the size, size distribution and dispersion of the supported particles, and often of the support particles, is of great importance, as is the evolution of the these parameters during thermal and chemical treatments. By recording HRTEM images and then identifying and measuring the size of the particles of interest, such information can be obtained. In order for the data to be representative and statistically meaningful, many images of many different regions of each sample must be obtained and several hundred particles must be measured. The measurements themselves may be made adequately by hand from micrographs. However, it is more convenient and more accurate to perform such measurements from digital images with the aid of appropriate software.

Particle size distributions are plotted in Figure 1.10(a) for Rh particles dispersed over the surface of the mixed oxide support in 1 0.78 wt.% $Rh/Ce_{0.68}Zr_{0.32}O_2$. Samples of the catalyst were reduced in H_2 for 1 h at different reduction temperatures. T_{red}, as indicated on the Figure. Sample size and mass-corrected mean particle size data are summarized in Table 1. The mass-correction is applied in order to weight the data in terms of the mass of particles of a particular size range, rather than simply in terms of the number of particles of that size range, as shown in Figure 1.10(a).

A procedure which is particularly useful to the study of heterogeneous catalyst materials is the ability to obtain dispersion data from the particle size distributions. Here, dispersion is defined as the ratio of total exposed particle surface area to total particle mass or, equally as the total number of metal atoms at an exposed surface divided by the total number of metal atoms. The relevant mathematical treatments are given and are discussed in detail elsewhere. Briefly, a morphology and an orientation with respect to the surface of the support are assumed for the particles. For the Rh particles studied here, a truncated cuboctahedral morphology was assumed

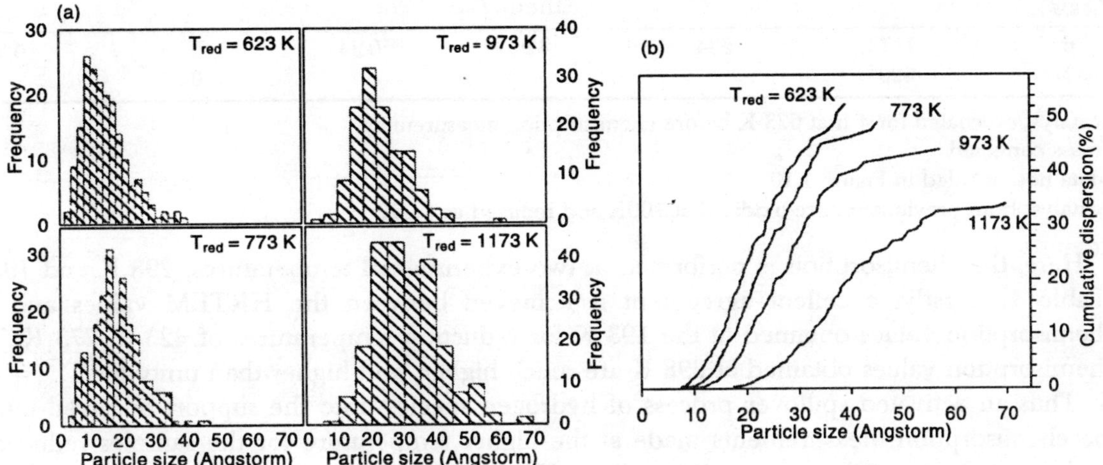

Fig. 1.10 *(a) Rh particle distribution and (b) Calculated cumulative dispersion plots for samples of 0.78 wt.% Rh/$Cr_{0.68}Zr_{0.32}O_2$ reduced for 1 h at reduction temperatures (T_{red}) of 623,773,973 and 1173 K.*

with {111}-type planes of both the metal and the oxide support in contact at their shared interface (assuming cubic crystal structures for both). These assumptions are based on detailed analysis of high resolution images of several representative particles. Based on this morphology, the exposed surface area and volume of each measured particle is estimated. Hence the contribution of the overall dispersion value of the sample made by each particle is also estimated and plotted against particle size in a graph of cumulative dispersion, a set of which are presented in Figure 1.10(b) for the Rh/$Ce_{0.68}Zr_{0.32}O_2$ samples. The dispersion value for the sample is given by the sum of all contributions, i.e. by the point on the extreme right of the curve, and these are given Rh/$Ce_{0.68}Zr_{0.32}O_2$ in Table 1. It can now be seen that mean particle size increases and the particle size distribution broadens on increasing the reduction temperature. Dispersion, as estimated from HRTEM measurements, decreases from a value of 52% (or Rh_S/Rh_T = 0.53 where S and T subscripts refer to surface and total, respectively) for a reduction temperature of 423K to 34% (0.34) for T_{red} = 1173K. In summary, the particles appear to undergo a classic evolution that can be explained by progressive sintering of the Rh particles.

TABLE 1 Comparison of Rh Particle Dispersion Data from Volumetric H_2 Chemisorption and from HRTEM measurements for Rh/$Ce_{0.68}Zr_{0.32}O_2$.

Run[a]	*Reduction temperature (K)*	*by HRTEM*			*by chemisorption*	
		Sample size	*Mean particle size*[b] *(Å)*	*Rh_S/Rh_T*	*H/Rh_T (193K)*	*H/Rh_T (289)*
1	423[c]	231	24	0.52	0.52	2.38
2	623	189	25	0.53	0.57	0.68
3	773	221	27	0.49	0.49	0.47
4	973	135	32	0.44	0.41	0.38
5	423[d]				0.47	

(Contd.)

(*Contd.*)

6	1173	234	42	0.34	0.23	0.20
7	423[d]				0.31	

[a] catalyst evacuated for 4 h at 673 K before chemisorption measurement
[b] mass corrected
[c] data not included in Figure 1.10
[d] catalyst from previous run, re-oxidised at 700K and reduced at 423K

Here, the chemisorption is performed at two experimental temperatures, 298 K and 193 K (Table 1). Firstly, excellent agreement is achieved between the HRTEM values and the chemisorption values obtained at the 193 K for reduction temperatures of 423 to 773 K. The chemisorption values obtained at 298 K are much higher, and higher than unity for $T_{red} = 423$ K. Thus an activated spillover process of hydrogen species onto the support occurred during the chemisorption measurements made at the higher temperature for the samples reduced at the lower reduction temperatures. Secondly, at $T_{red} > 773$ K, the chemisorption values (measured at 193 K) drop steadily below those obtained by HRTEM. However, after a mild re-oxidation and low temperature reduction, the higher dispersion values in the chemisorption experiment are recovered. So, a deactivation of the chemisorption of H_2 at the higher reduction temperatures is not related to the sintering of the particles, as detected by HRTEM.

SCANNING ELECTRON MICROSCOPE

Thus the **scanning electron microscope (SEM)** is a type of electron microscope capable of producing high-resolution images of a sample surface. Due to the manner in which the image is created, SEM images have a characteristic three-dimensional appearance and are useful for judging the surface structure of the sample. Unlike the TEM, where electrons of the high voltage beam from the image of the specimen, the Scanning Electron Microscope (SEM) produces images by detecting low energy secondary electrons which are emitted from the surface of the specimen due to excitation by the primary electron beam. In the SEM, the electron beam is rastered across the sample, with detectors building up an image by mapping the detected signals with beam position.

Generally, the TEM resolution is about an order to magnitude greater than the SEM resolution, however, because the SEM image relies on surface processes rather than transmission it is able to image bulk samples and has a much greater depth of view, and so can produce images that are a good representation of the 3D structure of the sample.

Detection of Secondary Electrons

The most common imaging mode monitors low energy (<50 eV) secondary electrons electrons originate with in a few nanometers from the surface. The electrons are detected by an detector which is a type of scintillato-photomultiplier device and the resulting signal is rendered into a two-dimensional intensity distribution that can be viewed and saved as a Digital images. The process relies on a raster-scanned primary beam. The brightness of the signal depends on the number of secondary electrons reaching the detector. If the beam enters the sample perpendicular

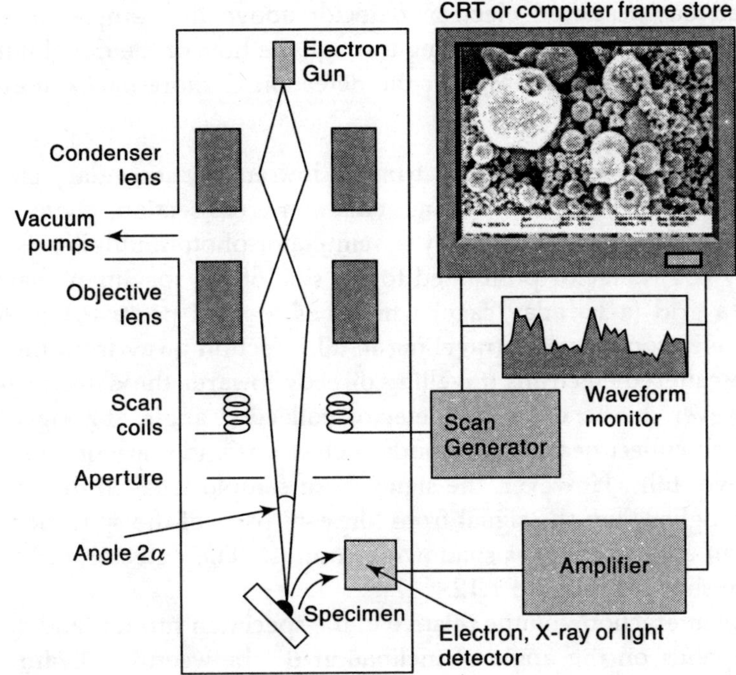

Fig. 1.11 *Schematic diagram of the layout of a scanning electron microscope*

to the surface, then the activated region is uniform about the axis of the beam and a certain number of electron "escape" from within the sample. As the angle of incidence increases, the "escape" distance of one side of the beam decreases, and more secondary electrons are emitted. Thus steep surfaces and edges tend to be brighter than flat surfaces, which result in images with a well-defined, three-dimensional appearance. Using this technique, resolutions less than 1 nm are possible.

Detection of Backscattered Electrons

Backscattered electrons consist of high-energy electrons originating in the electron beam, that are reflected or back-scattered out of the specimen interaction volume. Backscattered electrons are used to detect contrast between areas with different chemical compositions, especially when the average atomic number of the various regions is different, since the brightness of the BSE image tends to increase with the atomic number.

Backscattered electrons can also be used to form an **electron backscatter diffraction** (EBSD) image. This image can be used to determine the crystallographic structure of the specimen.

There are fewer backscattered electrons emitted from a sample than secondary electrons. The number of backscattered electrons leaving the sample surface upward might be significantly lower than those that follow trajectories toward the sides. The collection efficiency of backscattered electrons cannot be improved by detectors as they are positioned on one side of the sample; have a low collection efficiency for backscattered electrons due to small acceptance angles. The use of a dedicated backscattered electron detector above the sample in a "doughnut" type arrangement, with the electron beam passing through the hole of the doughnut, greatly increases the solid angle of collection and allows for the detection of more backscattered electrons.

Secondary Electron Imaging

During electron irradiation, a wide spectrum of low-energy secondary electrons are emitted from the surface of the specimen with a high emission cross section, giving images with a high signal-to-noise ratio. They are detected by a scintillator-photo-multiplier system known as an Everhart-Thronley (ET) detector positioned to the side of the specimen chamber. Surrounding the scintillator is a grid (a Faraday cage) which has a small positive bias to attract the low-energy secondary electrons that are travelling in all direction away from the specimen surface. High-energy backscattered electrons travelling directly towards the detector will also contribute to the image, however due to the small detector collection angle, this signal is relatively low. The ET detector will collect nearly all secondary electrons and therefore images appear to look down holes and over hills. However, the signal from sample areas in direct line-of-sight of the detector is sightly higher than the signal from 'unseen' parts of the specimen and consequently they will appear, an effect known as shadowing contrast. The detector and a typical secondary electron image are shown in Figure 1.12.

The shape of the interaction volume relative to the specimen surface, and hence the secondary electron yield, depends on the angle of inclination θ_{inc} between the beam and the specimen surface, varying as $1/\cos \theta_{inc}$. This inclination effect means that the edges of a spherical particle will appear brighter than the centre and, for studies of specimen topography, specimens are normally tilted 20–40° towards the detector to maximize the signal. Due to increased field strength at sharp edges and spikes on the specimen surface, the emission of secondary and

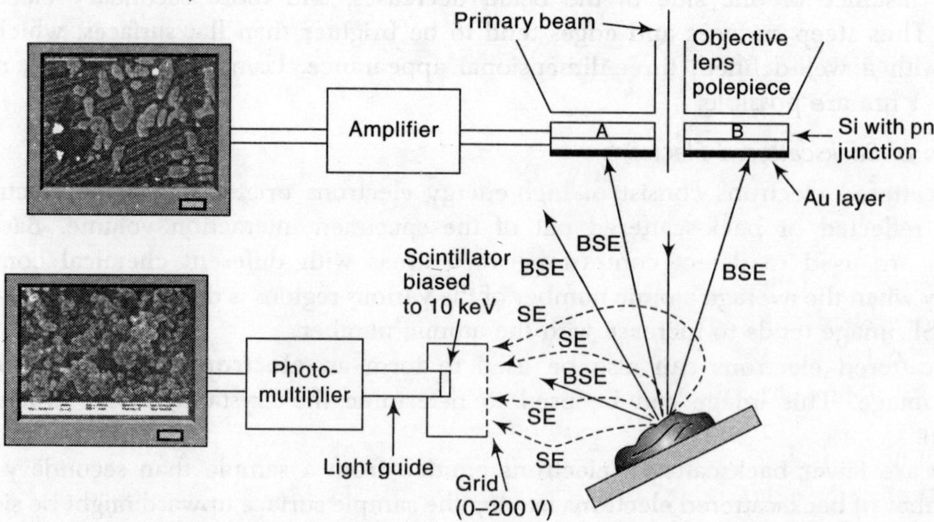

Fig. 1.12 *Schematic diagram showing a secondary electron detector and a solid-state back-scattered electron detector together with typical images.*

backscattered electrons increases, and such features will appear bright, known as edge contrast. Consequently, owing to the surface specificity and the contrast effects mentioned above, the major use of secondary electron imaging is for topographic contrast. It general, topographic images obtained with secondary electrons have the same appearance as an object illuminated with diffuse light, in that there are no harsh shadows. Quantitative information on surface topography may be obtained using stereomicroscopy techniques in which two images are taken under identical magnification conditions at slightly different tilt angles (e.g., –5° and +5°). The images may be viewed using a special stereo viewer (which allows the brain to interpret them as a 3D picture). For quantitative work, the relative displacement or parallax between two features on the pari of tilted allows their height separation to be determined.

Since the detected secondary electrons only originate from the top few nanometres of the specimen surface, the emission diameter will only be slightly bigger than the probe diameter the lateral resolution in secondary images is therefore of the order of 1–5nm for typical specimens but will obviously vary with probe size and signal-to-noise considerations. The depth resolution is a function of the accelerating voltage. Secondary electron imaging is therefore the most useful general signal for imaging nanostructures in the SEM due to the high signal level combined with high lateral and depth resolution.

Resolution of the SEM

The spatial resolution of the SEM depends on the size of the electron spot which in turn depends on the magnetic electron optical system which produces the scanning beam. The resolution is also limited by the size of the interaction volume, or the extent to which the material interacts with the electron beam. The spot size and the interaction volume are both very large compared to the distances between atoms, so the resolution of the SEM is not high enough to images down to the atomic scale, as is possible in the transmission electron microscope (TEM). The SEM has compensating advantages, though, including the ability to image a comparatively large area of the specimen; the ability to image bulk materials (not just thin films or foils); and the variety of analytical modes available for measuring the composition and nature of the specimen. Depending on the instrument, the resolution can fall somewhere between less than 1nm and 20nm. In general, SEM images are much easier to interpret than TEM images.

Environmental SEM

Conventional SEM requires samples to be imaged under vacuum, which means that samples that would produce a significant amount of vapor, e.g. biological samples, need to be either dried or cryogenically frozen. This means that processes involving transitions to or from liquids or gases, such as the drying of adhesives or melting of alloys, liquid transport, chemical reactions and solid-air-gas systems in general could not be observed.

Samples were observed in low-pressure gaseous environments (e.g. 1–50 Torr) and high humidity (up to 100 %), due to the development of a secondary-electron detector capable of operating in the presence of water vapour and by the use of pressure-limiting apertures with differential pumping in the path of the electron beam to separate the vacuum region around the gun and lenses from the sample chamber.

SEM picture of a bend in a seven-lobed polyester fiber. Note dotted line equals 20 micrometers.

An SEM image of various types of pollen.

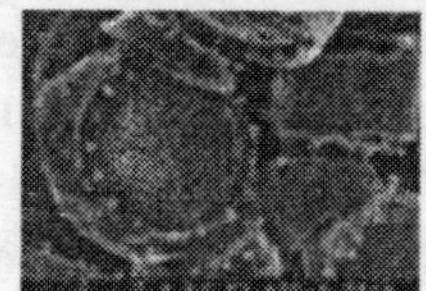
SEM Picture of a Diatom at magnification of 5000X.

False coloured SEM image of soybean cyst nematode and egg.

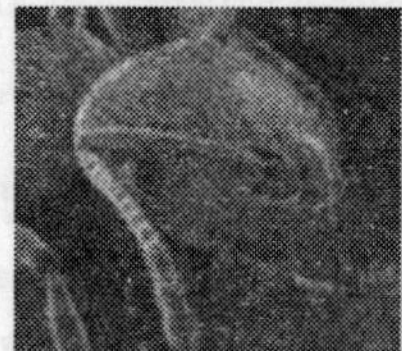
SEM image of an ant head.

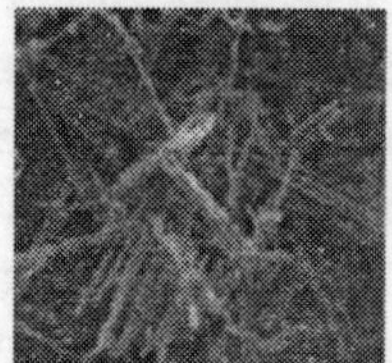
SEM image of asbestos fibres.

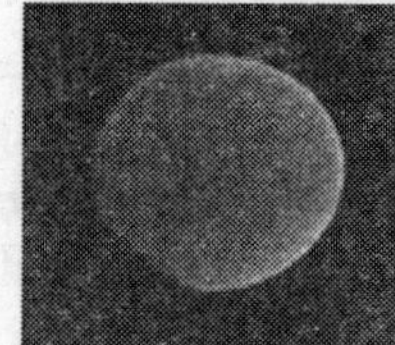
SEM Picture of a Diatom at magnification of 5000X.

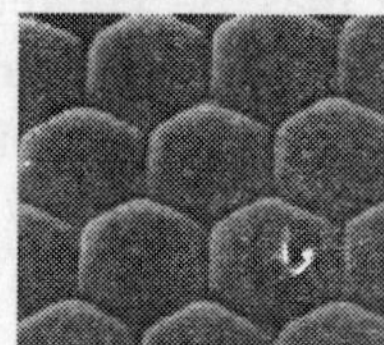
False coloured SEM image of soybean cyst nematode and egg.

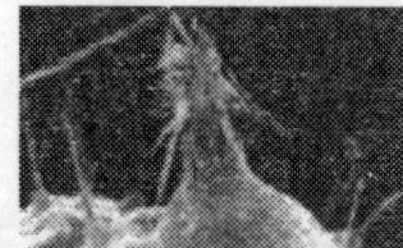
SEM image of discharged nematocyst.

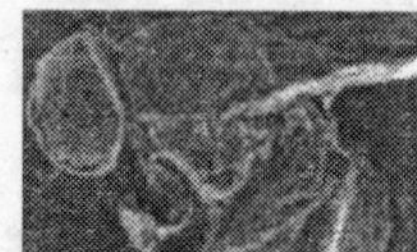
SEM image of the upper body of a Drosophila.

SEM images of the compound eye of a Drosophila

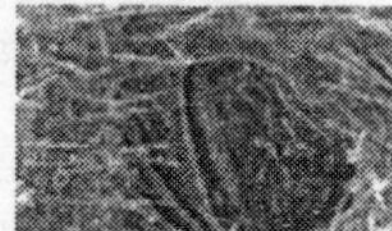
SEM image of the grasshopper's spiracle valve.

SEM image of normal circulating human blood.

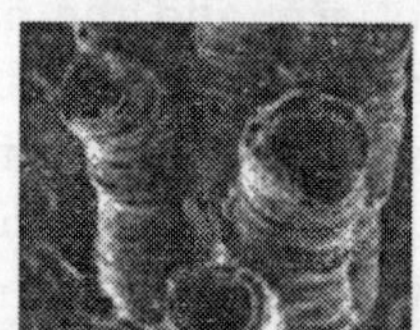
SEM image of a hederelloid from the Devonian of Michigan (largest tube diameter is 0.75 mm)

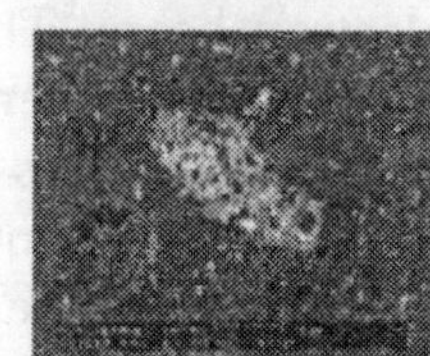
Backscattered Electron image of an Antimony rich region in a fragment of ancient glass.

SEM image of the corrosion layer on the surface of an ancient glass fragment, note the laminar structure of the corrosion layer.

Fig. 1.13

ESEM is especially useful for non-metallic and biological materials because coating with carbon or gold is unnecessary Plastics and Elastomers can now be routinely examined, as can animal and plant samples. Coating is irreversible, and reduces the value of the results obtained. Thus very small details on the surface of the sample may be concealed by the coating, let alone that coating is done under vacuum which drastically alters hydrated specimen in the first place.

SCANNING TRANSMISSION ELECTRON MICROSCOPY

A scanning transmission electron microscope (STEM) is a type of transmission electron microscope. With it, the electrons pass through the sample, but, as in scanning electron microscopy, the electron optics focus the beam into a narrow spot which is scanned over the sample in a raster.

The rastering of the beam across the sample makes these microscopes suitable for analysis techniques such as mapping by energy dispersive X-ray (EDX) spectroscopy, electron energy loss spectroscopy (EELS) and annular dark-field imaging (ADF). These signals can be obtained simultaneously, allowing direct correlation of image and quantitative data.

By using a STEM and high-angle detector, it is possible to form atomic resolution images where the contrast is directly related to the atomic number. This is in contrast to the conventional high resolution electron microscopy technique, which uses phase-contrast, and therefore produces results which need interpretation by simulation.

The general principle of STEM is similar to that of SEM, except that the electron detector is now located below a thin TEM specimen as shown in Figure 1.15. The small probe (of the order of angstroms), produced by a FEG and a condenser–objective lens system, is scanned across the specimen and the signal detected and imaged on a monitor or frame store. The resolution is determined mainly by the probe diameter, which is usually of the order of a nanometre. Many TEMs have a separate scanning attachment; however, STEMs operate entirely in scanning mode under UHVC conditions. Bright-field STEM imaging uses an axial detector to detect electrons scattered through relatively small angles, and images contain diffraction contrast. Dark-field STEM imaging is essentially incoherent and employs an annular detector to detect electrons scattered through higher angles. High-angle annular dark field (HAADF) image Fig. 1.14 intensity is roughly proportional to the square of atomic number (so-called Z contrast) and is therefore extremely useful for mapping phases at ultra-high resolution.

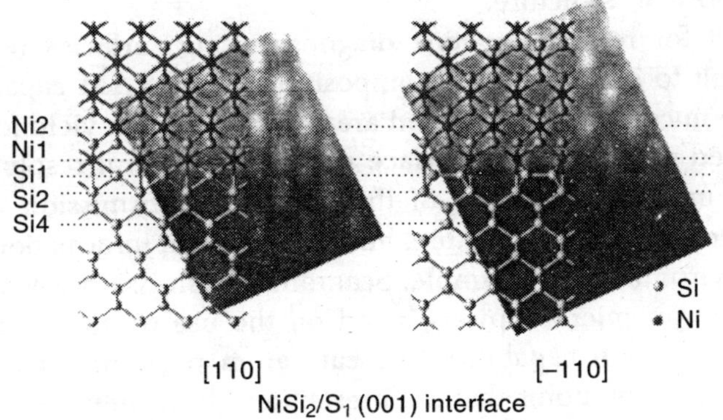

Fig. 1.14 *Layout of a STEM together with HAADF images of a silicon-nickel disilicide interface*

Scanning transmission electron microscopy is used to identify composition, bonding and electronic structure with atomic resolution in microelectronic device structures.

The mobility of electrons in a field-effect transistor depends on the density of defect states at the Si/SiO_2 interface next to the conduction channel. Intermixing of nitrogen within SiO_2 can

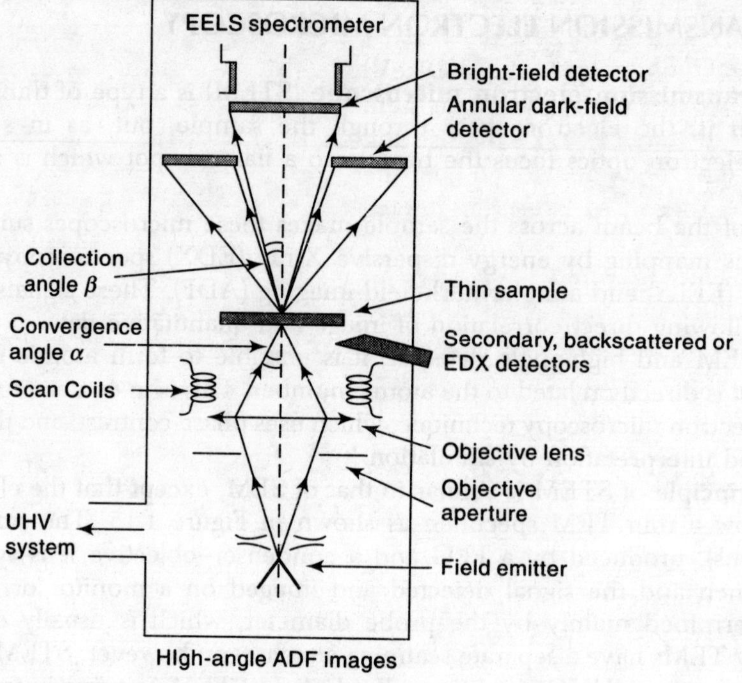

Fig. 1.15

produce better dielectric material for DRAM capacitor iso. Small, deliberately induced strain in a thin Si layer produces a high-mobility channel for very fast transistors, but misfit dislocations at the substrate interface of the channel can limit the practical use of such a layer. Layers of material only 2–3 atoms thick can produce a hundredfold change in the contrast resistance of a silicon/polysilicon (Si/poly-Si) structure.

Device are too small for reliable, routine diagnosis using surface-sensitive, low-resolution probes. It is very difficult to investigate the composition of a DRAM capacitor trench sidewall by means of SIMS. The uncertainties in probed area, secondary ion yield for non-normal probe incidence, and the buried nature of the interface all limit the lateral resolution to about 50 nm and the depth resolution to about 5 nm of the scanning transmission electron microscope (STEM) not only to determine atomic structure, but also to obtain information on the composition and atomic bonding environment of a sample. Scanning transmission electron microscopy, like SIMS and scanning electron microscopy, is based on the use of an electron probe. It shares with the TEM a high-resolution capability to locate atom positions; it shares with SIMS an ability to distinguish the kind of atoms that are present and how they are bonded; and it shares with the scanning electron microscope (SEM) a simplicity of interpretation grounded in its scanning nature. Under some circumstances it can also provide atomic-resolution electronic structure information similar to that obtained from large areas by optical absorption. It is also fundamentally a bulk-sensitive probe–it can, for example, be used to obtain information within a localized but subsurface region that controls the operation of a microelectronic device.

Annular dark-field imaging in the STEM

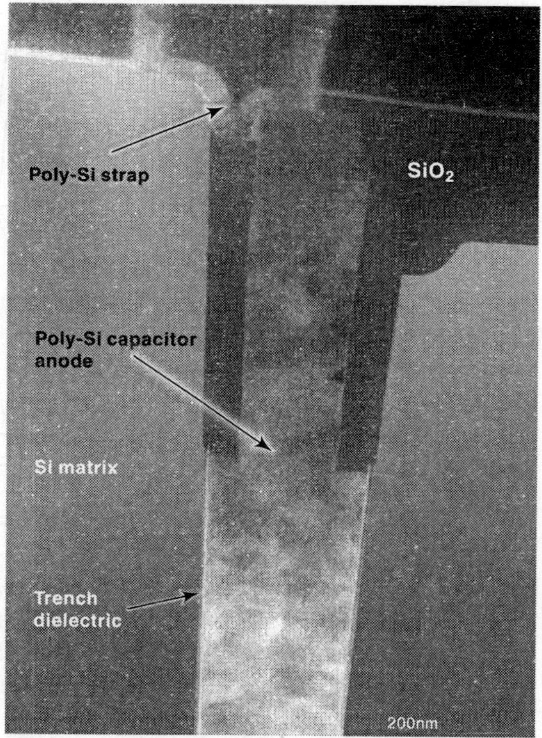

Fig. 1.16 *Low-magnification ADF images of an FET-trench capacitor structure under development for use in a DRAM chip. A poly-Si strap connects the Si matrix to the storage capacitor at the upper portion of the image, the capacitor extends several mm into the Si matrix; and the trench dielectric isolates the poly-Si capacitor element from the Si matrix.*

The STEM is a medium-energy (100–300 keV) instrument that works with standard Tem specimens. Therefore, it utilizes the expertise and infrastructure that have been developed over the years to integrate the TEM into the device development process. The instrument produces a 0.2-nm-diameter electron prove at an electron beam energy of 120 keV. Figure 1.16 shows a low-magnificent STEM image of a DRAM FET-trench capacitor structure. The annular dark-field (ADF) image shown is obtained using large-angle elastic scattering, which is sensitive to the atomic number of the material.

Figure 1.17 depicts the ADF imaging method. Shown there is a schematic illustration of the electron-scattering geometry in the STEM. Electrons are scanned in a raster across the specimen; they travel through the specimen and are caught by various detectors which feed video displays for viewing. The STEM is therefore a combination of a high-resolution TEM instrument with a low-resolution SEM. In ADF imaging, an electron which passes near an atom in a sample will suffer a large electrostatic deflection. If a detector is placed well away from the undeflected beam direction, a signal is obtained using only the electrons that have been deflected through large angles and which therefore pass close to an atom column. If the probe is smaller than the distance between the atom columns, about 0.3 nm for the [0 1 1] orientation of Si, this detection method yields a large change between the on-column and off-column positions of the probe. The detected intensity is roughly proportional to the number of atoms in the column times the square of their atomic number. Thus, a signal gives information about the locations of the atoms, their concentration, and their atomic numbers.

Recently, the technique has received extensive investigation for imaging interface and defects structure in crystal. Because the technique is sensitive to the atomic number (Z) of the atom, ADF imaging is sometimes referred to as "Z-contrast." Therefore, brighter areas are those composed of more atoms or toms having a higher atomic number. In Figure 1.16, the Si matrix is more dense and brighter than the SiO_2 dielectric. Because an ADF STEM signal averages over many scattering angles, there is less contrast arising from diffraction than is normally observed in a TEM. In the figure, the Si matrix surface is above the poly-Si-filled trench

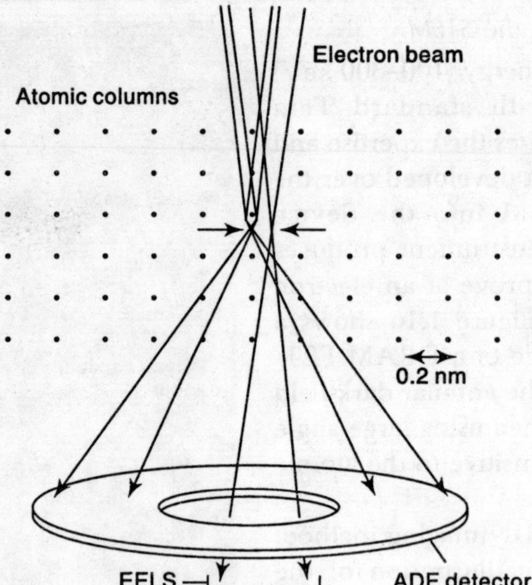

Fig. 1.17 *Scattering geometry for ADF imaging and spatially resolved FELS analysis. The STEM forms a 0.2-nm diameter prove of electrons which travels through a thin section of the specimen. If the probe travels a column of atoms, it is elastically deflected to large scattering angles, where it is intercepted by an annulus-shaped detector. An aperture located on the optic axis admits electrons undergoing small-angle scattering into an electron energy loss spectroscopy for FELS analysis. As the probe is scanned in a raster over the specimen, the ADF detector generates a view signal for display.*

capacitor extends several microns below into the substrate, and is bounded by an insulating layer along the trench sidewalls. There is a poly-Si strap connecting the top of the capacitor with the Si matrix region that would become the drain end of a FET in a completed device. The contrast in this image is primarily due to material density, with the most dense areas causing the most scattering into the high-angle detector.

Figure 1.18 is a composite of higher-magnification ADF images of the trench dielectric. Intensities are roughly proportional to atomic density, so that the Si matrix and poly-Si capacitor appear with the same brightness. Variation reflects grain boundaries and orientation differences in the poly-Si, SiO_2 appears darker than Si. Within the 15-nm dielectric, three regions are imaged–associated with the presence of SiO_2, Si_3N_4, and SiO_2. Bright-field (BG) imaging in this area shows no variation in contrast within the dielectric. There appears to be a 2-3 nm-thick region of brighter contrast following the edge of the Si matrix. Identification of this requires higher magnification and atomic-resolution analytical information.

Atomic Column Imaging

The diagram in Figure 1.17 suggests that if the probe is small enough and the crystal is oriented with the atom columns lined up with the electron beam direction, bright spots corresponds to the atom column positions are observed. In Figure 1.19, a $Ge_{30}Si_{70}$/Si interface is shown using (a) small-angle BF imaging, equivalent to a standard TEM result, and (b) ADF imaging. The BF image consists of many overlapping fringes that convey the projected symmetry of the lattice. Since their precise position is strongly dependent on focus conditions, they cannot be interpreted as positions of atoms. There is very little information about the atomic number. Therefore, the position of the composition interface is not visible in this image. In the AFG image, bright spots correspond directly to atomic columns, with their brightness indicating the average atomic number for equal column lengths, so the composition interface is clearly visible. The image is almost independent of focus and thickness, making ADF imaging a very useful and practical technique for imaging heterogeneous crystals.

In this image, each spot corresponds to a double column, with the column separated by 0.14 nm in the [1 1 0] projection. Since the STEM probe diameter is 0.2 nm, the columns are not resolved, appear as single spots elongated in the [1 0 0] direction (left-right) in the image. Figure 1.20 shows a further

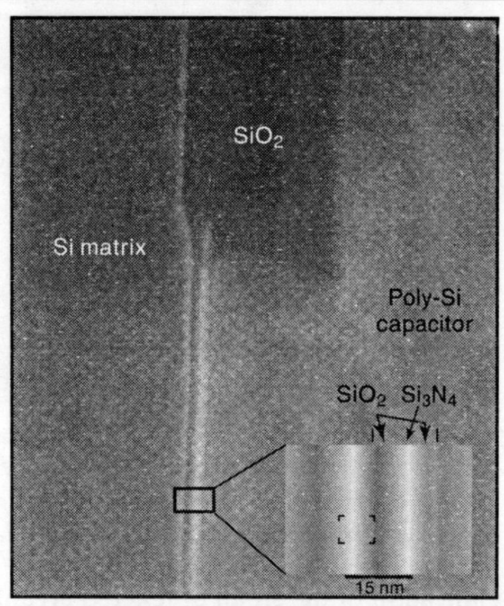

Fig. 1.18 *Trench dielectric region of the capacitor of Figure 1. The ADF imaging produces a strongly atomic-density-sensitive image, so that areas consisting of SiO_2 are clearly distinguished from areas of Si_3N_4. A bright region appears to lie within the Si matrix sidewall. The small corner brackets in the inset define this region for Figure 5.*

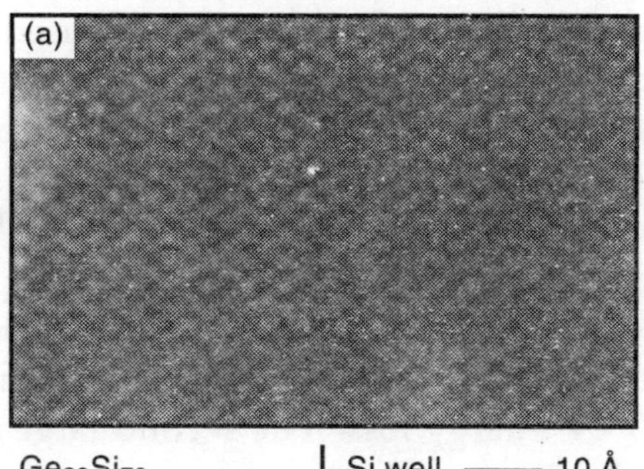

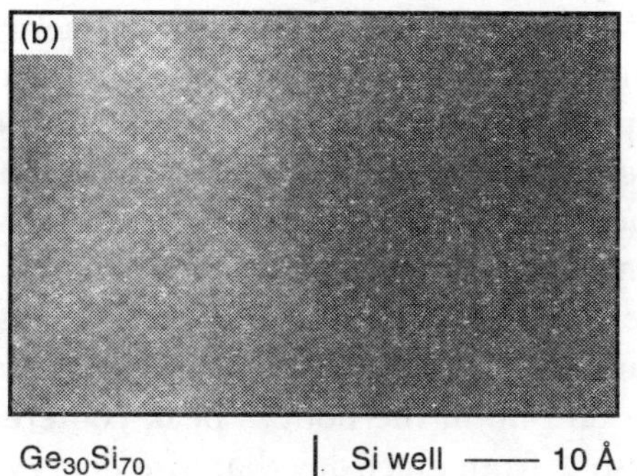

Fig. 1.19 *(a) Bright-field image of a $Si/Ge_{30}Si_{70}$ interface. The bright spots give correct atomic perodities but are not located correctly relative to corresponding atom positions. There is little atomic number sensitivity. (b) The AFF image gives spots elongated in the (0 0 1) direction, corresponding to the "dumbbell"-shaped double atom columns. There is a good registry of the image with the atomic lattice, and the brightness of the spots bears a close relationship to the atomic numbers of the atoms in each column.*

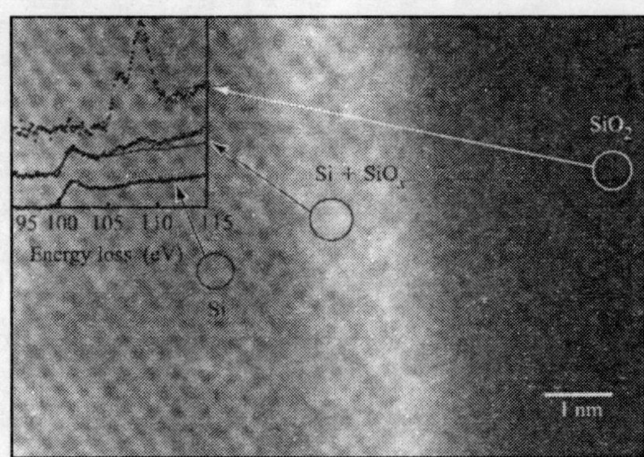

Fig. 1.20 *High-magnification ADF view of the bright region in the trench sidewall of Figure 1.22 This view shows that the region clearly lies within the crystalline region. EELS spectra in the inset suggest that the interface is rough on a scale of 1 nm, with SiO_2 and crystalline Si in intimate contact, or that O has been driven into the Si matrix during the cleaning and growth of SiO_2 on the trench sidewalls.*

magnification of the small bracketed area in the inset of Figure 1.18. This image shows elongated spots ending at the Si/SiO_2 interface at the trench sidewalls. the narrow region of bright contrast is seen in the lower-magnification images does indeed coincide with the Si lattice. The crystal structure appears relatively unmodified except for the ADF intensity variation.

Analytical Analysis

Now, Figure 1.17 also depicts an alternative analytical signal using electrons that do not suffer a large deflection. This is the area between atoms, and they scatter mainly by loss of energy to valence and conduction electrons within the specimen. They are therefore sensitive to its composition, bonding, and electronic structure. With an electron spectroscopy placed on the optic axis of the STEM, electron energy loss spectroscopy (EELS) can be used to obtain information about the energy lost by electrons in the probe as they traverse the sample. The most important loss features are absorption edges caused by electronic transitions from atomic core levels to the conduction band (CB). These identify the particular type of atom from their onset energy and summarize the symmetry-projected CB local density of states (LDOS) and local bonding configuration by their shape.

Figure 1.21 shows an illustrative EELs spectrum for Si in the 0–120-eV energy loss range. For a sample 50–100 nm thick, most of the incident electrons do not lose any energy and therefore end up in the no-loss peak centered at 0 eV energy loss. The second large peak near 17 eV results from electrons that lose energy to one collective electronic excitation, or plasmon. This excitation involves all of the valence electrons in the solid and so is characteristic of electron density, but is less informative about the bonding or electronic structure. Near 100 eV for Si, and absorption edge is evident for electronic transitions from the $2p$ core levels to the CB. This feature is only about 1% as strong as the plasmon peak. There is fine structure within 2 to 3 eV of the core loss onset energy. In the 1–3-eV region, near the no-loss peak, there is a direct interband absorption edge present which is caused by energy loss via direct excitation of single electrons across the band gap. This is only about 001% as large as the no-loss peak and is

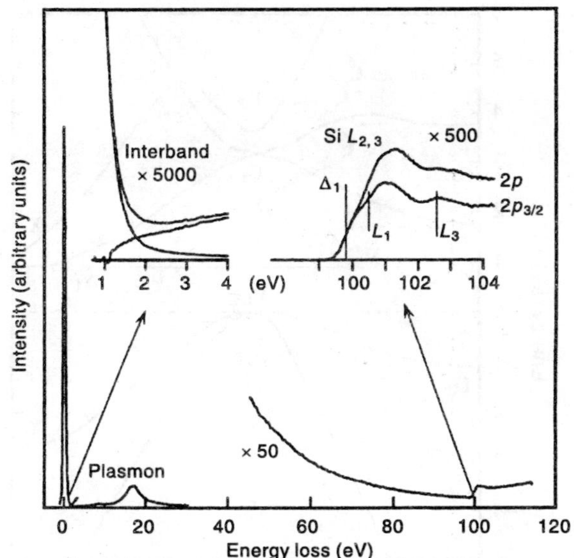

Fig. 1.21 *EELS spectrum for silicon. Effects can be seen due to no-loss electrons at 0 eV, plasmon losses at 17 eV the Si 2p core to CB absorption edge at 100 eV, and direct interband absorption near 1 eV. The fine structure on the core edges can be examined for structure in the CB LDOS. Direct interband information is difficult to obtain because the scattering cross section is only about % of the background intensity near 1 eV.*

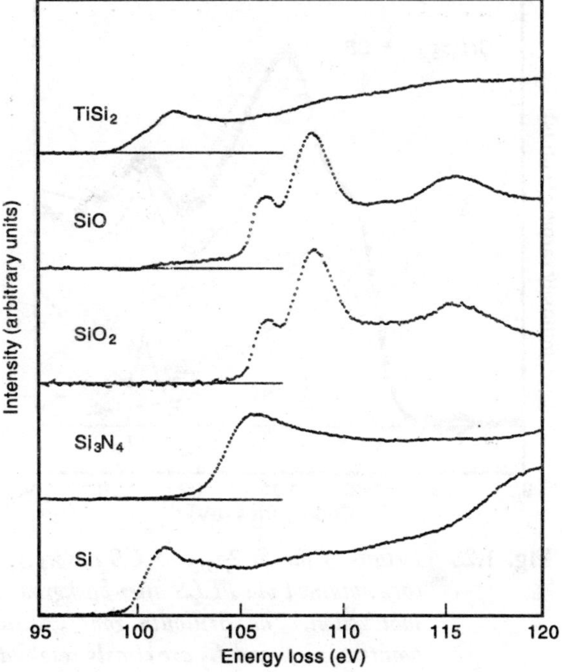

Fig. 1.22 *EELS spectra due to Si 2p core absorption edges, for several compounds. These show variation of structure on the 5-eV scale and can be used for "fingerprint" identification of unknown regions of a specimen.*

therefore very difficult to observe. High-energy-resolution optical spectroscopy is obtained in this energy range by cathodoluminescence (CL) using micron-sized electron beams, and energy-integrated CL is obtained in the STEM using small beams. However, EELS is the only STEP-based signal that promises to obtain both energy and spatial resolution to allow investigation of a semiconductor bandgap.

In Figure 1.22 several Si $2p_{3/2} \rightarrow$ CB spectra are shown for different Si bonding environments. These shapes can be used to identify the Si compounds with the lateral spatial resolution limited by the probs size even when analyzing single atom columns. In Figure 1.20, where EELS from three areas are displayed in the figure insert, shows clear signatures for the bulk Si and SiO_2 in regions a few atomic layers away from the interface. Within the narrow transition region, a composite shape is observed, one consisting partly of Si and partly of SiO_2, in spite of the fact that a clear Si atomic lattice is present. It seems likely, therefore, that this interface is not smooth in the [1 1 0] imaging projection. This might happen if oxygen is driven into the interface during cleaning, or possibly if the cross-sectional region did not precisely cut the curved trench sidewall perpendicular to its surface. This kind of imaging and spectroscopic

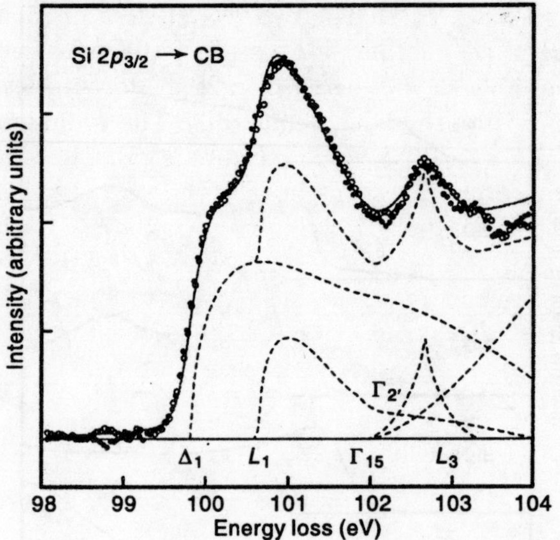

Fig. 1.23 *Details of the Si $2p_{3/2} \rightarrow$ CB absorption edge obtained via FELS after background subtraction. The Brillouin zone critical points Δ_1, L_1, and L_3 are clearly resolved. In addition, the edge shape may be modeled to obtain likely positions of $\Gamma_{1,5}$ and Γ_2. Two sets of experimental data are shown to illustrate the excellent reproducibility achievable using FELS. The edge position can be obtained with an accuracy of ±20 meV.*

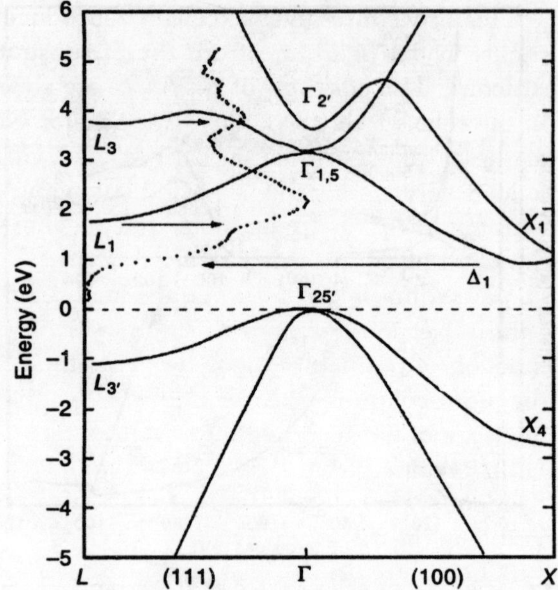

Fig. 1.24 *Interpretation of the Si $2p_{3/2} \rightarrow$ CB absorption. An experimental result is plotted over the Si band structure. Regions of the Brillouin zone that are diople-allowed find that have a large effective mass contribute to the spectrum. Final states at L_1, Δ_1, and L_3 are allowed. Scattering to final states at the Brillouin zone center at Γ is forbidden.*

behavior has also been observed at the Si/SiO_2 interface in thin gate dielectrics, where interface roughness has been cited as a possible cause.

The elemental Si absorption edge itself can be interpreted for CB information. In Figure 1.23 two Si $2p_{3/2}$ absorption edges are shown. These edges have been extensively processed by (1) subtraction of a slowly varying background, (2) deconvolution of an experimental instrumental resolution function, and (3) removal of structure derived from the spin-orbit splitting of the spin 3/2-1/.2 crore states. The data is compared with model spectra derived from trial *s,d*-symmetry-projected LDOS curves.

Figure 1.24 shows a typical band structure for Si. There are three main symmetry-allowed CB final states: (1) the *s*-like minimum at Δ_1, dispersing in the [1 0 0] direction; (2) the *s*-like minimum in the [1 1 1] direction L_1; and (3) the d-like saddle point at L_3 in the [1 1 1] direction. Data for the Si core absorption are plotted to show how the features of the spectra correspond to features of the bands. If a band is particularly flat in energy, an inflection point is generated in the LDOS.

This procedure gives excellent reproducibility (see Fig. 1.23). The two sets of experimental results in the figure are from the Si substrate and Si capping layer in $Ge_{30}Si_{70}$ quantum well structure. The positions of Δ_1, L_1, L_3 are clearly reproduced, with better than 20-mV accuracies. In the relaxed $Ge_{30}Si_{70}$ alloy, these major CB points have shifted systematically. The evolution of these values throughout the relaxed Ge_xSi_{1-x} alloy series is found (Figure 1.25). The CB follows very closely the predicted behaviour as a function of GE content. The Si 2*p* core level is constant in energy in the relaxed alloy, allowing the use of the absorption edge position to track the band offset in this system.

With some minor assumptions regarding strain, we can even use these results to obtain the band offset in a strained quantum well. Figure 1.26 shows, the heterojunction CB offset as a function of position across a 9-nm strained Si well. In the figure, ADF image of the well shows the atom columns and identifies the position of the well by the atomic number (*Z*-contrast). A line scan of the recorded intensity along the dotted line is shown in overlay. The position of the CB is obtained from EELS spectra and is also recorded on the image. This structure contains a deliberately introduced grading of the composition on the substrate side of the well, readily visible in the ADF line scan on the left. The band offer obtained clearly follows the composition variation on the left as well. At the upper interface (to the right), where quantum well conduction is desired, the composition is made as discontinuous as possible. A surprising result is that the EEL CB offset variation is sharper than the composition profile at the upper interface. This could be the result of "over-strain" at the interface.

Now, In the case of the DRAM structure of Figure 1.27 between the drain of a DRAM access transistor and a poly-Si region of a DRAM capacitor it is found that the resistance of the strap connection appears to be 100 times bigger than normal. TEM of the structure shows no reason for the behavior. In Figure 1.31(a), the Si/trap interface from the upper portion of Figure 1.20 is shown at high magnification in ADF imaging. The Si substrate is on the left; the poly-Si strap is on the right. Si 2*p* absorption spectra is shown at the lower right. It is clear from the image that there is a 1-nm-thick region of low density at the interface. The $2p \rightarrow CB$ absorption results show that the this region is largely composed of Si, although a minor amount of *O* may be present. The Si edge shape is distinctly bulklike on either side of the interface, but within the interface is less sharp, with the loss of fine structure attributable to the bulk bands. It appears that the poly-Si growth interface, although primarily Si in composition, is disordered on a 1-nm scale, leading to a lower density and to damping of the CB structure by electron confinement effects due to the ADF image, 1–2 nm deep into the Si substrate. Si stacking fault (SF) structures are visible (circled), suggest that the reactive ion etch (RIE) cleaning, used to prepare the Si surface for the growth of the poly-Si strap, caused subsurface damage.

Figure 1.27(b) shows an example of an analysis of a low-resistance strap interface. This interface has an almost uniform density with no discernible interface layer. In fact, the interface is rough at the 0.5-nm level, indicating intimate contact between the Si substrate and the poly-Si strap. The EELS spectra show bulklike fine structure throughout. No subsurface defects are visible. The interface show some blurring of the CB fine structure, suggesting that the interfacial quality is not yet as good as it could be, and that some improvement in electrical characteristics may therefore be possible with further work on this interface.

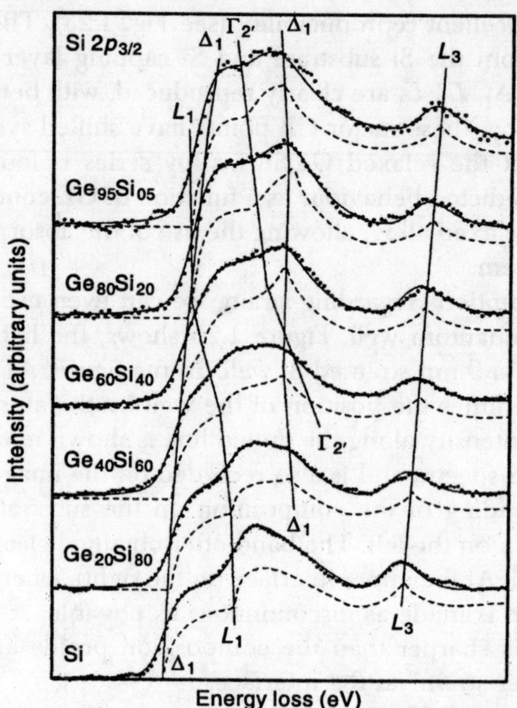

Fig. 1.25 *Summary of evolution Si $2p_{3/2} \rightarrow$ CB absorption edges as a function of alloy composition in $Ge_{1-x}Si_x$. Brillouin zone critical points are made obvious by their systematic shifts with alloy composition. Brillouin zone center movement can be inferred in spite of the dipole selection rules. Γ_1 denotes a saddle point which is absent in Si but appears in Ge as the critical point Γ_2. Descends and a becomes a local band minimum.*

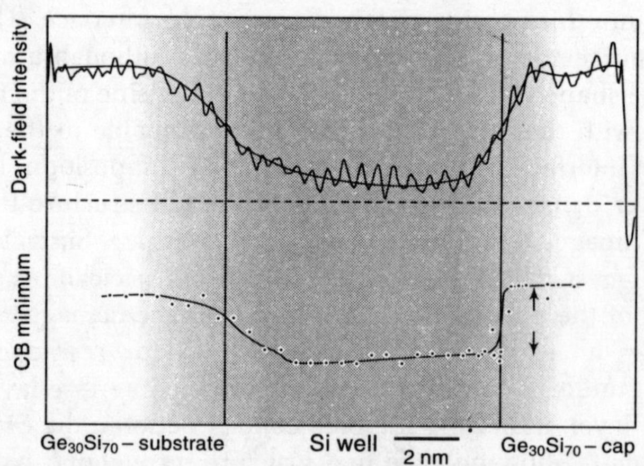

Fig. 1.26 *Strained Si quantum well confined between a $Ge_{10}Si_{70}$ substrate (left) and cap (right). The substrate composition was graded to reduce the impact of misfit strain. The upper curve is an ADF line scan along the dashed line. The lower curve is an EELS measurement of the CB minimum along a similar line. The heterojunction band offset can be accurately measured. Adapted from [30], with permission.*

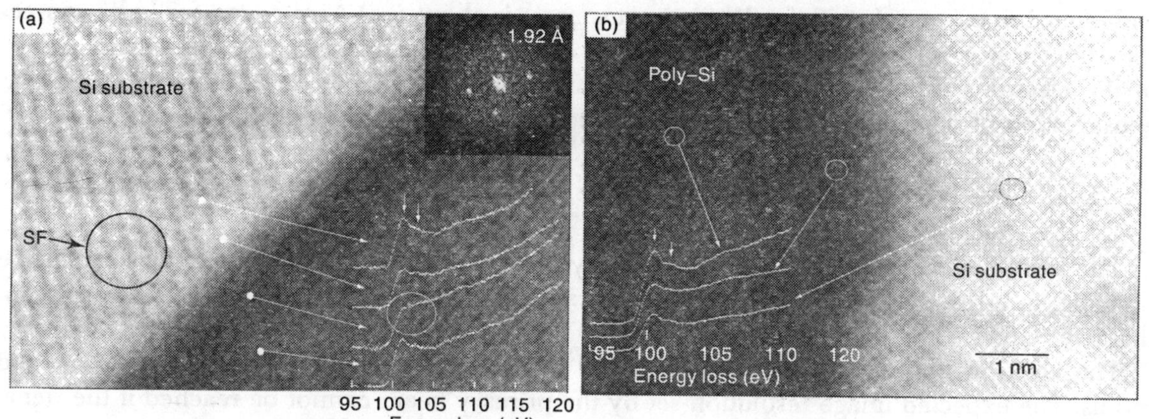

Fig. 1.27 *(a) ADF image and FELs spectra for a high-resistance Si/poly-Si interface, Si $L_{2,3}$ spectra show a modification of the near-edge structure that indicates the presence of a region of poor conductivity within 1 nm of the interface. (b) ADF image and FELs spectra for a low-resistance Si/poly-Si interface. The $L_{2,3}$ edges remain peaked, with no evidence for a heterogeneous layer at the interface.*

THE ELECTRON OPTICAL SYSTEM

For high resolution imaging, a well focused beam is required, just as in optical microscopy. Due to the short wavelength of electron beams with eV energies, as given by De Broglie wavelength, the properties of the electron optical system and the electron emitter mainly defines the limits on the achievable beam diameter. The current density in the electron beam can approximated by a Gaussion distribution of current density j [Am^{-2}] as function of radius, r, from the beam center

$$j(r) = j_0 e^{-\frac{r^2}{r_0^2}},$$

with radius determined by r_0, giving a the full width half maximum $FWHM = 2v(ln2)r_0$.

Total beam current, $I_b = \iint j(r)\, r dr d\theta$

i.e., $I_b = j_0 \pi$.

The electron optics impose a limit on the achievable beam current density and radius by the brightness of the electron emitter β_e, which is conserved throughout the system.

Brightness, β, is a measure of the current per area normal to the beam direction and per element of solid angle.

$$\therefore \quad \beta = \frac{j_0}{\pi\alpha^2}$$

i.e. the brightness is related to the current density. The emitter brightness β_e is determined by the type of electron emitter and the beam energy E_b

$$\beta_e = \frac{j_e E_b}{\pi \Delta E}$$

with emission current density for W-filament sources is about j_e~3 Acm^{-2}, and for LaB6 sources it is about 100 Acm^{-2}, while field emission guns (EFG) can reach 10^5 Acm^{-2}. The energy spread of the electrons from the sources are about ΔE~1 eV and slightly lower for FEGs. Due to conservation of the brightness in the system, the beam diameter depends on current as

$$r_0 = \frac{1}{\pi}\frac{r_{wd}}{r_a}\sqrt{\frac{I_b}{\beta_e}}$$

The ideal beam probe size is determined by the conservation of brightness that cannot be obtained in a real system. Effects such as aberration will make the minimum achievable beam diameter larger.

The electron optical system sets limitations to the achievable primary beam current and radius. The expected image resolution set by the primary beam cannot be reached if the signal detected for imaging is caused by electrons scattered far in the sample. The trajectory of an electron penetrating a bulk solid is a complex trajectory due to multiple elastic and inelastic collision events. As the primary electron (PE) penetrates into the sample it will gradually change direction and loose energy in collisions. The mean free path due to elastic and inelastic collisions, λ_{mfp}, depends on the atomic number of the material and the PE energy. At 100 keV λ_{mfp} = 150nm for carbon and 5 nm gold. For sample thinner than λ_{mfp} the main part of the PE will pass relatively unaffected through the sample, which is the basis for TEM

SEM can be used for thicker specimens. The electron that escape from the sample in a new direction compared to the PE due to elastic collisions are called backscattered electrons (BSE).

For sample thicker than λ_{mfp}, the volume interacting with the scattered PE defines the range of the electrons in the material, and this is considerably larger than the minimum achievable primary beam diameters.

AN ELECTROMECHANICAL MATERIAL TESTING SYSTEM FOR IN SITU ELECTRON MICROSCOPY AND APPLICATIONS

Nanochemical characterization of materials has attached significant attention because of the emergence of various novel nanoscale materials and structures. Nanotubes and nanowires have shown great potential in a range of applications from nanocomposites nanoelctronics to nanoelectromechanical systems. These nanostructures have presented new challenges for the identification of their properties and deformation mechanisms.

For this, quantitative mechanical testing of these structures by SEM, transmission EM (TEM), and scanning probe microscopes is essential. Progress in this area has been hindered by the lack of miniaturized material testing systems with adequate load an displacement accuracy. Existing measurement techniques can be categorized as follows:

- Techniques that measure load and displacement but cannot image the specimen during loading, e.g., force spectroscopy atomic force microscopy, nanoindentation, and no-chip testing;
- Techniques that image the specimen in real time but cannot measure load and deformation independently;

- Techniques that measure load and locally image deformation and failure by atomic force microscopy or by optical interferometry, although the resolution is somewhat limited for the study of nanostructures; and
- Techniques that can image defects and measure load and deformation but that require frequent switching of the electron beam, between specimen and a deflecting beam used as load sensor, during which local deformation events may go unobserved.

Therefore, nanomechanical testing requires a new material testing system with the capability to independently apply and measure load without the need of imaging, e.g., electronically, while exhibiting adequate resolution for nanostructure testing. In this manner, the imaging can be used exclusively for real-time continuous observation of the specimen at the desired magnetification. In this work, a previously undescribed microelectromechanical system (MEMS) is demonstrated.

The device includes an actuator (electromechanical or thermomechanical) and a load sensor based on capacitance measurement. The specimens can be thin films, 1D nanostructures, and biological structures. Thin films are cofabricated with the actuator and load sensor, while the nanostructures and biostructures can be placed across a gap between the actuator and the load sensor by a nanomanipulator or guided self-assembly. Because of its small size, the MEMS device is well suited for testing of nanostructures inside SEM, TEM, scanning probe microscopes, and x-ray synchrotron stages.

DEVICE DESIGN, FABRICATION AND CALIBRATION

Two types of actuators are used comb drive electrostatic actuator and in-plane thermal actuator. The comb drive actuator is force controlled, whereas the thermal actuator is displacement controlled, (i.e., it prescribes the displacement to the specimen on a given supplied voltage).

In the principle of the comb drive actuator, a thermal actuator, delivers a large force (up to tens of millinewtons) and a moderate displacement (up to tens of micrometers) (see Fig. 1.28(a)). It consists of inclined freestanding beams connected to a shuttle at one end and fixed to the substrate at the other end. When a voltage is applied across the inclined beams (V-shaped beams), the current flux causes joule-heating and thermal expansion. Because of the inclined configuration of the beams, the shuttle is pushed forward. In free motion (without connection to the specimen), the thermal actuator displacement is given as

$$U^{\Delta T} = \alpha \Delta T l \frac{\sin\theta}{\left(\sin^2\theta + \cos^2\theta \dfrac{12I}{Al^2}\right)} \tag{1}$$

where α is the thermal expansion coefficient of the beam material, ΔT is the average temperature change in the inclined beams, l is the beam length, I is the moment of inertia of one beam in the plane parallel to the substrate $(=(1/12)Eb^3h)$, E is the Young's modulus of the actuator material, b is the beam width, h is the beam height, A is the cross-sectional area of a beam, and θ is the angle between the beams and the transverse direction to the shuttle motion.

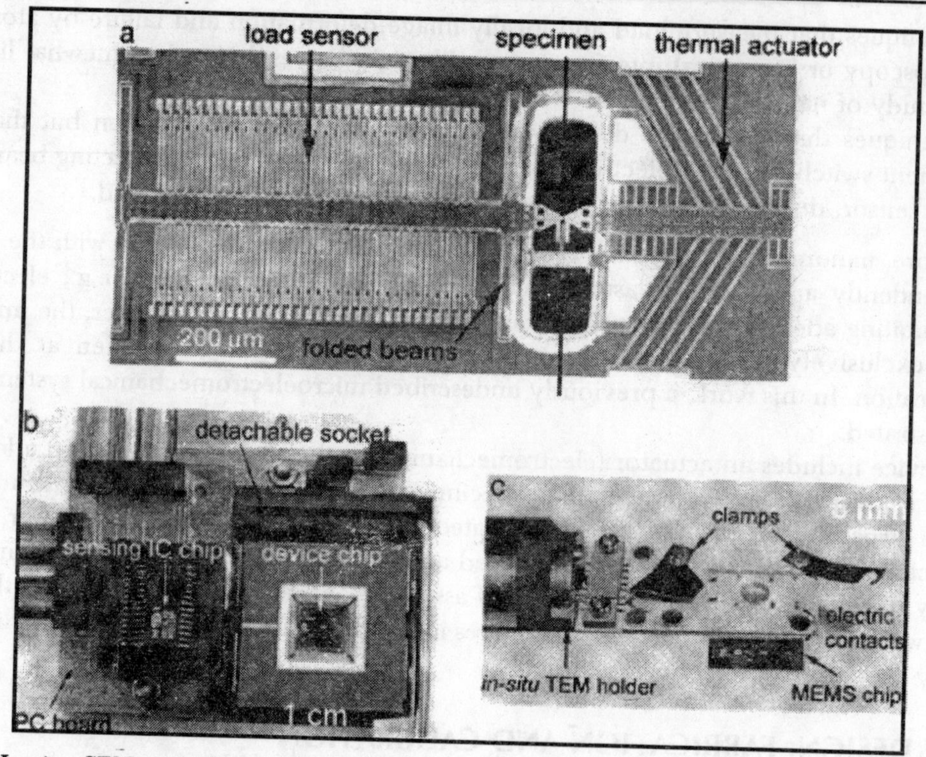

Fig. 1.28 *In situ SEM and Tem tensile testing device including actuator, load sensor, and specimen. Four folded beams support the load sensor. (a) Testing device used in situ TEM. (b) Experimental setup for in situ SEM testing. MEMS device chip is positioned near the MS3110 chip on a printed circuit board. The setup is connected to a power supply, a digital multimeter, and a computer outside the SEM by means of a chamber feedthrough. (c) In situ TEM holder (containing a feedthrough and eight electric contact pads), along with a 5 × 10 mm MEMS chip. In an actual experiment, the MEMS chip is flipped, placed in the Tem holder, and fixed by the left and right clamps.*

The load senior integrated in the testing platform is similar to the MEMS accelerometer in that they are both based on differential capacitive sensing and fabricated by surface micromachining. The load sensor consists of a rigid shuttle with one set of movable fingers and two sets of stationary fingers. A lumped model of the sensor consists of two capacitors in series. Shuttle displacement causes increases of one capacitance and decrease of the other. Within a moderate displacement range the capacitance change is proportional to the displacement, namely

$$\Delta C = C_1 - C_2 = N\varepsilon A\left(\frac{1}{d_0 - \Delta d} - \frac{1}{d_0 + \Delta d}\right) \approx \frac{2N\varepsilon A}{d_0^2}\Delta A \qquad (2)$$

where ε is the electric permittivity, N the number of units of differential capacitors, A and d_0 are the overlap area and initial gap between the movable finger and each stationary finger, respectively, and Δd the displacement of the load sensor. The load sensor is anchored to the substrate by four folded beams that are designed with a range of stiffness for testing various materials of interest.

The device are fabricated by using the multi-user MEMS process. For SEM testing, 20 devices with different of actuators and load sensors are arranged in 1 10 × 10 mm^2. To achieve electrical connection, there are 100 gold pads fabricated in the same chip around the periphery. The chip is glued to the cavity of a ceramic pin grid array package, and the gold pads are wire-bonded to the 100 leads around the cavity, as shown in Fig. 1.28(b). Corresponding pins in the back of the package make the electric connection to an outer circuit, which in turn is connected to electronic actuation and measuring instrumentation.

Measuring capacitance changes with subfemtofarad resolution, a charge-sensing measuring circuit that can mitigate the effect of parasitic capacitance with a commercially available integrated circuit based. The MEMS device chip is positioned very close to the sensing integrated circuit chip to minimize the stray capacitance and electromagnetic interference (Fig. 1.28(b)). The output voltage is proportional to the capacitance change.

In addition to the SEM measurements TEM testing of nanostructures is achievable by microfabrication of a window under the specimen area. The major challenge is to etch such a window, from the back of the silicon wafer, without damaging the previously fabricated structures. Such a task is accomplished by deep reactive ion etching of the window before device release. Fig. 1.28(c) shows a MEMS chip (5 × mm) containing four MEMS devices. The two devices in the center can be used to perform TEM testing, and the other two devices can be used in calibration tests. The chip has eight contact pads for electric actuation/sensing. Fig. 1.28(a) is a magnified view of one such device. The backside window, with dimensions of 130 × 300 μm, is clearly visible. The chip is designed to be directly mounted on a specially designed TEM holder containing a feed-through and interconnects to address electrically the devices (Fig. 1.28(c)). In this case, the sensing integrated circuit chip used in the capacitance measurements is located outside of the TEM.

A particular feature in the movable shuttle is selected as reference in the SEM image. The device is actuated ON and OFF sequentially four times during the SEM scan. The feature corresponding to the ON-OFF actuation cycles is captures in the SEM image (Fig. 1.29(a)), and simultaneously the capacitance change is recorded (Fig. 1.29(b)). Fig. 1.29(c) correlates the

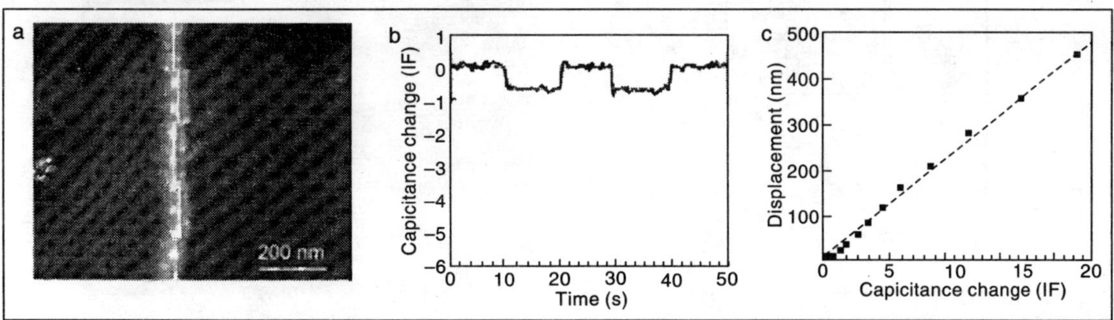

Fig. 1.29 *Calibration of the load sensor showing the relationship between capacitance change and measured displacement from SEM images at a series of actuation voltages. (a and b) Signatures when actuator is at 2 V; reference feature in the SEM image showing a motion of 15 nm due to four ON-OFF actuations (a) and plot corresponding to a 0.7 fF capacitance change resulting from the same actuation (b). Both raw data and fitted data are shown in the plot of capacitance measurements. (c) Plot of displacement vs. capacitance change resulting form the calibration.*

displacement (load) and the capacitance change. It follows a linear relationship, which agrees well with Eq. 2. The resolution of the measured capacitance change is 0.05 fF, and the corresponding displacement resolution is 1 nm. For a load sensor with stiffness of 11.8 N/m, designed for testing carbon nanotubes and nanowires, the load resolution is 12 nN, which is quite adequate. Depending on the application, the load resolution can be increased to ≈1 nN by decreasing the sensor stiffness.

Applications

The testing system shown in Fig. 1.28 was used to test three types of structures: freestanding polySi films and palladium nanowires inside an SEM and carbon nanotubes inside a TEM.

For the testing of freestanding thin films, polySi was selected because of its well-characterized Young's modulus and failure strength and ease of cofabrication with the device. Because the minimum feature size by standard photolithography is ≈2 μm, the cofabricated polySi specimen was further nanomachined by focused ion beam (FIB) to reduce the sample minimum dimension to 350–450 nm. A dog-bone-shaped polySi specimen with a trapezoidal cross-section was obtained (Fig. 1.30). Two platinum (Pt) lines (with a spacing of 2.5 μm) were deposited by electron beam-induced deposition (EBID) in a dual-beam FIB/SEM instrument for deformation measurement (Fig. 1.30(b)). Quantification of length increase between these two marks was done by means of image analysis with edge detection software. Two specimens are tested, resulting in the stress-strain curves shown in Fig. 1.30(c). The specimens exhibit nearly the same Young's modulus of 155 ± 5 Gpa and failure strengths of 0.7 and 1.42 Gpa. The results are consistent with the values reported for multi-user MEMS process polySi thin films.

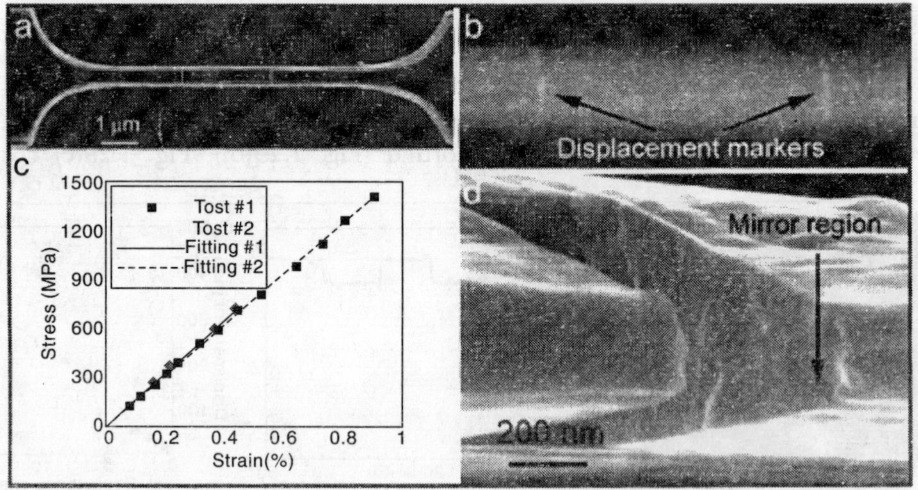

Fig. 1.30 *Quantitative in situ SEM test of a freestanding poly Si film. (a) Dog-bone-shaped tensile test specimen. (b) Magnified view of two Pt displacement markers deposited by EBID. (c) Stress-strain curves of two in situ SEM tests. Both specimens were 6 μm long and 1.6 mm thick. One had a top width of 0.34 μm and bottom width of 0.87 μm, whereas the other had a top width of 0.42 μm and bottom width of 1.04 μm. (d) Fracture surface of polySi beam showing a mirror region indicative of brittle fracture initiation.*

For the testing of nanowires, palladium (Pd) nanowires are synthesized by using a porous alumina. Membrane as template. The tested nanowires are ≈20–30μm long and 20 nm in diameter. They are polycrystalline with average grain size of ≈20 nm. To mount a single nanowire (specimen) onto the device, nanomanipulation is used consisting of the following steps: (i) nanowire dispersion in isopropyl by ultrasonication for 30 min; (ii) deposition of nanowire suspension on top of a TEM grid with a micropipette; (iii) manipulation of a protruding nanowire from grid edge; and (iv) mounting of the nanowire between the actuator and the load sensor. The last two steps are performed inside the same dual-beam FIB/SEM instrument. EBID of Pt was used for nanowelding. A nanomanipulator possessing 1-nm resolution in three orthogonal axes is used to pick up a protruding nanowire. After welding the nanowire to the tungsten tip of the manipulator, FIB is used to cut the nanowire and release it from the other end. Through x, y and z motion of the manipulator, the nanowire is brought close to the MEMS device (Fig. 1.31(a)). Contact is confirmed by electron beam and ion beam imaging through two different views. The free end in contact with the actuator shuttle is then welded (Fig. 1.31(b)). After that, FIB is used to cut the nanowire such that the other end is free (Fig. 1.31(c)). The manipulator tip then is used to push down this free end until it made contact with the sensor shuttle. It then is EBID nanowelded (Fig. 1.31(d)). Through this procedure, nanowires are successfully mounted on the devices.

A particular Pd nanowire is tested *in situ* the FIB/SEM. The length between the two welding locations is measured to be 3 ± 0.2 μm. The gap edges (2 μm apart) were used as displacement markers. Assuming no sliding between the nanowire and the two welded ends, the gap size increase during tensile testing is considered as the nanowire elongation. Quantification of the gap increase is done by means of image analysis with edge detection software. Hence, strain is computed as the ratio between gap size increase and distance between welding points. The stressstrain curve recorded during the experiment is plotted in Fig. 1.31(e). It is observed that the measured Young's modulus is 99.4 ± 6.6 Gpa, which is ≈20–30% lower than that of a polycrystalline Pd and in between the maximum and minimum values for single-crystalline Pd as a function of the crystal orientation. For single-crystalline Pd, the Young's modulus in the ⟨111⟩ direction is 191.5 GPa, in the ⟨100⟩ direction is 73.4 Gpa, and in the ⟨110⟩ direction is 136.5 GPa (23). Young's modulus of polycrystalline Pd is 132 Gpa. There is an ≈5° misalignment between the axial direction of the nanowire and the stretching direction, which causes $<1\%$ error of the Young's modulus. Reduction in the value of Young's modulus with respect to the bulk value is observed in nanocrystaline materials. Grain boundary compliance and creep are attributed to this phenomenon. Another feature revealed by the test is that the nanowire remainder elastic and did not fracture up to an applied stress of 1.5 GPa. This value is much higher than the yield stress of nanocrystalline Pd in bulk form. This finding confirms that when the material scales down, it strength increases and tends to approach the theoretical strength of the material (≈1/10 of the Young's modulus). This phenomenon can be attributed to absence of defects in the material atomic structure and the high stresses threshold for the nucleation of defects. Because the tested nanowire turned out to be much stronger then anticipated, the MEMS device designed with a relatively low load sensor stiffness was not able to break it within the range of admissible actuator voltages.

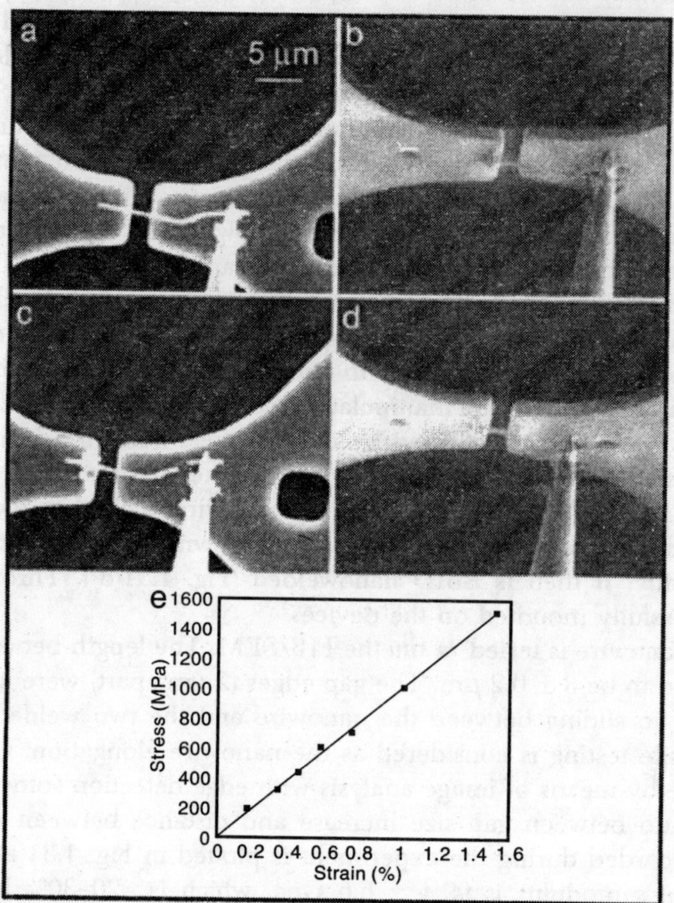

Fig. 1.31 *Nanomanipulation procedure to mount a Pd nanowire in the MEMS and test results (a) Approach and contact the nanowire to the MEMS device (b) Weld one end of the nanowire to actuator shuttle using EBID of Pt gas precursor. (c) Cut the other end using FIB. (d) Weld the free end to the load sensorshuttle. (e) Stress-strain curve of a PD nanowire exhibiting a Young's modulus of 99.4 ± 6.6 GPa and a strength higher than its bulk counterpart. A and c are SEM images; b and d are FIB images. There is a 52° view angle difference between SEM and FIB.*

Multiwalled carbon nanotubes (MWCNTs) are tested *in situ* the TEM. The specimens are mounted on the device by following the manipulation procedure described above. A particular MWCNT is tested inside a TEM by using the holder shown in Fig. 1.28(a). Its outer diameter was ≈130 nm and inner diameter was 99 mm as shown in Fig. 1.32(a). The length between the two welding locations is 2.57 ± 0.2 μm. The specimen is progressively loaded, and its deformation is continuously monitored. A movie showing the deformation of the MWCNT under tension until fracture is available from the authors upon request. A fracture strength of 15.84 GPa and a failure strain of 1.56% is measured. A previously undescribed phenomenon is observed during the *in situ* experiment. After failure, the graphite shell atomic structure disappeared. Instead, a large number of nanoparticles (with diameters ranging from 5 to 50 nm) embedded in an

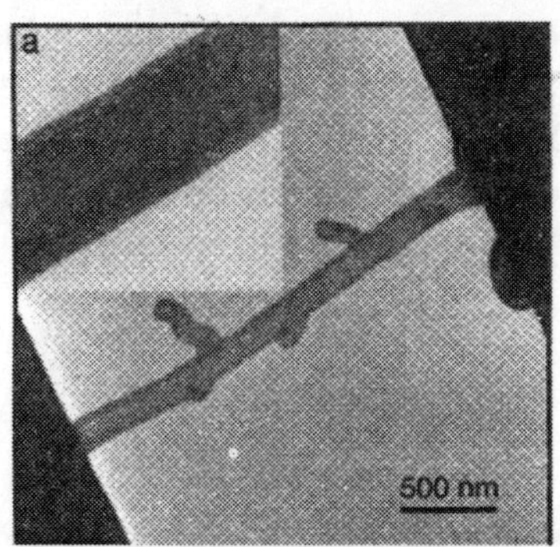

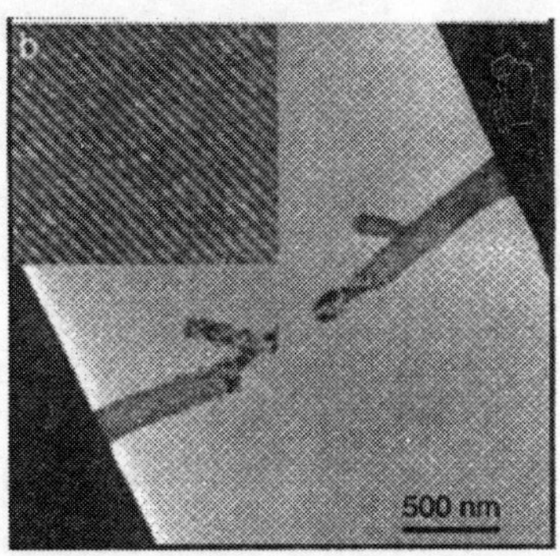

Fig. 1.32 *In situ TEM tensile test of an MWCNT. (a) TEM image before the test., Inset shows the multiple graphite shells. The MWCNT has outer diameter of 130 nm and inner diameter of 99 nm. (b) TEM image after the test. Post-failure crystallization is clearly seen. Inset is a high-resolution TEM image of a particualr nanograin at the fracture region showing that the grain is single crystalline. The three tiny tubes attasched to the main tube were present during nanomanipulation when the MWCNT specimen was taken out from a bundle. The procedure for specimen attachment to the MEMS testing system was the same to the one used in the testing of the Pd nanowire.*

amorphous matrix are observed (see Fig. 1.32(b)). Both high-resolution Tem and nanodiffraction confirmed that the nanoparticles are single-crystal Pt. Therefore with concluded that a thin layer of amorphous Pt coated the MWCNT during EBID and that this Pt layer is transformed to a crystalline phase during the straining and failure of the MWCNT. Furthermore, electron energy loss spectroscopy analysis revealed that the amorphous matrix is made of carbon. It is therefore inferred that the crystalline structure of the carbon nanotubes (graphite sheets) transformed to amorphous carbon upon failure.

Thus, as shown by the above tests, the MEMS-based material testing system with integrated electronic load applications and sensing greatly enhances the instrumentation capabilities for in situ atomic force microscopy, SEM and TEM testing of nanostructures. With this system, microscopy can be focused on capturing deformation mechanism while the corresponding load history is recorded electronically. Note that addition of a second differential capacitor sensor on the actuator side of the sample would directly provide an electronic measurement of average specimen elongation. In such a case, both load and average deformation would be measured electroncially. Furthermore, the MEMS-based material testing system potentially can be applied to characterize the mechanical and elctromechanical properties not only of nanowires and nanotubes but also of a large number of organic materials, including DNA, proteins and nanofibers.

Although the developed device has been discussed in the context of *in situ* SEM and Tem tensile testing, a large number of related applications are envisioned. To cite just one, let us highlight its applicability to *in situ* TEM nanoindentation experiments. Quantitative *in situ* nanoindentation in TEM has provided useful insight in the study of dislocation nucleation and propagation. In the dark-field condition, the image is formed by using a strongly diffracted electron beam from one grain only. MEMS-based material testing system with electronic load sensing can in principle be used to overcome this limitation. A low aspect ratio carbon nanotube or nanowire mounted on the actuator side, following a nanomanipulation procedure similar to that described above, would constitute the indenter. As a matter of fact, carbon nanotubes have been widely used as tips for high-resolution atomic force microscopy. Clearly, the V-shape of the actuator needs to be inverted such that the displacement vector produced by the actuator is in the direction of the specimen-load sensor.

LOW-POWER PHOTONIC DEVICES

Efficient Optical Coupling to the Nano-Scale

The ability to focus the optical field to deeply subwavelength dimensions opens the door to an entirely new class of photonic devices. If one could combine the imaging powers of X-ray wavelengths with the economy and maturity of visible light sources, one could greatly broaden the practical engineering toolbox. Imagine focusing visible photons to spatial dimensions less than ten nanometers. By doing so, *electron beam microscopy is immediately displaced by optical microscopy*, replacing expensive electron beam sources with inexpensive visible lasers. Beyond simple economics, through, this achievement would allow for the nanoscale imaging of living biological samples.

In addition to the goal of nano-scale optical energy concentration and focusing, we add the constraint of efficiency. It is not sufficient of simply deliver optical energy to a nanoscale spatial regime. This coupling must be done efficiently to launch a technology node. Such powerful focusing could then be used to generate optical nonlinearities with very low photon counts. Enabling a suite of novel non-linear devices in passive optical geometries, this leads to novel optical switches, all-optical logic and highly sensitive detector arrays. All of these reasons compel us to address the physical and engineering principles that determine the smallest volume to which light can be efficiently focused.

A simple thought experiment can clarify some of the ramifications of efficient optical coupling to the nanoscale. Consider the energy of a single visible photon. For the sake of quantitative discussion, a red photon is chosen at 2eV, yielding a free space wavelength of 620 nm. The reader is now asked to forego the objections of the Heisenberg Uncertainty Principle or the classical diffraction limit, and simply consider the energetic implications of confining the photon to a volume of 1nm^3. The details as to how we will arrive at such confinement are postponed until after the motivation has been established. This geometric scheme is illustrated below.

Computing the optical energy density is now trivial. The energy of the photon is a known quantity, as is the volume. Turning the crank on this transparent equation yields

$$\text{Energy Density} = \frac{2\text{ eV}}{(10^{-9}\text{ m})^3}\,\frac{106\times10^{-19}\text{ j}}{\text{eV}} = 3\times10^{8}\text{ j/m}$$

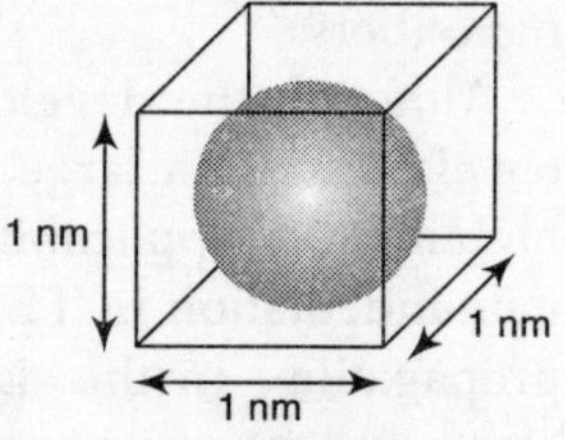

Fig. 1.33

Optical energy density is not a standard figure of merit of most photonics engineers. Serving as a comparison to a knon quantity, we now consider the sun. As the reader is well aware, all life on earth as well as most of the heat in our solar system, is powered by the optical energy radiating from the surface of the sun. The sun can be modeled as an ideal black body, which permits the use of a variation of the Stefan's law to determine the radiant intensity over all wavelengths emanating from the sun's surface. Dividing the intensity of velocity yields the optical energy density, which is computed below, assuming a black body temperature of 5,400K.

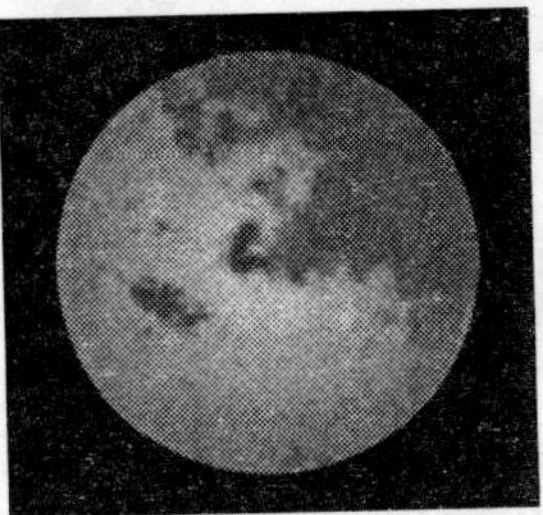

Fig. 1.34

The discrepancy between the energy densities of these systems is startling. The optical energy density, summed over all of the frequencies emanating from the sun, is more than ***a billion times smaller*** than that of a single red photon focused to the nanoscale. Clearly, compressed optical modes have enormous energy densities. In fact, the electric field of our nano-focused photon is on the order of 10^{10} V/m.

These levels of optical compression are great enough to achieve optical non-linearities in common materials with very low photon counts. For instance, Silica glass has a non-linear index coefficient (n_2) of approximately 6×10^{-23} m^2/V^2. Achieving a 1% change in refractive index, then, requires only 8 photons. Employing more exotic nonlinear materials, such as lead-silicate or chalcogenides, will increase the changes in refractive index by order of magnitude. Thus, focusing to the nanoscale will allow for a new regime of low-power nonlinear optics.

The Limits of Conventional Optics

There are, of course, several problems implicit in focusing light to deeply sub-wavelength dimensions. These limits for homogenous media come directly from Maxwell's Equations. We begin with Maxwell's Equations[i] in differential form.

$$\nabla \times E = -\mu \frac{dH}{dt}$$

$$\nabla \times H = J_f + \frac{dD}{dt}$$

These equations simplify at optical frequencies, as the optical magnetic response is negligible, making $\mu = \mu_0$. The free current, represented by J_f, cannot respond at optical frequencies, and they too are zero. The assumption of homogenous, isotropic linear media then allows us to eliminate the auxiliary field *H*. After some simplification, we arrive at:

$$-\nabla^2 E = \frac{n^2\omega^2}{c^2} E$$

In the equation above, *n* represents the index of reflection which is equal to the square root of the relative dielectric constant (e) of the medium. The electric field may now be decomposed into a complete basis set of plane waves, with k_i representing the spatial frequency in the direction *i.*

$$k_x^2 + k_y^2 + k_z^2 = \frac{n^2\omega^2}{c^2}$$

$$K_i \leq \frac{n\omega}{c} = \frac{2\pi n}{\lambda_0}$$

This puts a fundamental limit on the achievable spatial wave-vector, which is constrained principally by the low indices of refraction in conventional optical materials. In the regime of low loss in the visible band of the spectrum, indices of refraction top out around 1.9 with flint glass. This limits the maximum wave-vecto to a spatial frequency of $4\pi/\lambda_0$.

AS is known in the art, the spot focus in the image plane can be represented as a Fourier transform of spatial frequencies. The above equations then set an upper limit on the frequency which thereby determines the minimum pitch at the image to be greater than $\lambda_p/2$.

The use of focusing optics in the regime of Fraunhoffer diffraction drastically worsens the situation. As is typically the case with diffractive optics, a circular lens is used as the focusing element. This lens forms a circular aperture which acts as a low pass spatial filter with a maximum spatial frequency of $nD/2\lambda_0 d_i$. Here D represents the diameter of the lens and d_i is the distance from the lens to the image plane. With the high spatial frequencies cut-off, the circular optic creates an Airy disk in the image plane as illustrated in Figures 1.35 (a) and (b) The diffraction limited spot is now limited to a minimum diameter of approximately 1.22 λ_p/NA where the numerical aperture (NA) is defined a n sin θ. Although typically worse in practice, this then sets the minimum pitch to greater than $0.65\lambda_0$. Clearly, focusing to 1 nm spot sizes is not possible using conventional focusing techniques at visible frequencies. To fulfill the promise of advanced microscopy and low-power non-linear optics, then, a new solution is needed.

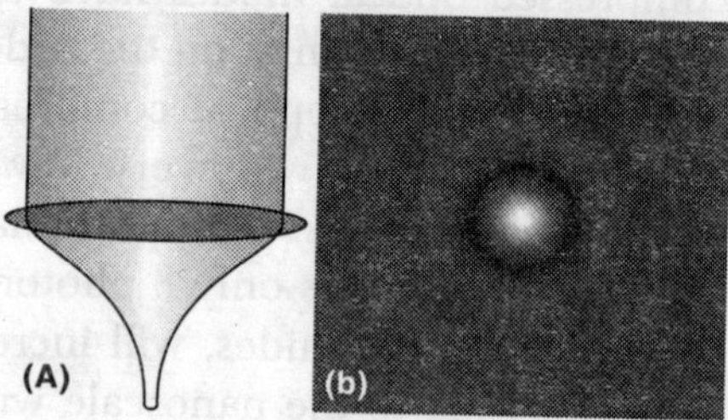

Fig. 1.35 *(a) Illustrates the diffraction limited spot size due to the clipping of the higher spatial frequencies. (b) shows the field pattern of an Airy disk.*

PLASMONS AND NANOFOCUSING

Surface plasmons have the unique ability to achieve very short wavelengths at modest optical frequencies. In addition to this, they can be stimulated with a purely optical field. They can thus mediate efficient optical coupling to the nanoscale.

It is evident that sub-wavelength focal spot cannot be formed through conventional focusing using a lens system or microscope objective. This is due, primarily, to the lack of high-index media at visible frequencies. What if, however, one was able to achieve a high *effective* index with conventional optical materials? That is the potential of surface plamsmon optics. By employing geometries of conductors (such as metal or doped semiconductor) with dielectrics (such as air or glass), modes at optical frequencies can be created with effective indices of refraction that are order of magnitude higher than those of the constituent materials. In fact, these indices can be so high as to create X-ray wavelengths (less than 10nm) with visible frequencies.

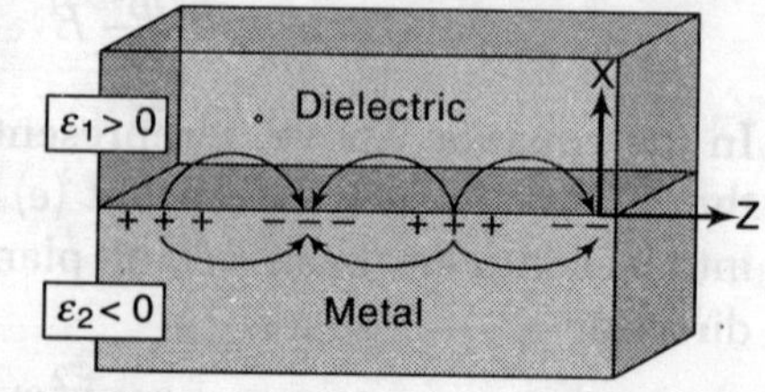

Fig 1.36

The reason surface plasmon modes can achieve anomalosuly high wave-vectors at visible frequencies is because they are mediated by electrons rather than free space optical fields. Surface plasmons are electron oscillations optical frequencies which are localized to the interface of a material with a positive dielectric constant and that of a negative dielectric constant. At wave-vectors much smaller than the Fermi wave-vector of the conductor, these modes can be well described by Maxwell's Equations. Quantum mechanical considerations only become necessary at very short plasmon wavelengths, beyond the scope of this dissertation and are not required for determining the limits of efficient focusing. At low wave-vectors, the behavior of these surface modes can be understood intuitively. In this regime, they are essentially transverse in character, although strictly speaking they have a small longitudinal component. These transverse fields generate a polarization in the dielectric which is aligned with the stimulating field. In the metal, however, the polarization will be in the opposite direction of the applied field owing to its negative dielectric constant. Now we have a situation where the stimulating field is creating equal and opposite electric displacements (*D*), in phase with each other across an interface. These opposing electric displacements serve to attract and confine the current to this interface, thus generating the collective electron oscillations of the surface plasmon.

Starting from Maxwell's Equations, it is valuable to derive the characteristics of this simple plasmonic system. Again we take the free current to be zero and the relative permeability (μ_r) of all media to be unity. Following Economou, we will assume translational symmetry and homogeneity in the direction and propagation with wave-vector *k* in the direction. The behavior in the direction is taken to be exponentially decaying away from the interface. Derived in the previous section, the wave equation tell us that the exponential decay constant in media *i* must be

$$K_i^2 = K^2 - \frac{\varepsilon_i \omega^2}{c^2}$$

We now have enough information to determine the field to with in a scale factor. In the dielectric (medium 1), we define

$$E_{1x} = Ae^{-K_1 x} e^{i(kz-\omega t)}$$

The scalar quantity *A* represent a scale factor to be determined. From Gauss's Law we may derive *Ez*

$$E_{1z} = \frac{K_1}{ik} E_{1x} = A\frac{K_1}{ik} e^{-K_1 x} e^{i(kz-\omega t)}$$

Likewise for the conductor, we amy derive E_x and E_z. In this case, we set the scale factor to unity, as only the relative scale factor carries importance

$$E_{2x} = e^{-K_2 x} e^{i(kz-\omega t)}$$

$$E_{2z} = \frac{K_2}{ik} e^{-K_2 x} e^{i(kz-\omega t)}$$

Applying the electromagnetic boundary conditions at the interface allows one to then solve for *A*. *The* continuity of E_z and D_z yield

$$A = \frac{\varepsilon_2}{\varepsilon_1} = -\frac{K_2}{K_1}$$

Some simple algebra may then be used to solve for *k*, finally generating the dispersion relation for these simple surface plasmon modes[v].

$$K = \frac{\omega}{c}\sqrt{\frac{\varepsilon_1\varepsilon_2}{\varepsilon_1 + \varepsilon_2}}$$

The wave-vector is no longer a linear function of permittivity as in standard dielectrics. Because we have the sum of dielectrics of opposite sign in the denominator, very large wave-vectors are possible.

These wave-vectors, of course, are intimately tied to the dispersion of the constituent materials. One cannot discuss the plasmon dispersion relations without a model for the relative permittivities. Dielectric materials tend to have fairly constant permittivities over large bandwidths, while conductors tend to be very dispersive. To illustrate the dispersion relations of a typical material system, a Drude model serves as simple approximation for a metal. Here we will assume a lossless Drude metal and denote the plasma frequency as ω_p.

$$\varepsilon_m = 1 - \frac{\omega_p^2}{\omega^2}$$

Taking free space as the dielectric material allows us to generate a dispersion relation, plotted in Figure 1.37. For generality, the frequency is plotted in units of the plasma frequency (ω_p) and the wave-vector in units of ω_p / c. As a point of reference, the light line is plotted as a grey dashed line and represents the dispersion relation of an optical field propagating in the dielectric medium along the same direction as the surface plasmon.

While this a very simplified material system, there are two important characteristics of surface plasmons which are evident in this model. The first is that the dispersion relation always lies at higher wave-vectors than the light line. Due to the difference in wave-vector, then, the plasmon field cannot efficiently couple to radiating modes. Conversely, free-space optical fields cannot directly stimulate surface plasmons unless a mechanism introduces additional momentum. Of course, this result is to be expected from our initial assumptions. By defining the mode to have an exponential decay normal to the surface, we assured an imaginary wave-vector in this dimension. The absolute square of this positive quantity then adds to the light-line wave-vector to determine k^2 (as in Equation), hence k must always be greater than that of the free space field.

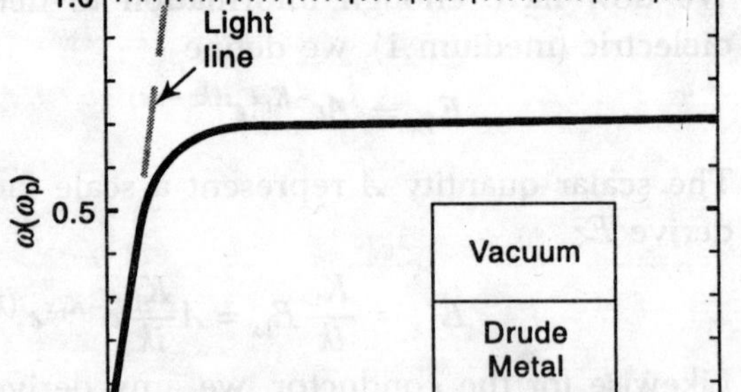

Fig. 1.37 *Dispersion relation for a Drude Metal plotted in units of the plasma frequency. The dashed line represents the light line.*

The second defining feature of this dispersion relation occurs as the frequency approaches $0.7\omega_p$. Here the wave-vector grows very large, orders of magnitude beyond the light line. It is precisely this feature which we will exploit to allow efficient focusing to the nanoscale. Although the permittivities of these materials may be modest, their geometry and interactions create an effective index much larger than that available from conventional transparent media. The reader may observe that these wave-vectors seem to violate Heisenberg's Uncertainty Principle for optical fields in these material. Because these modes are mediated by collective electron oscillations with sub-Angstrom wavelengths, they may achieve very large *optical* wave-vectors. In the parlance of plasma physics, the energy is in the from of oscillating charge separations between the negative electrons and the positive ionic background of the metal, waves that extend to large wave-vector. Various groups have reported low loss plasmons which involve clever schemes to keep a vast majority of the optical energy in the dielectric materials rather than in the metals. This however, makes those modes incapable of achieving very large wave-vectors.

Microscopy Techniques

Microscope Imaging

Some of the common techniques employed for the investigation of nanostructures physical analysis techniques, microscopical analytical techniques are currently in wide use.

There exist around ten possible primary probes (e.g., electrons, X-rays, ions, atoms, light (visible, ultraviolet, infrared), neutrons, sound) which may be used to excite secondary effects (electrons, X-rays, ions, light, neutrons, sound, heat, etc.) from the region of interest in the sample.

Many different techniques have been attempted, most of which use ions, electrons, neutrons or photons as the primary probes.

1. When *ions* are used as probes, generally ions and /or photons are emitted from the sample and analysed; example: secondary ion mass spectrometry (SIMS), Rutherford backscattering spectrometry (RBS) and proton-induced X-ray emission (PIXE).
2. When *electrons* are used as probe, generally electrons and/or photons are emitted and analysed; example: scanning electron microscopy (SEM), electron probe microanalysis (EPMA), analytical transmission electron microscopy (TEM) including energy-dispersive X-ray (EDX) analysis and electron energy loss spectroscopy (EELS), Auger electron spectroscopy (AES), low-energy electron diffraction (LEED) and reflection high-energy electron diffraction (RHEED).
3. When *photons* are used as probes, generally electrons and/or photons are emitted and analysed; examples include light microscopy, X-ray diffraction (XRD), X-ray fluorescence (XRF), X-ray absorption spectroscopy (XAS), infrared spectroscopy (IR), Raman spectroscopy and X-ray photoelectron spectroscopy (XPS).

In recent years there has been the development of proximal probe techniques such as atomic force microscopy (AFM), scanning tunnelling microscopy (STM) and scanning tunnelling spectroscopy (STS) which monitor the interaction between a localized probe and a sample surface.

Electron beam techniques

Transmission electron microscopy (TEM) is used to investigate the internal structure of micro-and nanostructures. It works by electrons through materials in conventional light microscopes. Because the wavelength of the electrons is much shorter than that of light, much higher spatial resolution is attainable for TEM images than for a light microscope. TEM can reveal the finest details of internal structure, in some cases individual atoms. The sample used for TEM must be very thin (usually less than 100 nm), so that many electrons can be transmitted across the specimen. However, some materials, such as nanotubes, nanocrystalline powders or small cluters, can be directly analysed by deposition on a TEM grid with a carbon support film. TEM and high-resolution transmission electron microscopy (HRTEM) is adequately equipped, chemical analysis can be performed by exploiting the interactions of the electrons with the atoms in the sample.

The scanning electron microscope (SEM) uses many of the basic technology developed for the TEM to provide images of surface features associated with a sample. Here, a beam of electrons is focused to a diameter spot of approximately 1nm in diameter on the surface of the specimen and scanned back and forth across the surface. The surface topography of a specimen is revealed either by the reclected (backscattered) electrons generated or by electrons ejected from the specimen as the incident electron decelerate secondary electrons. A visual image, corresponding to the signal produced by the interaction between the beam spot and the specimen at each point along each scan line, is simultaneously built up on the face of a cathode ray tube similar to the way that a television picture is generated. The best spatial resolution currently achieved is of the order of 1nm.

Scanning probe techniques

Scanning probe microscopy (SPM) uses the interaction between a sharp tip and surface to obtain an image. The sharp tip is held very close to the surface to be examined and is scanned back-and-forth. Here, a sharp conducting tip is held sufficiently close to a surface (typically about 0.5 nm) that electrons can 'tunnel' across the gap. The method provides surface structural and electronic information with atomic resolution. The invention of the STM led directly to the development of other 'scanning probe' microscope, such as the atomic force microscope (AFM). The AFM uses a sharp tip on the end of a flexible beam or cantilever. As the tip is scanned across the sample, the displacement of the end of the cantilever is measured, usually a laser beam. Unlike the STM, where the sample has to be conductive, an AFM can image insulating materials simply because the single corresponds to the force between the tip and sample, which reflects the topography being scanned across.

There are several different modes for AFM. In contact mode, the tip touches the sample; this is simple to implement but can lead to sample damage from the dragging tip on soft materials. Tapping mode mitigates this difficulty: the tip is oscillated and only touches intermittently, so that dragging during scanning is minimized Non-contact mode is where the tip senses only the attractive forces with surface, and causes no damage. It is technically more difficult to implement since these force are weak compared with contact force in non-contact model at larger tip-surface separation, the imaging resolution is poor. However, at small separation, true atomic resolution can be achieved in non-contact mode AFM.

Optical tweezers (single beam gradient trap)

Optical tweezers use a single laser beam (focused by a high-quality microscope objective) to a spot on a specimen plane. The radiation pressure and gradient forces from the spot creates an 'optical trap' which is able to hold a particle at its centre. Small interatomic forces and displacements can then be measured. Samples that can analysed range from single atoms and micrometre-sized spheres to strands of DNA and living cells. Optical tweezers are now a standard method of manipulation and measurement. Numerous traps can be used simultaneously with other optical techniques, such as laser scalpels, which can cut the particle being studied.

SCANNING PROBE MICROSCOPY (SPM)

Scanning probe microscopy cover the methods where a sharp top is scanned over a surface in a raster pattern and the interaction with the surface is recorded in each pixel to form an image of the interaction. There are a multitude of methods and interaction in SPM.

The main known SPMs are

1. Scanning Tunneling Microscopy (STM)
2. Atomic Force Microscopy (AFM).
3. Near-field Scanning Optical Microscopy (NSOM), or (SNOM),

Scanning Tunneling Microscopy

Scanning tunneling microscopy (STM) allows the real space imaging for surface with atomic resolution and has no lenses. The experimental arrangement is shown in Figures 2.1 and 2.2. A sharp conducting tip (often tungsten), acting as the anode, is brought close to the surface of the sample which is scanned by moving the probe with a piezo accuator (the cathode).

The tip is biased with respect to the sample as shown in the figure. The tip can be moved towards or away from the sample, i.e., in the –z or +z directions in the firm by means of piezoelectric actuators. At Ångstrom-scale distances, a quantum-mechanical effect, called tunneling, causes electrons to flow across the tip/sample gap, and a current can be detected. To first approximation, the tunneling current depends exponentially on the distance between tip and sample. This current is kept constant by a feedback circuit that controls the piezoelectric actuators. Because of the current/distance relationship, the distance is also kept

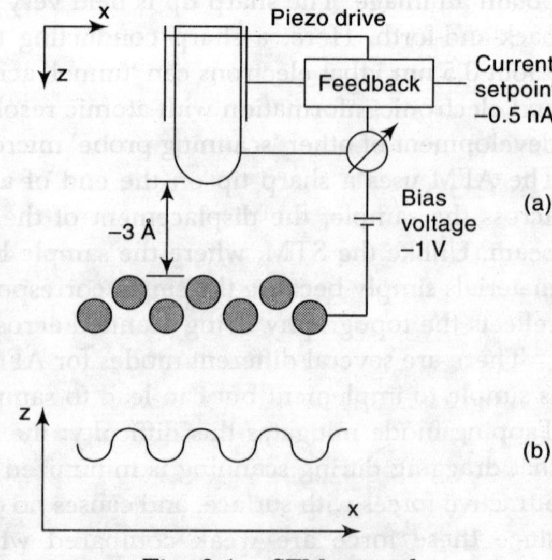

Fig. 2.1 *STM principles*

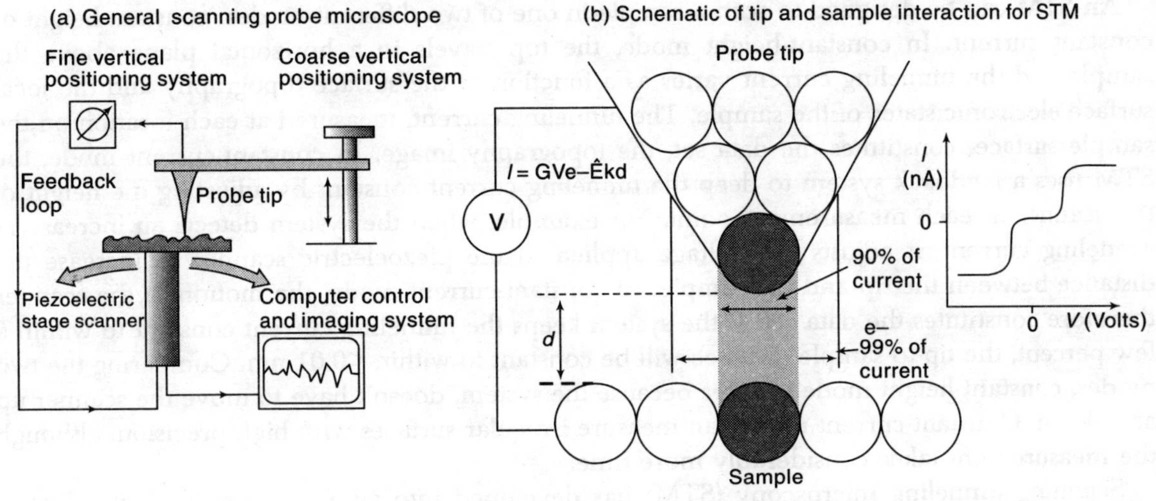

(c) 3D rendered image of a semiconductor surface

Fig. 2.2 *Schematic diagram of (a) a general scanning probe microscope, (b) the tip-sample interaction and I-V curve for STM, and (c) a typical STM image.*

constant, and the z accuracy is very high because an y small z variation causes an exponential error in the current. Additional piezo motors drive the tip in a xy scanning motion. Since the tip/sample gap is kept constant by the feedback, the scanning tip traverses a surface parallel to the sample surface. The result of a scan is a $z(x, y)$ terrain, map, with enough resolution to detect atomic-scale feature of the sample, as indicated diagrammatically in the figure (b).

A bias voltage ranging from 1 mV to 1 V is applied between the tip and the sample. Above the specimen surface there is an electron could due to surface atoms, and when the tip is brought within about 1 nm of the sample surface, electrons tunnel across the gap, causing a current to flow. The direction of electron tunneling across the gap depends on the sign of the bias voltage. It is this current that is used to generate an STM image. The tunneling current falls of exponentially with distance between tip and surface, and if the tip or sample is scanned laterally using piezoelectric drivers, the STM image reflects the variation in the sample surface topography. If the system is carefully damped from mechanical vibrations, the STM image will possess a sub-angstrom resolution vertically and atomic resolution laterally. The tunneling current also depends on the atomic species present on the surface and their local chemical environment. No vacuum is required for the adsorption of species on surfaces.

An STM can be designed to scan a sample in one of two different modes: constant height or constant current. In constant-height mode, the top travels in a horizontal plane above the sample and the tunneling current varies as a function of the surface topography and the local surface electronic states of the sample. The tunneling current, measured at each location on the sample surface, constitutes the data set, the topography image. In constant-current mode, the STM uses a feedback system to deep the tunneling current constant by adjusting the height of the scanner at each measurement point. For example, when the system detects an increase in tunneling current, it adjusts the voltage applied to the piezoelectric scanner to increase the distance between the tip and the sample. In constant-current mode, the motion of the scanner therefore constitutes the data set. If the system keeps the tunneling current constant to within a few percent, the tip-to-sample distance will be constant to within <0.01 nm. Comparing the two modes, constant-height mode is faster because the system, doesn't have to move the scanner up and down, Constant-current mode can measure irregular surfaces with high precision, although the measurement takes considerably more time.

Scanning tunneling microscopy (STM) has developed into an *imaging method* with diverse possibilities for *real-space imaging on a scale* which extends to *atomic dimensions.* A means of *imaging the structure and electrical properties of insulating layers thin enough* to permit electron tunneling and *also its local conductivity is developed.*

Electron Tunneling Microscopy

Electron Tunneling occurs between two conductors separated by a sufficiently thin insulating layer or, in physical terms, potential barrier. The tunneling current I is a measure of the overlap in the separating gap (or intervening insulator) of the wave functions of the two electrodes, I is a function of the electrode separation and the nature of the electronic states involved. The most common tunneling junction consists of a planar conductor-insulator-conductor sandwich in which the insulator is usually an oxide layer formed by oxidation of one of the electrodes [see Figure 2.2(a)]. For free electrons tunneling through a planar barrier at small voltages $V \ll \phi$, the current density, j, can be written as

$$j = (e^2/h) \cdot (\kappa_0/4\pi^2 s) \cdot V \cdot \exp(-2\kappa_0 s), \qquad \text{(i)}$$

where s is the effective tunnel distance in Å, κ_0 the inverse decay length of the wave-function density outside the surface, V the applied voltage in volts, and $(e^2/h) = 2.44 \times 10^{-4}\ \Omega^{-1}$; κ_0 is given by the expression $2\kappa_0$ (Å) $= 1.025\sqrt{\phi(\text{eV})}$, with ϕ the effective barrier height [$\approx(\phi_1 + \phi_2)/2$]. Here, the factor 1.025 is replaced by 1, distance and energies are given in units of Å and eV, respectively; and voltage (in volts) or energy (in eV) are designated as V. The effective barrier height of a vacuum gap between the two electrodes is, the average of the two respective work functions; most often a few eV. Oxide barrier heights are considerable smaller, usually below 1 eV. Figure 2.2(b) shows an energy-distance schematic of a typical metal-oxide-metal tunneling junction. At higher voltages, the effective barrier height in (i) becomes a function of V. In the case of non-free electrons and non-planar barriers, the total current I can no longer be expressed in such a simple and closed form, as there is an exponential dependence of I on some effective tip-surface separation and barrier height. The electrons close the Fermi level $E_F(1)$ [in conjunction with the empty states at $E_F(2) + V$] contribute most effectively to the tunneling current since they experience the smallest effective barrier height. The resulting energy is reflected in the current-voltage characteristic.

In STM, the signal is given by tunneling electrons of two surfaces brought at atomic scale distance. To a first approximation, the tunnelling current depends exponentially on the applied bias voltage and on the two conductors distance (d) according to the equation:

$$I = cVe^{-kd} \tag{1}$$

where c and K are constants.

A STM can operate in constant-height or in constant-current mode. In constant-height mode, the distance between the tip and the sample is kept constant and the variation of the tunnelling current is recorded. In the constant-current mode, the tunneling current is kept constant by adjusting the tip distance with a feedback loop system (the mapping signal is given by these adjustments). The constant height mode is usually employed to characterize the electronic properties of the surface. For example, a high resolution STM image of Highly-Oriented Pyrolytic Graphite (HOPG). The hexagonal close-packed structure of carbon atoms of graphite, with a periodicity of 2.5 Å (1 Å = 10^{-10} m), there typically, HOPG is chosen as test sample to validate the performance of STM, and in particular to calibrate the piezo-electric scanning tube.

The operation of the STM depends on tunneling distance of the tunneling current. If one electrode is formed into the shape of a tip, the tunneling current is confined to a filament between the apex of the tip and the surface or object under investigation [see Figure 2.2(c)]. The tunneling current decreases by roughly an order of magnitude for every distance increase by 1 Å, and the effective diameter of the filament, L_{eff}, can become extremely small for a pointed tip. In the case of a single atom at the apex, L_{eff} decreases to a magnitude of atomic dimension. By using vacuum, gas, or a liquid as the tunneling barrier between tip and surface, the tip and thus the tunneling current filament can be scanned freely across the surface. A possible STM, which uses a piezoelectric tripod for fine-positioning of the tip, is depicted in Figure 2.3(a). By adjusting the z-position of the tip, to keep the tunneling current constant, contours of constant tunneling current can be obtained, as illustrated in Figure 2.3(b). On a surface with uniform electronic properties, this procedure yields an image of the surface topography. Figure 2.4 shows an STM image of a clean Au(100) surface, exhibiting atomically flat terraces with monolayer steps lines.

Electronic and chemical surface properties manifest themselves primarily in the voltage dependence of the tunneling current. They appear as specific features in the local I-V, V-s, or I-s characteristics. In practice, electronic or chemical images are obtained by recording dI/dV or dI/ds while scanning and controlling the gap width by keeping the average current constant. Depicted in Figure 2.5 is an example of such scanning tunneling spectroscopic (STS) imaging. In (a), the strong peak at 0.8 V is attributed to surface nickel oxide. In (b), the surface is imaged with respect to that spectroscopic feature (top) and I = constant (bottom). Whereas the STS images differ dramatically, the STM images remain unchanged. Because of the usually close relation of the tip-to-sample spacing at I = constant with the topography, of dI/dV with the local density of states, and of dI/ds with the local barrier height (or work function), associated images are often referred to as topographical or STM images, spectroscopic or STS images, and work-function profiles, respectively.

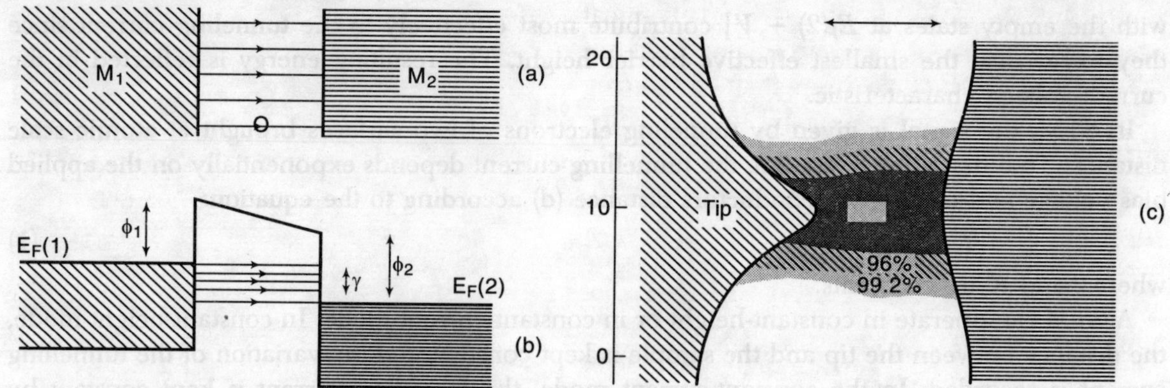

Fig. 2.3 *(a) Planar metal-oxide-metal junction (M_1-O-M_2),. The current density, is uniform. (b) a tunneling junction having a trapezoidal barrier (ϕ_1, ϕ_2) of thickness s. As indicated, only electrons tunneling from states between $E_F(1)$ and $E_F(1) - V$ in M_1 to unoccupied states between $E_F(2)$ and $E_F(2) + V$ in M_2 contribute to the tunneling current. The current density (per unit energy) is highest at $E_F(1)$. (c) Calculated current density distribution for tunneling from a tip to a corrugated surface, assuming an effective barrier height of 2.41 eV, a tip radius of 1.7 Å and s = 4 Å.*

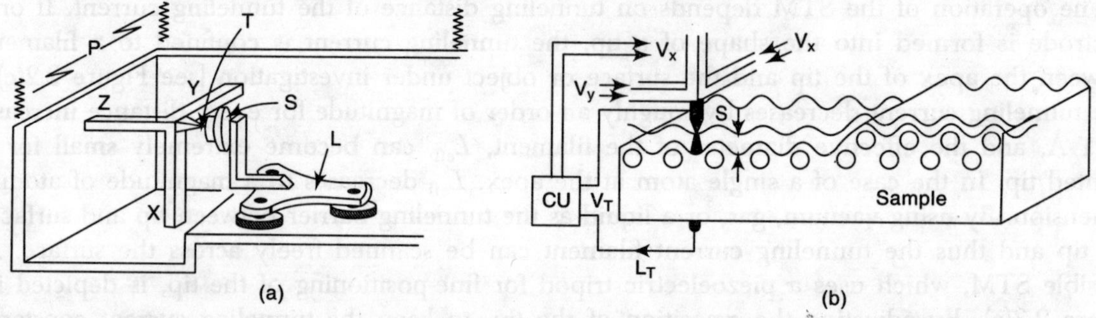

Fig. 2.4 *Scanning tunneling microscope and its operation. The tip T of the microscope depicted in (a) is scanned over the surface of a samples S with a piezoelectric tripod (X, Y, Z). The rough positioner L brings the sample within reach of the tripod. A vibration filter system P products the instrument from external vibrations. In the constant tunneling current mode of operation, a voltage V_z is applied to the Z piezoelectric element by means of the control unit CU depicted in (b) to keep the tunneling current constant while the tip is scanned across the surface by altering V_x and V_y. Electronic inhomogeneities also produce structure in the tip trace, as illustrated on the right above two surface atoms having excess negative charge.*

An important aspect of the STM is its nondestructive nature, viz., no perceptible irreversible damage occurs as a result of its use. Reversible local changes on the surface induced by the vicinity of the tip or by the high local electrical fields, take place. This is estimated by measuring at different currents or voltages. In addition, STM can be performed without appreciably altering surface electric fields. This simply requires application of a tunneling voltage which is equal and opposite to the contact potential [[$\phi_1 - \phi_2$ in Figure 2.2(b)]. STM can also be used to intentionally induce permanent local structural or chemical modifications.

Possibilities range from spurring chemical processes selectively by appropriate choice of electron energy or field strength to producing structural alternations caused by impact of the tip on the surface. Thus, STM can be used as a local position-control or displacement-monitoring system.

Various instruments analogues to the STM have been built. The exploit physical properties other than the tunneling effect on which the STM is based. The most common of these other instrument is the Atomic Force Microscope (AFM), which is based on interatomic forces. All of these instruments are collectively known as Scanning Probe Microscopes (SPMs).

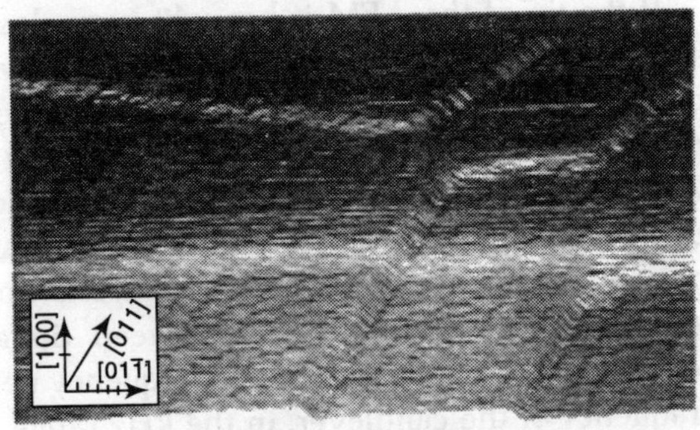

Fig. 2.5 *STM images of a clear Au(100) surface obtained at a constant tunneling current of 1 nA. The wavy structure is real and can be resolved into individual atomic rows. The divisions on the axis correspond to spacings of 5 Å.*

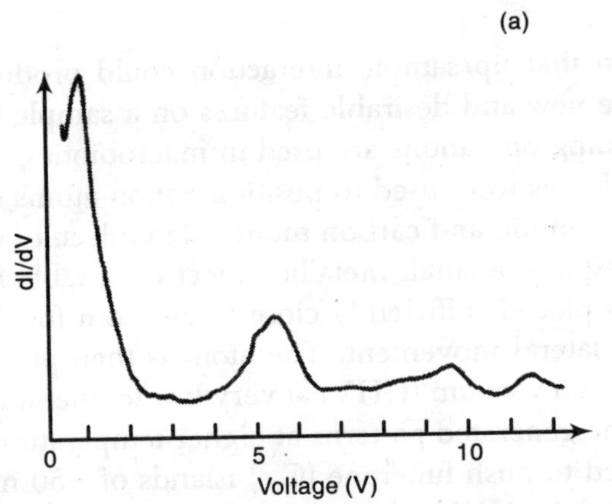

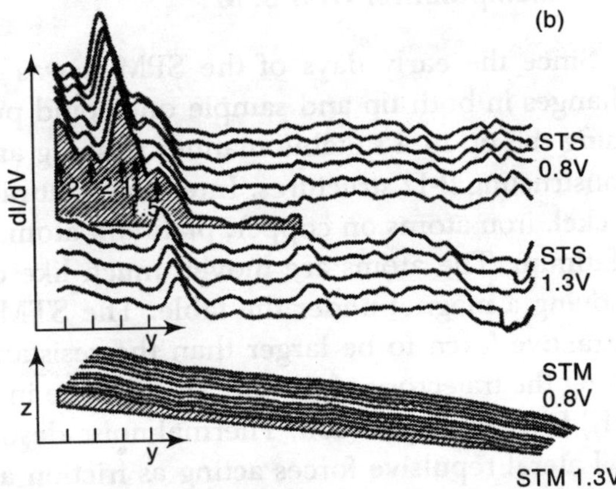

Fig. 2.6 *Spectroscopic and structural imaging of NiO on a Ni(100) surface. Shown in (a) is dI/dV vs. V from an oxide-covered region. The strong peak at 0.8 V is characteristic of NiO. The STS and STM images shown in (b) were obtained at indicated bias voltages of 0.8 and 1.3 V. An oxide island to the left is evident in the STS image obtained at 0.8 V. Spatial separations in units of the NiO lattice spacings are indicated at the left. The oxide island is hardly noticable in the STM images., The divisions on the y ands z axes correspond to spacings of 5 Å.*

The principles of operation of the AFM are shown in Figure 2.7. The forces between atoms in the tip and sample cause a deflection of the cantilever that carries the tip. The amount of deflection is measured by means of a laser beam bouncing off the top of the cantilever. (There are other schemes for measuring deflection.) The force depends on the tip/sample gap, and therefore serving on the force ensures that the distance is kept constant while scanning, as in the STM. The AFM does not require conducting tips and samples, and therefore has wider applicability than the STM.

If the tip of the AFM is brought very close to the sample, at distance of a few Å, repulsive forces prevent the tip from penetrating the sample. This mode of operation is called *contact mode.* It provides good resolution but cannot be used with delicate samples, *e.g.* biomaterials, which are damaged by the contact forces. Alternatively, the AFM tip can be placed at distances in the order of several nm or tens of nm, while the interatomic forces between tip and sample are attractive. The tip is vibrated at a frequency near the resonance frequency of the cantilever, in the kHz range. The tip/sample force is equivalent to a change in the spring constant of the cantilever, and causes a change in its resonance frequency. This change can be used as the error signal in the feedback circuit that controls the tip. (There are alternative detection schemes). This mode of operation if called *non-contact* mode. It has poorer resolution than contact mode but can be used with delicate samples.

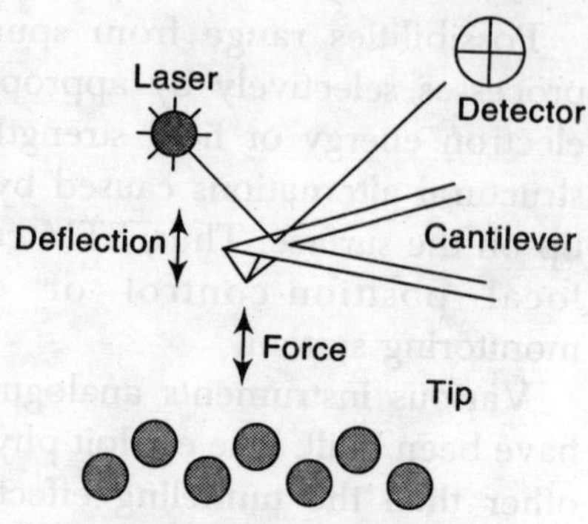

Fig. 2.7 **AFM principles**

Nanomanipulation with SPM

Since the early days of the SPM it was known that tip/sample interaction could produce changes in both tip and sample one could produce new and desirable features on a sample by using the tip in a suitable manner pushing and pulling operations are used in macrobiotics, for constructing 2-D structures. Interatomic attractive forces were used to position xenon atoms on nickel, iron atoms on copper, platinum atoms on platinum, and carbon monoxide molecules on platinum. The atoms are moved much like one displace a small metallic object on a table by moving a magnet under the table. The STM tip is placed sufficiently close to an atom for the attractive force to be larger than the resistance to lateral movement. The atom is then pulled along the trajectory of the tip. This is done in ultra high vacuum (UHV) at very low temperature (4K) for stable operation. Thermal noise destroys the generated patterns at higher temperatures.

Lateral repulsive forces acting as friction are used to push fullerene (C_{60}) islands of ~50 nm size on flat terraces of a sodium chloride surface, in UHV, at room temperature, with a modified AFM. The ability to move the islands depends strongly on species, substrate interaction, e.g., C_{60} does not move on gold and motion on graphite destroys the island pairs of antimonium atoms are rotated between two stable orientations 90 degrees apart. This is done in UHV at room temperature, on a silicon substrate, by scanning with an STM tip with a higher voltage than required for imaging. The rotation is reversible, although several scan are sometimes necessary to induce the desired motion.

Molecular Robotics is used to push colloidal gold nanoparticle with 15 nm diameters on a mica substrate at room temperature and in ambient air. The image is made in non-contact mode, then disable the feedback and push by moving the tip in a single line scan, without removing the top oscillation voltage. Figure 2.8 shows on the right a "USC" pattern written with gold nanoparticles. The z coordinate is encoded as brightness in this figure. On the left is the original random pattern, before manipulation with the AFM. Figure 2.9 shows the same "USC" pattern displayed in perspective.

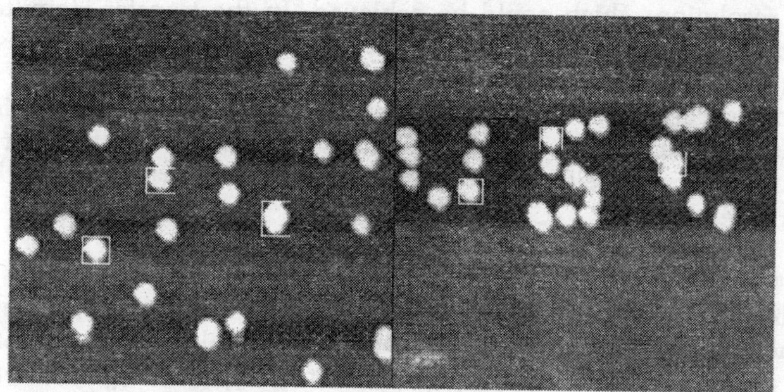

Fig. 2.8 *The initial pattern of 15 nm Au balls (left) and the "USC" pattern obtained by nanomanipulation (right).*

Smaller objects are arranged into prescribed patterns at room temperatures by pushing molecules at room temperature in UHV by using an STM. eg: prophyrin molecules are pushed on copper' bucky balls (i.e., C_{60}) are arranged in a linear pattern, using an atomic step in the copper substrate C_{60} molecules on silicon also have been pushed with an STM in UHV at room temperature. Here, the STM tip is brought closer to the surface than in normal imaging mode, and then scan across a rectangular region with the feedback turned off. This causes many probe crashed. The tip is brought close to the surface, the sweep is done with the feedback on and a high value for the tunneling current.

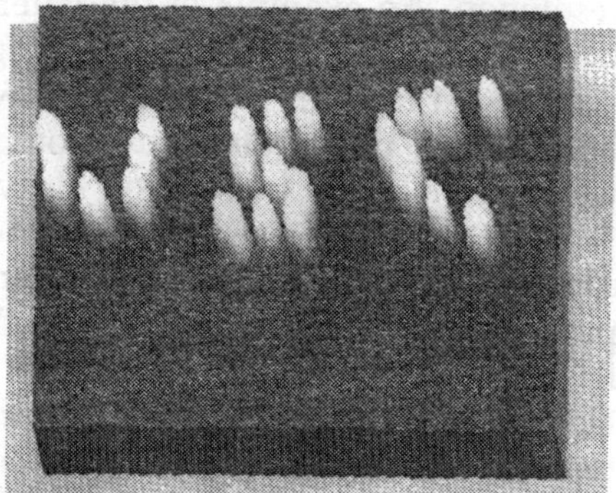

Fig. 2.9 *The "USC" pattern viewed in perspective*

Much of industrial macrorobotics is concerned with pick and place operations, which typically do not require very precise positioning or fine control. There are few examples of experiments in which atoms or molecules are transferred to SPM tips, these are moved, and the atoms transferred back to the surfaces.

Transferring xenon atoms from platinum or nickel surfaces to an STM to by moving the tip sufficiently close for the adsorption barriers of surface and top. An atom may leave the surface and become adsorbed to the tip, or vice-versa. Benene molecules also have been transferred to an from tips e.g. transfer xenon atoms between an STM tip and a nickel surface by applying voltage pulses to the tip. This is attributed to electromigration, caused by the electric current flowing in the tunneling junction in UHV at 4K. Transferring silicon atoms between a tungsten tip and a silicon surface in UHV at room temperature, by applying voltage pulses to the tip field-induced evaporation, aided by chemical phenomena at the tip/surface interface.

Imaging Structures

The two theoretical approaches, use of the transfer Hamiltonian and the direct-transmission approach provided the first insights regarding the physical basis for STM imaging and the associated lateral resolution which could be achieved. In the transfer-Hamiltonian approach, the tunneling tip is regarded as tracing contours of constant wave-function density, evaluated at the center of the (assumed) spherical tip, as it traverse a sample surface in the constant-current mode. The transmission approach provides an instructive view of the current distribution in the tunneling filament. Both approached lead to similar expressions for the lateral resolution. In particular, the minimum periodicity, a_m, of a sinusoidal surface corrugation with amplitude c_s to be resolved is, for free electrons.

$$a_m = \sqrt{\pi/\ln A} \cdot L_{\text{eff}} \leq \pi\sqrt{2/\ln A} \cdot \sqrt{r+s}/\phi^{1/4} \tag{i}$$

where r is the radius of curvature of the tip, s is the average width of the tunneling gap, ϕ is the effective barrier height, $A = c_s/c_d$, with c_d the smallest observable corrugation of the equicurrent surface traced by the tip (or, in effect, the gap-width stability), and L_{eff} is the effective diameter of the tunneling current filament. The dominant factors are tip radius, gap width, and barrier height; the vertical stability enters only logarithmically. Equation (i) is expected to be a good approximation for a tip for which the jellium concept is meaningful, i.e., a metal with a radius of curvature of, say, 7 Å or more. Not taken into account are image potential and nonuniform barrier heights. The former drastically decreases ϕ at small gap widths and thus affects the resolution. An example of the latter is the change of the barrier height at locations with strong curvatures, caused by the smearing out of the electronic charge. Enhanced focusing can be achieved if imaging is performed with a single, rather localized electron orbital such as could be provided by an adsorbate atom at the apex of a metal tip. Lateral resolutions less than 2 Å which have been obtained on cleaved graphite are believed to be due to such an effect.

STM experimentals have provided substantial insight into the structure of reconstructed surfaces. The general nature of such surfaces prepared by using standard surface-preparation procedures, are very often not as flat as assumed. They look quite rough, with "terraces" and flat "lakes" embedded in a hilly landscape, but, retain their "ideal" surface structure and electronic properties–even or quite restricted flat regions and in the immediate vicinity of defects, defects areas and step-lines.

The topographical imaging capability of STM is by no means restricted to atomic-resolution levels. Imaging with a lateral resolution in the nanometer scale is certainly also accessible and of broad importance. Applications thus far include studies of thin-film growth and technological preparation methods new roughness standards for industry important surfaces and imaging of biological matter. Lateral resolution at ambient pressure in the nanometer range with a vertical stability of about 1 Å is now standard with relatively simple instrumentation. Imaging is direct, without the need for decoration, staining, or replica technique which in nanometer-scale electron microscopy provide the mass contrast or transparency required. Chemical and electronic effects can dominate topographical STM image in the angstrom range, but should be much less important in nanometer imaging. Thus, the images, as obtained, are generally expected to be representative of the topographical structure being examined. Finally, adaptability to the environment is unique for imaging in both the angstrom and nanometer ranges, with fascinating prospects such as imaging of biological matter under living conditions.

A final point regarding the imaging of high-resistance materials in the constant-current mode: The potential drop in the sample, $\Delta V = IR_s$, where R_s is the spreading resistance, can vary across the surface. The applied voltage, therefore, must be larger than ΔV everywhere on the portion of the surface under examination when imaging in the constant-current mode. The relatively high voltage which was necessary in the first STM experiment on the 7 × 7 reconstruction of the Si(111) surface could be attributed to such a spreading resistance effect.

The SPM as a Robot

Assume an SPM is a 3 degree-of-freedom robot. It can move in *x*, *y* and *z*, but cannot orient its tip, which is the analog of a microbiotic hand. (How could a 6 degree-of-freedom SPM be built, and what could be done with it.) The vertical displacement is controlled accurately by a feedback loop involving tunneling current or force (or other quantities for less common SPMs). But nanocycle *x*, *y* motion over small (e.g., within a 5 micron square) is primarily open-loop, because of a lack of suitable sensors that can be used in a feedback scheme. Accurate horizontal motion relies on calibration of the piezoelectric actuators, which are known to suffer from a variety of problems such as creep and hysteresis. In addition, thermal drift of the instrument is very significant. At room temperature a drift of one atomic diameter per second is common, which mean that manipulation of atomic objects on a surface is not unlike picking parts from a conveyer belt. Thermal drift is negligible if the SPM is operated at very low temperatures, and all the experiments in atomic-precision manipulation to date have been done at 4K.

For room-temperature nanomanipulation, drift, creep and hysteresis must be taken into account automatically. Also, the SPM often must operate in a liquid environment, especially if the objects to be manipulated are biological.

Sensing

The SPM functions both as a manipulator and a sensing device. The lack of a direct and independent means of establishing "ground truth" while navigating the tip causes delicate problems. The SPM commands are issued in robot coordinates, and sensory data are acquired, whereas manipulation tasks are expressed in sample or *task* coordinates. These two coordinates systems do not coincide, and indeed are in relative motion due to drift. Accurate motion in task coordinates is achieved, by tracing features of the sample and moving relative to them.

A simple example of a practical implementation of these ideas is illustrated in Figure 2.10. To construct the pattern shown in Figure 2.8 and 2.9 we move the tip to the approximate position of a nanoparticle to be pushed, and then search for it through single-line scans.

Feature tacking assumes that features are stationary in task coordinates. The assumption may fail at room temperature if the features are sufficiently small, because of thermal agitation. Hence, atomic manipulation at room temperature require artificially introduced features that can be tracked to establish a task coordinates system. It must deal with the spatial uncertainty associated with the thermal motion of the atoms to be moved. Larger objects such as molecules and clusters have lower spatial uncertainty, and should be easier to handle.

The SPM output signal depends not only on the topography of the sample but also on the shape of the tip, and on other characteristics of the tip/sample interaction. For example, the tunneling current in an STM depends on the electronic wavefunctions of sample and tip. To

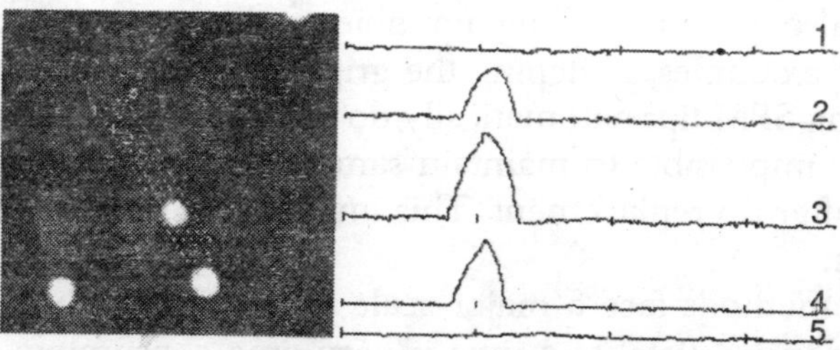

Fig. 2.10 *Searching for a nanoparticle with 5 line scans in the direction of motion*

first approximation one may assume that the tip and sample are in contact. Under this assumption one can use configuration-space techniques to study the motion of the tip. Figure 2.11 illustrates the procedure in 2-D. On the top of the figure consider a tip with triangular end and a square protrusion in the sample. In the bottom consider a tip with a semi-circular end and the same protrusion. Choose a reference point for the tip its apex, and reflect the tip about the reference point. The configuration-space obstacle that corresponds to the real-space protrusion obstacle is obtained by sweeping the inverted top over the protrusion so that the apex remains inside the obstacle in contact with the obstacle is the detected topographical signal. As shown in the figure, the sensed topography has been broadened by the dimensions of the tip. Also, the detected height is correct. Tip effects are called "convolution" and "deconvolution when" the image removed tip effects.

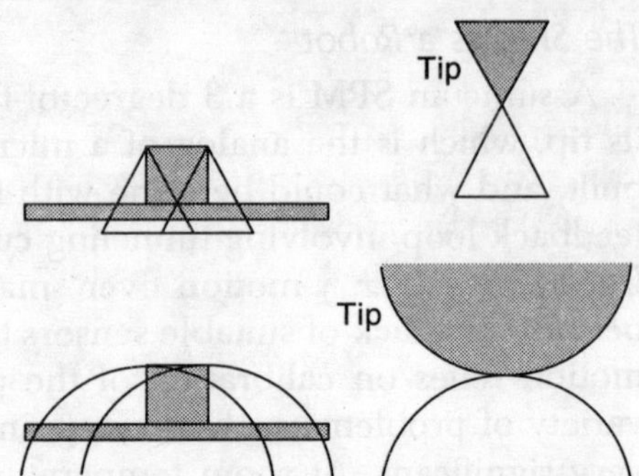

Fig. 2.11 *Configuration space obstacles for a triangular tip (top) and for a semi-circular tip (bottom)*

The configuration-space analysis outlined above is purely geometric and provides only approximation to the PSM output signal.

A major issue in tip effect compensation is that the shape of the probe is not known, and indeed may vary during operation. For example, atoms maybe adsorbed on the tip or lost because of the contact with the sample. The most promising approach for dealing with tip effects consists of estimating the tip shape by using it to image known features. If necessary, artificial features may be introduced into the scene for tip estimation purposes. The estimated tip shape can then by used to remove (at least in party) the tip effect from the image.

Sensor fusion techniques may be used, at least in principle, for increasing the quality of the sensory data, because it is possible to access several signals during an SPM scan. For example, vertical and lateral force (akine to friction) can be recorded simultaneously in typical AFMs.

The SPM tip is the primary end effector in nanomanipulation. A plain, sharp tip seems to be adequate for most pushing operations. In some cases it may also suffice for picking and depositing objects, especially in conjunction with electrostatic forces generated by applying a suitable bias to the tip. This requires both a conducting tip and a conducting substrate. Pick-and-place tasks, however, usually require the nanoscale analog of a *gripper*.

One can thick of a nanogripper as a molecular or cluster of molecules that are attached to a tip and capable of picking up and transporting other molecules or particles. These grippers are switchable, so as to pick and release objects on command, example molecules such as cyclodextrins, which have large cavities that can carry other molecules. Coating a tip with strands of DNA may also permit picking up objects that are coated with complementary strands. In both of these examples, switching the gripper on and off is not a solved problem.

Technique for changing SPM tips automatically do not exist. Changes must be done manually, and it is very difficult or impossible to maintain sample registration, i.e., to return to the same position on the sample after tip replacement. This implies that one often must image using a tip with an attached gripper.

SPM-based assembly methods face a major scale-up challenge. Building complex structures one atom (or even one nanoparticle) at a time is very time consuming.

ATOMIC FORCE MICROSCOPY

Introduction

Physicists are able to combine atoms, chemists create molecules and biologists create biomolecules with sizes on the nanometer scale. A greater challenge is to engineer structures in the mesoscopic scale of 10–1000 nm. Methods for creating mesoscopic structures include optical lithography, and beam lithography. Although successful, these techniques are limited in their use with Atomic Force Microscope (AFM), it is now possible to create mesoscopic sized devices of optical and ebeam techniques. ***The atomic force microscope is the most widely used scanning probe instrument in nanoscience and nanotechnology.*** The AFM combines surface profilometry and scanning tunneling microscopy and can operate under ambient conditions on both insulating and conducting materials. In general, the AFM can be used to measure, modify, or manipulate surface structures.

Theory

An atomic force microscope is optimized for measuring surface features that are extremely small, thus it is important to be familiar with the dimensions of the features being measured. An atomic force microscope is capable of imaging features as small as a carbon atom and as large as the cross section of a human hair. A carbon atom is approximately .25 nanometers (nm) in diameter and the diameter of a human hair is approximately 80 microns (4m) in diameter.

The common unit of dimension used for making measurements in an atomic force microscope is the nanometer. A nanometer is one billionth of a meter.

1 meter = 1,000,000,000 nanometers

1 micron = 1,000 nanometers

Another common unit of measure is the Angstrom. There are ten angstrom (*E*) in one nanometer:

1 nanometer = 10 Angstrom

Magnification in an atomic force microscope is the ratio of the actual size of a feature to the size of the feature when viewed on a computer screen. Thus when an entire cross section of a human hair is viewed on a 500 MM computer monitor (20 inch monitor)

Magnification = 500 mm/.08 mm = 6,250 X

In the case of extremely high resolution imaging, the entire field of view the image may be 100 nanometers. In this case the magnification on a 500 mm computer screen is:

Magnification = 500 mm/(100 nm*1 mm/1,000,000 nm) = 5,000,000 X

Piezoelectric Ceramic Transducer

Mechanical motion is created from electrical energy with an electromechanical transducer. Electrical motors such as used in washing machines are the most common types of electromechanical transducer. The electromechanical transducer most commonly used in an atomic force microscope is the piezoelectric ceramic. A piezoelectric material undergoes a change in geometry when it is placed in an electric field. The amount of motion and direction of motion depends on the type of piezoelectric material, the shape of the material, and the field strength. Figure 2.12 shows the motion of a piezoelectric disk when exposed to an electric potential:

A typical piezoelectric material will expand by about 1 nm per applied volt. Thus, to get larger motions it is common to make piezoelectric transducers with hundreds of layers of piezoelectric materials.

By using one thousand layers of piezoelectric material it is possible to get motions as large as 1000 nm per volt. Thus with volts it is possible to get 0.1 mm of motion with a multiple layer piezoelectric transducer.

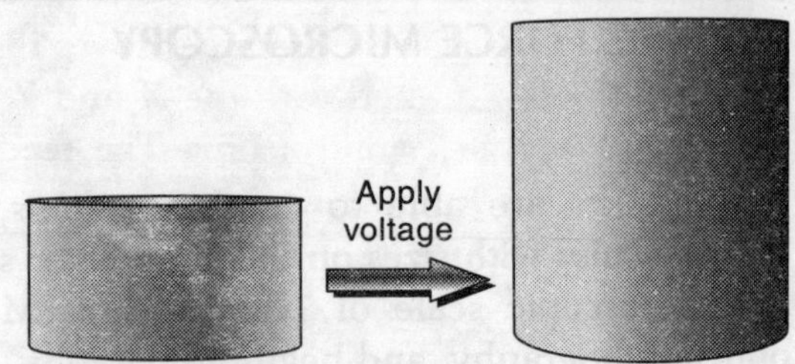

Fig. 2.12 *When a voltage is applied to the top and bottom surface of the piezoelectric disc, the disc will expand.*

AFM Instrumentation

In an AFM a constant force is maintained between the probe and sample while the prove is raster scanned across the surface. By monitoring the motion of the probe as it is scanned across the surface, a three dimensional image of the surface is constructed.

The constant force is maintained by measuring the force with the "light lever" sensor and using a feedback control electronic circuit to control the position of the Z piezoelectric ceramic. See Figure 2.13.

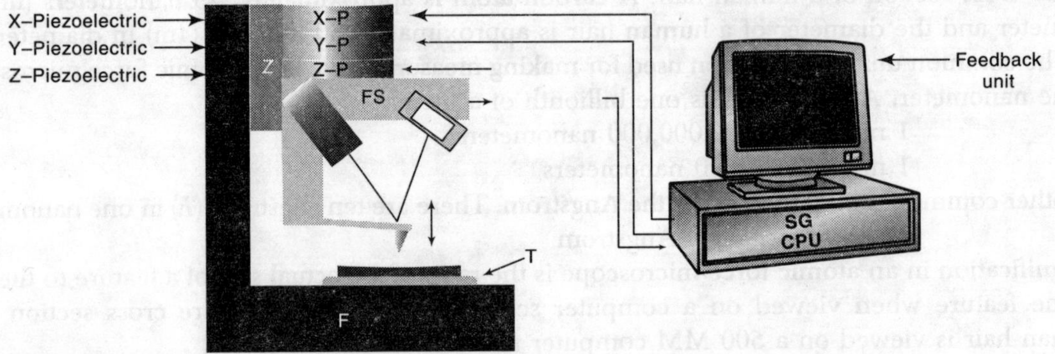

Fig. 2.13 *Components and subsystems of an atomic force microscope system.*

The motion of the probe over the surface is generated by piezoelectric ceramics that move the probe and force sensor across the surface in the X and Y directions.

(Z) Coarse Z motion translator–This translator moves the AFM head towards the surface so that the force sensor can measure the force between the probe and sample. The motion of the translator is usually about 10 mm.

(T) Coarse X-Y translation stage–The XY translation stage is used to place the section of the sample that is being imaged by the AFM directly under the probe.

(X-P) X and Y piezoelectric transducer–With the X and Y piezoelectric transducer the (Y-P) probe is moved over the surface in a raster motion when an AFM image is measured.

(FS) Force Sensor–The force sensor measures the force between the probe and the sample by monitoring the deflection of a cantilever.

(ZP) Z piezoelectric Ceramic–Moves the force sensor in the vertical direction to the surface as the probe is scanned with the X and Y piezoelectric transducers.

(FCU) Feedback control unit–The feedback control unit takes in the signal from the light lever force sensor and outputs the voltage that drives the Z piezoelectric ceramic. This voltage refers to the voltage that is required to maintain a constant deflection of the cantilever while scanning.

(SG) X-Y signal generator–The motion of the probe in the X-Y plane is controlled by the XY signal generator. A raster motion is used when an image is measured.

(CPU) Computer–The computer is used for setting and scanning parameters such as scan size, scan speed, feedback control response and visualizing images captures with the microscope.

(F) Frame–A solid frame supports the entire AFM microscope. The frame must be very rigid so that it does not allow vibrations between the tip and the surface.

Note–Not shown, is an optical microscope that is essential for locating features on the surface of the sample and for monitoring the probe approach process.

Atomic Force Microscope

Atomic Force Microscopes (AFM) was developed about five years after STM. While STM is limited to the surface of conducting specimens, AFM is used to study nonconducting materials, such as insulators and semiconductors, and electrical conductors. As shown in Figure 2.14, in AFM a sharp tip, about 2 μm long and down to a minimum of 20 nm in diameter is used, which is scanned closely over the specimen surface, but the magnitudes of atomic forces rather than tunneling currents are monitored as a function of the probe position on the sample surface. AFMs can operate under UHV, ambient air conditions or even with the solid sample and tip submerged in a liquid cell, which is highly useful for biological systems.

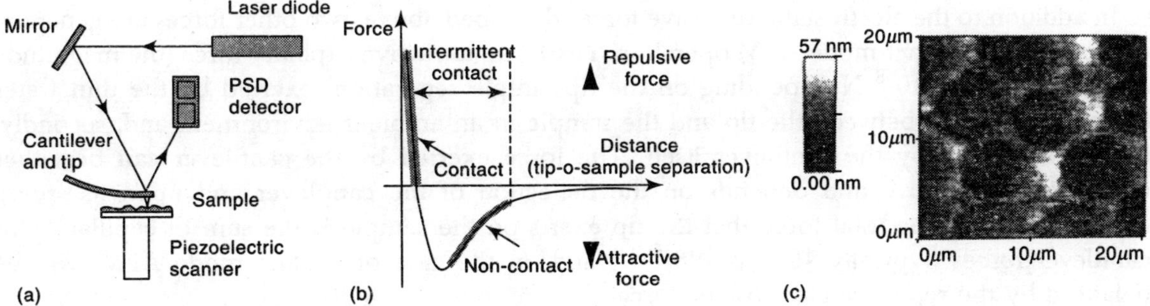

Fig. 2.14 *Schematic diagram of (a) an atomic force microscope, (b) the tip-sample interaction and (c) a typical AFM image of a carbon fibre-carbon composite.*

The AFM tip is located at the free end of a cantilever that is 010 to 200 μm long and forces between the tip and the sample surface cause the cantilever to bend, or deflect. A detector is used to measure the cantilever deflection (usually just the displacement in the *z* direction although lateral force microscopy is also possible) as the tip is scanned over the sample, or alternatively, as the sample is scanned under the tip. The measured cantilever deflections allow a computer to generate a map of surface topography. With AFM (Figure 2.14), samples a mapped according to the space dependence of the interaction forces between the tip and the sample surface. The mechanism of measurement is based on Hook's law

$$F = -kz$$

where k is the cantilever spring constant and depends on the material and dimensions of the cantilever and z is the measured deflection k is a constant. The interaction force can be calculated from the deflection of the cantilever if the cantilever stiffness is known. For a cantilever made of silicon that has dimensions of:

$L = 100$ microns; $W = 20$ min cross; $T = 1$ micron

The force constant, k is approximately 1 newton/meter. Thus if the cantilever is moved by 1 nm, a force of 1 nanonewton is required.

The forces experienced by the cantilever vary depending on the tip to sample separation. In Figure 2.15 shows, the typical dependence of the interatomic force on the distance. Thus, at short distance of the tip from the sample surface, repulsive forces dominate. These forces including exchange, hard-core, steric and Born repulsions. At large distance, attractive forces, such as Van der Waals, electrostatic or magnetostatic forces, prevail. Two different AFM regimes–i.e., contact or non-contact–correspond to these two different physical conditions. In the former, the cantilever is maintained at few Å from the sample surface, and the interaction is repulsive, whereas in the latter, the cantilever is held in a range of 10-100 of Å over the sample surface, and the intersection is attractive.

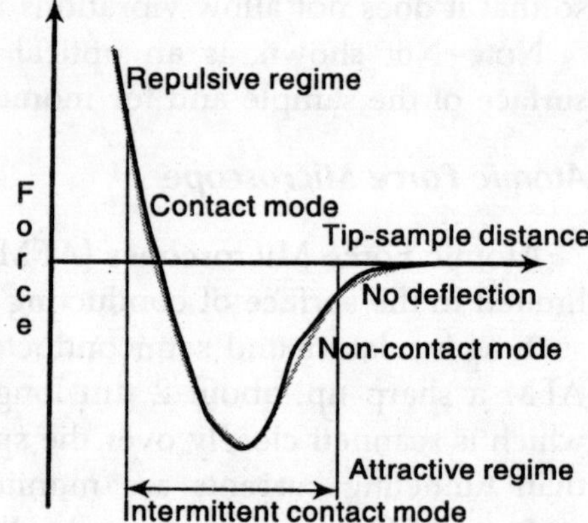

Fig. 2.15 *Typical force versus tip-sample distance curve.*

In addition to the electrostatic repulsive forces described above, two other forces are generally present during contact-mode AFM operation. Firstly, an attractive capillary force (the magnitude of which is about 10^{-8} N depending on the tip-sample separation) exerted by the thin water layer often present between the tip and the sample in an ambient environment and, secondly, the force exerted by the cantilever itself. The force exerted by the cantilever can be either repulsive or attractive and depends on the deflection of the cantilever and upon its spring constant. Overall the total force that the tip exerts on the sample is the sum of capillary plus cantilever forces (typically 10^{-8} to 120^{-6} N) and, in the case of contact mode AFM, will be balanced by the repulsive electrostatic force.

In non-contact AFM mode, a stiff cantilever is vibrated near the surface of a sample close to its resonant frequency. The spacing between the tip and the sample is of the order of tens to hundreds of Angströms and this is similar to the vibration amplitude of the tip. As the tip approaches the surface, this is detected as a change in the resonant frequency or vibration amplitude of the tip oscillation, and this can provide sub-Angström vertical resolution in the topographic image. Generally the detection system monitors the resonant frequency or vibrational amplitude of the cantilever and keeps it constant with the aid of a feedback system that moves the scanner up and down, ensuring that the average tip-to-sample distance also remains constant.

Non-contact AFM mode provides a means of measuring sample topography with minimum contact between the tip and the sample, thus removing any possibility of contamination or degradation of the surface by the tip, as can occur during contact AFM scanning. Furthermore, the total force between the tip and the sample in the non-contact regime is very low, approximately 10^{-12} N, which is beneficial for studying soft or elastic samples. Generally both contact and non-contact topographic images of rigid samples will appear similar. However, the presence of a few monolayers of liquid (e.g., water) on the surface of a sample will result in non-contact AFM mode imaging the surface of the liquid layer, whereas operating in contact mode will penetrate the liquid layer to image the underlying surface.

A further AFM variant is intermittent-content or 'tapping' mode. This is similar to non-contact mode except the vibrating cantilever tip is brought closer to the sample so that the bottom of its travel it just touches the sample. Tapping mode is less likely to damage the sample than contact AFM because it eliminates lateral forces (friction or drag) between the tip and the sample. Furthermore, in comparison with non-contact mode, tapping mode can image larger scan sizes that can include greater variations in sample topography.

The position of the cantilever is detected optically using a focused laser beam reflected from the back of the cantilever onto a position-sensitive (photo)-detector (PSD). Any bending of the cantilever results in a shift in the position of the focused laser spot on the detector and the PSD and measure such displacements to an accuracy of <1 nm. The ratio of the path length between the cantilever and the detector to the length of the cantilever, produces an additional mechanical amplification so, the overall system can detect sub-angstrom vertical movements of the cantilever tip. Other methods of detecting cantilever deflection relay on optical interference, or even an STM tip to read the cantilever deflection above a gold contact. If the cantilever is fabricated from a piezoresistive materials, this obviates the need for a laser and PSD system as any deflection will result in strain, thereby changing the cantilever's resistivity, which can be monitored using an integrated electrical circuit.

The AFM data sets are generated by operating in either constant-height or constant-face mode. In constant-height mode, the scanner height is fixed during the scan and the spatial variation of the cantilever deflection is recorded. Using the feedback control in the AFM, it is possible to scan a sample with a fixed cantilever deflection Fig. 2.16. Because the deflection of the cantilever is directly proportional to the force on the surface, a constant force is applied to the surface during a scan. This scanning mode is often called "contact" mode. However, because the forces of the probe on the surface are often less than a nano-newton, the probe is minimally touching the surface.

Fig. 2.16 *In contact mode AFM the probe directly follows the topography of the surfaces as it is scanned. The forces of the probe is kept constant while an image is measured.*

The cantilever in an AFM can be vibrated using a piezoelectric ceramic. When the vibrating cantilever comes due to a surface, the amplitude and phase of the vibrating cantilever may change Fig. 2.17. Changes in the vibration amplitude or phase are easily measured and the changes can be related to the force on the surface. This technique has many names including non-contact mode, and intermittent contact mode. It is important that the tip not "tap" the surface because the probe may be broken or the sample may be damaged. Constant-height mode is employed for recording real-time images of changing surface, (where high scan speeds are essential). It is also often used for taking atomic-scale images of atomically flat surfaces, where the cantilever deflections and thus variations in applied force are small. Conversely, in constant-force mode the deflection of the cantilever can be used as input to a feedback circuit that moves the scanner up and down in the *z* direction, responding to the topography by keeping the cantilever deflection (and thus also the total force applied to the sample) constant. Constant-force mode data sets are therefore generated form the scanner's motion in the *z* direction. In constant-force mode, the speed of scanning is limited by the response time of feedback circuit, but the total force exerted on the sample by the tip is well controlled. Constant-force mode is generally preferred for most applications.

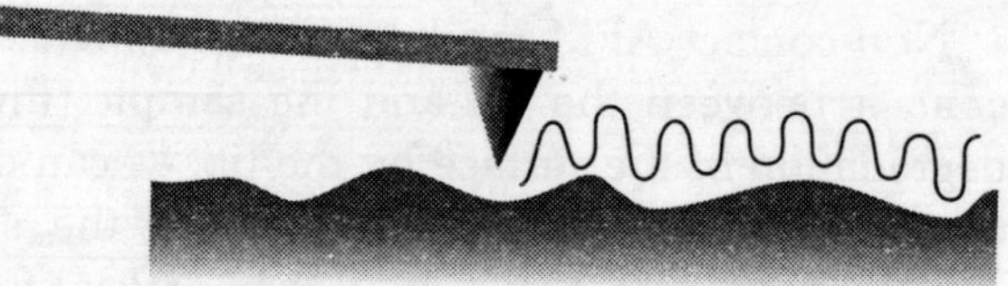

Fig. 2.17 *In vibrating methods, changes in probes vibrations are monitored to establish the force of the probe onto the surface. The feedback unit is used to keep the vibrating amplitude or phase constant.*

AFM may also be used to measure the local elastic properties of a point on the surface or even a whole surface. Here the force on an AFM cantilever tip in the *z* direction is monitored as a function of the *z* position of the piezoelectric scanner tube. This force versus distance curve will very with changes in local elastic properties of the surface, or the presence of contaminants and lubricants.

Generally AFM can only be used to provide information about the surface of a sample; and subsurface information about the morphology and structure must be obtained by scattering techniques (such as SEM or reflectometry) or sputter depth profiling (dynamic SIMS). It is often possible to reveal structural information of mixtures by selectively dissolving one component with an appropriate solvent. AFM is performed on the surface of, say, a film of a two component mixture before exposing that film to a solvent which dissolves only one componenty of the mixture. The change in morphology will tell us where that component resided in the film. However, on certain samples it is possible to reveal three-dimensional structural information with nanometre resolution. This technique, termed nanotomopgraphy, involves imaging the surface of a film before etching only a small amount in a plasma. The same part of the film is imaged again using AFM before another etching step is performed. The images can be stacked together to produce the final three-dimensional volume image. Image reconstruction is complicated by the difficulty in obtaining the same scan location *exactly* each time, and also by the fact that different components etch at different rates, hence the need for a careful alignment of images (registration) similar to that used in magnetic resonance imaging.

Imaging Modes

The AFM can be operated in a number of modes, depending on the application figure. In general, possible imaging modes are divided into static (also called contact) modes and a variety of dynamic modes.

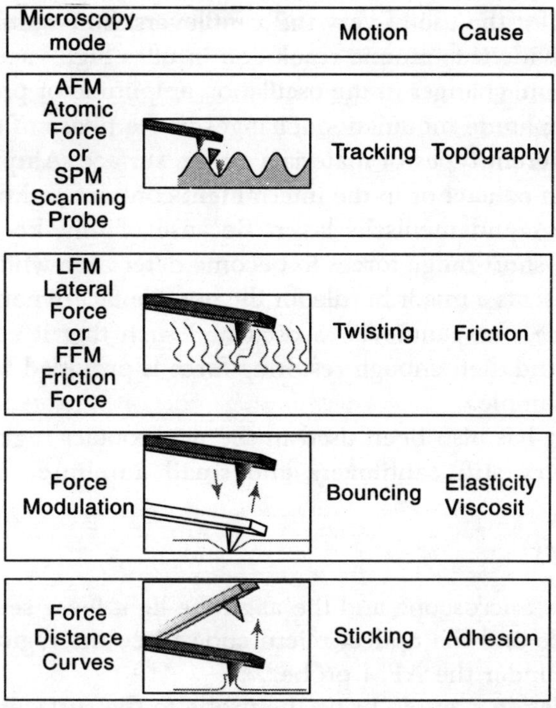

Fig. 2.18 *AFM instrumentation operates in several modes: contact, intermittent contact, noncontact and nonscanning. With both small and large samples stage, samples of many dimensions are accommodated nondestructively.*

The AFM can be used to image and manipulate atoms and structures on a variety of surfaces. The atom at the apex of the tip "senses" individual atoms on the underlying surface when it forms chemical bonds with each atom. Because these chemical interactions alter the tip's vibration frequency, they can be detected and mapped.

The primary modes of operation are contact mode and dynamic mode. In the contact mode operation, the tip deflection is used as a feedback signal. Because the measurement of a signal is prone to noise and drift, low stiffness cantilevers are used to boost the deflection signal. However, close to the surface of the sample, attractive forces can be quite strong, causing the tip to 'snap-in' to the surface. Thus in this mode AFM is always in contact where the overall force is repulsive. The force between the tip and the surface is kept constant scanning by maintaining a constant deflection.

In the dynamic mode, the cantilever is externally oscillated at or close to its resonance frequency. The oscillation amplitude, phase and resonance frequency are modified by tip-sample interaction forces; these changes in oscillation with respect to the external reference oscillation provide information about the sample's characteristics. Schemes for dynamic mode operation include frequency modulation and the more common amplitude modulation. In frequency modulation, changes in the oscillation frequency provide information bout tip-sample interactions. Frequency can be measured with very high sensitivity and thus the frequency modulation mode allows for the use of very stiff cantilevers. Stiff cantilevers provide stability to the surface and, thus provide true atomic resolution in ultra-high vacuum conditions.

In amplitude modulation, changes in the oscillation amplitude or phase provide the feedback signal for imaging. In amplitude modulation, changes in the phase of oscillation can be used to discriminate between different types of materials on the surface. Amplitude modulation can be operated either in the non-contact or in the intermittent contract regime. In ambient conditions, most samples develop a liquid meniscus layer. Because of this, keeping the probe tip close enough to the sample for short-range forces to become detectable while preventing the tip from sticking to the surface presents a major hurdle for the non-contact dynamic in ambient conditions. In dynamic contact mode, the cantilever is oscillated such that it comes in contact with the sample with each cycle, and then enough restoring force is provided by the cantilever spring to detach the tip form the sample.

Amplitude modulation has also been used in the non-contact regime to image with atomic resolution buy using very stiff cantilevers and small amplitude in an ultrahigh vacuum environment.

Measuring Images with AFM

- Place a probe in the microscope and the align the light lever sensing system.
- With the XY sample and the optical microscope place the region of the sample that will be imaged directly under the AFM probe.
- Engage the Z translation stage to bring the probe to the surface.
- Start the scanning of the prove over the surface and view thew image of the surface of the computer screen.
- Save the image on a computer disk.

Resolutions in an Atomic Force Microscope

An atomic force microscope has two measures of resolution the plane of the measurement and in the direction perpendicular to the surface.

In Plane Resolution: The in-plane resolution depends on the geometry of the probe that is used of scanning. The sharper the probe is the higher the resolution of the AFM image. Figure 2.19 below is the theoretical line scan of two spheres that are measured with a sharp probe and a dull probe.

Vertical Resolution: The vertical resolution in an AFM is established by relative vibrations of the prove above the surface. Sources of vibrations are acoustic noise, floor vibrations, and thermal vibrations. Getting the maximum vertical resolution requires minimizing the vibrations of the instrument.

Probe Surface Interactions

The strongest forces between the probe and surface are mechanical, which are the forces that occur when the atoms on the probe physically interact with the atoms on a surface. However, other forces between the probe and surface can have an impact on an AFM image. These other forces include surface contamination, electrostatic forces, and surface material properties.

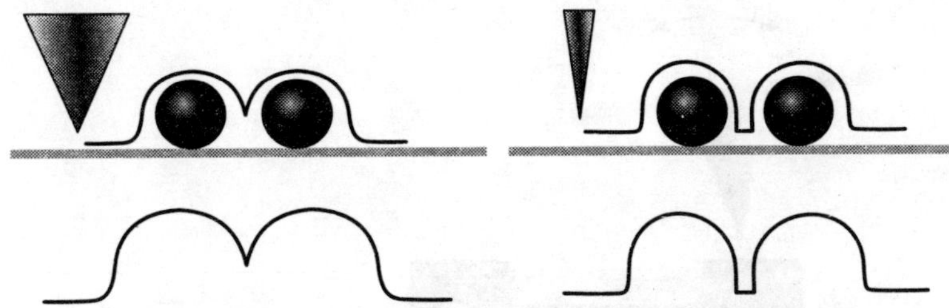

Fig. 2.19 *The image on the right will have a higher resolution because the probe used for the measurement is much sharper.*

Surface Contamination

In ambient air all surfaces are covered with a very thin layer, <50 nm, of contamination. This contamination is comprised of water and hydrocarbons and depends on the environment of the microscope. When the AFM probe comes into contact with the surface contamination, capillary forces pull the probe towards the surface.

Electrostatic Forces

Insulating surfaces can store charges on their surface. These charges can interact with charges on the AFM probe or cantilever. Such forces can be so strong that they "bend" the cantilever when scanning a surface.

Surface Material Properties

Heterogeneous surfaces can have regions of different hardness and friction. As the probe is scanned across a surface, the interaction of the probe with the surface can change when moving from one region to another. Such changes in forces can give "contrast" that is useful for differentiating between materials on a heterogeneous surface.

The interaction of the probe with the surface depends on the chemical and physical properties of the surface. It is possible to measure the interactions and thus "sense" the materials at a sample's surface.

Vibrating Sensing Mode

The AFM cantilever may be vibrated to measure the force between a probe and surface during on AFM scan. The magnitude of amplitude damping and the amount of phase change of the cantilever depends on the surface chemical composition and the physical properties of the surface. Thus, on an inhomogeneous sample, contrast can be observed between regions of varying mechanical or chemical composition. In vibrating material sensing mode, if the amplitude is fixed by the feedback unit, then the contrast of the material is observed by measuring phase changes.

Torsion Modes

In contact mode AFM it is possible to monitor the torsion motions of the cantilever as it is scanned across a surface. (See Fig. 2. 20)

Fig. 2.20 *Torsions of the cantilever are measured in the AFM. Changes in the torsion of the cantilever are an indication of changes in the surfaces chemical composition.*

The amount of torsion of the cantilever is controlled by changes in topography as well as changes in surface chemical properties. If a surface is perfectly flat but has an interface between two different materials, it is often possible to image the change in material properties on a surface. This technique is similar to lateral force microscopy (LFM).

Mechanical Surface Modification

The Atomic Force Microscope was initially developed to image the surfaces of insulating materials. As, AFM probe causes mechanical modification to a surface (see Fig. 2.21). So it is now used to alter surface topography.

Two types of mechanical surface modifications are possible. In the first, the probe is pushed into the surface. In the second, a line is scratched into the surface. The size of the feature depends on the following factors:

- Surface Material (hardness)
- Probe Diameter
- Probe Material (hardness)
- Force of the Probe on the Surface
- Probe Temperature

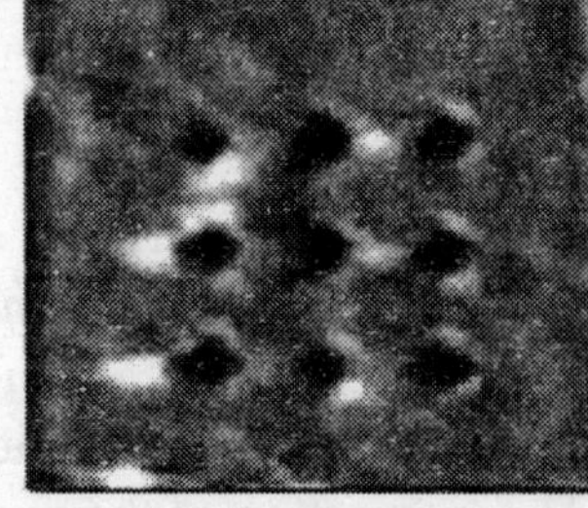

Fig. 2.21 *Mechanically modifying a surface. After indenting a 3 × 3 array of depressions ~0.4 mm in diameter and ~50 nm deep: image depth range: 103 nm. The image has an area 5.2 μm square.*

Figure 1 illustrates and example of an AFM used to indent ~0.4 μm diameter depressions in a material's surface.

Electrical Surface Modification

Electrically conducting AFM probes can be used to chemically modify a surface to "draw" an image. For example, applying an electrical bias between the conducting probe and a substrate can locally oxidize selected regions of the surface to form patterns. (See Fig. 2.22)

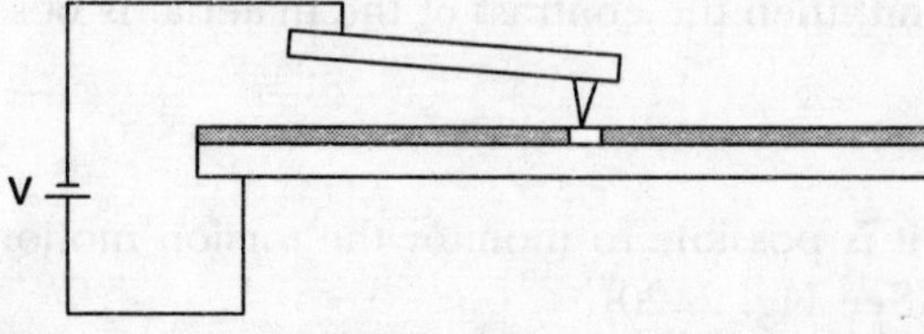

Fig. 2.22 *By placing a bias between an electrically conductive probe and a surface, the surface can be modified.*

The dimensions of the pattern drawn by electrically conducing techniques depends on:

- Diameter of the probe
- Potential between probe and surface

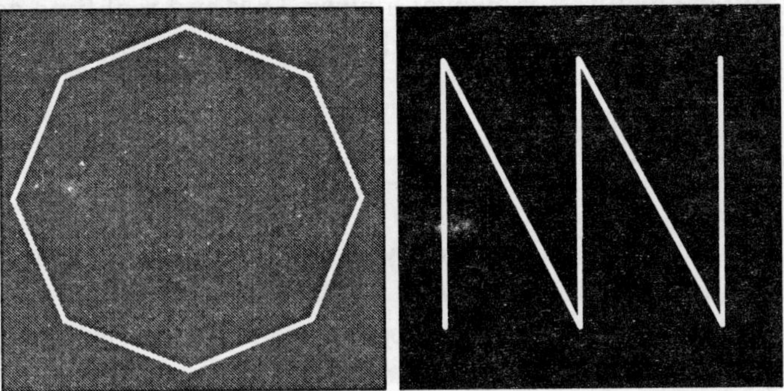

Fig. 2.23 *These patterns were drawn with an electrically conductive AFM probe using Anodic Oxidation. AT the left the, line widths are 50 nm. At the right, two lines widths were drawn using two different biases between the probe and sample.*

For growing an oxide on silicon, the line-width of the pattern ranges up to tens of nanometres. The thickness can be controlled in the range of 10–50 nm. When the writing is done in ambient air, the line width depends on the relative humidity, because water absorbed at the tip-substrate interface focuses the electric field and also acts as the anodization medium. The oxide thickness was found to depend on the electric field strength Figure 2.23 shows two patterns writing in silicon oxide with an AFM.

Advantages and Disadvantages

The AFM has several advantages over the scanning electron microscope (SEM). Unlike the electron microscope which provides a two-dimensional image of a sample, the AFM provides a true three-dimensional surface profile. The samples viewed by AFM do not require any special treatments (such as metal/carbon coatings) that can would change or damage the sample. Most AFM modes work perfectly well in ambient air or even a liquid environment. It is possible to study biological macromolecules and living organisms. AFM can provide higher resolution than SEM. It gives true atomic resolution in ultra-high vacuum (UHV). UHV AFM is comparable in resolution to Scanning tunneling Microscopy and Transmission Electron Microscopy.

A disadvantage of AFM compares with the scanning electron microscope (SEM) is the image size. The SEM can image an area on the order of millimetres by millimetres with a depth of field on the order of millimetres. The AFM can only image a maximum height on the order of micrometers and a maximum scanning area of ground 150 by 150 micrometres. At high resolution, the quality of an image is limited by the radius of curvature of the probe tip, and an incorrect choice of tip for the required resolution can lead to image artifacts. ALso, AFM can not scan images as fast as an SEM, requires several minutes for a typical scan, while and SEM is capable of scanning at near real-time (although at relatively low quality) after the chamber is evacuated. The relatively slow rate of scanning during AFM imaging often leads to thermal drift in the image, making the AFM microscope less suited for measuring accurate distances between artifacts on the image.

SFM images can be affected by hysteresis of the piezoelectric material and cross-talk between the (*x*, *y*, *z*) axes that may require software enhancement and filtering. Such filtering could "flatten" out real topographical features. However, newer AFM use real-time correction software or closed-loop scanners which practically eliminate these problems. Some AFM also use separated orthogonal scanners which also serve to eliminate cross-talk problems.

ATOMIC FORCE MICROSCOPY FOR NANOPARTICLES

The atomic force microscope (AFM) is ideally suited for characterizing nanoparticles. It offers the capability of 3D visualization and both qualitative and quantitative information on many physical properties including size, morphology, surface texture and roughness. Statistical information, including size, surface area, and volume distributions, can be determined as well. A wide range of particle sized can be characterized in the same scan, form 1 nanometer to 8 micrometers. In addition, the AFM can characterize nanoparticles in multiple mediums including ambient air, controlled environments, and even liquid dispersions.

AFM Capabilities in Nanoparticles characterization

Qualitative Analysis

(a) Using the AFM, individual particles and groups of particles can be visualized. It offers visualization in three dimensions. TEM provides images that are merely 2D projections, while generating 3D images from light microscopy and SEM are nonstandard and require rigorous calibration to a known standard.

In Figure 2.24, 73 mm NIST traceable microspheres are shown in both perspective view and top view. 3D information is incorporated in both views. In the perspective view, the 3D nature of the image and obvious in the top view, the intensity of the color reflects the height of the particle.

In material sensing mode, the AFM is distinguises between materials, providing spatial distributions information on composite materials. In Figure 2.25, material inhomogeneity is seen on a topographically flat organic film. Nanocomposites can be similar analyzed for dispersion of particulate matter.

(b) Software-based imaging processing of AFM data generates quantitative information on both individual nanoparticles and groups of nanoparticles. For individual particles, size information (length, width, and height) and other physical properties (such as morphology and surface texture) can be measured.

Groups of particles can also be measured through image analysis and data processing. Commonly desired ensemble include particle counts, particle size distribution, surface area distribution and volume distribution. With knowledge of the material density, mass distribution can also be calculated. Image processing of an AFM image are shown in Figure 2.26. A scan of colloidal gold and the particle size distribution calculated by image processing is shown in Figure 2.27.

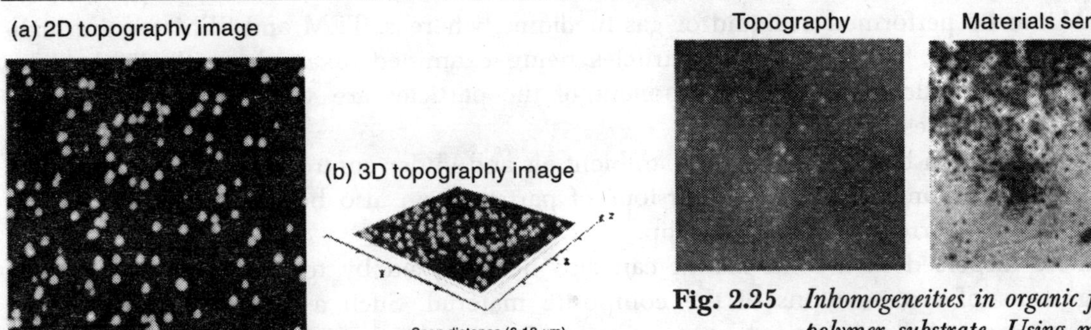

Fig. 2.24 *NIST traceable polystyrene microsphere scanned Mean W of microspheres is 73 nm. Scan size is 64 m × 64 m.*

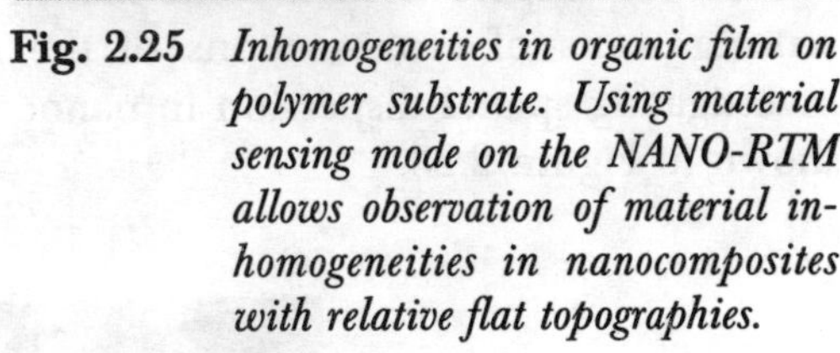

Fig. 2.25 *Inhomogeneities in organic film on polymer substrate. Using material sensing mode on the NANO-RTM allows observation of material inhomogeneities in nanocomposites with relative flat topographies.*

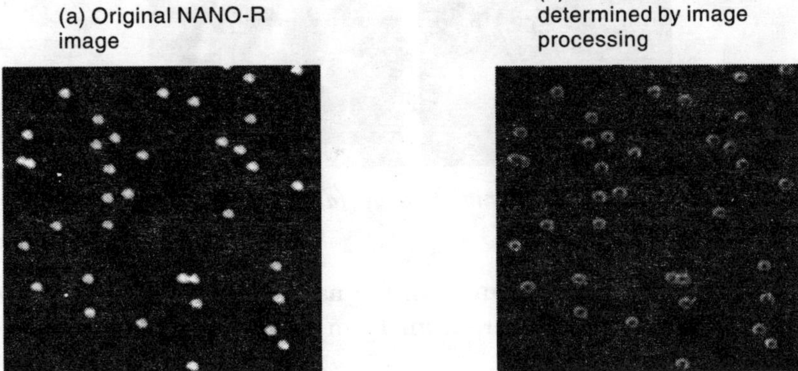

Fig. 2.26 *Polystrene micropspheres with mean diameter of 102 nm. (a) Topographical scan image. (b) Image processing used to determine boundaries of individual particles. Ensemble data can be further processed to provide statistical information such as particle size distribution.*

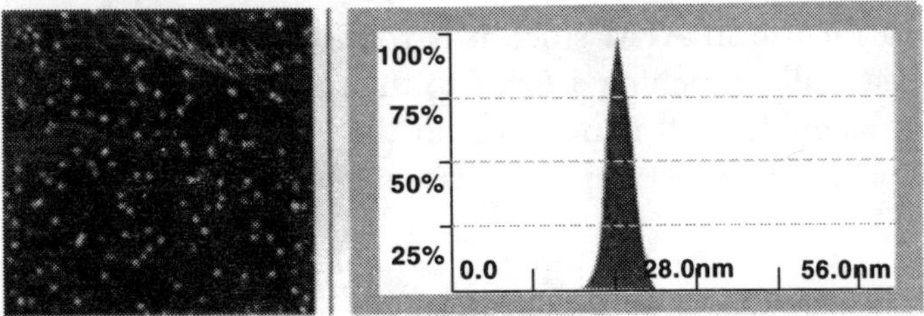

Fig. 2.27 *Scan of colloidal gold and associated particle size distribution. Mean W = 28 nm.*

Whenever data from single-particle technique is processed to provide statistical information, the concern over statistical significance exists. It is easy to attain greater statistical in AFM by combining data from multiple scans to obtain information on the larger population.

(c) AFM can be performed in liquid or gas mediums, whereas, TEM and SEM require high-vacuum conditions, which alter the particles being examined, example, with combustion-generated nanoparticles, a major component of the particles are volatile components that evaporate at low pressure.

Dry particles can be scanned in both ambient air and under controlled environments, such as nitrogen or argon gas. Liquid dispersions of particles can also be scanned, provided the dispersant is not corrosive to the probe tip.

Particles dispersed in a solid matrix can also be analyzed by topographical or material sensing scans of cross-sections of the composite material. Such a technique is useful for investigating spatial dispersion in nanocomposites. Precipitates in a nickel aluminum alloy are shown in Figure 2.28.

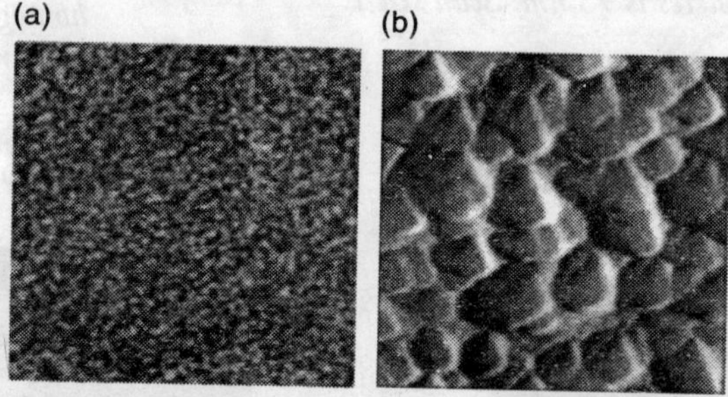

Fig. 2.28 *Ni3Al precipitates in a nickel aluminium alloy. (a) 27 mm × 27 mm topography image. (b) 3 mm × 3 mm shading image.*

(d) It many industries, the ability to scan from the nanometer range into the micron range is important. With AFM, particles anywhere from 1 nm to 8 mm in height can be measured in a single scan.

It is important to note that AFM scanning is done physically with a probe time in constant or intermediate contact. Therefore, particles must be pushed against a 2D surface during the scan.

Atomic Force Microscope with CNT Tips

CNT is an ideal tip for use in AFM since it is robust and can be just a few nanometers in diameter. However, manually attaching a CNT to the tip of a cantilever can be arduous. CVD is used to directly grow single wall nanotubes on the AFM cantilever. Figure 2.29 shows an example of this. An electrochemical approach is developed to attach a multiwall nanotube to the AFM cantilever.

The AFM with nanotube tip is used to image Mars dust. Figure 2.30 shows an image of Red Dune Sand which is a Mars analog. The CNT tip-only a few nanometers in diameter–is able to reside the image the dust particles. In addition, the resolution, the robustness of CNT provides long lasting tips in contrast to the quick wearing of silicon tops. With CNT probes metrology applications (for example, a profilometer) and nanoscale imaging of semiconductor surfaces is also possible. Figure 2.31 shows an AFM image of a thin indium film obtained using a single-walled nanotube tip. The resolution is much better then conventional silicon tips. These CNT

TEM Micrograph of Electro-deposited catalyst on tip of AFM Cantilever

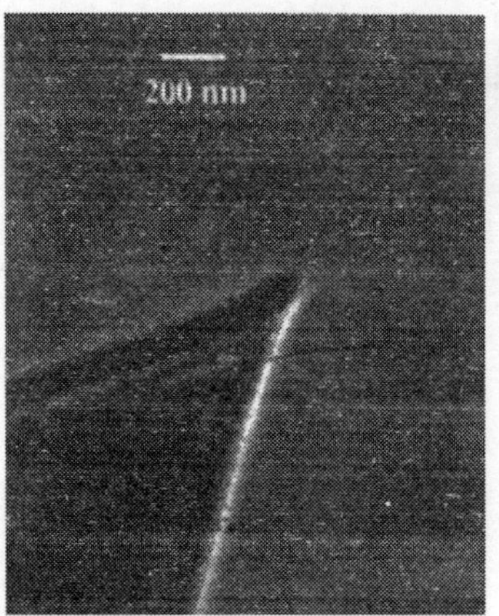

TEM Micrograph of Shortened NT Synthesized by CVD

Fig. 2.29 *Nanotube probe directly grown on AFM cantilever.*

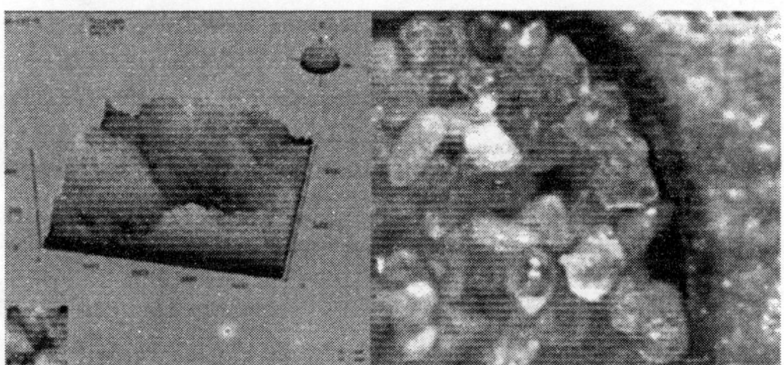

Fig. 2.30 *AFM Image of Red Due Sand (Mars Analog) obtained using a nanotube tip (left) compared to an image using optical microscope (right).*

probes are also have also been used to image biological samples both in the dry from and in aqueous environment. The latter is accomplished by coating the nanotube probe with ethylene diamine to render the probe hydrophilic in order to be able to image liquid samples.

Techniques

Because of their flexibility, nanotubes could also be used in scanning probe instruments. Since MWNT-tips are conducting, they can be used in STM and AFM instruments (Figure 2.31). Advantages are the improved resolution in comparison with conventional Si or metal tips and the tips do not suffer from crashes with the surfaces because of their high elasticity.

Nanotubes tips can be modified chemically through attachment of functional groups to be used as molecular probes.

First CNTs are picked up from a cartridge with an AFM tip (Figure 2.32) by attraction due to van der Walls forces. When a single CNT is picked up the AFM tip moves the CNT towards

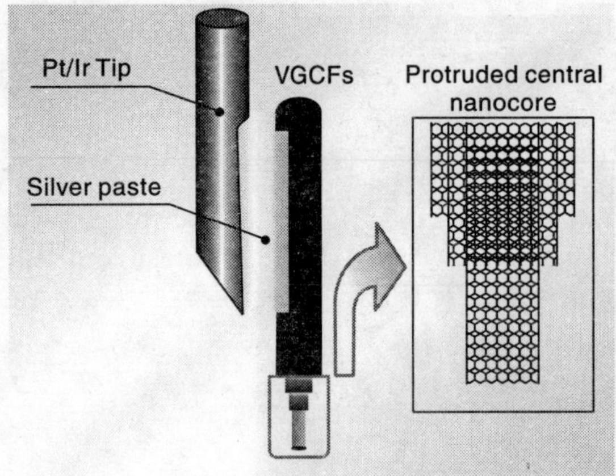

Fig. 2.31 *Use of a MWNT as AFM tip. VGCF stands for vapour Grown Carbon Fibre. At the centre of this fibre the MWNT forms the tip.*

Fig. 2.32 *(a) The AFM tip touches a single CNT in an attempt to pick it up. (b) After the CNT sticks to the tip, the tip is moved away from the cartridge towards the MEMS.*

the MEMS. Near the MEMS, the real positioning starts. In four steps (Figure 2.33) the CNT is placed across the gap and is welded to the system by Electron Beam Deposition. This is a technique that uses a focuses SEM electron beam to dissociate organic species in a specific area and deposit the residual ionized organic gas molecules on the junction of the CNT with the measuring system. It takes approximately 15 minutes to weld a single side of a CNT.

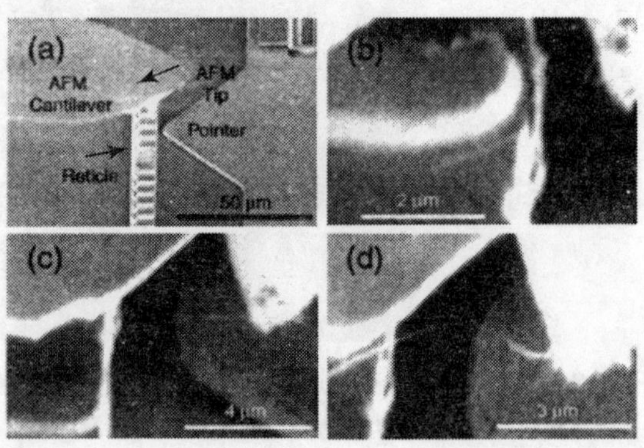

Fig. 2.33 *(a) The MEMS. AN AFM tip brings the CNT is its position. (b) When the CNT makes contact with the site it is seen by the sudden change in shape of the CNT. Now it can be welded to the site. (c) After welding the CNT is stretched. (d) The CNT is placed across the gap and also welded to the site.*

After positioning and welding, electric and/or mechanic measurements are carried out. Currents and stress strain relationships using the AFM tip to apply a strain to the CNT can also be measured.

Direct growth by Chemical Vapour Deposition of AFM Tips

Atomic Force Microscope (AFM) is an important technique in biology and chemistry for imaging and characterizing structures in liquid, ambient, and vacuum. The level of information obtained from AFM images, depends on the size, shape, and terminal functionality of the probe tips used for imaging.

An AFM probes consist of microfabricated silicon or silicon nitride cantilevers with integrated pyramidal tips. These tips can have radii of curvature as small as 5 nm but are often larger, and have pyramid half cone angles of about 35°. The larger cone angle limits the effectiveness of these tips in imaging deep and narrow features. Super sharp silicon tips with radii of curvature as small 2–5 nm and half cone angles of about 10° are also available and these tips exhibit limited stability, due to brittleness of the fine silicon tip. Alternatively approach for fabricating sharp and robust probes has involved depositing a spike of carbonaceous material onto the end of a regular pyramidal AFM and probe. These amorphous carbon tips have end radii of 5–12 nm and high aspect ratios. An advantage of these tips is that the resolution determined from images of isolated biological macromolecules reflects their radii.

Carbon nanotube tips have several advantages, including (i) high aspect ratio for imaging deep and narrow feature, (ii) low tip-sample adhesion for gentle imaging, (iii) the ability to elastically buckle rather than break when large forces are applied making them highly robust and (iv) the potential to have resolution <0.5 nm in the case of individual single-walled carbon nanotubes (SWNTs) Molecular Structures of nanotubes under controlled synthesis can make every nanotube tips with an identical structure. This feature enables nanotube probes to be characterized with a well defined transfer function. Multiwalled carbon nanotube (MWNT) tips and SWNT bundle tips have shown to image with high lateral resolution and have selectively modified at their ends with organic and biological specie allow functionally sensitive imaging. So, nanotube tips will have a large impact on imaging of biological structures and macromolecular assemblies which X-ray crystallography and electron microscopy cannot do.

As, nanotube AFM tips are fabricated by direct mechanical assembly and in this process micromanipulators are used to control the positions of a cantilever-tip and bundle of nanotubes. While viewing with an optical microscope, there are several limitations. First it selects thick bundles of nanoιubes sizes that are easiest to observe in the optical micro-scope. Mechanical assembly of nanotube tips also has been performed inside a scanning electron microscope (SEM), with diameters greater than 5–10 nm; thus increasing the overall time required to make one tip. Second, well defined and reproducible tip etching procedures to expose individual nanotubes do not exist. Third, a relatively long time is required to attach nanotubes to cantilevers. To overcome these limitations nanotubes tips (Fig. 2.34) are "grown" directly by chemical vapor deposition.

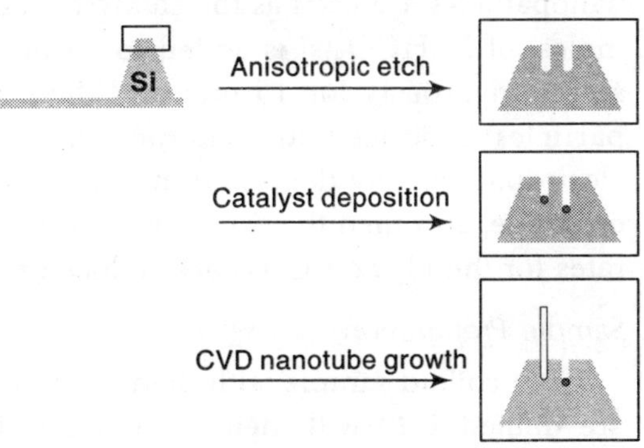

Fig. 2.34 **CVD nanotube tip preparation approach**

Here a flattened area is created at the tip pores are etched into the flat area, and catalyst is deposited into the pores. CVD then allowed nanotubes to grow from the tip aligned by the pores. This method of produces thin, individual MWNT tips, which are images biological structures with high resolution. CVD could ultimately be used to mass produce tips. The detailed studies of CVD growth of MWNT and SWNT tips, the characterization of the structural and mechanical properties of these tips, and high resolution imaging of individual IgG and GroES proteins are given below.

CVD MWNT Tip Preparation

Silicon AFM $k = 0.5–5$ N/m, are flattened by scanning on silicon or CVD diamond surfaces in contact mode at high load (≈1 μN), high scan speed (30 Hz), and a large scan size (40 μm) for several minutes. Thus creating flat area of 1–5 μm^2 at the tip. Under the view of an optical microscope, the last 50 μm of the cantilever/tip assembly are placed in an ≈10μl drop of 2.5% HF (aq) supported by two parallel grounded platinum wires acting as counter electrodes. To create ≈100-nm-diameter, 1-μm-deep pores, the tips are anodized at 2.1 V for 100 sec. rinsed with ethanol, and then dried in air; the end of the cantilever is anodised to reduce signal-to-noise. Anodized tips are etched in 0.03% KOH (aq.) for 50 sec, rinsed with ethanol, and dried in air to remove excess silicon and open up the pores at the surface iron catalyst is electrochemically deposited into the pores, under the view of an optical microscope, from an aqueous 0.01 M $FeSO_4$/0.05 M H_2SO_4 solution at –1.4 V versus the platinum counter electrodes. The tips are rinsed in ethanol to remove residual iron solution. They are oxidized in a flow of pure O_2 at 288 K/min to 873 K held for 20 min, and are cooled at 288 K/min to room temperature. For CVD nanotube growth, the tips the were heated at 268 K/min to 1073 K in a flow of 950 standard cubic centimeters per minute (sccm) argon (Ar) and 40 sccm hydrogen (H_2). At 1073 K 10 sccm of ethylene (C_2H_4) is added for 10 min, and then the furnace is cooled at 288 K/min in 1,000 sccm of Ar.

CVD SWNT Tip Preparation

SWNT tip are prepared in a similar manner to MWNT tips, except that iron oxide (FeO_x) nanoparticles are used as the catalyst. The FeO_x colloid catalyst is synthesized. A portion (0.033 moles) of $NaHCO_3$(s) is added to 50 ml of 0.0165 M $Fe(NO_3)_3$ (aq). The brick red solution is stirred vigorously for 15 min and then diluted 1:5 times. The 1.5 nm to 4nm colloidal FeO_x particles produced are characterized by transmission electron microscopy (TEM, and electrophoretically deposited into the pores of an HF-anodized tip at –0.5 V versus a platinum electrode, and then the tip is rinsed with water to remove the excess catalyst solution. The flow rates for the H_2 and C_2H_4 are reduced to 8 and 2 sccm, respectively.

Sample Preparation

Gold colloid sample with diameters of 5.7 nm the nanotube tips in imaging. The Au colloids are diluted 1:20 with deionized water. Five microliters of 1% poly-L-lysine is deposited on freshly cleaved mica for 10 sec, and then rinsed with deionized water and dried with a stream of nitrogen. Next, 5 μl of the diluted gold colloid solution is deposited on the mica for 1 min, rinsed with deionized water, and then dried with nitrogen.

IgS (Protein) sample are prepared from bovine IgG (>95% purity) by diluting to 2 μg/ml in0.05 M PBS solution at pH 7.0. A 25-μl drop of this solution is placed onto freshly cleaved mica for 10 sec, rinsed with deionized water, and dried with nitrogen.

GroES (Protein) samples are prepare from *Escherichia coli* GroES (>90% purity) by diluting the lyophilized powder to 6.25 μg/ml in a 25 mM buffer solution containing 75 m KCl, 0.5 mM DTT, and 1.25% trehalose at pH 7.5. Ten microliters of this solution is dropped onto freshly cleaved mica for 10 sec, rinsed with deionized water, and dried with nitrogen.

All AFM images are recorded in tapping mode in air. Free oscillation amplitudes of the cantilever are 10–20 nm. As-grown CVD tips are generally too long for higher-solution AFM imaging and must be shortened; They are shortened using a controlled electrical etching procedure where the nanotube tip is brought into contact with a rough niobium sample in force-calibration mode, and then a negative 5–20 V bias is applied to the sample with the tip grouped. This allows the controlled etching of the nanotubes; lengths of 50–300 nm are used. Tapping mode images of the rough niobium surface can be used to evaluate qualitatively the sharpness of the tip. Typically, the etching procedure continues until feature with <10 nm are observed in the AFM image of the niobium surface.

TEM experiments are performed at 100 kV. The entire AFM cantilever/tip assembly with nanotube tip is mounted on a custom TEM holder for imaging. The field emission scanning electron microscopy (SEM) images of CVD nanotube tips are obtained with a Electron Microscope operated with a 3kV accelerating voltage.

Growth of MWNTs from a porous substrate by CVD results in nanotubes oriented perpendicular to the substrate plane. This strategy is used to fabricate aligned carbon nanotube AFM tips by CVD with the proper orientation for AFM imaging. After reaction of the electrodeposited iron catalytic tips with ethylene at elevated temperature, individual multiwalled nanotubes grow from the pores etched in the flattened area at the tip. SEM images demonstrate that these nanotube tips grow in an ideal orientation for AFM imaging (Fig. 2.35(a)). But the nanotube diameters are, less than the limit of what can be accurately measured by the SEM. TEM images of the nanotube tips confirm that they are multiwalled nanotubes with well graphitized walls and radii ranging from 3–8 nm, as shown in Fig. 2.35(b). In addition, the as-grown nanotube tips have typical lengths of several micrometers, and are too long to be used for high-resolution AFM imaging the nanotube length is optimized for imaging by using an electrical etching process.

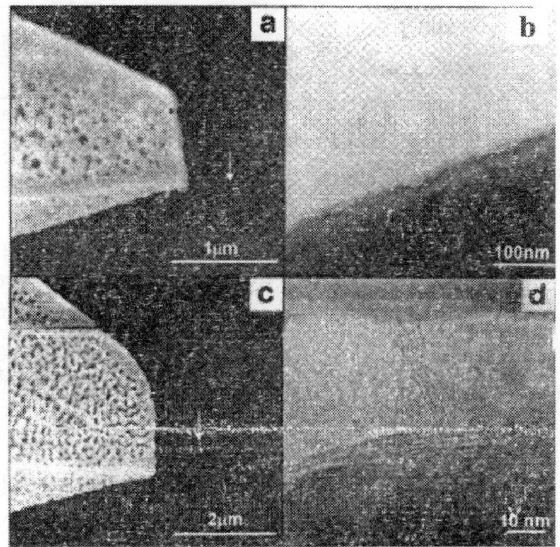

Fig. 2.35 *Electron microscopy of CVD nanotubes tips. (a) FESEM image of MWNT tubes grown from a Si cantilever/tip assembly. (b) TEM of a CVD MWNT tip. (c) field emission/SEM image of SWNT bundles grown from a Si cantilever/tip assembly. (d) TEM of several CVD SWNTs comprising a bundle tip.*

Since SWNT have the potential to produce robust probes with a resolution ≤ 0.5 nm; their growth is investigated, by using colloidal FeO_x CVD catalyst that has particle sizes comparable to the desired nanotube diameter. The gas flow is reduced to favor SWNT growth in bulk catalyst powder. SEM images of the nanotube tips prepared by using these conditions reveal that the nanotubes grow from the tips aligned perpendicular to the flattened area and are well positioned for AFM imaging (Fig. 2.35(c)). TEM images show that these nanotubes are SWNTs, and thus demonstrate that choice of catalyst and reaction conditions favor efficient SWNT growth from the tip ends (Fig. 2.35(d)). Analysis of the TEM images also shows that the individual nanotubes have radii of 1–2 nm, which is comparable to the starting catalyst size, and that most of the tips consist of several SWNTs in a bundle.

The force calibration mode of the AFM is used to characterize the mechanics of the nanotube tips. In this mode, raster scanning is disabled so that the tip oscillates near its resonance frequency over one point on the sample. The *z*-piezo is used to bring the sample into and out of contact with the tip, while the tip oscillation amplitude and deflection are plotted versus the sample Z position. Typical plots obtained for a CVD nanotube tip are shown in Fig. 2.36a and b. Both MWNT and SWNT tips show similar characteristics. As the tip approaches (from right to left), it begins to contacts intermittently the sample and the amplitude decreases as expected (region A). Upon further approach, the nanotube buckles as seen in both curves; the amplitude begins to increase again and the deflection increases to the Euler buckling force and then remains relatively constant with further approach (region B).

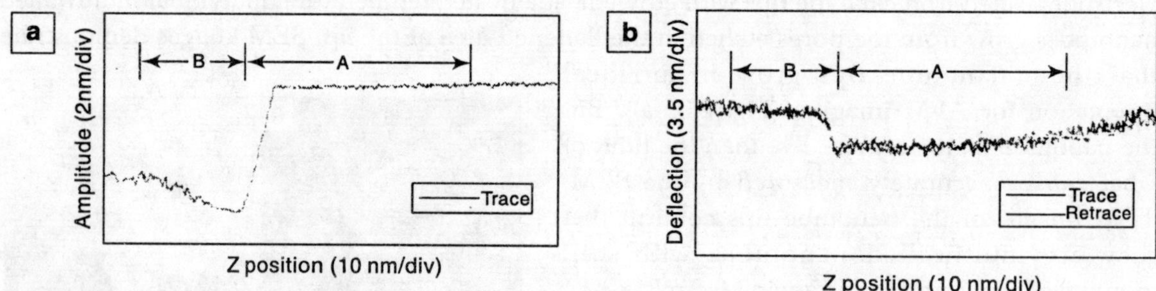

Fig. 2.36 *Force calibration plot for a CVD nanotube tip. (a) Tip amplitude oscillation vs. height above the sample. (b) Tip deflection vs. height above the sample. Regions A and B highlight pre and postbuckling areas.*

The tip adhesion is measured from these curves as the hysteresis in the deflection plot; it is below the noise level of the AFM, which ≈100 pN. This is smaller than that obtained with silicon tips.

Gold nanoparticles are imaged to evaluate the resolution of CVD nanotube tips. These nanoparticles are well defined spherical structures and are relatively incompressible MWNT tips produced by CVD are used to image 5.7-nm gold nanoparticles. By measuring the width of the particles in the image, the effective tip radii can be determined. That the radii are is the range 3-6 nm agrees with TEM results.

Analysis of AFM images of gold nanoparticles taken with SWNT tips shows smaller 2- to 4-nm tip radii of curvature. This size range coincides with the SWNT bundle diameters as a measured by TEM. (Table 1).

Table 1. Summary of CVD Nanotube Tip Resolution

MWNT tip number	0C	20E	38A	31G	38B
Radii ± 1σ, nm	3.7 ± 0.7	4.4 ± 0.5	5.7 ± 1.7	3.2 ± 0.7	5.3 ± 0.7
SWNT tip number	51A	58B	61A		
Radii ± σ, nm	3.8 ± 0.7	3.0 ± 1.3	2.0 ± 0.4		

The tips are defined by a specific number and a capital letter. The number designates a particular tip and the letter the growth cycle, where A = 1. The resolution was calculated using a two-sphere model using the full-width at half-maximum determined from the experimental images of 5.7-nm gold nanoparticles.

The decrease in tip radius in going from individual MWNT tips to SWNT bundle tips is due to catalysis. Moreover, the observed image resolutions obtained with nanotube tips are in good agreement with those from TEM measurements made on these tips. Those results show that it is possible to make very high-resolution AFM probes that have predictable resolution. In addition, as smaller diameter, higher resolution nanotube tips become available, the control of their length remains an important issue. Specifically, an individual 0.5-nm radius SWNT will need to be of a nm to avoid compromising the resolution due, to thermal vibrations.

The small radii of curvature and low adhesion forces of nanotube tips make them ideal for imaging soft biological structures. To evaluate the potential of CVD nanotube tips in biological AFM, two proteins with known structures are imaged in tapping mode IgG and GroES. IgG is a 150-kDa antibody molecule that consists of four polypeptide chains arranged in a Y shape. Low and high-resolution images of IgG (Fig. 2.37(a) and 2.37(b)) show examples of the characteristic Y shape. Because of the intrinsic conformational flexibility of IgG, all molecules absorbed from solution do not exhibit the Y shape; the high resolution obtained with these tips can readily observed, by comparing the image with the crystal structure (Fig. 2.37(c)); that is, there is little tip-induced broadening. IgM, which consists of five IgG-like monomers arranged

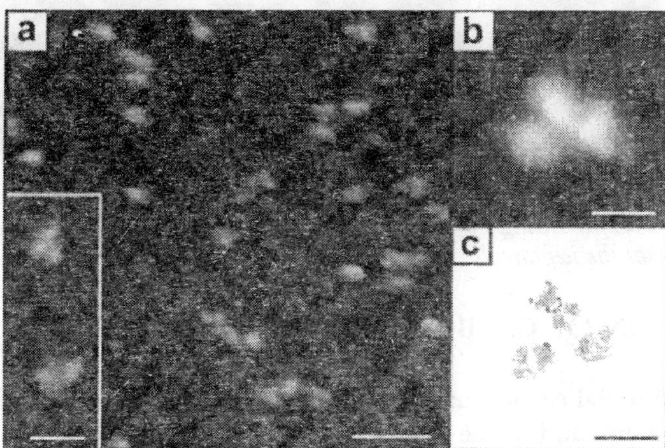

Fig. 2.37 *IgG imaged by a CVD nanotube tip. Large scan area in a shows several molecules that exhibit the characteristics Y shape. (b) High resolution images of IgG reveal very little tip-induced broadening when compared with the crystal structure in c. [Bars = 50 nm (a), 20 nm(Inset), 10 nm (b), and 10 nm(c).]*

as a pentamer is also imaged. The images resolve pentameric structure and show little broadening on comparing with structural models based on electron microscopy and x-ray scattering.

GroES, which is component of the GroEL/GroES chaperonin system involved in protein folding, is a hollow dome shaped heptamer that is approximately 8 nm in outer diameter. The seven 10-kDa subunits each consists of a core β-barrel with a βhairpin loop at the top and bottom. The top β-hairpins point inward to form the top of the dome, while the bottom hairpins are disordered when not in contact with GroEL. Low-resolution AFM images of individual, well separated GroES molecules on mica reveal two conformations (Fig. 2.38(a)). One conformation shows a ring-like structure with an 11nm outer diameter and some features on the ring. The other conformation simply looks like a dome of the same diameter. These two conformations are two sides of the GroES heptamers. The domed image is the side with the ordered β-hairpins creating the dome roof, and the ringed image is the disordered β-hairpins open to the hollow cavity. The disordered hairpins have conformational flexibility, but it is possible to observe the seven-fold symmetry, as shown in Fig. 2.38(b). Fig. 2.38(c) shows the bottom view of the GroES crystal structure with the top of the dome removed to make the pore visible. Hence CVD nanotube tips achieve submolecular resolution on isolated protein assemblies.

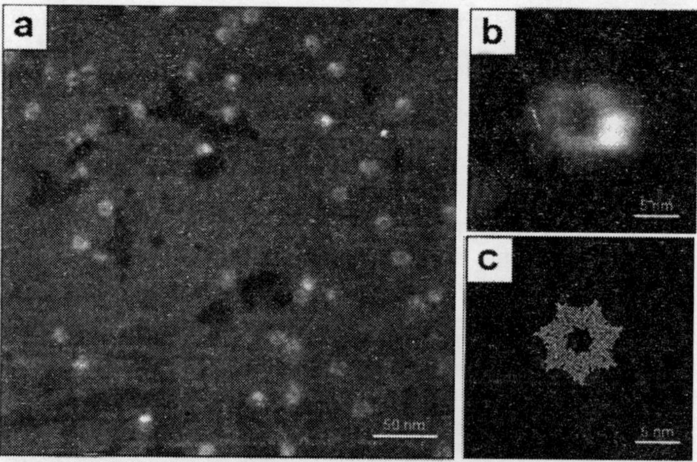

Fig. 2.38 *GroES imaged by a CVD nanotube tip. Large scan area in a shows both "come" and "pore" conformations, representing the two sides of GroES facing up. (b) A higher resolution image of the pore side shows the heptameric symmetry which matches well with the crystal structure (c).*

AFM MANIPULATION OF CARBON NANOTUBES

In order to use individual carbon nanotubes to build nanostructures and devices one must be able to manipulate them and place them in predefined positions at well. The interaction between the nanotubes and the substrate on which they are dispersed suggests that their position at room temperature by applying lateral forces of the appropriate magnitude with the tip of an AFM can be found. As the shear stress on surfaces of H-passivated silicon is high, of the order of 10^7 N/m, only one position of the nanotube and its shape can be controlled.

To manipulate the nanotubes, the mode of operation of the AFM is changed. For tube imaging the AFM in the non-contact mode is used, with very low forces applied by the tip (in the range of pN), but in the contact mode normal forces of 10–50 nN is employed. In Fig. 2.39, shows a sequence of manipulation steps. A single nanotube originally on an insulating substrate (SiO_2) is manipulated in a number of steps (not all shown) onto a tungsten thin film wire (~80 Å high), and, finally, is stretched across an insulating WO, barrier (itself made by AFM tip-induced oxidation at an earlier stage).

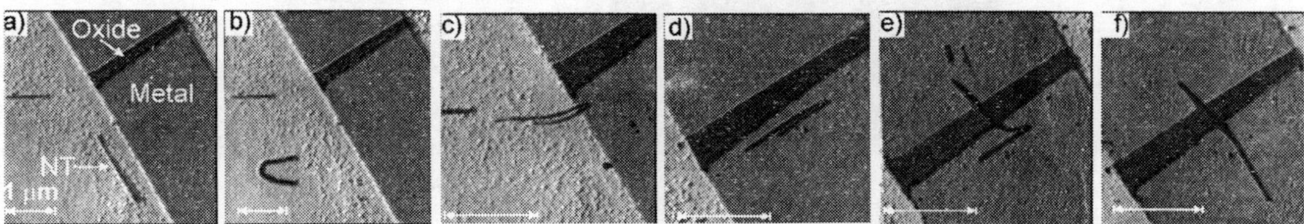

Fig. 2.39 *AFM manipulation of a single multi-wall nanotube such that electrical transport through it can be studied, Initially, the nanotube (NT) is located on the insulating (SiO_2) part of the sample. In a stepwise fashion (not all steps are shown) it is dragged up the 80 Å high metal thin film wire and finally is stretched across the oxide barrier.*

The tip of an AFM can be used to manipulate multiwalled carbon nanotubes on a passivated silicon surface. The interaction between nanotubes and the surface is crucial for such manipulations. How the strength of nanotube–surface interactions can be determined from the observed elastic deformations of tubes on the surfaces is given below:

The Experiment

The multiwalled nanotubes used are dissolved in methylene chloride using an ultrasonic bath. The solution is dispersed on a H-passive Si(100) surface. Imaging and manipulation is performed in air using an atomic force microscope.

Most of the structure and associated properties of nanotubes refer to isolated (unsupported) nanotubes. Supported nanotubes, intertact strongly with the substrate, and as a result a number of deformations arise.) The AFM images of nanotubes shown in Figure 2.40 illustrate different types of elastic deformations observed on the freshly covered surface and after manipulation with the AFM tip. Figure 2.40(a) shows how nanotubes crossing one another are deformed elastically due to their interaction with the surface–despite their exceptionally large Young's modulus of about 1.5 TPa.

The attractive interaction with the substrate forces the upper nanotube along profile 2 in Figure 2.40(a) to maximize its area of contact with the surface at the expense of strain energy building up when it flexes around a lower nanotube. Profile 1 illustrates how a nanotube can also be compressed elastically under the load of another nanotube lying on top of it. The strength of the attractive forces between nanotubes and the surfaces are estimated using a quasi one-dimensional model where the profile $d(x)$ along the tube principal axis is determined by the balance of strain and adhesion energies $E_\varepsilon + E_{ad}$. Now, the profile $d(x)$ is calculated by minimizing the total energy of a non-dimensional tube, that is crossing a pointlike obstacle of

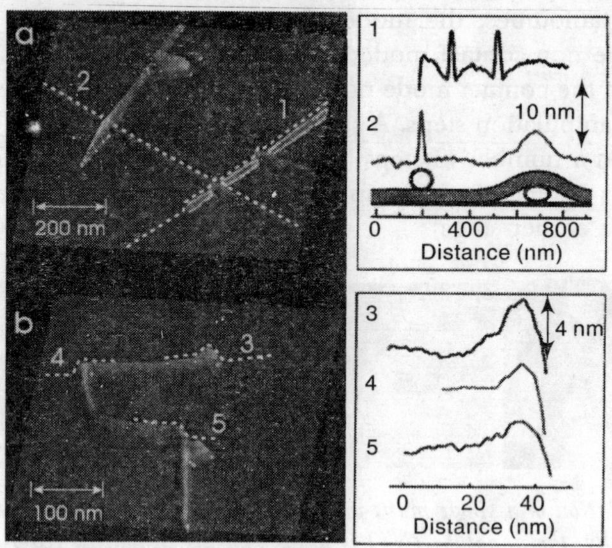

Fig. 2.40 *Nanocontact AFM images of elastically deformed nanotubues on a hydrogenated Si(100) surface. For imaging in the noncontact mode the tip is scanned across the sample at a distance of about 30 Å so that vertical forces between tip and surface are extremely small (typically about 5 pN). (a) The height of the nanotube along profile 1 is reduced by at least 30 Å in the vicinity of the contact with another nanotube crossing on top. The height at the crossing point is also found to be reduced almost 60 Å with respect to the sum of the individual nanotube diameters. (b) Strongly bent nanotube with buckling of 15 Å in heavily strained regions.*

given height & obtain the strain strategy E_ε by integration of $E\pi(a^4 - b^4)/8\int r(x)^{-2}\,dx$ over the length of the tube where E is Young' modulus, a and b give the outer and inner nanotube radii, respectively and $r(x)$ is the local radius of curvature along the principal axis (b^4 is assumed to be significantly smaller than a^4 and thus can be neglected). The outer tube radius a is determined from apparent height of nanotubes in AFM images. The binding energy is determined by short-ranged chemical and long-ranged van der Walls interactions, and the corresponding forces may be modeled using appropriate potentials. The profiled $d(x)$ depends only on the depth of the potential and is independent of the particular shape. Figure 2.41a, provides a unique way to determine the nanotube binding energy. The profile $d(x)$ is determined experimentally for various tube diameters and obstacle heights are compared to a number of different configuration as shown in Figure 2.40(a). For nanotubes with a diameter of about 100 Å, a binding energy of 0.8 ± 0.3 eV/Å is obtained. This large binding energy of nanotubes to distorts them.

The experimentally obtained binding energy of 0.8 eV/Å allows to estimate the forces at nanotube/nanotubesubstrate contacts as in Figure 2.40(a) by using the calculated change of the total energy when the height of the lower tube is reduced. The resulting force acting at the point of contact in Figure 2.40(a) is about 35 nN. Such strong forces are due to a substantial increase of the contact area of the upper tube with the substrate, and thus there is a gain in the binding energy, when the obstacle height is reduced. Thus, the pressure between nanotubes at these contacts is about 10 GPa, which is evident from the strong compression of the lower nanotube seen in profile 2.41 of Figure a. Such enormous pressures are known to lead to drastic modifications of the bonding and electrical properties of carbon compounds. For singlewalled nanotubes with a diameter of 13 Å (e.g. a (10, 10) nanotube), the force to compress them by 30% is of the order of 10 nN so that the tip of an AFM is used to locally perturb their geometric and electronic structure.

The observed strong interaction of nanotubes with the surface is determined with both the normal force that binds them to substrate and the pinning forces that are responsible for frictional properties when nanotubes are moved laterally across the surface. This interaction is, crucial for any attempt to manipulate nanotubes with the AFM tip and for stabilization of nanotubes in strained configurations. Evidence for the strength of the frictional forces can be obtained from images of nanotubes after they have been perturbned buy scanning the surface with the AFM tip in contact. In this mode, the loading force between tip and surface of 20 nN provides is an lateral forces strong enough to cause bending and displacement of nanotubes. Figure 2.40(b) example where a nanotube has been bent so strongly that curved segments start to buckle and form kinks. The correlation between bending angle and buckling height in Figure 2.40(b) suggests a successive vertical compression of the nanotube in the kinked region. Such kinked nanotube can be straightened by controlled manipulation without structural damage. High-resolution transmission electron micrographs, indicate that some of the strain in strongly bent nanotube segments are also relieved by strain fracture in addition to elastic deformations.

When scanning the surface in the contact AFM mode, nanotubes can be displaced on the surface in a controlled manner i.e., the nanotubes could be bent and realigned for shape and position. The area of images in the nanocontact mode and then the contact mode, is employed to apply higher vertical and lateral forces. The images in Figure 2.41 demonstrate how the nanotube can be realigned vertically, translated, bent, and then rotated (around the surface

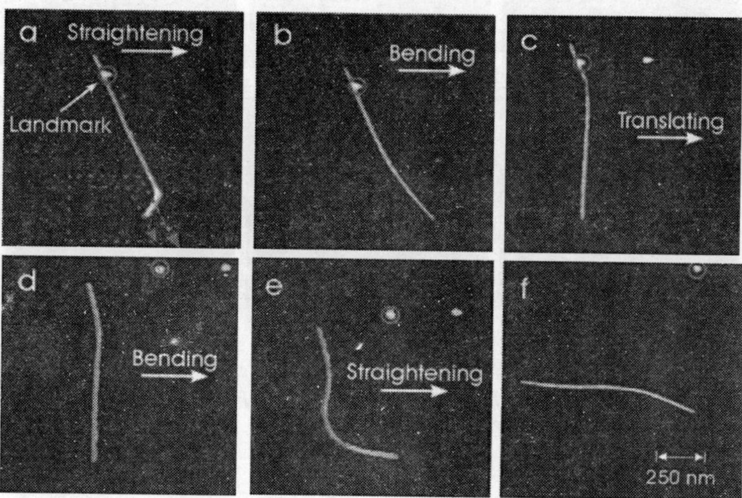

Fig. 2.41 *Manipulation of a 0.9 mm long and 80 Å nanotube at a vertical load of 20 nN. (a, b) A strongly nanotube segment can be straightened by repeated movement of the tip parallel to the nanotube axis (dashed line). Every more brings the nanotube successively closer to the desired geometry. (c-f) The nanotube can also be realigned vertically, translated, bent and rotated around the surface normal.*

normal). Figure 2.44a-f shows the realigning of a nanotube bundle in several steps to form the Greek letter Θ. Thus, the ability of the AFM tip as a tool to manipulate individual nanotube or bundles in a controlled way is achieved.

Quantitative information on lateral (frictional) forces is obtained from images of bent nanotubes as shown in Figure 2.41e. These configurations posses strain energies of several hundred electronvolts and are pinned in this shape by surface potential or by interaction with defects. The force *F* required to keep the nanotube of Figure 2.41e in place is about 5 nN. As a result of the strong nanotube–surface interaction, the nanotube flattens on the surface side thus increases the effective area of contact. Molecular mechanics calculations indicate the widths of the contacts region lie between 5% and 50% of the nanotube diameter, depending on the tube radius and number of shells. Thus, using the estimated contact area A of 5×10^{-6} m^2 for the lower tube segment in Figure 2.41e a shear stress $\sigma = F/A$ of 10^7 N/m^2 is obtained. This compares well, with the shear strength of graphite of 1.03 × 107 N/m^2.

To obtain control over the length of nanotubes, long nanotubes having a small diameter (~40 Å) are disected by pulling the tip contact across the nanotube on the surface. Images are taken during a successful cutting experiment (Figure 2.42) which show that the nanotube is stretched before breaking. In this particular case, the nanotube is contaminated with amorphous carbon, as seen from the spotty line left behind at its original position. The contamination may have increased its adhesion to the substrate. Type of contamination is not found in other experiments as seen from absence of such spotty lines in Figure 2.41 and 2.42.

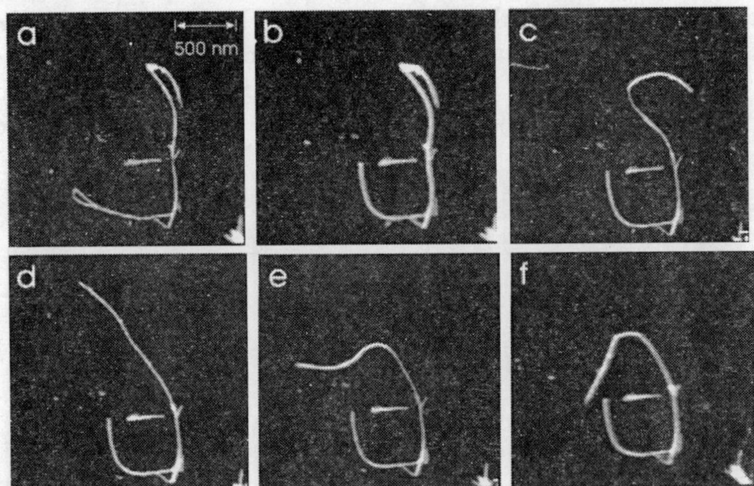

Fig. 2.42 *Fabrication of a more complex pattern (the Greek letter Θ) from a nanotube structure with a total length of about 2.5 mm, possibly consisting of more than one individual nanotube (not all steps are shown).*

Lateral Force Microscopy

There are an increasing number of variations of the standard STM and AFM techniques that can provide additional information to surface topography. However, these may well be at a lower spatial resolution depending on the particular interaction involved. Force modulation microscopy works in contact mode and involves applying a periodic oscillation to the tip or the sample and monitoring the amplitude and sometimes phase lag of the cantilever oscillation; it provides information on the elastic properties of the sample. Lateral force microscopy (LFM) detects lateral deflections such as twisting of the cantilever that arise from force on the cantilever parallel to the plane of the sample surface thus the frictional forces on the surface can be revealed. The lateral twist of the cantilever can arise from changes in surface friction or in surface slope. It is able to acquire both topographical and friction information of the surface.

LASER SCANNING MICROSCOPY

Confocal Laser Scanning Microscopy (*CLSM or LSCM*) is a technique for obtaining high-resolution optical images. The key features of confocal microscopy is its ability to produce in-focus images of thick specimens, a process known as *optical sectioning*. Images are acquired point-by-point and reconstructed with a computer, allowing three-dimensional reconstructions of topologically-complex objects.

In a confocal laser scanning microscopy, a laser beam passes through a light source aperture and then is focused by an objective lens into a small (ideally diffraction limited) focal volume within a fluorescent specimen Fig. 2.43. A mixture of emitted fluorescent light as well as reflected laser light from the illuminated spot is then recollected by the objective lens. A beam splitter separates the light mixture by allowing only the laser light to pass through and reflecting the fluorescent light into the detection apparatus. After passing a *pinhole*, the fluorescent light is detected by a photodetection device (a photomultiplier tube (PMT) or avalanche photodiode), transforming the light signal into an electrical on that is recorded by a computer.

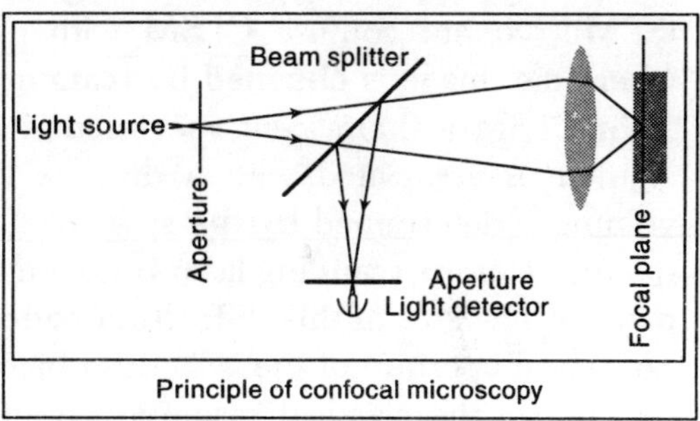

Fig. 2.43 *Principle of confocal microscopy.*

The detector aperture obstructs the light that is not coming from the focal point, as shown by the dotted gray line in the image. The out-of-focus light is suppressed: most of their returning light is blocked by the pinhole, resulting in sharper images than those from conventional fluorescence microscopy techniques, and permits one to obtain images of various *z* axis planes (also known as *z* stacks) of the sample.

The detected light originating from an illuminated volume element within the specimen represents one pixel in the resulting image. As the laser scans over the plane of interest, a whole image is obtained pixel-by-pixel and line-by-line, whereas the brightness of a resulting image pixel corresponds to the relative intensity of detected fluorescent light. The beam is scanned across the sample in the horizontal plane by using one or more (servo controlled) oscillating mirrors. This scanning method usually has a low reaction latency and the scan speed can be varied. Slower scans provide a better signal-to-noise ratio, resulting in better contrast and higher resolution. Information can be collected from different focal planes by raising or lowering the microscope stage. The computer can generate a three-dimensional picture of a specimen by assembling a stack of these two-dimensional images from successive focal planes.

Confocal microscopy also provides a substantial improvement in lateral resolution and the capacity for direct noninvasive, serial optical sectioning of intact, thick, living specimens with a minimum of sample preparation. Because CLSM depends on fluorescence, a usually needs to be treated with fluorescent dyes to make objects visible. However, the actual dye concentration can be low to minimize the disturbance of biological systems: some instruments can track single fluorescent molecules. Also, transgenic techniques can create organisms that produce their own fluorescent chimeric molecules (such as a fusion of GFP, green fluorescent protein Fig. 2.44 with the protein of interest).

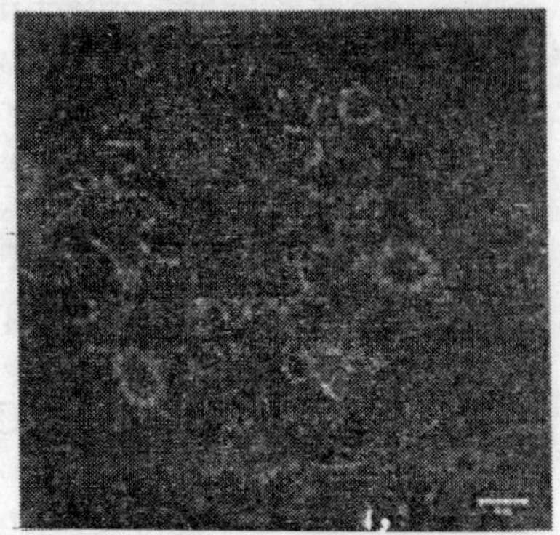

Fig. 2.44 *An example of a GFP fusion protein.*

Resolution Enhancement

CLSM is a scanning imaging technique in which the resolution obtained is best explained by comparing it with another scanning technique like that of the scanning electron microscope (SEM). Do not confuse CLSM with phonograph-like imaging–AFM or STM, for example, where the image is obtained by scanning with an atomic tip over a conducting surface.

In CLSM a fluorescent specimen is illuminated by a point laser source, and each volume element is associated with a discrete fluorescence intensity. Here, the size of the scanning volume is determined by the spot size (close to diffraction limit) of the optical system because the image of the scanning laser is not an infinitely small point but a three-dimensional diffraction pattern. The size of this diffraction pattern and the focal volume it defined is controlled by the numerical aperture of the system's objective lens and he wavelength of the laser used. This can be seen as the classical resolution limit of conventional optical microscopes using wide-field illumination. However, with confocal microscopy it is even possible to overcome this resolution limit of widefield illuminating because only light generated in a small volume element is detected at a given time. Here the effective volume of light generation is usually smaller than the volume of illumination; that is, the diffraction pattern of detectable light creation is sharper and smaller than the diffraction pattern of illumination. The resolution limit in confocal microscopy depends not only on the probability of illumination but also on the probability of creating enough detectable photons, so that the actual addressable volume being associated with a generated light intensity is smaller than the illuminated volume. Depending on the fluorescence properties of the used dyes, there is a more or less subtle improvement in lateral resolution compared to conventional microscopy. However, with light creation processes with much lower probabilities of occurrence such as second harmony generation (SHG), the volume of addressing is reduced to a small region of highest laser illumination intensity, substantially improving lateral resolution. Unfortunately, the probability decrease in creation of detectable photons negatively affects the signal-to-noise ratio. One can compensate for this effect by using more sensitive photodetectors or by increasing the intensity of the illuminating laser point source Increasing the intensity of illumination later risks excessive bleaching or other damage to the specimen of interest, especially for experiments in which comparison of fluorescence brightness is required.

Uses

CLSM is widely-used in numerous biological science disciplines, from cell biology and genetic to microbiology and developmental biology.

Clinically, CLSM us used in the evaluation of various eye diseases, and is particularly useful for imaging, qualitative analysis, and quantification of endothelial cells of cornea.[4] It is used for localizing and identifying the presence of filamentary fungal elements in the corneal stroma in cases of keratomycosis, enabling rapid diagnosis and thereby early institution of definitive therapy. Research into CLSM techniques for endoscopic procedures is also showing promise.[5]

CLSM is also used as the data retrieval mechanism in some 3D optical data storage systems and has helped determine the age of the Magdalen papyrus.

X-RAY MICROSCOPE

An X-ray microscope uses electromagnetic radiation Fig. 2.45 in the Soft X-ray band to produce images of very small objects.

Unlike visible light microscopes, X-rays do not reflect of easily, and they are invisible to the human eye. Therefore the basic process of an X-ray microscope is to expose film or use a charge-coupled device (CCD) detector to detect X-rays that pass through the specimen, rather than light which bounces of the specimen. It is a contrast imaging technology using the difference in absorption of soft x-ray in the water window region (wavelength region: 2.3-4.4 nm, photon energy region: 0.28-0.53 keV) by the carbon atom (main element composing the living cell) and the oxygen atom (main element for water).

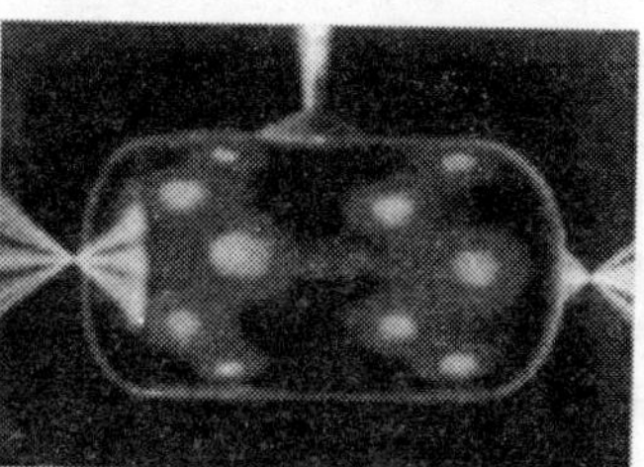

Fig. 2.45 *Indirect drive laser inertial confinement fusion uses a "hohlraum" which is irradiated with laser beam cones from either side on its inner surface to bathe a fusion microcapsule inside with smooth high intensity X-rays. The highest energy X-rays which penetrate the hohlraum can be visualized using an X-ray microscope.*

Early X-ray microscopes used grazing-incidence reflective optics to focus the X-rays, which grazed X-rays off parabolic curved mirrors at a very high angle of incidence. An alternative method of focusing X-rays is to use a tiny fresnel zone plate of concentric gold or nickel rings on a silicon dioxide substrate.

Sources of soft X-rays suitable for microscopy, such a synchrotron radiation sources, have fairly low brightness of the required wavelengths, so an alternative method of image formation is scanning transmission soft X-ray microscopy. Here the X-ray are focused to a point and the sample is mechanically scanned through the produced focal spot. At each point the transmitted X-rays are recorded with a detector such as a proportional counter or an avalanche photodiode.

The resolution of X-ray microscopy lies between that of the optical microscope and the electron microscope. It has an advantages over conventional electron microscope in that it can view biological samples in their natural state. Electron microscopy is widely used to obtain images with nanometer level resolution but the relatively thick living cell cannot be observed as the sample has to be sliced thinly and then dried to get the image. However, it should be mentioned that cryoelectron microscopy allows the observation of biological specimens in their hydrated natural state. Until now, resolutions of 30 nanometer are possible using the Fresnel Zone plate lens which forms the images using the soft X-rays emitted from a synchrotron. Recently, there has been active research on using the soft x-rays emitted from the laser-produced plasma rather than synchrotron radiation light.

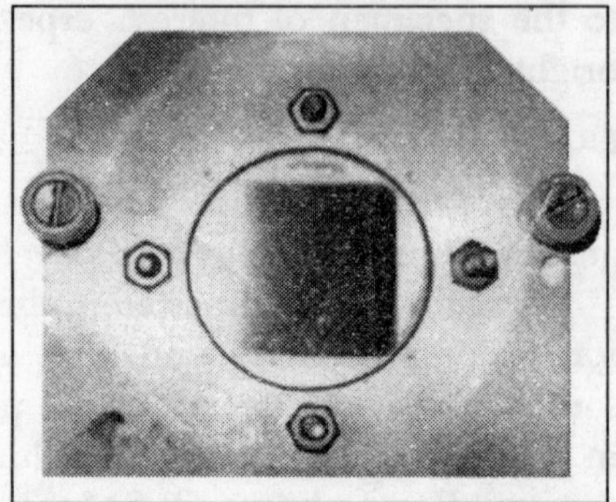

Fig. 2.46 *A square beryllium foil mounted in a steel case to be used as a window between a vacuum chamber and an X-ray microscope. Beryllium, due to its low Z number is highly transparent to X-rays.*

Additionally, X-rays cause fluorescence in most materials, and emission can be analyzed to determine the chemical elements of an imaged object Fig. 2.46. Another use is to generated diffraction patterns, a process used in X-ray crystallography. By analyzing the internal reflections of a diffraction pattern (usually with a computer program), the three-dimensional structure of a crystal can be determined down to the placement of individual atoms within its molecules. X-ray microscopes are sometimes used for these analyses because the sample are too small to be analyzed in any other way.

MAGNETIC FORCE MICROSCOPE

A magnetic force microscope (MFM) is a scanning probe microscope (SPM) that can map the spatial distribution of magnetism by measuring the magnetic interaction between a sample and a tip. As magnetic devices have become smaller and smaller, an evaluation technique with nanoscale spatial resolution has become necessary. To meet this need, the MFM was developed.

The principle of MFM measurement is based on noncontact atomic force microscopy (NC-AFM). Unlike NC-AFM, magnetic materials are used for the sample and tip, so that not only the atomic force but also the magnetic interaction are detected. Many kinds of magnetic interaction are measured by MFM, including magnetic dipolar interaction.

Because the magnetic stray field from the sample will effect the magnetized state and vice versa, in most cases it is difficult to obtain quantitative information from the MFM measurement, a typical resolution of 30 nm can be achieved. The best resolution obtained so far is 18 nm.

A potential method of increasing the resolution would involve using an electromagnet on the tip instead of a permanent magnet. By only enabling the magnetic tip when place over the pixel being sampled could increase the resolution.

With AFM, magnetic domains can be revealed and mapped by coating the tip of the cantilever with a thin film of magnetic material (e.g. Fe-Ni/Cr, Co/Cr, or CoSm/Cr). The cantilever oscillates near the surface sample, as in non-contact mode, but at a distance where the magnetic interaction becomes dominant on Van der Waals force. The amplitude of the oscillation is modified by the magnetic field. A map of magnetic domains can be achieved by detecting the variation during the scan. Topographic and magnetic images can be acquired simultaneously rising or lowering the tip on the surface during the scan in a way that the variation of the amplitude of the cantilever is due to magnetic force or to Van der Waals force. High resolution image of dominas written by a focused laser beam in a magnetic thin film has been reported. A spatial resolution of 10 nm resolution was achieved.

MAGNETIC RESONANCE FORCE MICROSCOPY

Overview

Magnetic resonance force microscopy (MRFM) is a new microscopic imaging technique that combines aspects of atomic force microscopy (AFM) with magnetic resonance imaging (MR). Interest in this technique is driven by the possibility or reaching the "Holy Grail" of microscopy: true three-dimensional' sub-surface imaging with atomic resolution and chemical specificity. This goal has not yet been reached but good progress is being made.

A future MRFM imaging apparatus suitable for molecular imaging look like Fig. 2.47:

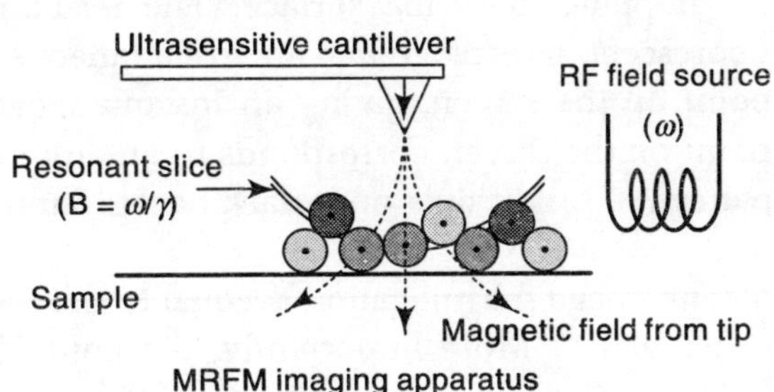

Fig. 2.47 *MRFM imaging apparatus.*

A molecule whose structure is to be determined is placed below a sharp magnetic tip. The magnetic tip is attached to a sensitive micromechanical cantilever that will bend in response to small forces, including the magnetic forces due to the magnetic nuclei in the sample. (Many common elements, such as hydrogen, fluorine, carbon-13, phosphorus, etc., have magnetic nuclei.) Nuclear magnetic resonance is used to manipulate individual nuclei that are just the right distance from the tip (i.e., within the "resonant slice"). By applying suitably modulated radiofrequency (rf) magnetic field using a small coil, the magnetic moment of the nucleus within the resonant slice can be flipped up, down, up, down, etc., thus generating an alternating force on the tip that causes that cantilever to vibrate slightly. This vibration is detected using a sensitive fiber-optic interferometer (not shown). By scanning the sample with respect to the tip in a threedimensional raster pattern, an image of the atomic structure of the molecule can, in principle, be obtained.

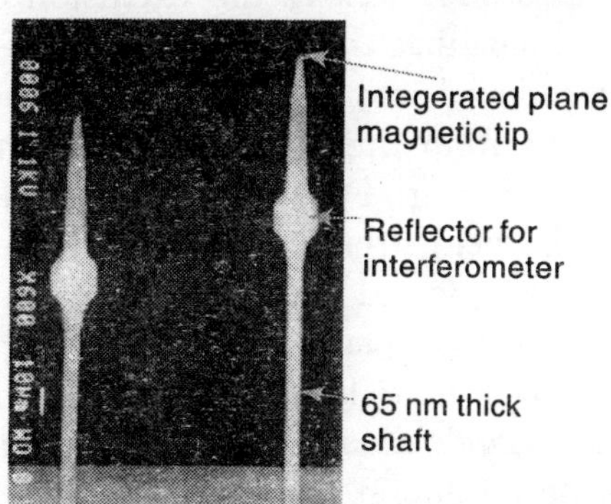

Fig. 2.48 *Ultrasensitive micromechanical cantilevers.*

For detecting extremely small force; ultrasensitive micromechanical cantilevers as shown below in Fig. 2.48 are fabricated. These cantilevers are made from single crystal silicon. They are about 200 microns long, but only 65 nm thick. A liquid helium temperatures, force as small as 3 attonewtons (3×10–18N) can be detected in a 1 Hz bandwidth. This level of sensitivity should be sufficient for the detection of individual electron spins, provided that the electron spin has a relaxation time greater than 100 milliseconds. Besides MRFM applications, these cantilevers can be used for super sensitive electrometry and magnetometry.

FIELD ION MICROSCOPY

In the field ion microscope (FIM), the specimen consists of a conducting or semiconducting fine needle-tip of radius *r* (usually between 5 and 100nm) prepared by etching or electropolishing techniques, such that individual atoms stand out in a terrace of atomic-scale ledges. The tip is held at a positive potential (between 5 to 20 keV), with respect to a negatively charged conductive fluorescent screen of radius *R*, in a high vacuum and is maintained at cryogenic temperatures. A low-pressure inert gas, such as hydrogen or helium, is introduced into the chamber, the atoms of which become polarized in the electrostatic field between the screen and the tip and are therefor attracted to the tip by the increased field intensity. The polarized atoms are then ionized and lose electrons (which channel into the tip) either at one point of contact or by 'hopping' over the surface. The resulting positive ions are then accelerated towards the fluorescent screen, giving an instantaneous fluorescent signal. In principle, each fluorescent point on the screen, giving an instantaneous fluorescent signal. In principle, each fluorescent point on the screen corresponds to an atom in the tip and the high-contrast image consists of a pattern of bright dots on a dark background, which reproduces the atomic arrangement of the tip.

The image magnification is equal to the ratio of the screen to that of the tip (*R/r*) and is of the order of 10^6. More importantly, the limit of resolution is that of an individual atom and the FIM was the first instrument to provide images of atoms and their arrangement at the tip, as well as the location of vacancies and line imperfections. Since no lenses are involved, there are no depth-of-field or diffraction limitations to be considered.

So much for the (potential) advantages of FIM. A part from the difficulties associated in the preparation of such a fine tip (and the geometrical constraints on a specimen which this imposes) as well as the exclusion of contamination or adsorbed surface species, there is the problem that the field strength at the tip must be sufficiently high to ionize the gas atoms introduced into the chamber, but not so high as to result in field evaporation or emission of atoms from the tip. In practice only those high melting point elements with high cohesive energies (e.g., Fe, Mo, W, Rh and Pt) and their compounds, which can resist field evaporation at the field strengths needed or inert gas ionization, fulfil this requirement. No breakthrough has been found in the case of gases which ionize more easily. On the other hand, this problem is turned to advantage in the atom probe FIM (APFIM), which involves the controlled progressive removal via field evaporation of (ionized) atoms from the tip. Moreover, the desorbed atoms can be individually chemically identified using a time-of-flight mass spectrometer giving atomic-scale resolution maps of elemental distributions. A recent variant, the position-sensitive atom probe (POSAP), allows a three-dimensional reconstruction of the atomic structure in the tip across an area of about 20 nm^2.

In a sense, APFIM has come full circle from the field emission microscope (FEM), invented prior to the FIM, in which the polarities of the tip and the screen are reversed and which, of course, from the basis of the fild emission gun in electron microscopy. In principle the emitted electrons recorded on the conductive fluorescent screen arise from individual atoms projecting from the tip surface as in FIM, but in practice the images suffer from noise and perturbations in the electron trajectories.

NEAR-FIELD SCANNING OPTICAL MICROSCOPE

Scanning Near Field Optical Microscopy

In near field scanning optical microscopy (NSOM) a very small light source very close to the sample is scanned. Light passes though a sub-wavelength diameter aperture and illuminates the sample that is placed within its near field, i.e. at a distance much less than the wavelength of the light. The resolution achieved is far better than which is attainable in conventional optical microscopes.

Near field scanning optical microscopy (NSOM) was developed as a means to break the diffraction limit on spatial resolution attainable with optical measurement. Traditional NSOM has a tapered, metal-coated optical fiber with a small aperture as either an excitation source or collection device. The spatial resolution is defined by the size of the aperture itself. With rough metallic coatings on glass fibers, coupled with the finite skin depth of metals at optical frequencies, there is a limit on the spatial resolution attainable with fiber probes. Spatial resolution for the glass fiber approach to NSOM is ~50–100nm, though careful and sample preparation have yielded ~30nm resolution.

The field of electromagnetic wave consists of two components: a propagating and an non-propagating wave. The propagating field is the light collected by the optical microscopy. The non propagating component, decays exponentially to negligible values within a wavelength from its source. It is characterized by high spatial frequency that the optical lens system filters out. High spatial frequencies correspond to the fine details of the object, and are not collected by lens. The region of space where the non propagating field is collectable defines the near field region. The Near-Field Scanning Optical Microscopy (NSOM, or SNOM) is based upon the detection of nonpropagating waves in the near-field region.

If light passes through a sub-wavelength diameter aperture and illuminates a sample placed within the near field from the aperture, the light coming from the surface sample can be collected in the far without loosing the near field resolution, overcoming the diffraction limit—also known as Abbe's limit—of optic microscopy. By moving the aperture over the surface in a raster scan mode, as in a SPM, and collecting and detecting the light coming from the sample, a SNOM high resolution optical image can be acquired. Simultaneously to the optical map, a morphological image can be also acquired.

In the SNOM, an aperture with a diameter about of 20–100nm can be formed at the end of a tapered metal-coated optical fiber. The tip can be fabricated starting from and optical fiber by heating and pulling method or by chemical etching method. Optical probes have been fabricated, within Si AFM cantilever, and use the SNOM probe as an AFM probe is made in contact mode.

Other crucial elements in the SNOM apparatus are the local illumination and the constancy of the distance between the tip and the surface. In order to guarantee operation in the near field, the shear-force detection mode has been so far used: a lateral oscillation is forced to then tip, and the damping of the oscillation due to shear-forces is monitored to keep constant this distance. The increasing interest for SNOM has triggered the development of various techniques to monitor and maintain a constant tip-sample distance. In the first SNOM set-ups, a laser spot focused and diffracted from the oscillating fiber tip, or interferometric techniques Fig. 2.49 is used. The methods are difficult to use as they require a supplementary detection optical system. Non optical methods a thus developed. The use of quartz resonator tuning fork, to generate a piezoelectric voltage, proportional to the amplitude of the oscillation, appears to the best option: the fiber is attached to one leg on a high Q quartz resonator tuning fork, and oscillates at the resonance frequency of the resonator (typically at 33 kHz). As the probe is brought to about 10 nm from the surface, a feedback loop system monitors the damping in the oscillation amplitude due to shear force. As a variation is recorded, during the scanner of the surface sample, the feedback reports the oscillation to the original value, moving up or down the z-piezo scanner in relation to the corrugation of the surface sample. By recording this adjustment, a 3D morphology map is obtained see Fig. 2.50, During scanning, the excitation light from the small aperture interacts locally with the surface. It can be absorbed, backscattered, or it can induce emission from the sample. To collect the absorbed, backscattered or emitted light, conventional optical systems placed in the far field are used. Coupling a spectrometer system to the SNOM output, and using the SNOM in a static mode, optical information resolved in wavelength can be achieved.

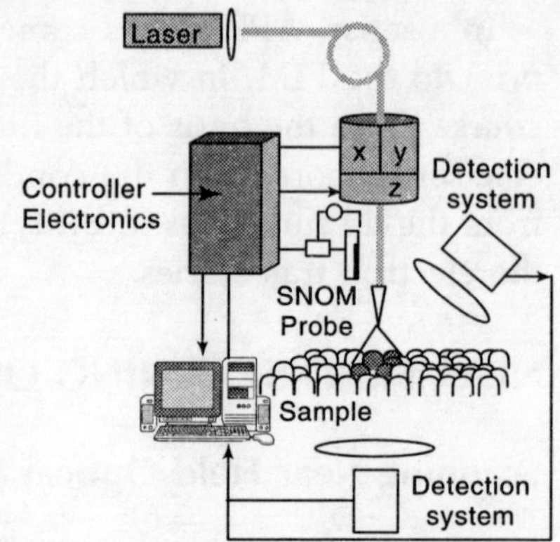

Fig. 2.49 *SNOM operation scheme.*

Typically, the SNOM can operate in different modes depending on the probe, the nature of the sample, and the direction mode (Fig. 2.51). Thus, the options are illumination mode, collection mode, transmission mode, and reflection mode.

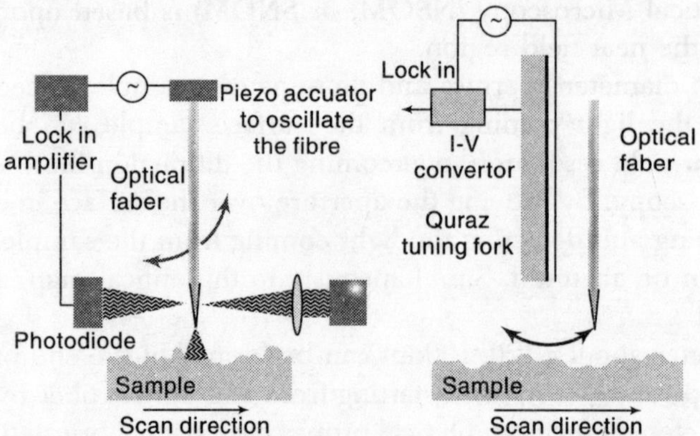

Fig. 2.50 *Optical (left) and tuning fork (right) method for Shear-Force detection.*

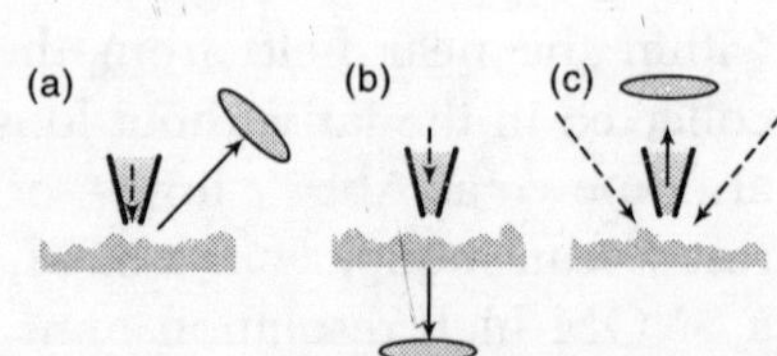

Fig. 2.51 *Operation modes of SNOM: (a) reflection mode; (b) transmission mod; (c) collection mode; the probe collects the light.*

In non contact mode, it vibrates at its resonant frequency at a distance where the electrostatic forces dominate. The voltage applied to the tip is

$$U = U_{dc} + U_{ac} \sin(\omega t),$$

thus the electrostatic force measured is given by

$$F = \frac{1}{2}\frac{dC}{dz}(F_{dc} + F_{\omega} + F_{2\omega})$$

where C is the tip-sample capacitance and ω is the voltage frequency. The three separate components (F_{dc}, F_{ω} $F_{2\omega}$) can be measured selectively by lock-in-amplifiers. So, it is possible to obtain information about locally charged domains, capacitance between the tip and sample surface simultaneously to the topography. Thus, one can characterize dielectric or dopoant surfaces, and locally measure the work function at the surface. The presence of dielectric material over Si wafer is detected and measured in the voltage in a *p-n* junction with sub-micron spatial resolution. Electric forces as small as 10^{-10} N are measured. The contact potential difference between different materials such as gold, platinum, and palladium surfaces with high resolution and fluctuations of thickness with single atomic layer resolution in In GaN/GaN quantum wells grown by metallorganic chemical vapor depositon are also measured. Direct imaging of charge distributions by SCM has been demonstrated in thin SiO_2 layers after electric stress and in presence of Ge-and Si-nanoclusters.

RHEED

RHEED stands for Reflection High Energy Electron diffraction. It is quite similar to Low energy electron diffraction (LEED).

RHEED uses an electron gun and a display or recording device (e.g., a fluorescent screen) Fig. 2.52 for typically obtaining information on the surface symmetry. The surface structure can be determined using RHEED however the process is a time consuming one, involving trial-and-error methods.

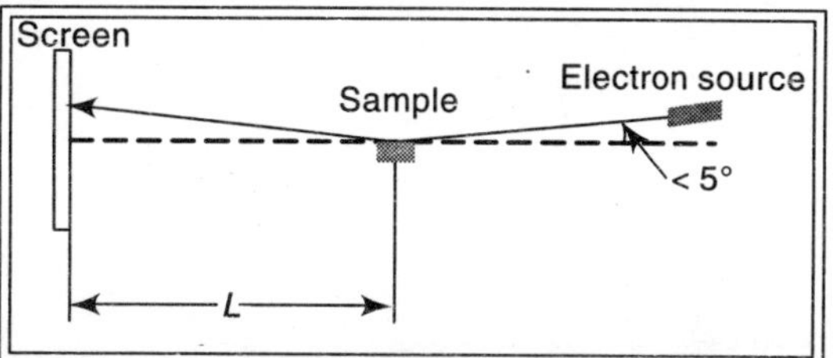

Fig. 2.52 *Schematic of a RHEED setup. The distance from sample to screen, L, and the energy of the electron beam must be known to derive the lattice spacing of the crystal.*

RHEED uses much higher electron energy (typically 5–100keV) and lower impact angles (<5°). The higher energy sharpens the picture, while the lower angle means electrons penetrate just a few atomic layers into the crystal. Consequently RHEED patterns represent the surface symmetry and not the bulk crystal.

Due to this surface sensitivity, single crystal materials that have a flat surface on the nanometre scale produce a RHEED pattern with lines instead of dots. Rough surfaces produce a RHEED pattern with dots which indicate transmission electron diffraction.

Good RHEED patterns consist of well defined streaks which follow the shape of a circle. In RHEED, "view" the reciprocal lattice rods (which are perpendicular to the surface) and rods have a non-zero width and so the observed pattern is a streak.

RHEED is often used for real-time analysis and monitoring of crystal growth, for example in Molecular-beam epitaxy growth systems. The technique is particularly suited to this application as it does not block the direction vertical to the surface of the crystal which is observed, and is particularly sensitive to surface roughness, down to monolayer sensitivity. This allows growth rate of a layers of atoms on the surface to be monitored, by analysing the 'RHEED oscillations'—the change from a RHEED pattern consisting of streaks to one of dots, and back.

By knowing both the distance from the sample to the screen or recording medium and the energy of the electron source, it is possible to calculate the lattice spacings of the sample analysed.

Spectroscopy

SPECTROSCOPY TECHNIQUES

Generally in all types of spectroscopic measurements (e.g., optical, UV, IR, X-ray and electron spectroscopies) there are three types of processes: absorption (transmission). reflection or emission (luminescence). The different experimental methods are shown in Figure 3.1. Such experiments probe the energy differences between electronic, vibrational and, rotational quantum energy states of atoms. A pulsed radiation source is also used.

The wavelength or energy resolution of both the incident radiation and the detected radiation, to determine the overall spectral resolution besides spatial resolution are used. The intensity of the reflected, transmitted (absorbed) or emitted radiation is recorded and analysed as a function of the wavelength of the incident radiation.

Another variable in such spectroscopic experiments is the polarization of the incident and/or reflected, transmitted or emitted radiation. The absorption of electromagnetic radiation involves the resonant response of electrons in an atom or molecule to the oscillating electric field of the radiation, which is a vector quantity and is orthogonal to the propagation direction and the oscillating magnetic field vector of the radiation. If the radiation is linearly polarized, the electric field vector of the radiation to one plane (plane polarized radiation) is restricted and is used to probe the directional properties of, the electronic transitions of atoms within the sample and hence the structure of the sample with respect to the plane of polarization. There are other forms of polarization, known as circular and elliptical polarization, which possesses a polarization vector that describes either a left- or right-handed helical path as a function of time or position and are used for optically active materials or magnetic transitions. Circular polarization involves the electric and magnetic fields of the polarization each being described by two equal orthogonal components which are out of phase (by 90°), while elliptical polarization involves the electric and magnetic fields of the polarization each being described by two unequal orthogonal components which are out of phase. In addition to electromagnetic radiation, it is also sometimes possible to perform similar experiments with electrons.

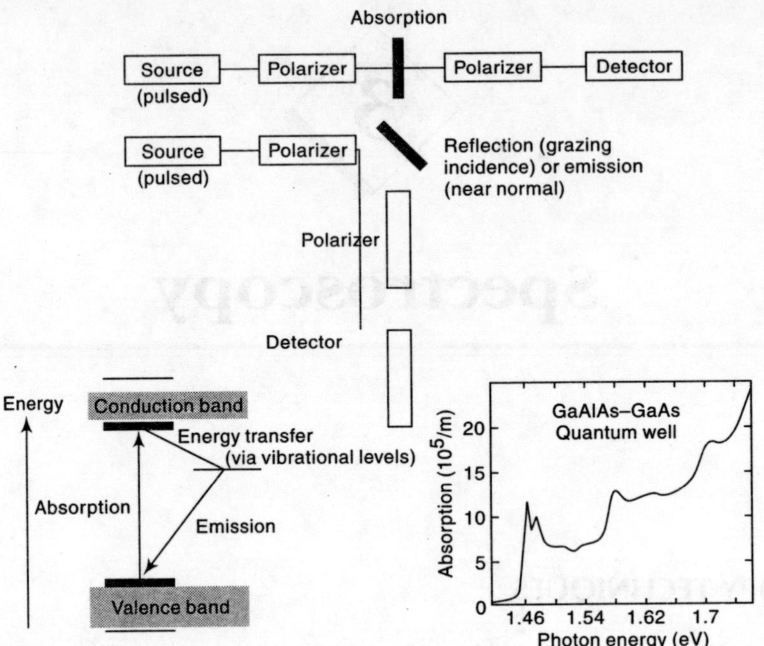

Fig. 3.1 ***Diagram of absorption, reflection and emission spectroscopic experiments together with an energy level diagram and a typical absorption spectrum from a quantum well.***

Photon Spectroscopy

Generally the lateral spatial resolution of photon-based spectroscopies is limited by diffraction, and can attain probe sizes of at best a few micrometres, unless near field techniques are used whereas X-ray synchrotron sources have demonstrated spot sizes of 50–100 nm by focusing using Fresnel Plates. Lateral resolution is therefore limited and essentially the techniques proved an ensemble-averaged signal from a nanostructured sample. Furthermore, most photon-based spectroscopies probe bulk rather than surface properties, however there do exist various techniques for localizing the signals from surface which are of use for analysing thin film nanostructures. Photon-based spectroscopies have been revolutionized in the X-ray regime by the use of synchrotron radiation and, in the optical/UV regime, by the use of lasers that provide coherent, intense and often tunable radiation (over a large range of wavelengths). The lasers can be pulsed over sub-picosecond timescales.

Despite the limited spatial resolution, from the photon-based spectroscopies, all electronic energies in nanostructures affected by quantum confinement is obtained, in addition to the effects of the increased contribution of surfaces. These changes are most apparent for transitions between the highest occupied molecular orbitals (HOMO) in the valence band and the lowest unoccupied molecular orbitals (LUMO) in the conduction band of a solid. Such transitions often lie in the optical region of the electromagnetic spectrum, where λ is typically 400-700 nm (sometimes expressed as a wavenumber $(1/\lambda)$ in cm^{-1}). Photon-induced transitions follow a set of dipole selection rules for the quantum numbers of the initial and final electronic states of the transition, which causes only selected transitions.

Spectroscopy and Spectroscopic Imaging

Tunneling spectroscopy gives access to the chemical and electronic properties of the surface or objects under investigation. It is concerned with density-of-state effects on the tunneling current. These effects manifest themselves in an anomalous voltage dependence of the tunneling current-voltage characteristics or derivatives thereof. In conventional tunneling spectroscopy, the tunneling barrier is a solid insulator of given width, and the applied voltage is the only external experimental parameter on which the tunneling current experimental parameter on which the tunneling current depends (aside, of course, from parameters such as temperature and magnetic field, which are of no concern here). In STM, however, the gap width is adjustable and its experimental control is crucial to obtaining meaningful tunneling spectra. The main merits of tunneling spectroscopy performed with an STM are the following:

- It is *local.*
- It is preformed at *preselected positions.*
- It is performed under *well-defined conditions.*
- It can be combined with *other methods.*
- It can provide *spectroscopic images.*

When spectroscopy is carried out with the STM, the gap width is controlled by the tunneling current itself, either in the constant-current mode or constant-resistance mode. Therefore, the time scales of the current changes of spectroscopic interest should be either slow or fast with respect to the cutoff frequency of the tunneling-current control loop. Tunneling spectra are usually taken by superposing a small, fast modulation on the tunneling voltage while sweeping it slowly, as in conventional tunneling spectroscopy. Alternately, *I-V* curves obtained with fast voltage sweeps or *s-V* characteristics obtained with slow sweeps are also used to obtain spectroscopic information. Advantages of using fast sweeps are that the shapes of the spectroscopic features which are seen are not distorted by a changing gap width and are thus more familiar in their appearance, and that STM spectroscopy can be rendered insentive to mechanical gap-width instabilities. An advantage of using slow sweeps is that the electric fields at the surface are prevented from becoming too high since, under the imposed condition of constant tunneling current, the gap width increased with tunneling voltage. This can be particularly important in systems with loosely bound absorbates of delicate molecules.

Currently achievable gap-width stabilities of about 0.1 Å make it possible to carry out a large variety of spectroscopic studies, even without the necessity of very high sweep frequencies. The first examples of STM spectroscopy, indicate good promise for the use of the STM for this purpose, even in this moderate gap-width stability range. Many spectroscopic features can be much more pronounced in STM spectroscopy because of its local character. For instance, resonant tunneling states of the Gundlach type or image states are readily seen in (s vs. V) and ($\partial I/\partial V$) vs. V) STM, spectra, but only under special conditions in solid junctions. Additionally, as in inelastic tunneling spectroscopy, other aspects such as image force lowering of the tunneling barrier and reduced dielectric screening enhance spectroscopic feature. Finally, STM spectroscopy might relieve us, in many cases, of the necessity of observing very week spectroscopic features. Adsorbate atoms or molecules, for instance, and their electronic states might instead be recognizable via local density-of-state effects in elastic STM tunneling spectroscopy or energy dissipation.

STM imaging should be viewed in a broad contact. The scanning tunneling microscope provides images in a five-dimensional space, namely, in the three-dimensional real space (x, y, z), the tunneling current I, and the tunneling voltage V. It is the selection of the dimensions of interest which determines the "imaging mode." For the contours of constant wave-function overlap, we measure z as $z(c, y)$ at constant I and V, and call it a "topographical" image. If we measure $(\partial I/\partial V)(x, y)$ at constant V and with I as gap-width control parameter, we obtain an STS image, and $(\partial I/\partial V)(x, y)$ gives a "work-function" profile. An STS image or a work-function profile can also be obtained from the subtraction of two topographical images taken at different voltages or different currents, respectively.

One of the first STS images which was obtained pertained to the spatial variations in the superconductivity of Nb_3Sn, with the superconducting energy gap as the spectroscopic parameter of interest. STS of NiO was mentioned above (in discussion of Figure 4). In both cases, the "topographical" images are representative of the topography, and the "spectroscopic" images of the spatial variation of the spectroscopic feature chosen. In other situations, however, we do not have such a close correspondence between image and imaging mode. The following into figures illustrate that caution is required when interpreting the topographical or electronic significance of STM and STS images.

Figure 3.2 shows an STM image of graphite, taken with a tunneling voltage of 50 mV. The surface atomic structure of graphite is a flat, honeycomb arrangement of carbon atoms. The STM image, however, exhibits an appreciable corrugation, and two inequivalent carbon sites are found, in accord, with the presence of carbon atoms having two different electronic structures. When the tunneling voltage is increased to 1.5 V, the STM image becomes flat. That obtained at $V = 50$ mV reflects the corrugated state density close to the Fermi level, which is very different from the integrated state density or the corrugated of the surface-atom positions. It is thus in effect a typical spectroscopic image.

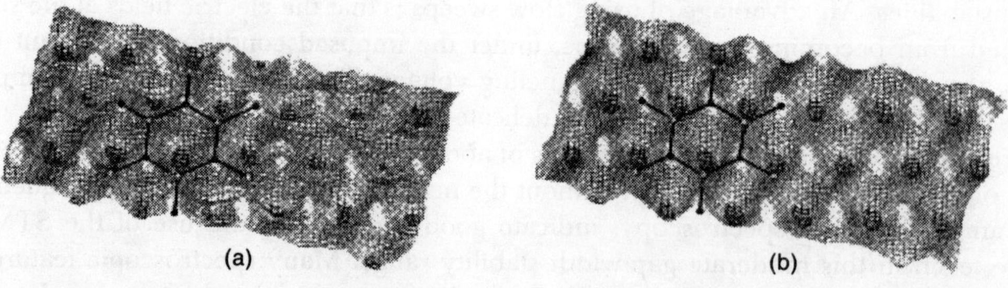

Fig. 3.2 *Constant-current STM image of a graphite surface: (a) raw data, (b) periodic continuation of an averaged unit cell (for the image processing method used. The solid and open circles indicate the positions of the two electronically different types of carbon atoms, which are arranged in a flat honeycomb lattice. The corrugation measured is not that of the surface carbon atoms but corresponds to that of the local density of states at the Fermi level.*

Ultraviolet-visible Spectroscopy

Ultraviolet-visible spectroscopy or ultraviolet-visible spectrophotometry (UV/VIS) involves the spectroscopy of photons and spectrophotomoetry. It uses light in the visible and adjacent near ultraviolet (UV) and near infrared (NIR) ranges. In this region of energy space molecules undergo electronic transitions.

UV/VIS spectroscopy is routinely used in the quantitative determination of solutions of transition metal ions and highly conjugated organic compounds.

Solutions of transition metal ions can be colored (i.e., absorb visible light) because d electrons within the metal atoms can be excited from one electronic state to another. The color of metal ion solutions is strongly affected by the presence of other species, such as certain anions or ligands. For instance, the color of a dilute solution of copper sulphate is a very light blue; adding ammonia intensifies the colors and changes the wavelength of maximum absorption.

Organic compounds, with a high degree of conjugation, also absorb light in the UV or visible regions of the electromagnetic spectrum. The solvents for these determinations are often water for water soluble compounds, or ethanol for organic-soluble compounds. (Organic solvents may have significant UV absorption; not all solvents are suitable for use in UV spectroscopy. Ethanol absorbs very weakly at most wavelengths.)

Charge transfer complexes also give rise to colors, the colors are often too intense to be used for quantitative measurement.

The Beer-Lambert law states that the absorbance of a solution is directly proportional to the solution's concentration. Thus absorbance changes with concentration. This can be taken from references (tables of molar extinction coefficients), or more accurately, determined from a calibration curve.

The method is most often used in a quantitative way to determine concentrations of an absorbing species in solution, using the Beer-Lambert Law:

$$A = -\log_{10}(I/I_0) = E \cdot C \cdot L,$$

where A is the measured absorbance, I_0 is the intensity of the incident light at a given wavelength, l is the transmitted intensity, L the pathlength through the sample, and C the concentration of the absorbing species. For each species and wavelength, E is a constant known as the molar absorptivity or extinction coefficient. This constant is a fundamental molecular property in a given solvent, at a particular temperature and pressure, and has units of $1/M\ cm$ or often $AU/M\ cm$.

The absorbance and extinction E are sometimes defined in terms of the natural logarithm instead of the base-10 logarithm.

A UV/VIS spectrophotometer may be used as a detector for HPLC. The presence of an analyte gives a response which can be assumed to be proportional to the concentration. For accurate results, the instrument's to the analyte in the unknown should be compared with the response to a standard; this is very similar to the use of calibration curves. The response (e.g., peak height) for a particular concentration is known as the response factor.

Spectrophotometry

The instrument used in ultraviolet-visible spectroscopy is called a UV/VIS spectrophotometer. It measures the intensity of light passing through a sample (I_0). The ratio I/I_0 is called the *transmittance*, and is usually expressed as a percentage (%T). The absorbance, A, is based on the transmittance:

$$A = -\log(\%T)$$

The basic parts of a spectrophotometer Fig. 3.3 are light source (often an incandescent bulb for the visible wavelengths, or a deuterium are lamp in the ultraviolet), a holder for the sample, a diffraction grating or monochromator to separate the different wavelengths of light, and a detector. The detector is typically a photodiode or CCD. Photodiodes are used with monochromators, which filter the light so that only light of a single wavelength reaches the detector. Diffraction gratings are used with CCDs, which collects light of different wavelengths on different pixels.

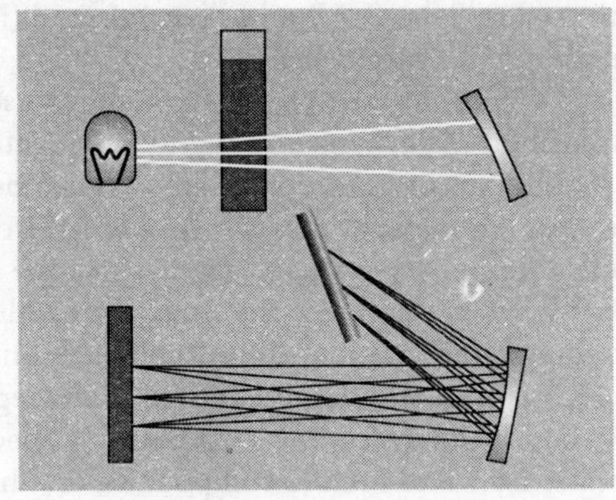

Fig. 3.3 *Diagram of a single-beam UV/vis spectrophotometer.*

A spectrophotometer can be either *single beam* or *double beam.* In a single beam instrument, all of the light passes through the sample cell. I_0 must be measured by removing the sample.

In a double-beam instrument, the light is spit into two beams before it reaches the sample. One beam is used as the reference; the other beam passes through the sample. Some double-beam instruments have two detectors (photodiodes), and the sample and reference beam are measured at the same time. In other instruments, the two beams pass through a beam chopper, which blocks one beam at a time. The detector alternates between measuring the sample beam and the reference beam.

Sample for UV/VIS spectrophotometry are most often liquids, although the absorbance of gases and even of solids can also be measured. Sample are typically placed in a transparent cell, known as a **cuvette**. Cuvettes are typically rectangular in shape, commonly with an internal with an internal width of 1 cm. (This width becomes the path length, L, in the Beer-Lambert Law.) Test tubes can also be used as cuvettes in some instruments. The best cuvettes are made of high quality quartz, although glass or plastic cuvettes are common. (Glass and most plastics absorb in the UV, which limits their usefulness to visible wavelengths.)

Ultraviolet-visible Spectrum

An ultraviolet-visible spectrum is essentially a graph of light absorbance versus wavelength in a range of ultraviolet or visible regions. Such a spectrum can often be produced directly by a more sophisticated spectrophotometer, or the data can be collected one wavelength at a time by simpler instruments. Wavelength is often represented by the symbol λ. Similarly, for a given substance, a standard graph of the extinction coefficient (ε) *vs.* wavelength (λ) may be made. For the given substance, the wavelength at which maximum absorption in the spectrum occurs is called λ_{max}. A set of empirical observations are used to predict λ_{max}, the wavelength of the most intense UV/VIS absorption, for conjugated organic compounds such as dienes and ketones.

This spectrum can be used qualitatively to identify components in a sample as each component has their own unique absorbance spectrum (like a fingerprint).

X-ray Spectroscopy

X-ray from a high-intensity source with a controllable wavelength, such as a synchrotron, is used to perform absorption or reflectivity. X-ray absorption spectroscopy (XAS) involves the ionization of atoms by high-energy photons causing electron transitions to occur from deep inner shell atomic levels to the empty conduction ban. X-ray ionization edges occur at energies characteristic of both the element and the inner shell involved and show two types of fine structure: X-ray absorption near-edge structure (XANES), which involves strong intensity oscillations within about 40 eV of the absorption edge onset, and weaker oscillations at higher energies known as extended X-ray absorption near-edge structure (EXAFS). XANES is directly related to the unoccupied density of electronic states and has characteristic features and energy positions for different bonding environment of the atoms in the solid, whereas EXAFS oscillations can be used to determine bond lengths and coordination numbers in terms of a radial distribution function around the ionized atom. This is shown in Figure 3.4. Both these techniques have electron equivalents in the technique of EELS in the TEM atoms in sample. This technique is restricted to elements possessing isotopes with suitable nuclear transitions such as ^{57}Fe, ^{199}Sn and ^{197}Au.

Electron paramagnetic resonance (EPR) is a microwave spectroscopy which employs a variable magnetic field to analyse crystals containing molecules or ions with unpaired electronic spins. The magnetic field causes Zeeman splitting of electronic energy levels, and transitions between these levels are induced by the absorption of microwave radiation.

ELECTRON SPECTROSCOPY

Electron-induced spectroscopies in the electron microscope, is done by SEM or TEM, by analysis the X-rays emitted by the sample as given below:

X-ray emission in the SEM and TEM

Following ionization of atoms in a sample by an electron beam, is one possible de-excitation process of X-ray emission. The energy of the X-ray photon emitted when a single outer electron drops into the inner shell hole is given by the difference between the energies of the excited states involved. A set of dipole determines the transitions to be observed. The energies and associated wavelengths of the set of emitted X-ray have a characteristic value for each of the atomic species present in the specimen. By measuring the energies or wavelengths of the X-ray emitted from the top surface of the sample, it is possible to determine which elements are present at a particular position of the electron probe. This is the basis for energy-dispersive and wavelength-dispersive X-ray analysis (EDX and WDX). WDX spectrometers use crystal monochromators to disperse the emitted X-ray spectrum in terms of diffraction angle and hence wavelength. Detectors then move along the arc of a circle centered on the specimen and collect the spectrum serially. EDX detectors collect X-rays in a near-parallel fashion and rely on the creation of electron hole pairs in a biased silicon crystal. The number of electron-hole pairs and hence current is directly proportional to the energy of the incident X-ray. Pulses of X-ray are seperated and measured electronically. EDX detectors depending on the material,

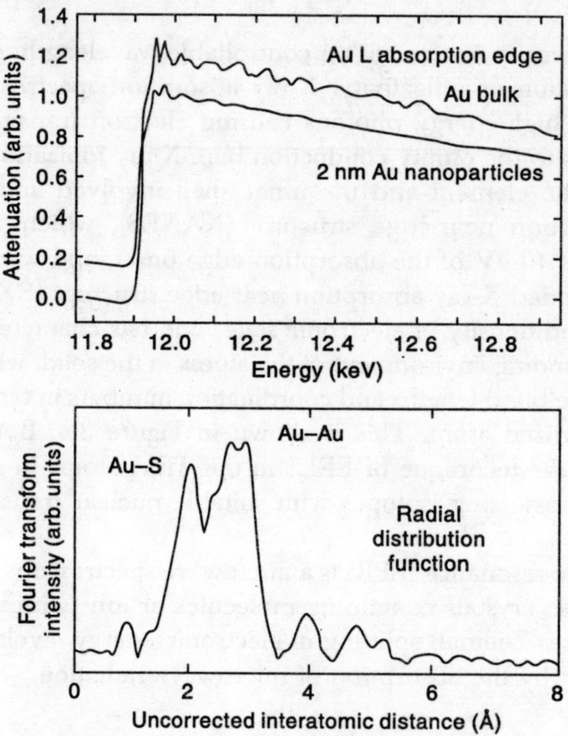

Fig. 3.4 *Schematic comparison between the L X-ray absorption edge in bulk gold and nanoparticulate gold together with the radial distribution function obtained via Fourier transformation of the EXAFS oscillations.*

reduce sensitivity to light elements ($Z<11$). WDX is slow but has increased resolution and is sensitive to all elements. It is confined to dedicated analytical SEMs known as electron probe microanalysers (EPMA), while EDX detectors are may be fitted as an add-on attachment to most SEMs and TEMs.

Fig. 3.5c shows a typical electron-generated X-ray emission spectrum from molybdenum oxide. The Mo Kα, Kβ and Lα X-ray lines as well as the O Kα line are superimposed. These X-rays are not characteristic of any particular atom but depend principally on specimen thickness. To a first approximation, peak intensities are proportional to the atomic concentration of the element and, with careful measurement, EDX and WDX detect levels of elements down to 0.1 at %.

When EDX is used in the TEM, the reduced sample thickness and probe size lead to much higher spatial resolution. At 100 kV, a 100nm thick sample typically gives a beam broadening of the order of 10 nm. EDX analysis collect all X-rays produced isotropically within the beam-broadened volume, for the analysis of nanostructures. EDX maps of multilayers and nanoscale precipitates have demonstrated resolutions of a few nanometres in ultrathin sample areas. Impurity levels or defect concentrations modify the band gap and thus the CL wavelength. Spectroscopic analysis of the CL emission therefore provide information on the microstructure

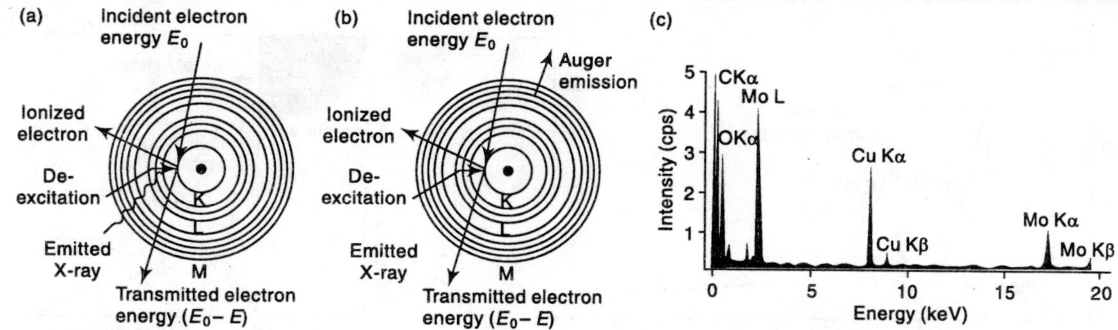

Fig. 3.5 *Diagram of (a) de-excitation by X-ray emission, (b) de-excitation by Auger electron following ionization by an electron of incident energy E_0. (c) Energy-dispersive X-ray emission spectrum from a thin specimen of molybendium, oxide on a carbon support film; the peaks are labelled with standard X-ray notation; the peaks due to copper are from the specimen holder*

of the sample. The emitted photon intensity is low and therefore low scan rates and high probe currents (large probe diameters) are required, limiting the resolution to 1–10 mm in the SEM.

Electron Energy Loss Spectroscopy

Electron energy loss spectroscopy (EELS) in a TEM or STEM involves analysis of the inelastic scattering suffered by the electron beam via measurement of the energy distribution of the transmitted electrons. This allows high-resolution elemental mapping and the measurement of local electronic structures for the determination of the local chemical bonding, as that present at an interface or defect.

The transmitted electrons are dispersed according to their energy losses using a magnetic sector spectrometer (see Figure 3.6). EELS spectra can be recorded in parallel using a scintillator and diode array detection system, with the TEM in either diffraction or image mode: spatially resolved spectra are obtained using a small, highly focused probe in a STEM. The transmitted EELS signal is highly forward-peaked and the collected signal is well defined (using a suitable aperture) thus it significantly improves the ultimate spatial resolution of the analysis compared to EDX, where beam broadening will limit the ultimate (lateral) spatial resolution for microanalysis. EELS spectra can be obtained from individual atomic columns.

The various energy losses observed in a typical EELs spectrum are shown in Figure 3.6a, which gives the scattered electron intensity as function of the decrease in kinetic energy (the energy loss E) of the transmitted fast electrons and essentially represents the response of the electrons in the solid to the disturbance introduced by the incident electrons. In a specimen of thickness less than the mean free path for inelastic scattering (roughly 100 nm at 100 keV), by far the most intense feature is the zero-loss peak at 0 eV energy loss, which contains all the elastically and quasi-elastically (i.e., phonon) scattered electron components. Neglecting the effect of the spectrometer, the full which at half maximum (FWHM) of the zero-loss peak is usually limited by the energy spread inherent in the electron source. In a TEM the energy spread generally lie between ~0.1 and 3eV, depending on the type of emitter, and this energy often determines the overall spectral resolution.

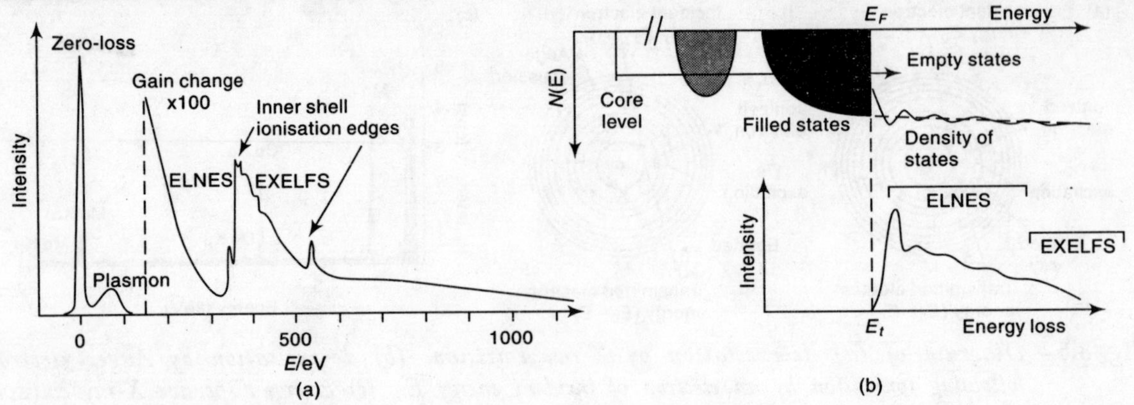

Fig. 3.6 *Diagram of (a) an EELS spectrum and (b) the ELNES intensity which reflects the unoccupied DOS above the Fermi level*

The *low-loss* region of the EELs spectrum, extending from 0 to about 50 eV, corresponds to the excitation of electrons in the outermost atomic orbitals which are often delocalized due to interatomic bonding and extend over several atomic sites. This region reflects the solid-state character of the sample. The smallest energy losses (10–100 meV) arise from phonon emission, but these are usually subsumed in the zero-loss peak. The low-loss spectrum is from collective, resonant oscillations of the valence electrons known as plasmons. The energy of the plasmon peak is governed by the density of the valence electrons, and the width by the rate of decay of the resonant mode. In a thicker specimen (≥100nm) there are additional peaks at multiples of the plasmon energy. Also, in the low-loss spectra of insulators are peaks, known as interband transitions, which correspond to the excitation of valence electrons to low-energy unoccupied electronic states above the Fermi level. These single-electron excitations lead to a shift in the energy of the plasmon resonance. The low-loss region is used mainly to determine the specimen thickness and to correct for the effects of multiple inelastic scattering when performing quantitative microanalysis on thicker specimens. The overall shape of the low-loss region is also related to the dielectric response function of the material, which allows a correlation with reflectivity and band gap determination in insulators and semiconductors.

The *high-loss* region of the EELs spectrum extends from about 50 eV to several thousand electron volts and corresponds to the excitation of electrons from localized orbitals on a single atomic site to extended, unoccupied electron energy levels just above the Fermi level of the material. This region reflects the atomic character of the specimen. As the energy loss increases, this region exhibits steps or edges superimposed on the monotonically decreasing background intensity; the intensity at 2000 eV is typically eight orders of magnitude less than at the zero-loss peak and so, in Figure 3.6 a gain change is inserted in the linear intensity scale at 150 eV. These edges correspond to excitation of inner shell electrons and are known as ionization edges. The various EELS ionization edges are classified using the standard spectroscopic notation; e.g., K excitation for ionization of 1s electrons and are known as ionization edges threshold is determined by the binding energy of the particular electron subshell within an atom. The signal under the ionization edge extends beyond the thershold, since the amount of kinetic energy given to the excited electron is not fixed. The intensity or area under the edge is proportional to the number of atoms present and this allows the technique to be used for quantitative analysis. EELS is particularly sensitive to the detection and quantification of light elements ($Z <$ 11) as well as transition metals and rare earths.

If electrons are scattered via inelastic collisions with K-shell electrons of free atoms (e.g., gases) the core-loss edges are sharp, sawtooth-like steps displaying no features. Other core-loss excitations in free atoms display a variety of basic edge shapes that are essentially determined by the degree of overlap between a variety of basic edge shapes that are essentially determined by the degree of overlap between the initial and final state wavefunctions. In solids, however, the unoccupied electronic states near the Fermi level, are modified by chemical bonding, leading to a complex density of stated (DOS), and this is reflected in the electron energy loss near-edge structure (ENES), which modifies the basic atomic shape within the first 30–40 eV above the edge threshold. The ELNES effectively represents the unoccupied DOS (above the Fermi level) in the environment of the atom(s) being ionized, as shown in Figure 3.5b, hence it gives information on the local structure and bonding. Beyond the near-edge region, superimposed on the gradually decreasing tail of the core-loss edge, a region of weaker, extended oscillations is observed, known as the extended energy loss fine structure (EXELFS). The period of the oscillation may be used to determine bond distances, while the amplitude reflects the coordination number of the particular atom.

Energy-filtered TEM (EFTEM) involves the selection of a specific energy loss value, or narrow range of energy losses, from the transmitted electron beam via use of an energy-selecting slit after the spectrometer. Using only zero-loss (elastically) scattered electrons to form images and diffraction patterns increases contrast and resolution. Chemical mapping is achieved by acquiring and processing images formed by electrons which have undergone inner shell ionization events. An alternative approach, is to raster the electron beam across the specimen and record an EELS spectrum 'spectrum imaging' at every point (x, y). The complete date set is then processed to from a 2D quantitative map of the sample using either standard elemental quantification procedures or the position and/or intensity of characteristic low-loss or ELNES features. This inherently high spatial resolution of the EELS technique has allowed maps of elemental distributions to be formed at subnanometre resolution.

RAMAN SCATTERING

Raman scattering or the **Raman effect** is the inelastic of a photon.

When light is scattering from an atom or molecule, most photons are elastically scattered (Rayleigh scattering). The scattered photons have the same energy (frequency) and, therefore, wavelength, as the incident photons. However, a small fraction of scattered light (approximately 1 in 10 million photons) is scattered from excitations with optical frequencies different from, and usually lower than, the frequency of the incident photons. In a gas, Raman scattering can occur with a change in vibrational, rotational or electronic energy of a molecule. Chemists are concerned primarily with the vibrational Raman effect.

In 1922, Indian physicist Chandrasekhara Venkata Raman published his work on the "Molecular Diffraction of Light," the first of a series of investigations with his collaborators which ultimately led to his discovery (on 28 February 1928) of the radiation effect which bears his name. The Raman effect was first reported by C.V. Raman and K. S. Krishnan, and independently by Grigory Landsberg and Leonid Mandelstam, in 1928. Raman received the Nobel Prize in 1930 for his work on the scattering of light. In 1998 the Raman Effect was designated an ACS National Historical Chemical Landmark in recognition of its significance as a tool for analyzing the composition of liquids, gases, and solids.

It is also possible to observe molecular vibrations by an inelastic scattering, an absorbed photon is re-emitted with lower energy. The difference in energy between the absorbed and re-emitted photons corresponds to the energy required to excite a molecule to a higher vibrational mode.

Typically, in Raman spectroscopy high intensity laser radiation with wavelengths in either the visible or near-infrared regions of the spectrum is passed through a sample. Photons from the laser beam are absorbed by the molecules, exiting them to a virtual energy state. If the molecules relax back to the vibrational state, the reemitted photon has the same energy as the original photon. This leads to scattering of the laser light, but with no change in energy between the incoming photons and the reemitted/scattered photons. This type of scattering is known as Rayleigh scattering.

THEORY OF RAMAN SCATTERING

Degrees of Freedom

For any given chemical compound, there are a total of 3N degrees of freedom, where N is the number of atoms in the compound This number arises from the ability of each atom in a molecule to move in three different direction (x, y and z). When dealing with molecules, it is more common to consider the movements of the molecule as a whole. Consequently, the 3N degrees of freedom are partitioned into molecular translation, rotational, and vibrational motion. Three of degrees of freedom correspond to translational movement of the molecule as a whole (along each of the three spatial dimensions). Similarly, three degrees of freedom correspond to rotations of the molecules about the x, y, and z-axes. However, linear molecules only have two rotations because correspond to molecular vibrational modes. These modes include stretching and bending motions of the chemical bonds of the molecule. For a linear molecule, the number of vibrational modes is:

$$3N - 3 - 2 = 3N - 5$$

whereas for a non-linear molecule the number of vibrational modes are

$$3N - 6$$

Molecular Vibrations and Infrared Radiation

The frequencies of molecular vibrations are in range of 10^{12} to 10^{14} Hz. These frequencies correspond to radiation in the infrared (IR) region of the electromagnetic spectrum. At any given instant, each molecule in a sample has a certain amount of vibrational energy. However, the amount of vibrational energy that a molecule has continually changes due to collisions and other interactions with other molecules in the sample.

At room temperature, most of molecules will be in the lowest energy states, which is known as the ground state. A few molecules will be in higher energy states, which are known as excited states. The fraction of molecules occupying a given vibrational mode at a given temperature can be calculated using the Boltzmann distribution. Performing such a calculation shows that, for relatively low temperatures (Such as those used for most routine spectroscopy), most of the molecules occupy the ground vibrational state. Such a molecule can be excited to higher vibrational mode through the direct absorption of a photon of the appropriate energy. This is the mechanism by which IR spectroscopy operates: infrared radiation is passed through the sample, and the intensity of the transmitted light is compared with that of the incident light. A reduction in intensity at a given wavelength of light indicates the absorption of energy by a vibrational transition. The energy, E, of a photon is

$$E = hv$$

Where h is Planck's constant and v is the frequency of the radiation. Thus, the energy required for such transition may be calculated if the frequency of the radiation is known.

Stokes and Anti-stokes

The interaction of light with matter in a liner regime allows the absorption or simultaneous emission of light precisely matching he difference in energy levels of the interacting electrons.

The Raman effect corresponds, in perturbation theory, to the absorption and subsequent emission of a photon via an intermediate electron state, having a virtual energy level. It is also possible for the molecules to relax back to a vibrational state that is higher in energy. Here, the original photon and the reemitted photon differ in energy by the amount required to vibrationally excite the molecule. The difference in energy is recorded as the difference in wavenumber ($\Delta \bar{v}$) between the laser light and the scattered light. A Raman spectrum is generated by plotting the intensity of the reemitted light versus ($\Delta \bar{v}$). If the reemitted radiation is lower in energy than the incident laser light, the change in wavenumber is positive and results in a series of peaks in the Raman spectrum known as Stokes lines.

A Raman spectrum also exhibits peaks that correspond to negative values of $\Delta \bar{v}$. These peaks are due to re-emitted photons that are higher in energy than the incident photons. This occurs when molecules that initially are in an excited vibrational state absorb the laser light and relax back to the lower vibrational state when they reemit the photon. These lines in the Raman spectrum are known as anti-Stokes lines. Since the Stokes lines and anti-Stokes lines gain and lose the same amount of energy, they are symmetric with respect to the peak due to elastic (Rayleigh) scattering ($\Delta \bar{v} = 0$). The anti-Stokes lines are appreciably less intense than the corresponding Stokes lines. This is because initially very few molecules are in excited vibrational states compared to the number in the ground state. Since anti-Stokes lines arise from the former and Stokes lines arise from the latter, the Stokes lines are much more intense. However, it should be noted that in molecules which exhibit fluorescense, the Stokes lines may be obscured while the anti-Stokes lines remain unaffected. In Such cases, it is necessary to use the anti-Stokes lines despite their lower intensity.

Thus there are three possibilities:

No energy exchange between the incident photon and the molecules (and hence no Raman effect)

Energy exchanges occur between the incident photons and the molecules. The energy differences are equal to the differences of the vibrational and rotational energy-levels of the molecule. In crystals only specific photons are allowed (solutions of the wave equations which do not cancel themselves) by the periodic structure, so Raman scattering can only appear at certain frequencies. For amorphous materials like glasses, more photons are allowed and thereby the discrete spectral lines become broad.

Molecule absorbs energy: Stokes scattering Fig. 3.7. The resulting photon of lower energy generates a Stokes line on the red side of the incident spectrum.

Molecules loses energy: anti-stokes scattering Fig. 3.7. Incident photons are shifted to the blue side of the spectrum, thus generating an anti-Stokes line.

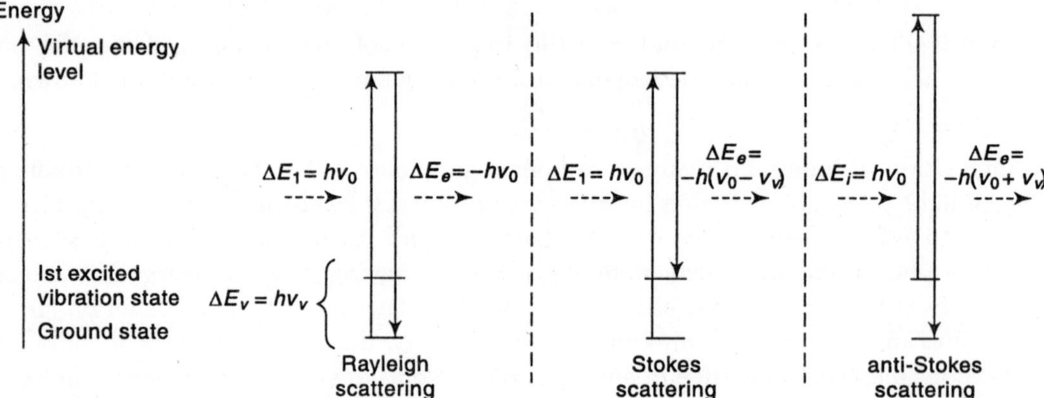

Fig. 3.7 *The different possibilities of visual light scattering: Rayleigh scattering (no Raman effect), Stokes scattering (molecule absorbs energy) and anti-Stokes scattering (molecule loses energy)*

These differences in energy are measured by subtracting of the mono-energetic laser light from the energy of the scattered photons. The absolute value, however, doesn't depend on the process (Stokes or anti-Stokes scattering), because only the energy of the different vibrational levels is of importance. Therefore, the Raman spectrum is symmetric relative to the Raleigh band. In addition, the intensities of the Raman bands are only dependent on the number of molecules occupying the different vibrational states, when the process began. The Boltzmann distribution shows that more molecules occupy the lower energy levels.

$$\frac{N_1}{N_0} = \frac{g_1}{g_0} e^{-\frac{\Delta E_v}{kT}}$$

with:

N_0: amount of atoms in the lower vibrational state
N_1: amount of atoms in the higher vibrational state
g_0: degeneracy of the lower vibrational state (number of orbitals of the same energy)
g_1: degeneracy of the higher vibrational state (number of orbitals of the same energy)

ΔE_v: energy difference between these two vibrational states
k: Boltzmann's constant
T: temperature in kelvin

Thus the Stokes spectrum is more intense than the anti-Stokes spectrum.

Distinction with Fluorescence

The Raman effect differs from the process of fluorescence. Here, the incident light is completely absorbed and the system is transferred to an excited state from which it can go to lower states only after a certain resonance. The result of both processes is essentially the same: A photon with the frequency different from that of the incident photon is produced and the molecule is brought to a higher or lower energy level. But the major difference is that the Raman effect can take place for any frequency of the incident light. In contrast to the fluorescence effect, the Raman effect is therefore not a resonant effect.

Transition

The distortion of a molecule in an electric field, and therefore the vibrational Raman cross section, is determined by its polarizability.

A Raman transition from one state to another, and therefore a Raman shift, can occur only when the polarizability changes during the process under consideration (that is, during the vibration or rotation). The key quantity is the derivative of the polarizability with respect to the normal mode excited during the transition.

Stimulated Raman Scattering and Raman Amplification

Raman amplification is obtained by using Stimulated Raman Scattering (SRS), which actually is a combination between a Raman process with stimulated emission. It is used for application in telecommunication fibers to amplify inside the standard material with low noise for the amplification process. The process requires significant power and thus imposes more stringent limits on the material. The amplification band can be up to 100 nm broad, depending on the availability of allowed photon states.

For high intensity CW (continuous wave) lasers, SRS can be used produce broad bandwidth spectra. This process can also be seen as a special case of four wave mixing, where the frequencies of the incident photons are equal and the emitted spectra are found in two bands separated from the incident light by the phonon energies. The initial Raman spectrum is built up with spontaneous emission and is amplified later on. At high pumping levels in long fibers, higher order Raman spectra can be generated by using the Raman spectrum, thereby building a chain of new spectra with decreasing amplitude. The disadvantage of intrinsic noise due to the initial spontaneous process is overcome by seeding a spectrum at the beginning. This technology is used ni the fiber optic laser field and there is demand for transversal coherent high intensity light sources (i.e. broadband telecommunication, imaging application).

Applications

Raman spectroscopy employs the Raman effect for materials analysis. The frequency of light scattered from a molecule may be changed based on the structural characteristics of the molecular bonds. A monochromatic light source (laser) is required for illumination, and a spectrogram of the scattered light then shows the deviations caused by sated changes in the molecule.

Raman spectroscopy is also used in combustion diagnostics. Being a completely non-intrusive technique, it permits the detection of the major species and temperature distribution inside combustors and in flames without any perturbation of the (mainly fluid dynamic and reactive) processes examined.

Stimulated Raman transitions are also widely used for manipulating a trapped ion's energy levels, and thus basis quit states, in ion trap quantum computing.

RAMAN SPECTROSCOPY

Raman spectroscopy is a spectroscopic technique used in condensed matter physics and chemistry to study vibrational, rotational, and other low-frequency modes in a system. It relies on inelastic scattering, or Raman scattering of monochromatic light, usually from a laser in the visible, near infrared, or near ultraviolet range. The laser light interacts with phonons or other excitations in the system, resulting in the energy of the laser photons being shifted up or down. The shift in energy gives information about the phonon modes in the system. Infrared spectroscopy yield similar, but complementary information.

Typically, a sample is illuminated with a laser beam. Light from the illuminated spot is collected with a lens and sent through a monochromator. Wavelengths close to the laser line (due to elastic Rayleigh scattering) are filtered out and those in a certain spectral window away from the laser line are dispersed onto a detector.

Spontaneous Raman scattering is typically very weak, and as a result the main difficulty of Raman spectroscopy is separating the weak inelastically scattered light from the intense Rayleigh scattered laser light. Raman spectrometers typically use holographic diffraction gratings and multiple dispersion stages to achieve a high degree of laser rejection. Use of CCD detectors is made in dispersive Raman spectral acquisitions.

Raman spectroscopy has a stimulated version, and is called stimulated Raman scattering.

RAMAN MICROSPECTROSCOPY

Raman spectroscopy offers several advantages for microscopic analysis. Since it is a scattering technique, specimens do not need to be fixed or sectioned. Raman spectra can be collected from a very small volume (< 1 μm in diameter); these spectra allow the identification of species present in that volume. Water does not interfere very strongly. Thus, Raman spectroscopy is suitable for the microscopic examination of minerals, materials such as polymers and ceramics, cells and proteins. A Raman microscope begins with a standard optical microscope, and adds an excitation laser, a monochromator, and a sensitive detector (such as a charge-coupled device (CCD) or photomultiplier tube (PMT)). FT-Raman has also been used with microscopes.

In *direct imaging*, the whole field of view is examined fro scattering over a small range of wavenumbers (Raman shifts). For instance, a wavenumber characteristic for cholesterol could be used to record the distribution of cholesterol within a cell culture.

The other approach is *hyperspectral imaging*, in which thousands of Raman spectra are acquired from all over the field of view. The data is then used to generate images showing the location and amount of different components. For, the cell culture, a hyperspectral image shows the distribution of cholesterol, as well as proteins, nucleic acids, and fatty acids. Sophisticated signal-and image-processing techniques are used to ignore the presence of water, culture media, buffers, etc.

Raman microscopy, and in particular confocal microscopy, has very high spatial resolution. For example, the lateral and depth resolutions of 250 nm and 1.7 μm, respectively are obtained using a confocal Raman microspectrometer with the 632.8 nm line from a He-Ne laser with a pinhole of 100 μm diameter.

Since the objective lenses of microscopes focus the laser beam to several micrometres in diameter, the resulting photon flux is much higher. This has the added benefit of enhanced fluorescence quenching. However, the high photon flux also causes the sample degradation, and for this reason some setups require a thermally conducting substrate (which acts as a heat sink) in order to mitigate this process.

By using Raman microspectroscopy, *in vivo* time-and space-resolved Raman spectra of microscopic regions of samples can be measured. As a result, the fluorescence of water, media, and buffers can be removed. Consequently *in vivo* time-and space-resolved Raman spectroscopy is suitable to examine proteins, cells and organs.

Raman microscopy for biological and medical specimens generally uses near-infrared (NIR) lasers (785 nm diodes and 1064 nm Nd: YAG). This reduces the risk of damaging the specimen by applying high power. However, the intensity of NIR Raman is low (owing to the ω^{-4} dependence of Raman scattering intensity), and most detectors required very long collection times. Raman microscopy of inorganic specimens, such as rocks and ceramics and polymers, can use a broader range of excitation wavelengths.

Applications of Raman Spectroscopy

Raman spectroscopy is commonly used in chemistry, since vibrational information is very specific for the chemical bonds in molecules. It provides a fingerprint by which the molecule can be identified. The fingerprint region of organic molecules is in the range 500–2000 cm^{-1}. With it changes in chemical bonding, e.g. when a substrate is added to an enzyme can be known.

Raman gas analyzers have many practical applications. They are used in medicine for real-time monitoring of anaesthetic and respiratory gas mixtures during surgery.

In solid state physics, spontaneous Raman spectroscopy is used to, characterize materials, measure temperature, and find the crystallographic orientation of a sample.

Raman spectroscopy can also be used to observe low frequency excitations of solids, such as plasmons, magnons, and superconducting gap excitations.

The spontaneous Raman signal gives information on the population of a given phonon mode in the ration between the Stokes (downshifted) intensity and anti-Stokes (upshifted) intensity.

Raman scattering by an anisotropic crystal gives information on the crystal orientation. The polarization of the Raman scattered light with respect to the crystal and the polarization of the laser light can be used to find the orientation of the crystal, if the crystal structure is known.

Raman active fibers, such as aramid and carbon, have vibrational modes that show a shift in Raman frequency with applied stress. Polypropylene fibers also exhibit similar shifts.

The radial breathing mode is a commonly used technique to evaluate the diameter of carbon nanotubes.

Spatially Offset Raman Spectroscopy (SORS), which is less sensitive to surface layers than conventional Raman, can be used to discover counterfeit drugs without opening their internal packaging, and for non-invasive monitoring of biological tissue.

Raman spectroscopy can be used to investigate the chemical composition of historical documents and contribute to knowledge of the social and economic conditions.

INFRARED (IR) AND RAMAN VIBRATIONAL SPECTROSCOPY

The energy of most molecular vibrations corresponds to the infrared region of the electromagnetic spectrum and these vibrations may be detected and measured in an infrared (IR) or Raman spectrum, which typically covers a wavelength range of 2–16 mm (more commonly expressed as 400–5000cm$^{-1)}$.

For IR the spectrometer system consists of an IR source emitting throughout the whole frequency range; the beam is split, with one beam passing either through the sample (transmission) or alternatively reflected (for studies of surfaces) while the other beam is employed as a reference beam. Samples may be a vapour or solution contained in a special NaCl cell, a liquid between NaCl plates, or a solid ground together with excess KBr and pressed into a disc. Fourier transform IR (FTIR) spectroscopy works by allowing the transmitted (or reflected) beam to recombine with the reference beam after a path difference has been introduced. The interference pattern so produced is Fourier transformed to produce the spectrum of material being analysed as a function of wavelength or wavenumber of the incident radiation.

If a molecular vibration is excited in the sample, the molecule absorbs energy of the particular frequency and this is detected as absorption relative to the reference beam. As shown in Figure 3.8, different functional groups have characteristic vibration frequencies arising due to stretching, bending, rocking and twisting of bonds which allows the particular functional groups to be identified. These may change slightly when the functional group is incorporated into a solid. However, not all possible vibrations are excited in IR spectroscopy, as a dipole moment must be created during the vibration. In molecules possessing a centre of symmetry, vibrations symmetrical about the centre of symmetry are IR-inactive.

The attenuated total reflection spectrometer (ATR-FTIR) is an extension of the FTIR technique and consists of an infrared-transparent block; silver chloride, zinc selenide and germanium are suitable materials. An infrared beam passes through tis block at a glancing angle and is totally reflected several times at the top ad bottom before exiting at the other end. The samples can be adhered to the block, changing the IR spectrum. In total internal reflection there is no power transmission outside of the block but this odes not imply that the IR wave is completely

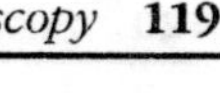

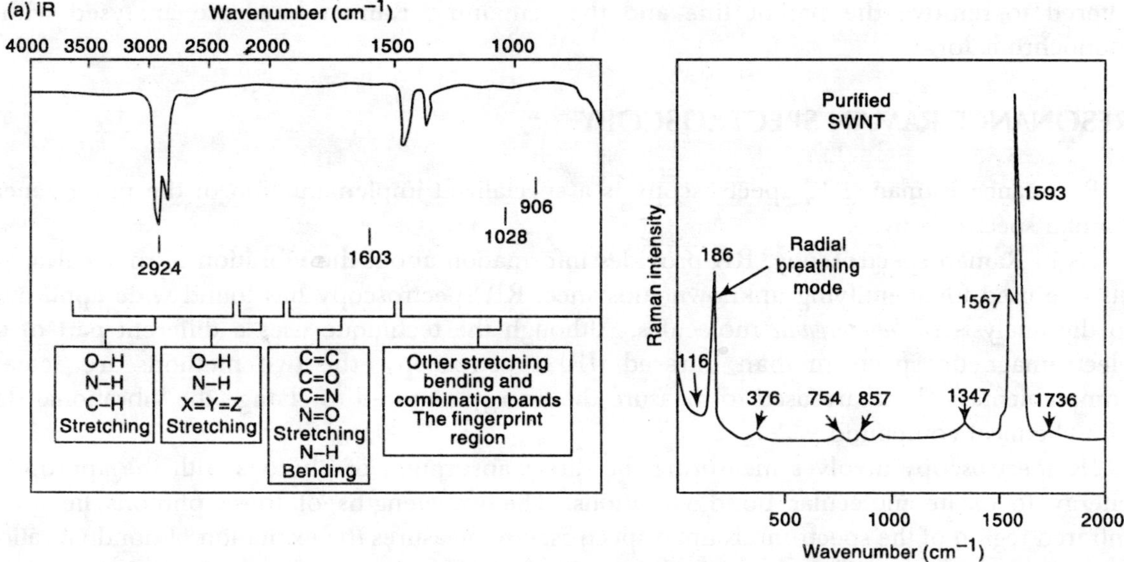

Fig. 3.8 *(a) Diagram of the IR spectral regions for different chemical bonds in organic material. (b) A schematic example of a Raman spectrum from single-walled carbon nanotubes: the radial breathing mod at 186 cm^{-1} is characteristic of the entire nanostructure.*

confined within the block; in fact, the beam penetrates a very small distance into the contacting sample layer (this is the evanescent wave) and this can be utilized for infrared absorption. In the IR case, the evanescent wave penetrates about a micron into any sample clamped onto the sample has strong IR absorption lines, these will be revealed in the final spectrum. In fact, ATR-FTIR is not restricted t solid materials attached to the ATR block; the block can also be immersed in solution to study adsorption to the block. This is a particularly useful way of studying nanoscale conformational events such as protein folding and unfolding, which might take place at surfaces.

A complementary vibrational spectroscopic technique is Raman spectroscopy; this usually employs a laser source and here the scattered light is analysed in terms of its wavelength, intensity and polarization. The parent line is simply formed by absorption and re-emission of the light, whereas other weaker lines involve not only absorption/emission but also vibrational excitation or de-excitation (Raman lines). The vibrational frequency is just the difference between the parent line and the Raman line. Generally IR-inactive vibrations are Raman active, which makes the techniques complementary for analysis of functional groups or bonds (chemistry). Additionally, low-frequency Raman scattering can probe elastic vibrations of an entire nanoparticle, which most often shows a distinct size dependence; it is used extensively to analyse single-and multi-walled carbon nanotubes.

Both IR and Raman microscopy are possible using a specially designed confocal scanning laser microscope (CSLM) that provides a laser spot size, hence lateral spatial resolution, of around 1 mm. The confocal nature of the microscope allows different depths of up 2 μm below the surface to be sampled. Typically the laser produced polarized red light, the scattered light is filtered to remove the parent line and the remaining Raman lines are analysed with a monochromator.

RESONANCE RAMAN SPECTROSCOPY

Resonance Raman (RR) spectroscopy is a specialized implementation of the more general Raman spectroscopy.

As in Raman spectroscopy, RR provides information about the vibration of molecules, and also be used for identifying unknown substance. RR spectroscopy has found wide application to the analysis of *bioinorganic* molecules. Although the technique uses a different part of the electromagnetic spectrum than infrared (IR) spectroscopy, the two methods are actually complementary. Both are used to measure the energy required to change the vibrational state of a chemical compound.

IR spectroscopy involves measuring the direct absorption of photons with the appropriate energy to excite molecular bond vibrations. The wavelengths of these photons lie in the infrared region of the spectrum. Raman spectroscopy measures the excitation of bond vibrations in an indirect manner. The two methods are complementary because some vibrational transitions that are observed in IR spectroscopy are not observed in Raman spectroscopy, and *vice versa.* RR spectroscopy is an improvement of traditional Raman spectroscopy that has increased sensitivity and is better suited for the study of complicated systems.

Raman spectroscopy can be used to identify chemical compounds because the values of $\Delta \overline{v}$ are indicative of different chemical species. This is because the frequencies of vibrational transitions depend on the atomic masses and the bond strengths. (Heavier atoms correspond to lower vibrational frequencies, while stronger bonds correspond to higher vibrational frequencies.) Thus, armed with a database of spectra from known compounds, one can unambiguously identify many different known chemical compounds based on a Raman spectrum. The number of vibrational modes scales with the number of atoms in a molecule, which means that the Raman spectra from large molecules will be very complicated. For example, proteins, typically contain thousands of atoms, and will therefore have thousands of vibrational modes. If these modes have similar energies ($\Delta \overline{v}$), then the spectrum may be incredibly cluttered and complicated.

Not all vibrational transitions will be "Raman active", i.e., some vibrational transitions will not appear in the Raman spectrum. This is because of the spectroscopic selection rules for Raman. As opposed to IR spectroscopy, where a transition can only be seen when that particular vibration causes a net change in dipole moment of the molecule, in Raman only transitions where the polarizability of the molecule changes can be observed. This due to the fundamental difference in how IR and Raman spectroscopy access the vibrational transitions. In Raman spectroscopy, the incoming photon causes a momentary distortion of the electron distribution around a bond in a molecule, followed by reemission of the radiation as the bond returns to its normal state. This causes temporary polarization of the bond, and an induced dipole that disappears upon relaxation. In a molecule with a center of symmetry, a change in dipole is accomplished by loss of the center of symmetry, while a change in polarizability is accomplished by preservation of the center of symmetry. Thus, in a centrosymmetric molecule, asymmetrical stretching and bending will be IR active and Raman inactive, while symmetrical stretching and bending will be Raman active and IR inactive. Hence, in a centrosymmetric molecule, IR and Raman spectroscopy are mutually exclusive. For molecules without a center of symmetry, each vibrational mode may be IR active, Raman active, both, or neither. Symmetrical stretches and bends, however, tend to be Raman active.

In resonance Raman spectroscopy, the energy of the incoming laser energy is adjusted such that it or the scattered light coincide with an electronic transition of the molecule or crystal. In most materials the incoming and outgoing electronic resonances are sufficiently broad that they can bot be distinguished. So, rather than exiting the molecule to a virtual energy state, it is excited to near one of its excited electronic transitions. Since the energy of these transitions differ form one chemical species to the next, this technique did not become applicable until the advent of tunable lasers. (Tunable lasers are those where the wavelength can be altered within a specific range.) When the frequency of the laser beam is tuned to be near an electronic transition (resonance), the vibrational modes associated with that particular transition exhibit a greatly increased Raman scattering intensity. This usually overwhelms Raman signals from all of the other transitions. For instance, resonance with a π-π^* transition enhances stretching modes of the π-bonds involved with the transition, while the other modes remain unaffected.

This aspect of Raman spectroscopy becomes especially useful for large biomolecules with chromophores embedded in their structure. In such chromophores, the charge-transfer (CT) transitions of the metal complex generally enhance metal-ligand stretching modes, as well as some of modes associated with the ligands alone. Hence, in a biomolecule such as hemoglobin, tuning the laser to near the charge-transfer electronic transition of the iron center results in a spectrum reflecting only the stretching and bending modes associated with the tetrapyrrole-iron group. Consequently, in a molecule with thousands of vibrational modes, RR spectroscopy allows us to look at relatively few vibrational modes at a time. This reduces the complexity of the spectrum and allows for easier identification of an unknown protein. Also, if a protein has more than one chromophore, different chromophores can be studied individually if their CT bands differ in energy. In addition to identifying compounds, RR spectroscopy can also supply structural identification about chromophores in some cases.

The main advantage of RR spectroscopy over traditional Raman spectroscopy is the large increase in intensity of the peaks in question (by as much as a factor of 10^6). This allows RR spectra to be generated with sample concentrations as low as 10^{-8} M. This is in stark contrast to conventional Raman spectra, which usually requires concentrations greater than 0.1 M. Also, as previously mentioned, RR spectra usually exhibit only a few peaks, and different peaks can be selected for by targeting specific electronic transitions. This main disadvantage of RR spectroscopy is the increased risk of fluorescence and photodegradation of the sample due to the increased energy of the incoming laser light. Both of these factors can be minimized by using an infrared laser instead of visible light.

Instrumentation

In RR spectroscopy, the light source consists of tunable laser, whose radiation lies in the near-infrared, visible, or near-ultraviolet regions of the spectrum. In creating a sample handling system, RR spectroscopy offers an advantage over IR spectroscopy in that glass can be used for window, lenses, and other optical components. Another advantage over IR spectroscopy is that whereas water absorbs strongly in the IR, it si completely Raman inactive. Therefore, water can easily be used as a solvent. Since lasers can be easily focused on small surface areas, the risk or sample heating and photodegradation is diminished, and the emitted radiation can be focused more efficiently. Typically, the sample is placed into a tube, which can then be spun to further decrease the sample's exposure to the laser light, further diminishing the threat of photodegradation. Gaseous, liquid, and solid samples can all be analyzed using RR spectroscopy. Gas and liquid samples can be put directly into the sample chamber whereas solid samples must first be ground into a power. With gaseous and solid samples, Raman scattering may still be too weak to easily detect. For these samples, the holder is placed between two mirrors that reflect that laser beam multiple times through the sample.

Since scattered light leaves the sample in all directions the probes that carry the scattered light back to the detector in Raman spectroscopy may be placed at any angle. Usually, the detector probes are most placed at an angle of 135° to the path of the exiting laser light beam. Two other common arrangements position the detector probe at 90° or 180° with respect to the incident light. Detection angles greater than 90° are generally called back-scattering detectors because they are oriented in the same direction as the incident laser light so the radiation must scatter back to the probes. In transmitting the incident laser light to the sample and the scattered light *back* to the detector, fiber-optic cables may be used. Such cables can transmit light 100 m or more, thereby allowing the analysis of samples under relatively rough experimental/environmental conditions.

After the scattered radiation exits the sample, it is through a monochromator. Typical monochromators consist of a diffraction grating mounted on a rotating platform. A diffraction grating causes light dispersion. Rotating the grating controls which wavelengths of scattered radiation reach the exit slit leading to the detector. The detector itself is usually a charge-coupled device (CCD), which allows the entire spectrum to be recorded simultaneously. Consequently, multiple scan can be acquired in a short period of time, which can drastically increase the signal-to-noise ratio of the spectrum.

4

Lithography

The term lithography generally means the transfer of structures of an electronic or an image pattern into a thin radiation-sensitive layer called photoresist, by means of electromagnetic waves or particle beams. The process flow is given in Fig. 4.1.

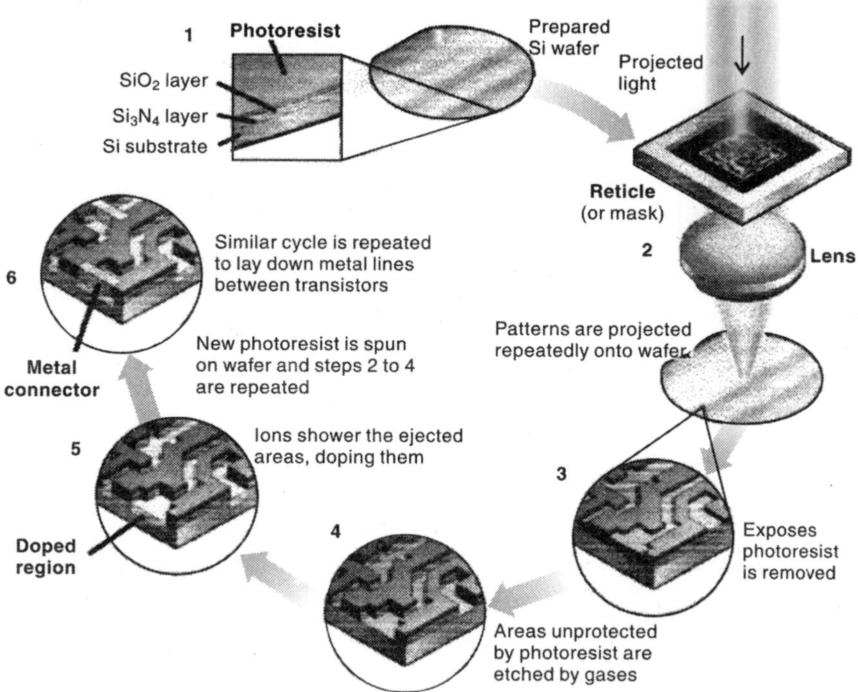

Fig. 4.1 *Lithographic Process flow*

The requirements of a lithographic system are:

1. Small dimensions (linewidth)
2. Small variations in dimensions (linewidth control)
3. Large depth of focus (tolerance of non-planar substrates and thick resists)
4. Accurate alignment of subsequent patterns (registration)
5. Low distortion of image and sample (high quality masks, projection systems)
6. Low cost (high throughput)
7. High reliability (high yield)
8. Tolerance of contamination particles on mask and sample (clean room requirements)
9. Uniformity over large areas (large wafers)

Lithography Techniques

Following Lithographic techniques are mainly used:

- Photolithography
- Electron beam lithography
- X-ray lithography
- Extreme Ultraviolet lithography
- Ion beam lithography
- Imprint lithography
- Contact lithography
- Atomic lithography

For ethching techniques used are (chemical plasma and reactive-ion), liftoff, electroplating, sputter deposition e-beam evaporation.

The execution of the lithography method involves a series process consisting of deposing the *photoresist*, exposure and development of the radiation-sensitive layer.

The *photoresist* deposition on the substrate takes place via *spin-coating* in which the resist is given on a rotating plate (approximately 3000 rpm).

In the lithography technique, developing the resist means removing the exposed or unexposed areas in abase solution. Development takes place in NaOH or TMAH solution by dipping. Alternatively, the spray development offers the highest reproducibility.

NANOLITHOGRAPHY

Nanolithography—or lithography at the nanometer scale—refers to the *fabrication of nanometer-scale structures*, meaning patterns with at least one lateral dimension between the size of an individual atom and approximately 100 nm. Nanolithography is the key technology driving technological progress in electronics, photonics information technology and biotechnology. Nanolithography is used during the fabrication of leading-edge semiconductor integrated circuits or nanoelectromechanical systems (NEMS). The fabrication of stable structures requires changes in bonds between atoms of a solid or a resist layer. Therefore, the locally introduced energy has to overcome the activation barrier of the respective chemical reactions. If this activation relies just on temperature, only the total energy or the energy per area and time (power density) are of interest, but not the energy of the individual particles. Locally introduced thermal energy is easily dissipated on the micro-and nanoscale, so the required power density increased with the decreasing diameter of a structure of interest. High power densities are realized by lasers that are used routinely for structure fabrication in microtechnology. Therefore, processes with particles each providing sufficient energy for bond activation are preferred on the nanoscale rather than methods based on the integral effect of many particles.

Typical bond energies of material that are stable at room temperature are in the order of magnitude of 100 kJ mol^{-1} (about 1 eV per bond). For bond cleavage by individual activated particles or photons, these particles have to exhibit at least the same energy. To achieve chemically selective effects, the introduction of resonant energy at the bond is advisable. This phenomenon is utilized in photochemical processes, but their resolution in optical projection is restricted by the diffraction limit. Fabrication techniques also apply particles of higher energies, so that one particle is able to cleave a group of bonds.

The energy required for bond cleavage can be provided as photons. To achieve activation energies of about 1 eV, photons of wavelength 1.2 mm and below are needed. This energetic reason limits the range of the electromagnetic spectrum that is usable in nanostructure technology to lower wavelengths. For particularly stable bonds, the required energies reach the range of 1 MJ mol^{-1}. Hence light from the far UV into the near vacuum UV is applied.

To fabricate nanostructures with typical dimension below 100nm using photons, a wavelength below 200 nm is required. Nanotechnology with optical projection requires the use of vacuum UV or even shorter radiation, such as X-ray or extreme UV (EUV). Soft X-ray radiation is especially suited for structure fabrication. A wavelength of about 1 nm is sufficient to achieve structure in the medium nanometer range. For energies in the MeV range, the wavelengths are in the pm range, therefor of subatomic dimension. Millions of bonds are activated by a single particle yielding activation volumes with diameters of 10 nm and more. Secondary emission of energetic electrons and hard X-ray radiation lead to the growth of such activated regions of up to 1 μm^3. Thus, not all bonds in this volume are activated but changes to the bonds distributed in this volume occur. Therefore, an upper-energetic limit for photons regarding applications in nanotechnology should be defined. This value can be found in the medium keV range.

One solution to the diffraction limit of resolution is the structure fabrication in the optical near field. Near-field processes rely on very small distances between the volume element of the target and the light source of light-forming element such as the edges of masks, tools or structures. These critical dimensions should be below half of the applied wavelength. Optical near-field probes reach distances of 1 nm.

A special method of direct-writing structure generation utilizes focused atom beams in standing electromagnetic waves. The standing wave acts as a periodic lens for the beam, so that the beam is projected with a period of half the wavelength onto a substrate with a radiation-sensitive layer. Line gratings with a line width of 28 nm and a period of 213 nm were fabricated using a light wavelength of 425nm[(1)].

Standard pattern generators in microtechnology apply two apertures arranged at an angle to each other. The form of the structural elements is controlled by changing the aperture size if possible—the angle. Visible light of shorter wavelengths or UV is used to project the aperture onto the light sensitive layer; the resulting diffraction limit hampers any application in nanolithography.

To transfer the principle of the pattern generator for use in nanotechnology, radiation with shorter wavelength (such as extreme UV or X-ray) is applied. Large complexes of structures are transferred together in the size of a chip or larger areas. Such group transfer processes require a high optical precision and highly homogeneous illumination. Again, the wavelength limits the resolution.

Mass production of 130 nm integrated solid-state circuits is possible. Assuming a minimum structure width at half wavelength for optical projection, far UV radiation in the vacuum UV)190 nm) region would enable structures of below 100nm width to be mass produced. The further development of high resolution and high contrast resists, on the one hand, and the introduction of exposition equipment working in the vacuum UV will result in structure dimensions considerably below 100nm in the microelectronic industry within the next few years. So a process of continuous stepwise improvement of equipment and technologies ranging from visible through the near, medium and far UV to the vacuum UV is leading from a micro- to a true nanotechnology. Such a development was unbelievable several years ago. It seems possible, that with the extension of optical projection of small structures into the extreme UV range that further development will addresses the medium nonometer range (20–50 nm) and so the optical lithography will reach a seamless transition into EUV and X-ray.

Other Nanolithography Techniques

- X-ray lithography can be extended to an optical resolution of 15 nm by using the short wavelength of 1 nm for the illumination. This is implemented by the proximity printing approach. The technique is developed to the extent of batch processing. The extension of the method relies on Near Field X-rays in Fresnel diffraction: a clear mask feature is "demagnified" by proximity to a wafer that is set near to a "Critical Condition". This Condition determines the mask-to wafer Gap and depends on both the size of the clear mask feature and on the wavelength. The method is simple because it requires no lenses.
- Double Patterning is a method of pitch resolution enhancement. This technique increases feature density by printing new features in between pre-printed features on the same layer. It is flexible because it can be adapted for any exposure or patterning technique. The feature size is reduced by non-lithographic techniques such as etching or sidewall spacers.
- An optical maskless lithography tool, uses a digital micro-mirror array to directly manipulate reflected light without the need for an intervening mask. Throughput is inherently low, but the elimination of mask-related production costs–which are rising exponentially with every technology generation–means that such a system might be more cost effective in the case of small production runs of state of the art circuits, such as in a research lab, where tool throughput is not a concern.

- The most common nanolithographic technique is Electron-Beam Direct-Write Lithography (EBDW), the use of a beam of electrons to produce a pattern–typically in a polymeric resist such as PMMA.
- Extreme ultraviolet lithography (EUV) is a form of optical lithography using ultrashort wavelengths (13.5 nm). It is the most popularly considered NGL technique.
- Charged-particle lithography, such as ion–or electron-projection lithographies (PREVAIL, SCALPEL, LEEPL), are also capable of very-high-resolution patterning.
- Nanoimprint lithography (NIL), and its variants, such as Step-and-Flash Imprint Lithography, LISA and LADI are promising nanopattern replication technologies. This technique can be combined with contact printing.
- Scanning probe lithography (SOL) is a promising tool for patterning at the deep nanometer-scale. For example, individual atoms may be manipulated using the tip of a scanning tunneling microscope (STM). Dip-Pen Nanolithography (DNP) is the first commercially available SPL technology based on atomic force microscopy.
- The furthest developed NGL remains X-ray lithography which is extensible to 15 nm resolution by use of "demagnifcation" in the Near Field.
- Chemomechanical Surface Patterning using an atomic force microscope is another type of nanolithography.

Structural Generation and Fabrication of Lithographic Masks

Adhesive Mask Technique

The transfer of a small structure onto a substrate or a functional layer is carried out with a wide variety of different sources, ranging from 1:1 initial images to digital data sets. The various techniques applied in this context are combined in the term microlithography, with photolithography being a widely utilized method. The largest proportion of microlithographic technique apply a resist layer to create adhesion masks on the substrate or on the functional layes.

The adhesion mask technique is a key aspect of planar technology. In this way the structure of every functional layer is patterned into a mask layer (the resist) prior to pattern transfer in the functional layer. A resistance of the mask layer to the applied etch media is required. This requirement often narrows the choice of mask material.

Resists based on organic polymers are preferred as adhesion masks. These resists show a local change in solubility in a certain solvent (development) due to local exposure to electromagnetic radiation. The change to higher or lower solubility differentiates the resists into positive and negative ones, respectively (Fig. 4.2). Steep edges in the resist are a prerequisite for the fabrication of small structures and high aspect ratios. Therefore, the resist should exhibit a high gradation, so that a small difference in dose results in a larger change in solubility (Fig. 4.3)

The requirements of high gradation is of particular importance in nanotechnology due to the proximity effects that are observed. These effects are usually associated with external parameters, such as wavelength of electromagnetic radiation (a proximity effect due to diffraction) or the

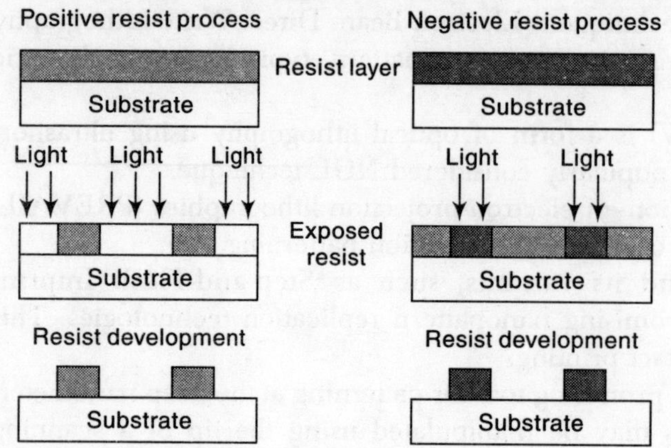

Fig. 4.2 *Principles of positive and negative resist processes.*

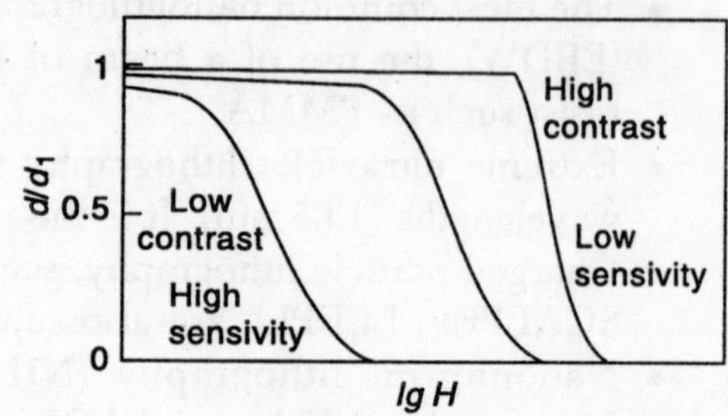

Fig. 4.3 *Gradation curve for positive photoresist. Remaining resist thickness d related to the initial thickness d_0 dependent on the exposure dose H.*

energy of fast particles (a proximity effect due to ionization and X-ray emission). These effects result in a distribution of the locally deposited energy of a structural element instead of a sharp transition along the image boundaries. Hence the transformation product, which is generated locally by the exposure, also exhibits no sharp distributions (Fig. 4.4). This product dispersion results in a lateral gradient of the solution rate, which in turn increases the deviation from the upright flanks. This blurring effect decreases with a steeper gradation curve of the resist material.

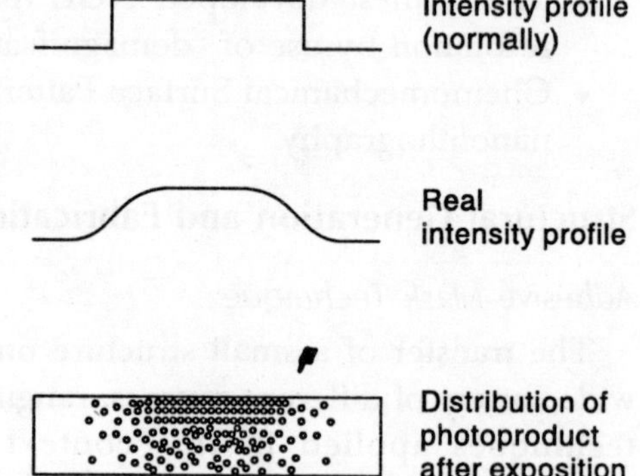

Fig. 4.4 *Distribution of the photoproduct due to the optical proximity effect in photolithography (scheme).*

There is a variety of demands on resist structures. The resist must be stable against liquid process media with respect to dissolution and rate of chemical reaction. As well as the resist, the interface between the resist and the substrate also has to show a high stability. This requirement is not trivial, because the interfaces are usually quite different from the resist regarding their composition. If the process solution reacts with chemical groups in finest cracks or dissolves components of the interface, a shift of structures or even an under cutting (and therefore a loss of the mask) is observed.

Adhesive masks should therefore establish a stable bond with the substrate. This is achieved by a certain density of strong covalent bonds or by a higher density of coordinative, dipole-dipole, hydrogen bridge or Van der Waals interactions. As long as established bonds are not opened by the process solution, weaker polyvalent and unspecific interactions between resist molecules and surface groups of the substrate can be as efficient as individual strong bonds. Van der Walls interactions between polymers and hydrophobic surfaces are very capable of avoiding the intrusion of aqueous etch baths between the substrate and resist.

The stability of the interface is of lesser importance for pattern transfers in the gas phase, especially for the case of the interaction of directed particles with the surface. On the other hand, this situation requires a higher resistance of the layer to the impact of particles on the surface. This requirement is hard to accommodate, because the kinetic energy of the particles exceeds the binding energy of the resist often by some order of magnitude.

The choice of resist has to fulfill three conditions.

The first point is the stability of the material with respect to the process that includes the pattern transfer. For all lithographic processes that utilize some type of radiation, the resist has to be sensitive to this radiation, which means that the exposed part undergoes a local chemical change, but the remaining area is unaffected.

Secondly, this local chemical change has to result in the development of a relief in the resist. Therefore, the change has to significantly affect the removal rate in an appropriate development media. In the simplest case, exposed areas are removed completely (positive resist) in the development process compared with the preserved remaining parts, or the surface areas without exposure are removed by the development and the exposed parts remain (negative resist).

Thirdly, in subsequent processes the remaining parts have to exhibit a protective property with regard to the underlying material, so that the functional layers can be etched by dry or wet techniques, or so that the process for lifting of the layers deposited on the adhesion mask are possible.

Also the integration of solvent by insoluble structural elements of the adhesion mask affects the fabrication of small structures. Swelling results in failed geometries, especially in the case of negative resists. Negative resists are usually based on cross-linking of linear or slightly branched polymers by an irradiation step. The resulting network in the exposed areas significantly decrease the solution rate in a solvent, but the affinity of the resist to the solvent is preserved to a large extent. The solvent also diffuses into the exposed regions, increases their volume and so deforms the preserved structures. Negative resists are therefore in most cases not suitable for the realization of high aspect ratios and very small structures.

The molecular length is significant in the case of nanotechnology application of long-chained polymers. A molecular mass of 10^6 result in a length of several micrometers for an extended molecule, and a length in the upper nanometer range is still observed for a molecular mass of 10^5. Although the molecules are usually in a coiled state, the ends are exposed in the development process and generate a gel-like zone in the solvent. This zone also induces a smoothing of the structures and an increase in the radius of curvature of the edges.

The majority of the effects discussed induce a flattening of the resist and so a decrease in resolution especially for thicker resist layers. Therefore, thinner resist layers are preferred for nanostructures. Besides the protective function required in particular in plasma processes, there is another negative aspect for an extremely this resist layer. For the typical case of multilayer arrangement, resists must often planarize previously structures surfaces or at least cover the edges of the structures in a subsequent step. This function requires an adequate resist thickness.

The so-called reversal processes counteract the effect of flattened resist edges. These processes invert the contrast values of the resist. For positive resist processes in particular, the common orientation (perpendicular to the substrate) of decreasing locally deposited energy and solvent attack can be transformed into opposite processes which compensate each other. Positive resist-based reversal processes apply a fixation step after the masked exposure, so that the exposed regions are stabilized. Then the whole substrate is irradiated, so that the previously unexposed regions can be dissolved (Fig. 4.5). In addition to the simple conversion of the contrast values of masks, this procedure is ideal for the realization of steep or even overhanding flanks, as if often required for dry etch processes or an additive generation of structures.

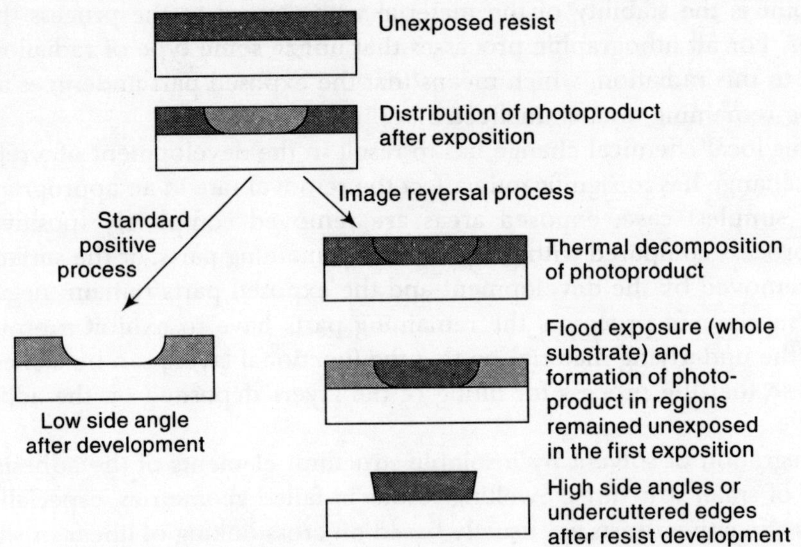

Fig. 4.5 *Comparison of positive and image reversal process.*

The choice of resist thickness is influenced by the level required for the sufficient protection, the topological relief of the substrate and the expected size of particles form the environment. Although media and air are filtered, this cleaning procedure has a cutoff and the number of smaller particles is reduced but not negligible. So there is a given probability of smaller particles on the substrate. The negative effect of this event increases with the density of nanostructures and the miniaturization of the structures. Although the probability of a defect in a given nanostructure is minimal, for higher coverage of substrate with nanostructures the probability of a defect structures increases. Defects not covered by the resist are usually failures, so a decreased resist thickness results in increased malfunctions due to particles.

PHOTOLITHOGRAPHY

Photolithography (also optical lithography) is a process used in microfabrication to selectively remove parts of a thin film (or the bulk of a substrate). It uses light to transfer a geometric pattern from a photomask to a light-sensitive chemical (photoresist, or simply "resist") on the substrate. A series of chemical treatments then engraves the exposure pattern into the material underneath the photoresist. In a complex integrated circuit (for example modern CMOS), a wafer will go through the photolithographic cycle up to 50 times.

Photolithography is the classical method for the fabrication of reliefs on substrates by pattern-transfer processes, and will therefore serve as an example for a discussion of the generation and function of adhesion masks. Pattern transfer occurs by light. Light transfers the pattern from the original, e.g., a shadow mask, onto the substrate which incorporates a light sensitive layer. For a photolithographic process, this layer is a photoresist, which means it must adsorb at the particular applied wavelength and undergo a photochemical reaction. A transfer without optics occur either direct (contact lithography) or in close proximity (proximity lithography) of the mask and substrate. The use of an optical system allows the scaled or distorted projection of the pattern onto a resist layer. With respect to the optical image transfer, photolithography corresponds to the principles known from photolithographic techniques. The different lies in the result, which is a relief and not a change in colour or brightness. Because photolithographic transfer does not involve single structures but whole image or at least groups of structure, photolithography represents a process of group transfer.

Ideal materials for photolithography are resists based on relatively short-chained phenol resins. The phenolic OH groups lead to stable networks with dyes such as naphtochinondiazid, which nitrogen under irradiation prior to a reaction with water to give carbonic acids. This reaction increases the solubility of the resist in weak basic solutions by several orders of magnitude. Moreover, the phenolic OH groups can be involved in interactions leasing to hydrogen bridges to protic groups on the substrate surfaces, so that substrates with surface OH groups (as are often found in inorganic materials) show a good resist binding. Another point is the potential for interactions with organic surfaces due to the aliphatic substituents in the phenol cores. The chemical protection of phenol resins extends down to very low pH values, allowing the use of processes with strong acids. The resist is only unstable in the higher pH range. Phenol resins have advantages in the protection against plasma and energetic particles compared with resists based on aliphatic polymers. The stability can be explained by the relatively low hydrogen content and the aromatic groups.

Photolithography resembles the conventional lithography used in printing, and shares some fundamental principles with photography. It is used because it affords exact control over the shape and size of the objects it creates, and because it can create patterns over an entire surface simultaneously. Its main disadvantages are that it requires a flat substrate to start with, it is not very effective at creating shapes that are not flat, and it can require extremely clean operating conditions. Optical lithography, is capable of producing sub-100-nm patterns with the use of very short wavelengths say, 193 nm. It requires the use of liquid immersion and a host of photomask enhancement technologies phase-shift masks (PSM), optical proximity correction (OPC) at the 32 nm node.

Optical lithography with light in the wavelength range of 465 nm down to 193 nm is used for the structural transfer from the mask onto the photoresist in all micro techniques.

A single iteration of photolithography combines the use of atuomated, robotic wafertrack systems. The wafer is initially heated to a temperature sufficient to drive off any moisture that may be present on the wafer surface. Wafers that have been in storage must be chemically cleaned to remove contamination. A liquid or gaseous "adhesion promoter", such as hexamethyldisilazane (HMDS), is applied to promote adhesion of the photoresist to the wafer.

It is covered with photoresist ("RR") by spin coating. A viscous, liquid solution of photoresist is dispensed onto the wafer, and the wafer is spun rapidly to produce a uniformly thick layer. The spin coating typically runs at 20 to 80 Hz for 30 to 60 seconds, and produces a layer between 2.5 and 0.5 micrometres thick.

The photoresist-coated wafer is then "soft-baked" or "prebaked" to drive off excess solvent, typically at 333 to 373 K for 5 to 30 minutes. Sometimes a nitrogen atmosphere is used.

After prebaking, the photoresist is exposed to a pattern of intense light. Optical lithography uses ultraviolet light. Positive photoresist, becomes less chemically robust when exposed; negative photoresist becomes more robust. This chemical change allows some of the photoresist to be removed by a special solution, called "developer". A post-exposure bake is performed before developing, to help reduce standing wave phenomena caused by the destructive and constructive interference patterns of the incident light. Developers originally contain sodium hydroxide (NaOH). However, sodium is considered an extremely undesirable contaminant in MOSFET fabrication because it degrades the insulating properties of gate oxides. Metal-ion-free developers such as tetramethylammonium hydroxide (TMAH) are now used.

The resulting wafer is then "hard-baked", typically at 393 to 453K for 20 to 30 minutes. The hard bake solidifies the remaining photoresist, to make a more durable protecting layer in future ion implantation, wet chemical etching, or plasma etching.

Wet Etching Processes

Wet etching processes include a transfer of the removed solid state material into the liquid phase. Therefore, a chemical species is generated that is soluble (and therefore mobile) in the liquid phase. Metals and semiconductors are transformed into cations or soluble coordinated compounds of their cations, through the involvement of local electrochemical processes. The standard electroless etching process compensates the local currents by reduction of oxidizing agents at the metal or semiconductor surface (Fig. 4.6).

For isotropic wet etching processes, the local removal rate is equal for all spatial directions. It is determined by diffusion, and therefore the minimal structure size correlates strongly with

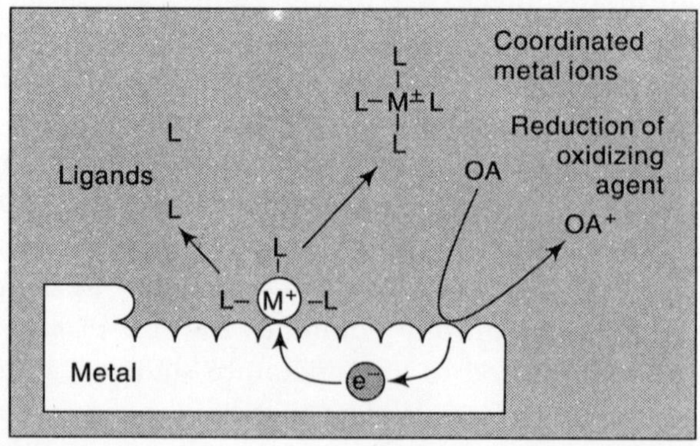

Fig. 4.6 *Anodic and cathodic process during the electrodeless etching of metals and semiconductors (M, metal; L, ligand; OA, oxidizing agent).*

etching depth. Ideal isotropic transport means equal rates of material removal in both the lateral and vertical direction. Underetching of the mask edges occurs to the same extent as the layer material is removed in vertical direction. Therefore, the maximum aspect ratio for structures realized through an opening in a mask is 0.5 (related to the dimensions of the structure). Individual upright structures overcome this limitation in the case of precisely controlled underetching of the mask edges on both sides. This process is very sensitive to changes in the process, and is therefore barely reproducible. Moreover, such process are only applicable for isolated structures. For periodical arrangement, the distance between two structures should be more than twice the etch depth.

This disadvantage of isotropic etching processes becomes critical in the case of nanometer structures in relatively thick layers. Thicknesses of 100 nm, for example, only allow structures down to 200 nm width. The limitation is less critical for very thin layers, because the lateral etch width (the underetching along the mask edges) is minimal. The extreme selectivity of wet etching processes therefore allows the application of monomolecuarl layers as a mask. Patterns generated in such masks can be transferred with minimal variation into a functional layer, as long as this layer is sufficiently thin. An alkythiol mask with a thickness of typically 2 nm, for instant, allows wet etching of thin gold layers resulting in etch depths significantly greater thickness compared with the mask thickness. An Au thickness of 10 nm and a desired structures width of 40 nm require an alkylthiol mask opening of 20 nm in order to compensate for isotropic underetching on both sides.

Anisotropic wet etching procedures need an isotropy of the etch rate in the material to be removed. This effect occurs at wet chemical procedure, which are controlled by chemical surface processes instead of diffusion. Such a procedures is utilized in the crystallographic etching of monocrystalline semiconductor, especially in silicon micromechanics (Fig. 4.7).

The basic chemical etch principle, which differentiates between the various crystalographic planes, can be transferred into nanometer dimensions. A prerequisite is a nanostructures mask with sufficient resistance towards the anisotropic etch baths. In nanotechnology, wet etching is utilized to structure ultrathin layers in a highly selective manner. Isotropic wet etching is not applicable for nano-trenches at a thicknesses of several tens of nm, due to underetching on both edges of dimension of the etch depoths, so that the final structure has dimensions of a twofold of the etch depth. Elevated structures, such as thin lines, are only achieved by isotropic wet etching when the

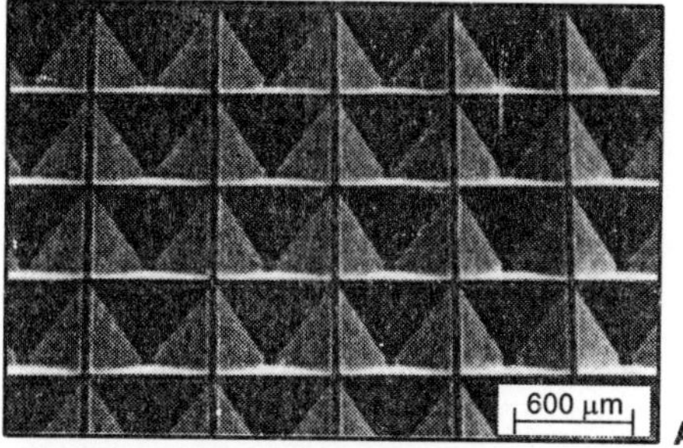

Fig. 4.7 *Anisotropic etched pits in single-crystalline silicon structures A, Pyramidal pits in Si(100). B, Parallel channels with upright side walls in Si(110).*

distance to neighbouring structures is larger than the horizontal etch component. So, controlled underetching yields steep edges (Fig. 4.8). As prerequisite is control of the process time, because underetching is determined by the etch time, but not by selectivity. Anisotropic wet chemical etching, e.g., on monocrystalline substrate, allows the realization of nanostructures with greater aspect ratios.

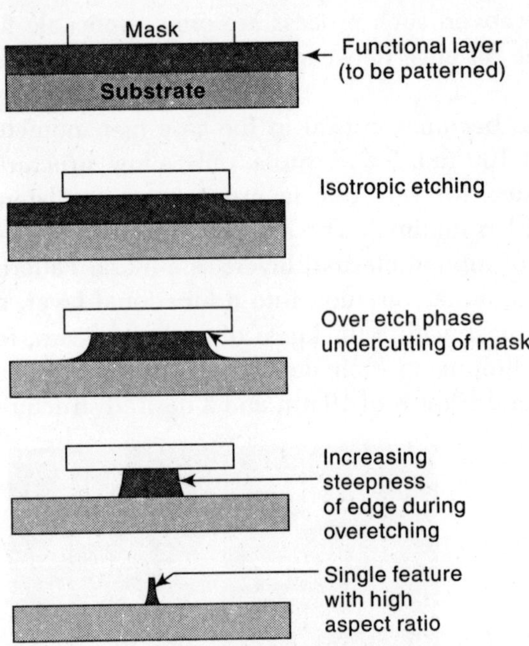

Fig. 4.8 *Preparation of embossed discrete features with steep flanks by time-controlled isotropic overetching.*

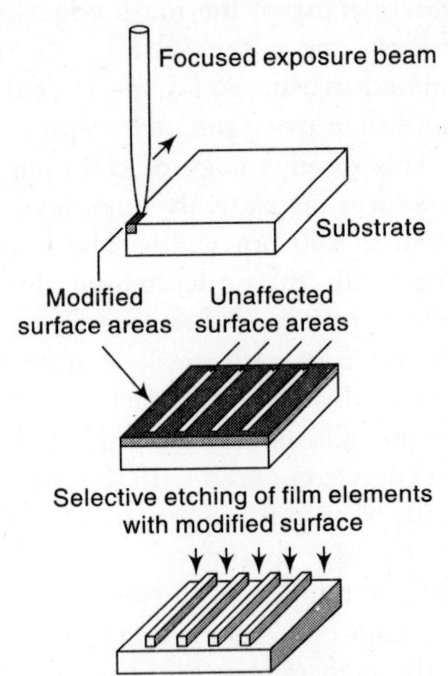

Fig. 4.9 *Direct-writing fabrication of nanolithographic structures by local modifications of the material by focused ion beam and subsequent selective etching.*

Wet etching is preferred for the transfer of pre-differentiated structure into a relief. Local implantation, electron or beam damages change the wet-chemical etch rate significantly. Based on this effect, focused particle beams induce a locally (lower nanometer range) confined implantation or damage prior to removal of the modified regions by a highly selective etchant (Fig. 4.9). So the basically isotropic etching combined with the previously generated anisotropy yields an anisotropic process for nanofabrication.

Dry Etching Processes

Etched nanostructures of high are complexity with aspect ratios larger than 0.5 are realized by dry etching. This process transfers the removed material from the surface of a solid-state substrate into a gaseous chemical species, e.g., by highly reactive vapors or by plasmas[(1)]. Therefore, the transport of the etching species to the and of the etch product form the surface is diffusion-limited. Such processes are (comparable to the isotropic wet etching) thus limited by a maximum aspect ratio of 0.5.

Dry etching procedures with a preferred orientation regarding the removal of material are of particular interest in nanotechnology. This effect is realized, when the etch rate is determined by the density of particles with higher energy reaching the surface in a preferred orientation.

Non-reactive dry etching methods, such as physical sputtering (PE, physical etching) and ion beam etching (IBE) utilizing inert ions, are able to structure any material. Ions with a kinetic energy many times that of the binding energy are accelerated onto the surface. The local transfer of the mechanical impulse knocks individual atoms or clusters out of the surface and they are transferred into the gas phase prior to transport by convection (Fig. 4.10).

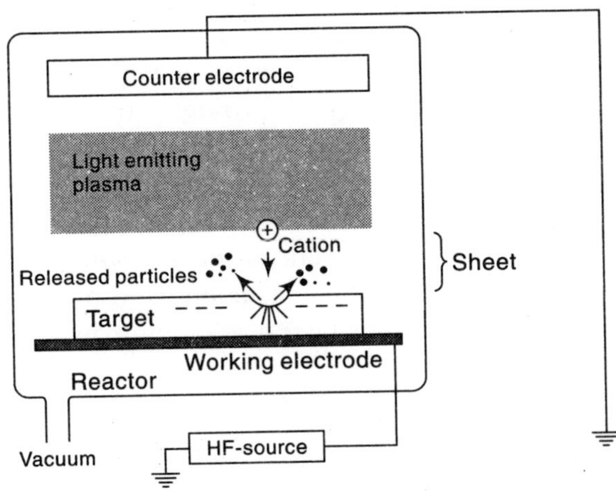

Fig. 4.10 *Reactor for sputter etching and scheme of sputtering.*

Standard sputter etching generates and applies energetic ions in the same reactor chamber. Therefore, an inert gas under reduced pressure (typically argon at 1–10 mirror) is exposed to high frequency source, so that a plasma is generated. Because of the much higher mobility of electrons compared with cations, spatial charging zones are generated, originating from the plasma and reaching the electrodes and the walls (sheet zone). Here the cations are accelerated towards the electrodes (and thus the substrate to be etched). This so-called self bias effect can be enhance by an external DC field. Typical potentials are some hundred volts, so that monovalent cations are accelerated to energies of several hundred electron volts.

As the lines of force of electrical fields arrive approximately perpendicular to the surface of the substrate, the sputtering effects is induced in the same direction. Therefore, sputter etching is ideal for anisotropic etching. Upright mask structures can be transferred into functional layers with high precision and usually minimal undercutting. Tilted mask planes are problematic, being enhanced by faceting or leading to the reflection of particles into narrow cleft structures and thereby to an underetching with bent edge profiles. Because the mask material in sputter etching usually shows non-negligible etch rates compared with the etch rates of the functional layer, tilted mask edges result in a projection of their angle into the edge angle of the functional layer. However, sputter etching is well suited for lithographic masks with steep edges, to facilitate small structures with high aspect ratios and are therefore often used for he fabrication in the medium nanometer range.

Sputter etching as well as reactive ion etching are usually conducted in reactors with two parallel electrodes. This arrangement ensures a relatively homogeneous distribution of the plasma. The ion energy and the density of effective sputtering particles depend in particular on the pressure and the introduced power.

The conditions for etching by energetic particles can be controlled more effectively when the ion source is separated from the ion generation. This principle is realized in ion beam etching (IBE). An ion source creates and accelerates the ions, and the rector is separated by at least on electrode. Creation, acceleration and the movement of the energetic particles onto the target are controlled separately. The spatial separation allows for a precise control of the ion energies and a high density of ions being achieved without thermal or plasma to the substrate. Moreover, the energetic ions can be neutralized in order to etch with neutral energetic particles. Ion etching allows may orientation of the target substrate with respect to the direction of the particles, so that structures with defined edge angles are possible.

For lower and medium aspect ratios, the achievable resolution of sputter and ion etching is mainly determined by the dimensions and the etch stability of the mask. The latter is the key problem in non-reactive dry etching. As a result of the variability of the processes, practically all mask materials exhibit a significant removal. This limits in particular the application of thin mask layers, resulting in limitations to the lateral resolution in the direct lithography of the mask layer.

Reactive dry etching processes are applied to achieve high etch rates with high anisotropy but also sufficient selectivity. They are realized either by replacement of the noble gases by reactive gases (reactive ion etching–RIE) or by the use of reactive gases for the ion generation (reactive ion beam etching–RIVE). Another possibility is the insertion of highly energetic inert particles (such as anoble gas ions) into the reaction zone with the simultaneous delivery of reactive thermalized particles (chemical assisted ion beam etching–CAIBE).

High-resolution Dry Etching Techniques

Dry etching processes are ideal for the realization of ultra-small structures, due to their anisotropic effect on homogeneous and isotropic materials. Thus upright or other chosen flank angles can be realized. They are required for higher integrated small structures with aspect ratios below 0.5. While wet etch techniques dominate in micro mechanics and for structures in the medium and upper micrometer range, dry etching becomes interesting with decreasing structure sizes. Therefore, dry etching processes play a key role in microelectronics today. They are also the most important group of techniques in nanotechnology. The directed acceleration of energetic particles towards the substrate surface is required for an anisotropic etch process with a high precision of pattern transfer, even for small dimensions. Sputter and ion etching along with the respective techniques with reactive gases, RIE and RIBE, are the most important techniques for the etching of nanometer structures.

To etch with a high rate and simultaneously achieve a high selectivity of the dry etch process with respect to mask and other materials, a reactive etch process above a certain gas pressure is required. Thus the non-specific sputter effect is minimized compared with the reactive removal. Under these conditions, the isotropic component in the process is hardly avoidable. It leads to an undereteching below the mask edges and prevents uprights flanks being obtained. To overcome this problem, a reactive etch gas with a component that induces a stable chemical passivation of the flank is applied. The passivation by the deposition of a thin and etch-resistant film prevents the lateral removal of material. However, the etch parameters have to ensure that this passivation layer is limited to flanks. The flank passivation can be controlled by the composition of the etch gas. An increase in the percentage of trifluortrichlorethane in an etch gas mixture of SF_6 and $C_2Cl_3F_3$ from 20% to 27% and then to 32% leads to transformation of isotropic to partial anisotropic and to anisotropic etching of silicon. An increase in etching time was needed for formation of an efficient flank protection. If necessary, a periodic change in etch parameters in required to induce alternating etch and protection steps.

Depending on the material and the required etch depth, ion etching allows the transfer of very small structures from the mask into functional layers. Hence lines with less then 10 nm width were transferred from a PMMA mask into a GaAs substrate in a reactive ion etching with an SiCl4/Ar mixture. Using an AuPd mask for a silicon substrate, individual structures with diameters as low as 7 nm were obtained with a mixture of $SiCl_4$ and CF_4.

In addition to mask-based etching techniques, direct etching methods with focused probes are also applied. Ion beams in particular can be well focused and controlled by electronic optics. Therefore, focused ion beam (FIB) processes are especially well suited for to obtaining small structure by probe-induced etching. Instead of ions, molecular clusters and noble has atoms can also be applied in nanostructure fabrication[(3)]. Clusters of gas mixtures are ideal, when they or their products are easily transferrable into the gas phase and when no secondary mask effects occur. The gas cluster are created by condensation of the respective atomic or molecular disperse gases through expansion, prior to an ionization of the cluster and the beam generation using an accelerating electrode with an aperture and additional electrode is required.

The clusters usually include some hundreds up to some thousands of atoms or individual molecules. Apart from clusters of inert gases such as Ar or mainly inert molecules such N_2, N_2O or CO_2, reactive gas clusters such as SF_6 or O_2 can also be utilize(4).

Damage to the mask or damage of the layers or interface below the removed material are a problem of etching techniques using energetic particles. This makes the application of monomolecular layers or single supermolecules as the etching mask difficult. Subsequent layers are hit by the energetic particles when the functional layer is removed. Because of local variation in the layer thickness and etching rate, there is no absolute control of the etching process, and the subsequent layer is at least partially exposed to the etching process. In the case of sensitive surfaces, RIE and RIBE processes result in significant damage or even removal. However chemically rather inert structures, such as inorganic tunneling barriers, are also affected due to their lack of thickness. Additionally, the impact of energetic particles results in failures of the solid substrate, which can change the electronic, such as the local conductivity.

In the etching step, a liquid ("wet") or plasma ("dry") chemical agent removes the uppermost layer of the substrate in the areas that are not protected by photoresist. In semiconductor fabrication, dry etching techniques are generally used, as they can be made anisotropic, in order to avoid significant undercutting of the photoresist pattern. This is essential when the width of the features to be defined is similar to or less than the thickness of the material being etched. Wet etch processes are generally isotropic in nature, which is often indispensable for microelectromechanical system (MEMS), where suspended structures must be "released" from the underlying layer.

If a photoresist is no longer needed, it must be removed from the substrate, by a liquid "resist stripper", Alternatively, photoresist may be removed by a plasma containing oxygen, which oxidizes it. This process is called ashing, and resembles dry etching.

Exposure systems typically produce an image on the wafer using a photomask. The light shines through the photomask, which blocks it in some areas and lets it pass in others. (Maskless lithography projects a precise beam directly onto the wafer without using a mask). Exposure systems may be classified by the optics that transfer the image from the mask to the wafer.

Contact Mode

Optical lithography in contact mode uses masks from glass on whose surface' the desire structures are available on a 1:1 scale in the form of a thin chromium film as absorber. Boron silicate or quartz glass is used depending on the selected wavelength.

During the contact exposure the photomask is in direct contact with the photoresist film at the surface of the substrate so that during irradiation of the mask the structures are transferred on a 1:1 scale. For the contact improvement of the resolution the substrate is pressed against the mask before the exposure. Additionally, vacuum is applied between mask and substrate.

The resolution is limited only by the diffraction effects at the structure edges so that minimum structural widths of about 0.8 μm for 436 nm wavelength down to about 0.4 μm for 248 nm wavelength are possible on plane surfaces as a function of the photoresist thickness and the used wavelength. By decreasing the wavelength to 220 nm line widths of about 100nm are obtained. Presumably, the procedure can also be extended to structural widths below 100 nm by further reduction of wavelength.

The disadvantage of this procedure is the unavoidable position-dependent adjustment error of already manufactured structures on the substrate, resulting from temperature gradients and mechanical stresses, as well as the strong load of the expensive mask by direct contact between mask and substrate surface. The contact leads to a fast contamination of the mask, possibly existing particles between photoresist and mask prevent a conclusive contact and thus worsen the quality of the imaging. Moreover, the close contact can cause scratching of the photoresist film on the substrate or the photomask itself.

A contact printer, the simplest exposure system, puts a photomask in direct contact with the wafer and exposes it to a uniform light. A proximity printer puts a small gap between the photomask and wafer. In both cases, the mask covers the entire wafer, and simultaneously patterns every die.

Contact printing is liable to damage both the mask and the wafer. Both contact and proximity lithography require the light intensity to be uniform across an entire wafer, and the mask to align precisely to features already on the wafer. As modern processes use increasingly large wafers, these conditions become increasingly difficult.

Contact lighography is used because it can achieve high optical resolution. The resolution is approximately the square root of the wavelength and the gap distance. Hence, contact printing offers the best resolution, because its gap distance is approximately zero (neglecting the thickness of the photoresist itself).

Non-contact mode

With this procedure the disadvantage of the close contact between substrate and mask is eliminated by which the wafer is kept reproducibly at 20–30 μm away from the mask by means of defined spacers. Therefore few errors or contaminations occur both in the resist layer and at the mask.

The UV exposure delivers a shadow image of the mask in the photoresist. However, the resolution clearly decreases as a result of the proximity-distance; due to the diffraction effects at the chromium edges of the mask only structures with smallest dimensions down to about 2.5 μm are resolved. For the nanometer lithography these devices are completely unsuitable. In the industrial production the proximity exposure is also only rarely in use because of the insufficient resolution. An improvement of the resolution by advancement of the devices does not take place.

Very-large-scale integration lithography uses projection systems. Unlike contact or proximity masks, which cover an entire wafer, projection masks show only one die. Projection exposure system (steppers) project the mask onto the wafer to create the complete pattern.

The image for the mask originates from a computerized data file. This data file is converted to a series of polygons and written onto a square fused quartz substrate covered with a layer of chrome using a photolithographic process. A beam of electrons is used to expose the pattern defined in the data file and travels over the surface of the substrate in either a vector or raster scan manner. Where the photorsist on the mask is exposed, the chrome can be etched away, leaving a clear path for the light in the stepper/scanner systems to travel through.

RESOLUTION IN PROJECTION SYSTEMS

The ability to project a clear image of a small feature onto the wafer is limited by the wavelength of the light that is used, and the ability of the reduction lens system to capture enough diffraction orders from the illuminated mask. Current state-of-the-art photolithography tools use deep ultraviolet (DUV) light with wavelengths of 248 and 193 nm, which allow minimum feature sizes down to 50 nm.

The minimum feature size that a projection system can print is given approximately by:

$$F = k \cdot \frac{\lambda}{N_A}$$

where

F is the minimum feature size
k is a coefficient that encapulates process-related factors, and typically equals 0.5
λ is the wavelength of light
N_A is the numerical aperture of the lens

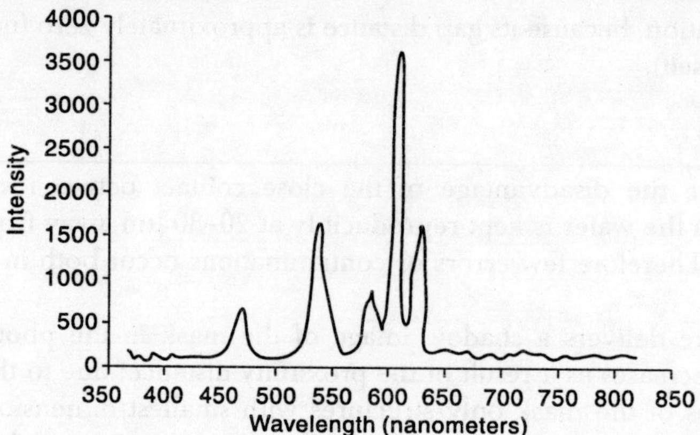

Fig. 4.11 ***The filtered fluorescent lighting in photolithography cleanrooms contains no ultraviolet or blue light in order to avoid exposing photoresists.***

According to this equation, minimum feature sizes can be decreased by decreasing the wavelength, and increasing the numerical aperture, i.e. making lenses larger and bringing them closer to the wafer. However, this design method runs into a competing constraint. In modern systems, the depth of focus is also a concern:

$$D_F = 0.6 \cdot \frac{\lambda}{N_A^2}$$

The depth of focus restricts the thickness of the photoresist and the depth of the topography on the wafer. Chemical mechanical polishing is used to flatten topography before doing high-resolution lithographic.

In photolithography ultraviolet light is used from gas-discharge lamps using mercury, sometimes in combination with noble gases such as xenon. These lamps produce light across a broad spectrum with several strong peaks in the ultraviolet range. This spectrum is filtered to select a single spectral line, usually the "g-line" (436 nm) or "i-line" (365 nm).

Now, lithography has moved to "deep ultraviolet", that is produced by excimer lasers. (In lithography, wavelengths below 300 nm are called "deep UV".) Krypton fluoride produces a 248-nm spectral line, and argon fluoride a 193-nm line.

Optical lithography can be extended to feature sizes below 50 nm using 193 nm and liquid immersion techniques. Also termed immersion lithography, this enables the use of optics with numerical apertures exceeding 1.0. The liquid used is typically ultra-pure, deionised water, which provides for a refractive index above that of the usual air gap between the lens and wafer surface. This is continually circulated to eliminate thermally-induced distortions. Water will only allow *NA*'s of up to ~1.4, but materials with higher refractive indices will allow the effective *NA* to be increased further.

Tools using 157 nm wavelength DUV in a manner similar to current exposure systems have been developed. These were once targeted to succeed 193 nm at the 65 nm feature size node but have now all but been eliminated by the introduction of considerations that provided strong incentives for the continued use of 193 nm technology. High-index immersion lithography is the newest extension of 193 nm lithography to be considered. Features less than 30 nm have also been demonstrated.

Limitations

Line widths of 100 nm, 70 nm or even 50 nm, can be transferred by optical lithography. Further resolution up to 35 nm structure width may be possible.

An increase of the resolution is aimed at by reduction of the wavelength down to 156 nm (F_2 Laser) and further down into the x-ray regime (EUV, extreme ultraviolet). For this, new optics have to be developed for the projection lithography since quartz lenses age or lose transparency due to radiation stress (production of color centers, missing transparency of the materials for this wavelength).

Perspectives

EUV lithography is seen as a continuation of the optical method for further reduction of the wavelength to approximately 13 nm. Due to the wavelength within the X-ray regime, operation can be done only with reflecting optics.

The structures which have to be transferred are produced by a reduced image of a reflecting mask Fig. 4.12 in the photoresist applying wavelengths around 11–14 nm. Reflecting optics in the from of multilayer mirrors are used as optical elements.

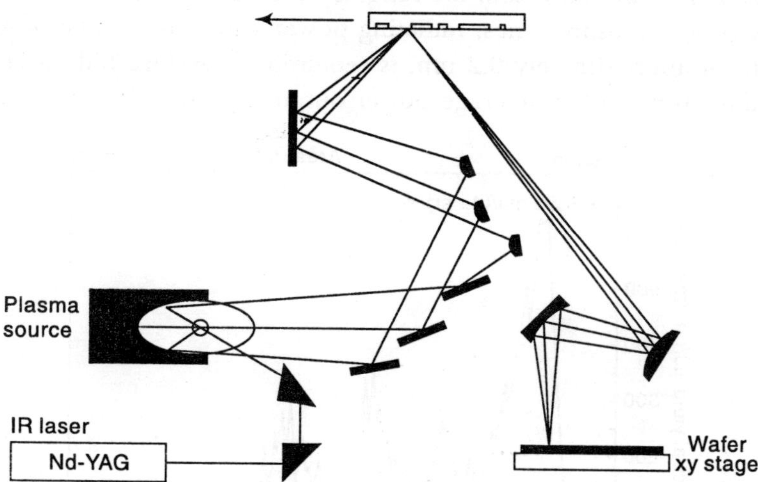

Fig. 4.12 *Structure of EUV step-scan exposer with plasma source and reflex mask.*

Presently, multilayer systems from silicon-molybdenum films are suitable as mirrors with reflectivities about 70 %; this means a remaining intensity of maximum 8 % at the substrate surface for an optical system of 7 elements. The technological hurdles of this procedure exist essentially in the guarantee for surface quality of the optics over larger areas and the availability of efficient radiation sources in this wavelength range. Since the masks are to be produced as a reflecting element, therefore bragg reflection can be used on a series of thin layers. The thicknesses of the respective layers must be controlled. An example of a mask is shown in Fig. 4.13.

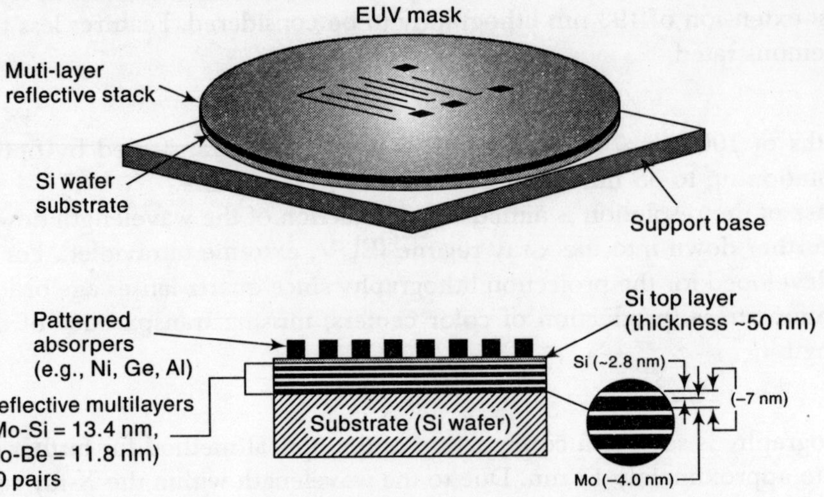

Fig. 4.13 *Structure of multilayer reflex mask for EUV exposure.*

Also, a laser-generated plasma is favored as radiation source with which pulsed laser light is focused on a collection of cooled xenon clusters. The xenon atoms are so strongly heated that characteristic radiation is emitted within the range of interest between 11 and 14 nm (Fig. 4.14).

For sufficiently short exposure times, radiating power within the range of approximately 20 W in a bandwidth of approximately 0.2 nm, is required. Therefore pulsed lasers with a pulse duration of typically 10 ns and an average power output of some kilowatts are used.

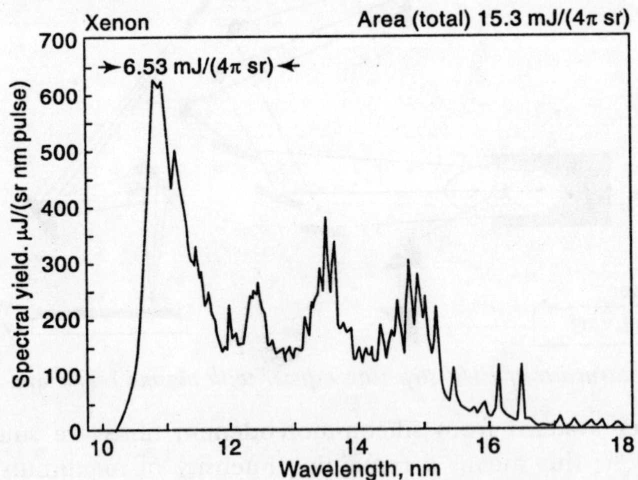

Fig. 4.14 *Spectrum of a xenon plasma source for EUV radiation.*

But as these sources are not available so interest in gas discharge-based EUV radiation sources increases. With these sources the plasma emitted within the EUV area is produced by stored electrical energy in form of a pulsed discharge. Hence there is direct and more effective transformation of the electrical energy into light.

A new gas discharge radiation source xenon clearly promises higher operating life and repeating rates within the range of some kilohertz. Such repeating rates are necessary in order to ensure a sufficiently high average radiating power. Figure 4.14 shows an emission spectrum operated with xenon as discharge gas.

In comparison to the laser-produced plasma this radiation source is more simple, more compact. The spectral characteristics are comparable with those of other radiation sources and fulfill the requirements of EUV lithography. The life time limiting erosion of the electrodes does not occur here. With already currently attainable radiating power within the range of several 100 mW this source lies at the apex of gas discharge-based sources.

ELECTRON BEAM LITHOGRAPHY

Electron-Beam Lithography (EBL) is a specialized technique for creating extremely fine patterns. The practice of using a beam of electrons to generate patterns on a surface is known as Electron beam lithography. The technique consists of scanning a surface with an electron beam. The surface is covered with a resist film sensitive to electrons. By the electron beam, the resist film undergoes structural changes. After development (of the resist film) in a liquid solution, part, of the resist is dissolved away and the wafer is further processed with etching or metallization procedures.

Thus the electron beam technique applies to fast electrons for the generation of structures. Because of their electrical charge, electrons can be easily accelerated and deflected. The two-directional deflection of the beam, is also applied in lithography to project structures. The energies of the electrons are above the binding energies, so that an individual electron induces a number of reactions after hitting the target. The beam voltage is determined by the geometrical requirements of the beam.

Due to a cloud of negatively charged particles of the beam, interparticle electrostatic repulsion is observed. High acceleration voltages and focusing only in the proximity of the substrate surface minimize this effect. High beam voltages suppress the beam divergence due to the short time interval the electrons are at high density. An electron beam is always subject to inference by external fields. The beam width is of the order of a few nanomaters. Accelerated electrons are sensitive both to electric and magnetic. The higher the beam voltage the lower the influence of external fields. High voltages result in a relatively tight beam. Typical values for the beam voltage are about 20 kV, and some times also 20–40 kV.

Due to their large energy, the beam electrons do not interact with the outer electrons, but knock electrons out of the inner shells. Because the beam electrons lose only part of their energy, one particle is able to initiate several such ionization processes. The removed electrons have a sufficiently high energy to ionize the material in their surroundings.

After the ionization of target atoms, an electronic relaxation occurs by transfer of electrons from the outer shells to the free places of the inner shells and an energy emission as radiation. Owing to the large differences in energy, X-ray radiation is generated. This radiation is readily absorbed by the adjacent material. However the expansion distance is much larger than the diameter of the electron beam. Under the influence of direct electronic excitation and the secondary effect of the X-rays, an excitation volume element develops in the target material around the impact area of the electron beam, which extends to a larger penetration depth and is therefore pear-shaped (Fig. 4.15). Because X-rays induce ionization and therefore chemical reactions, the resist solubility in this volume changes. The lithographic resolution of electron beam lithography is mainly determined by the dimensions of this excitation pear, which is based on the target material and beam parameters especially the acceleration voltage, the resolution of electron beam lithography (EBL) does not really depend on the beam diameter. The deposition of beam energy around the actual beam area causes the so-called proximity effect of electron beam lithography. This region can reach dimensions of 0.5–1 μm for low atomic number elements, as in the case of organic resist.

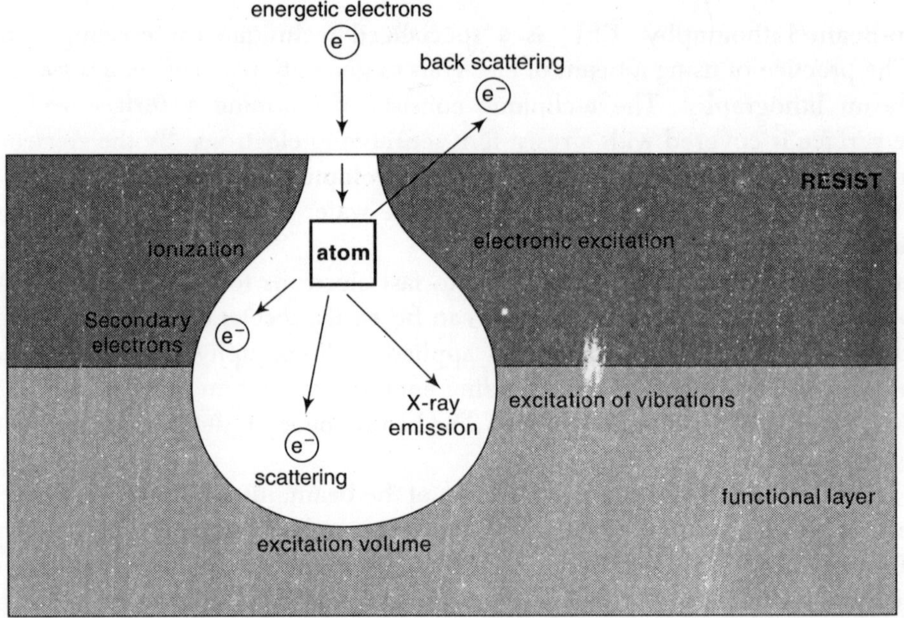

Fig. 4.15 *Proximity effect by formation of an excitation volume in electron beam lithography.*

Now, in electron beam lithography a finely focused computer controlled electron beam is scanned over the substrate coated with a special electron beam sensitive resist. The areas which are not to be exposed are blanked i.e., they are illuminated with electrons.

After irradiation the semiconductor wafer coated with electron-sensitive resist is transferred into the high vacuum system. There, scanning can take place line by line (line scan procedure). Since not only chip for chip must be written but also each structure of each chip, as the exposure procedure is time consuming. In order to minimize the writing time, the beam width in the point of focus can be continuously varied during writing (variable beam shape).

This form of lithography has found wide usage in mask-making (for masks used in photolithograph), low volume production of semiconductor components, the beam be scanned across the surface to be patterned. This makes for very slow pattern generation compared with a parallel technique like photolithography (the current standard) in which the entire surface is patterned at once. As an example, to pattern a single wafer it would take an electron beam system approximately ten hours; compared with less than a few minutes.

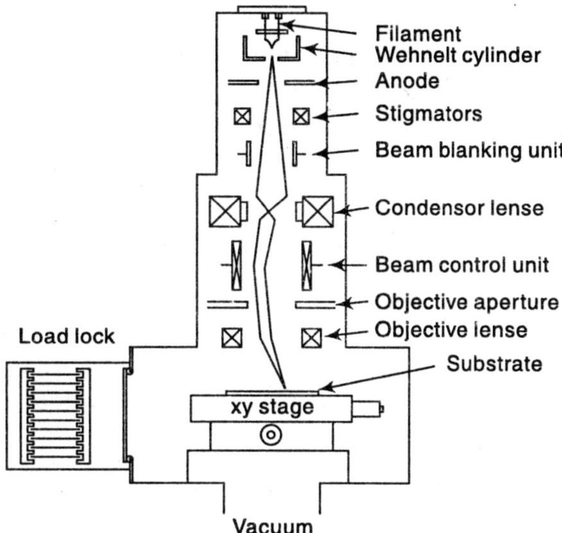

Fig. 4.16 *Cross-sectional view of the electron-optical of an electron-beam writer.*

The electronic beam technique is used mainly for mask manufacturing for the optical lithography because of small throughput. It is only profitable in special cases for direct writing on substrates Fig. 4.16 , e.g., for mask-programmed circuits with small number of elements.

The resolution of the electron beam lithography procedure of modern devices with finely focused beam is clearly smaller than 30 nm in line width; 5 nm structure widths are partly achieved. However, the writing time increases strongly with the required resolution, so that for direct writing of very fine structures irradiation times of some hours/substrate are expected.

The electron beam exposure offers the possibility of exposing the individual layers of a chip with structure sizes down to 30 nm width fast and differently from wafer to wafer without taking the expensive route of mask manufacturing, particularly within the area of the specific applications in integrated circuits (ASIC). Thus, small chip numbers can also be manufactured relatively low-priced despite the high devices costs.

In order to compensate the disadvantage of the long writing time for each wafer respective the small throughput, electron-beam writers with several independently controllable beams are currently being developed. Both multi-beam writers (multi beam) and writers with many micro electron columns (multi-column) Fig. 4.17 are being developed.

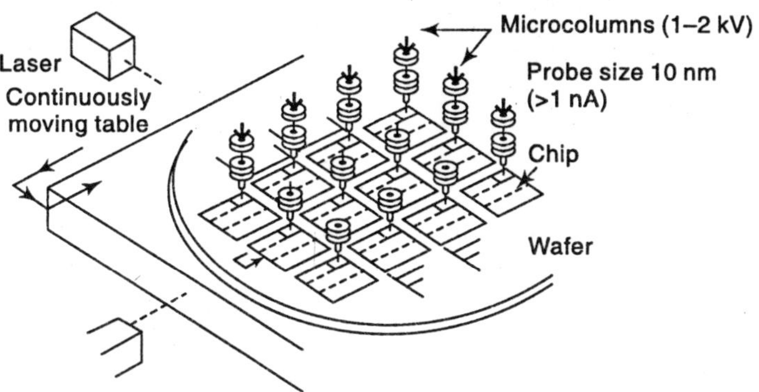

Fig. 4.17 *Multi-column electron beam exposure.*

The most important application for electron beam lithography is the generation of masks of or photolithography with structure dimensions below 200 nm. The electron point beam method is a key technology in nanolithography to fabricate small structures on the substrate directly.

Development of negative resists with high resolution have enabled the fabrication of electron beam lithographically prepared nanostructures with high aspect ratio. Acid-induced cross-linking of oligo cresols with tris (hydroxyethyl) benzene and a 50 kV exposure yielded 80 nm structures with aspect ratios of nearly 4, in the resist ZEP520 for a thickness of 50 nm in structures of only 10 nm width. Under optimized conditions, periodic structures are prepared in this resist structures with repetitions down to 20 nm, e.g., by point beam exposure of line gratings and two-dimensional arrays in PMMA (Fig. 4.18). These structures are transferable in metal films (e.g., Au, AuPd, Pd) by lift-off techniques fig. 4.19. A highly focused electron beam (diameter below 1 nm) enables the smallest structures with extreme aspect ratios to be created, such as structures of less than 30 nm width in PMMA of 400 nm thickness, resulting in an aspect ratio of up to 15. Similar structural periods and line widths of up to 10 nm a realized in calixaren-based resists.

The highly sensitive and chemically enhanced electrons beam resist (NEB 22a) has demonstrated aspect ratios up to 7 for a structure width of 75, and better than 3 for a resolution of 36 nm, respectively. Addition of fullerenes to EBL resists increased the thermal stability.

Also, metal fluorides such as AIF_3, FeF_2, CrF_2, LaF_3, BaF_2 and SrF_2 are inorganic electron beam resists. Here, aluminium fluoride exhibits a good resolution in combination with the

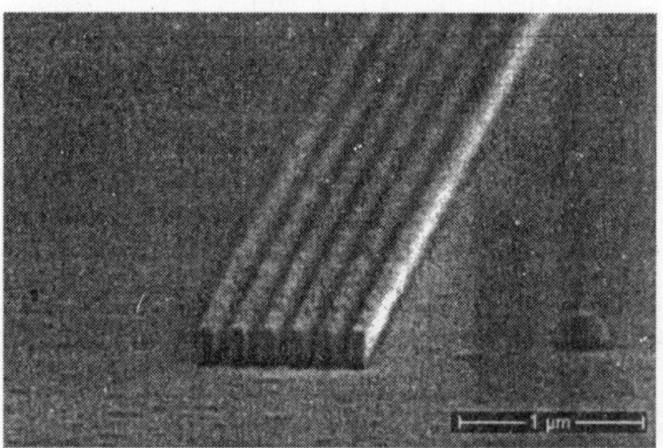

Fig. 4.18 *Nanometer resist structures (negative resist ma-N 24,000m Microresist Technology: 80 nm lines, aspect ratio 2.25) fabricated by electron beam lithography (courtesy of H. Elsner)*

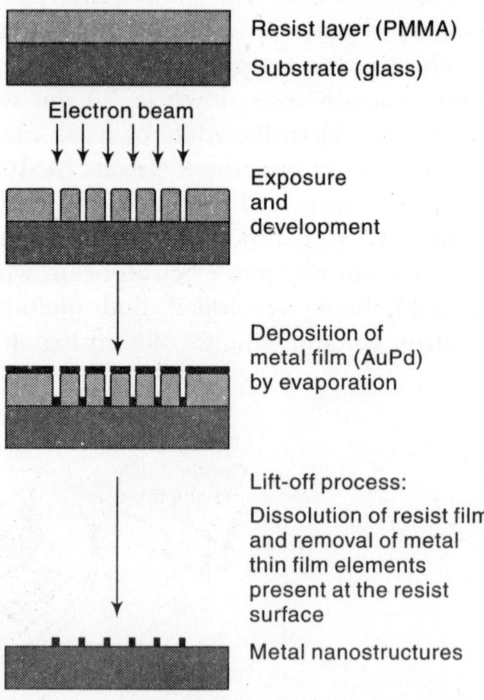

Fig. 4.19 *Electron beam lithographic fabrication of metal nanostructures by a lift-off process.*

resist stability in plasma and an acceptable development method. After exposure, the layers are developed in water. Structural periods of down to 50 nm were realized in 50 nm thick layers, so that a minimum aspect ratios of 2 was achieved.

To avoid large aspect ratios in the lithographic masks, very thin spin-on layers and monomolecular layers can be made as an electron beam resist. On metal surfaces, especially on metals with easily polarizable electron shells (e.g. noble metals and heavy elements of the Main Groups (III to V), the spontaneous assembly of closed monolayers of alythiols and dithiols can be used for such a mask technique. On oxidic surfaces, OH groups are created or existing OH groups are used to fabricate monolayers to act as a resist via a siloxane coupling, e.g., starting with trichlorosilanes or trialkoxysilances. Also a perfluroalkyl monolayer can be used as the electron beam sensitive mask layer. Using an octadeclythiol layer on GaAs and an octadecylsiloxane layer on SiO_2, EBL structures of 25 nm width and 50 nm periodicity were realized and transferred by wet etching into the functional layer.

Regular structures such as line gratings with a constant periodicity realized without a mask or position-controlled beam: in analogy with the lithography with electromagnetic radiation, EBL can use the interference of electrons for structure generation. Applying this approach, gratings with line widths of 0.1 μm are structured into a 30 nm thick PMMA layer.

Electron Energy Deposition in Matter

Primary electron in the incident beam lose energy upon entering a material through inelastic scattering or collisions with electrons. In such a collision the momentum transfer form the incident electron to an atomic electron can be expressed as $dp = 2e^2/bv$, where b is the distance of closest approach between the electrons, and v is the incident electron velocity. The energy transferred by the collision is given by $T = (dp)^2/2m = e^4/Eb^2$, where m is the electron mass and E is the incident electron energy, given by $E = (1/2)mv^2$. By integrating over all values of T between the lowest binding energy, Eo, and the incident energy, one obtains the result that the total cross section is inversely proportional to the incident energy E, and proportional to $1/Eo - 1/E$. Generally, $E >> Eo$, so the result is essentially inversely proportional to the binding energy.

By using the same integration approach, but over the range $2Eo$ to E, one obtains by comparing cross-sections that half of the inelastic collisions of the incident electrons produce electrons with kinetic energy greater than Eo. These secondary electrons are capable of breaking bonds (with binding energy Eo) at some distance away from the original collision. Additionally, they can generate additional, lower energy electrons, resulting in an electron cascade. Hence, it is important to recognize the significant contribution of secondary electrons of the spread of the energy deposition.

Spatial-phase-locked Electron-beam Lithography (SPLEBL)

It reduces pattern-placement errors in electron-beam-lithography systems to the 1 nm level.

As shown in Figure 4.20, SPEEBL uses a periodic signal, derived from the interaction of the scanning e-beam with a fiducial grid on the substrate, to continuously track the position of the beam while patterns are being written. It senses deviation of the beam from its intended location on the substrate, and corrects them and feeds them back to the beam-control electronics;

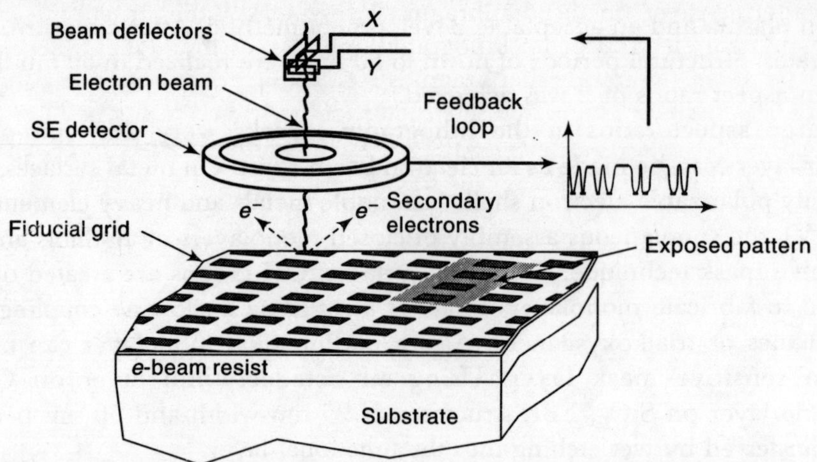

Fig. 4.20 *Fiducial-grid mode of spatial-phase-locked electron-beam lithography. The periodic signal detected from the fiducial grid which include both X and Y components, is used to measure placement error, and a correction signal is fed back to the beam deflection system.*

to cancel errors in the beam's position. Thus, the locations of patterns are directly registered to the grid on the substrate.

SPLEBL in its continuous-feedback mode is then implemented on a scanning e-beam lithography system which provides a sub-20-nm patterning resolution. A thin (<10 nm) metallic fiducial grid Fig. 4.20 is placed on top of the e-beam resist. During exposure, a periodically varying secondary electron (SE) signal is produced from the interaction between the electron beam and the metal reference grid. Its limited thickness makes the grid layer transparent to the primary electron-beam.

Here, silicon nitride is used as a membrane, transparent for electrons with strengthens in the masking area with a metallic scattering layer. Electrons, which hit this scattering layer, are strongly deflected while electrons which hit the membrane only slightly modify their direction of propagation.

After focusing of all electrons, the aperture diaphragm fades out the strongly scattered electrons and only the particles with small diversion pass through the aperture and lead to exposure (Fig. 4.21). The heating up of the mask is relatively small because the electrons are only scattered but not absorbed.

A resolution of 30 nm line width has already been demonstrated.

The mask for the SCALPEL technique consists of a silicon wafer which is coated as diaphragm with silicon nitride. The scatterers made of tungsten are vapor-

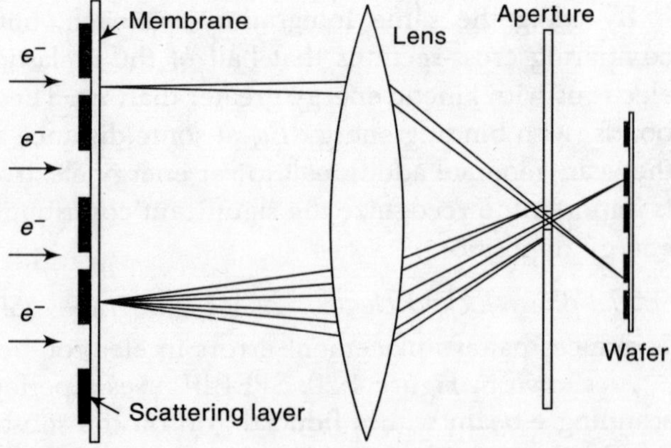

Fig. 4.21 *SCALPEL method for electron beam expo.*

deposited as scattering layer on the diaphragm. Subsequently, the structuring of the tungsten layer with conventional electron beam lithography and dry etching takes place.

Now, the silicon under the nitride diaphragm is removed by anisotropic wet-chemical etching with KOH or EDP solutions. The resulting structure is presented in Fig. 4.22.

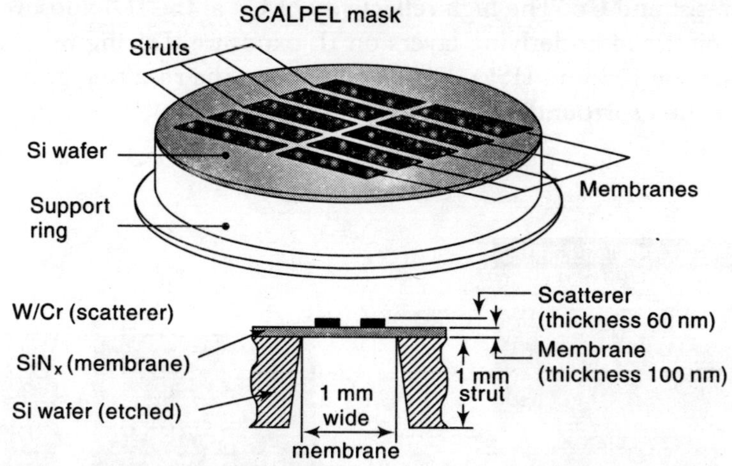

Fig. 4.22 *Masks for the exposure according to the SCALPEL method.*

A much simplified grid-fabrication process based on interference lithography (IL) has been developed, as shown in Fig. 4.23.

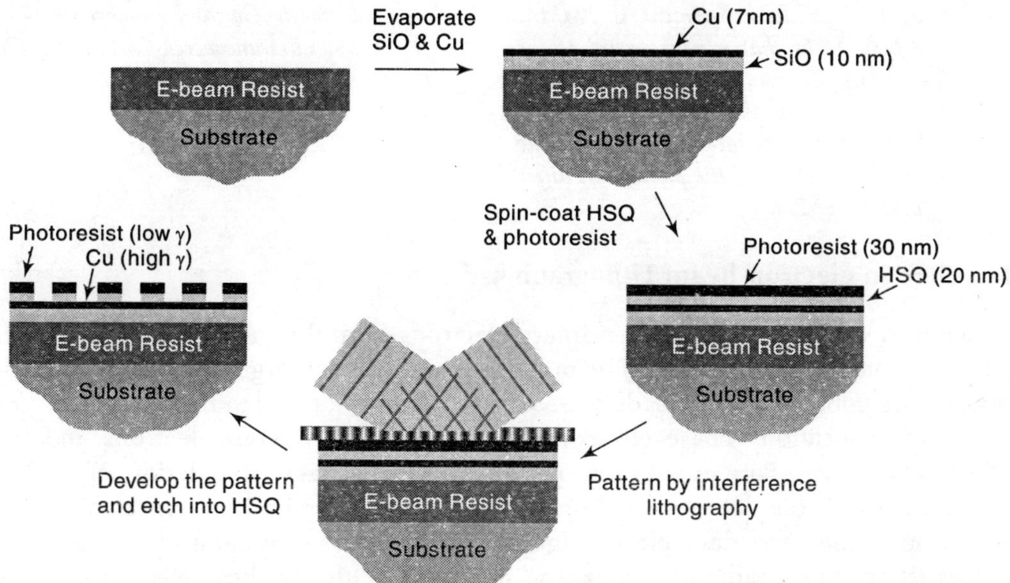

Fig. 4.23 *Process diagram for grid fabrication. The process starts with a sample coated with e-beam resist. Then, 10 nm of SiO_2 and 7 nm of Cu are evaporated on the e-beam resist. The SiO_2 layer is placed under the CU layer to reduce the stress in Cu. A phase-shifting layer (PSL) of 20nm HSQ and 30 nm photoresist (PR) are then spun onto the sample. After IL exposures and development, the pattern is directly transferred into HSQ. Areas covered by HSQ will appear darker than other areas.*

After spin-coating the e-beam resist (e.g. PMMA), 10 nm of SiO_2 and 7 nm of Cu are evaporated onto the sample. The SiO_2 layer is necessary to reduce the stress in the CU film. To reduce the effect of the vertical standing waves (Fig. 4.24), the photoresist layer is kept very thin, and a 'phase-shifting' layer (PSL) of 20-nm hydrogen silsesquioxane (HSQ) is added between the photoresist and Cu. The high reflectivity of Cu at the IL exposure wavelength (325 nm) eliminates the effects of underlying layers on IL exposure. During resist development, the pattern in directly transferred into HSQ. Figure 4.25 shows the resulting grid. Areas covered by HSQ appear darker than surrounding areas.

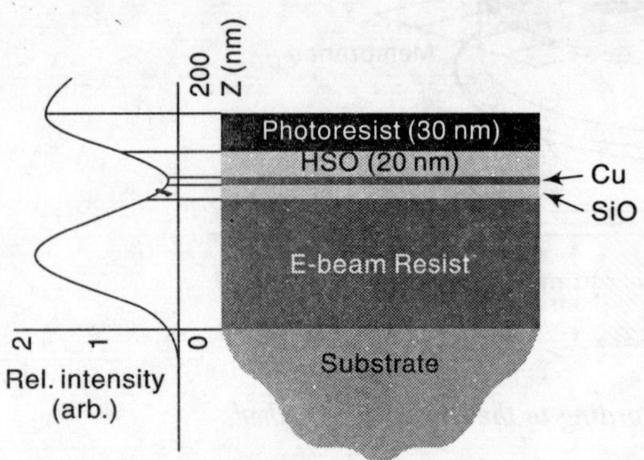

Fig. 4.24 *Relative intensity distribution in the resist stack during IL exposures. The e-beam resist used in the experiment is PMMA. The plot shows that a null occurs at the Cu surface. To achieve uniform exposure, 20 nm HSQ was placed between Cu and the photoresist. The intensities between the top and the bottom of the photoresist differ by a factor of 2.4.*

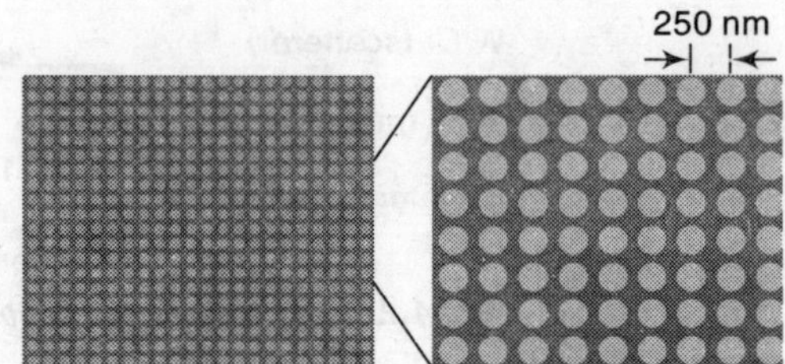

Fig. 4.25 *SEM images of Cu grid. The signal contrast is due to the differently yield between Cu and photoresist. The bright holes are uncovered Cu, and the dark areas are photoresist.*

New Frontiers in electron-beam Lithograph

In addition to secondary electrons, primary electrons from the incident beam with sufficient energy to penetrate the photoresist can be multiply scattered over large distances from underlying films and/or the substrate. This leads to exposure of areas at a significant distance from the desired exposure location. These electrons are called backscattered electrons and have the same effect as long-range flare in optical projection systems. A large enough dose of backscattered electrons can lead to complete removal of photoresist in the desired pattern area.

To get around the secondary electron generation, it will be imperative to use low-energy electrons as the primary radiation to expose photoresist. Ideally these electrons should have energies on the order of not much more than several eV in order to expose the photoresist without generating any secondary electrons, since they will not have sufficient excess energy. Such exposures has been demonstrated using a scanning tunneling microscope as the electron beam source. The data suggest that electrons with energies as low as 12 eV can penetrate 50 nm thick polymer photoresist. The drawback to using low energy electrons is that it is hard to prevent spreading of the electron beam of the photoresist. Low energy electron optical systems are also hard to design for high resolution. Coulomb inter-electron repulsion always becomes more severe for lower electron energy.

Another alternative in electron beam lithography is to use extremely high electrons energies (at least 100 keV) to essentially "drill" or sputter the material. The phenomenon has been observed frequently in transmission electron microscopy. However, this is a very inefficient process, due to the inefficient transfer of momentum from the electron beam to the material. As a result it is a slow process, requiring much longer exposure times than conventional electron beam lithograhy. Also high energy beams always bring up the concern of substrate damage.

Interference lithography using electron beam is another possible path for patterning arrays with nanometer-scale periods. A key advantage of using electrons over photon in interferometry is the much shorter wavelength for the same energy.

Despite the various intricacies and subtleties of electron beam incident at different energies, it remains the most practical way to concentrate the most energy into the smallest area.

X-RAY LITHOGRAPHY

X-ray lithography is developed for the semiconductor industry. Batches of microprocessors have been produced.

The short wavelengths of 0.8 nm X-rays overcome diffraction limits in the resolution of optical lithography. The X-rays illuminate a mask placed in proximity to a resist-coated wafer. No lenses are used; only rudimentary collimating mirrors are used. The X-rays are broadband, typically from a compact synchrotron radiation source, allowing rapid exposure.

X-ray lithography uses shorter wavelengths, about 0.1 nm to fabricate three dimensional structures with reduced resolution.

The mask consists of an X-ray absorber Fig. 4.26, made from gold or compounds of tantalum or tungsten, on a membrane that is transparent to X-rays, typically of silicon carbide or diamond. The pattern on the mask is written by direct write electron beam lithography onto a resist that is developed by conventional semiconductor processes. The membrane can be stretched for overlay accuracy.

X-ray lithography is performed by copying with image fidelity, i.e., without magnification, 1x, on the line of fuzzy contrast as illustrated in the Figure 4.27. But with the increasing need for high resolution, X-ray lithography is now performed on the Sweet Spot, using local "demagnification by bias".

Dense structures are developed by multiple exposures with translation. Many advantages accrue from the application of 3x "demagnification": the mask is more easily fabricated; the mask to wafer gap is increased; and the contrast is higher. The technique is extensible to dense 15 nm prints. The resulting printing has high contrast.

X-rays generate secondary electrons as in the cases of extreme ultraviolet lithography and electron beam lithography. While the fine pattern definition is due principally to secondaries

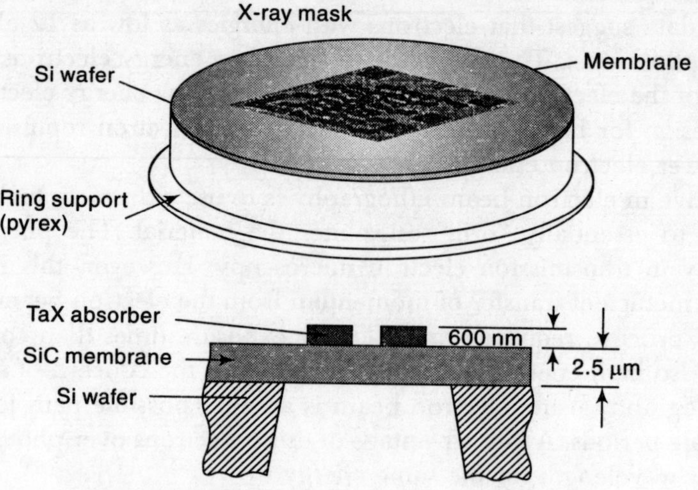

Fig. 4.26 *Design of a mask for x-ray lithography.*

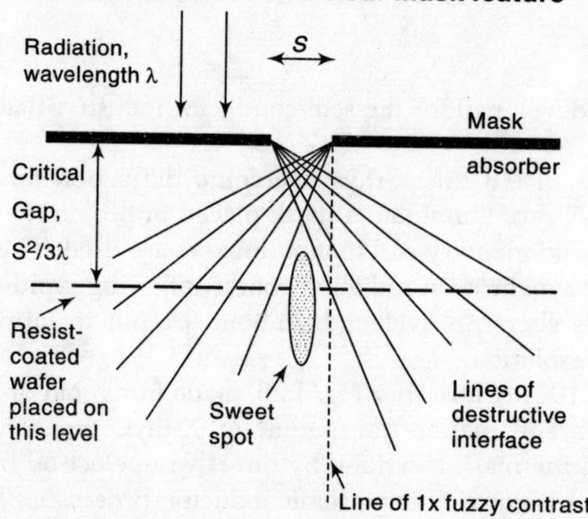

Fig. 4.27

from Auger electrons with a short path length, the primary electrons will sensitize the resist over a larger region than the X-ray exposure. While this does not affect the pattern pitch resolution (determined by wavelength and gap), the image exposure contrast (*max* – *min*)/(*max* + *min*) is reduced since the pitch is on the order of the primary photo-electron range. Several prints at about 20 nm have been published.

Another manifestation of the photoelectron effect is exposure to X-ray generated electrons from thick gold films used for making daughter masks.

By simulations, photoelectron generation from the gold substrate effect dissolution.

Advantages of X-ray Lithography:

- Large depth of focus
- Excellent resist profiles (pillars of resist)
- Large process latitude
- Line width independent of substrate topography
- Relatively immune to low atomic weight contaminants

Disadvantages of X-ray Lithography:

- 1 X mask technology (gold on 1–2 µm thick silicon)
- Defects, aspect ratio, bending, and heating are problems
- Source cost and/or complexity
- Alignment/registration is nontrivial

X-ray and Synchrotron Lithography

Because of the substantially smaller wavelength in comparison to the optical lithography, diffraction patterns at structure edges occur only with structural widths well below 100 nm when using x-rays. Therefore finer structures can be imaged with X-ray lithography than with optical procedures. The wavelength of approximately 1 nm promises a considerably higher resolution. However, due to the Fresnel diffraction and because of the generated photo electrons, limiting effects occur for the minimum attainable structure width, so that the limit of the resolution as a local masking cannot be implemented by chromium layers., Here, heavy elements such as gold, tantalum or tungsten are required for the adsorption of radiation. An intensity ratio of 10:1 is achieved between the permeable and the impermeable mask areas.

The structuring of the masks of the X-ray lithography takes place with the help of the electron beam technique. The absorber layer is deposited galvanically on then thin carrier membrance, whereby the exposed resist mask in negative technique releases only the desired structures for coating.

The mask making is very complex and the necessary size accuracy is not satisfactory.

X-ray lithography has been unsuccessful because the optical lithography is substantially more simple to execute with the previous line widths. X-ray lithography may be a solution for structure width between 70 and 40 nm. Finer structures cannot be imaged.

X-Ray Nanolithography (XNL)

XNL is a reliable and simple means of replicating patterns with feature sizes down to 20 nm. Typically the X-ray mask is made with scanning-electron-beam lithography (SEBL), although employ a combination of interference lithography, photolithography, SEBL, and XNL is employed to fabricate the mask. Once the mask it fabricated it can be replicated repeatedly. XNL has been used in the fabrication of a large variety of structures and devices, including: photonic bandgap devices, short-channel MOSFETs, and optical channel-dropping filters.

Source for XNL are electron-bombardment targets. The L line of copper is $\lambda = 1.32$ nm is used. The sources are separated from a helium-filled exposure chamber by a 1.5 µm-thick SiN_x vacuum window.

Although the wavelength used is very short (1.32 nm) compared to the minimum sizes of interest (e.g., 20 nm) diffraction in the gap between the mask and the substrate can be determined. For example, with a CuL source, a 50 nm feature must be exposed at a mask-to-substrate gap of less than about 4 μm in order to maintain good process latitude. A 25 nm feature would require a gap of 1 μm. For such very small features, eliminate the gap and use contact between the substrate and the flexible membrane mask; essentially the same as in imprint lithography. This technique has enabled to replicate features as small as 20 nm.

EXTREME ULTRAVIOLET LITHOGRAPHY

Extreme Ultraviolet Lithography (also known as EUV or EUVL) is a next-generation lithography technology using the 13.5 nm wavelength. EUV is a significant departure from the deep ultraviolet lithography used today. All matter absorbs EUV radiation. Hence, EUV lithography needs to take place in a vacuum. All the optical elements, including the photomask, must make use of defect-free Mo/Si multilayers which act to reflect light by means of interlayer; any one of these mirrors will absorbs around 30% of the incident light. This limitation can be avoided in maskless interference lithography systems. However, the latter tools are restricted to producing periodic patterns only.

EUV and X-ray Lithography

The wavelength of electromagnetic radiation can be decreased until it limits no longer the resolution. Extreme UV radiation (EUV) includes the medium nanometer range. Values around 10 nm have been reached, the region of the transition to soft X-rays. The lower nanometer and sub-nanometer range is addressed by X-ray lithography (XRL). Using X-rays, sharp local distributions of photoproducts are realized for structure dimensions in the medium and lower nanometer range, which are not accessible even for multilayers mask processes with UV radiation (Fig. 4.28). Such radiation transfers structures in the medium and lower nanometer range, so that this technology is applied for the fabrication of quantum dot.

The transition from UV/VIS to EUV and XRL transfers the limitations due to diffractions to those for materials. Consider the optical aspects. Materials with extremely large differences in optical transmission exist i.e. the absorption for a nanometer thick metal layer is the same as for a kilometer long glass fiber. Also, the absorption coefficients in the soft X-ray region exhibit only small differences at a maximum two orders of magnitude. The fabrication of transmissive and mechanically stable lithographic masks is therefore difficult. Also the realization of high-resolution optics for lithographic projection systems is difficult.

To achieve structures with high aspects ratios, X-ray lithography is already in use. X-ray radiation is extremely parallel, especially for synchrotron sources. An exceptionally parallel beam and large depth of focus allows the penetration of even thick layers of resist. Therefore, X-ray radiation is suitable for the direct writing of nanometer structures with high aspects ratios into X-ray sensitive material. Using a special plasma source with wavelengths of 1.2 and 1.7 nm, columns of 70 nm diameter and an aspects ratios of 4 have been structured directly into the radiation resist. Aspect ratios of up to 13 are achieved for structure widths of 20 nm) in PMMA (layer thickness of 250 nm) or a PMMA/MMA layer (layer thickness of 265 nm) using

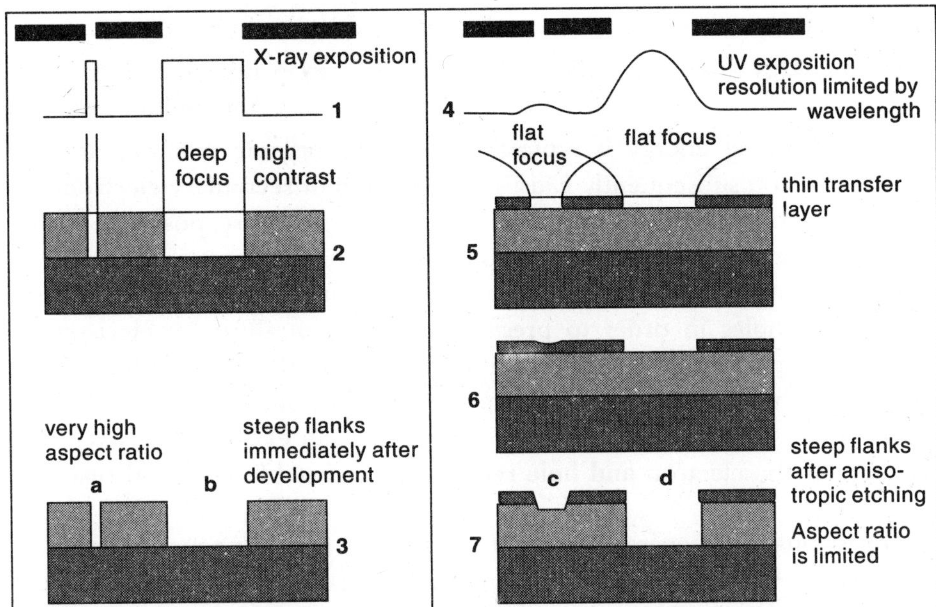

Fig. 4.28 *Comparison of the principle possibilities for nanostructure generation by X-ray lithography (left) and diffraction-limited UV-lithography in multilayer resist technology (right).*

Au and W absorption layers. High aspect ratios for the EUV range, for a wavelength of 13.4 nm, line-space patterns for 80 nm structure widths have been generated in a resist layer of 600 nm.

EUV Tools

The pre-production EUV systems being built to date are expected to contain at least two condenser multilayers mirrors, six projection multilayers mirrors, and a multilayer object (mask). Since the optics already absorbs 96% of the available EUV light, the ideal EUV source will not be sufficiently bright. EUV source development has focused on plasmas generated by laser or discharge pulses. The mirror responsible for collecting the light is directly exposed to the plasma and is therefore vulnerable to damage from the high-energy ions and other debris. This damage associated with the high-energy process of generating EUV radiation precluded the successful implementation of practical EUV light sources for lithography

EUV Absorption in Matter

When an EUV photon is absorbed, photoelectrons and secondary electrons are generated by ionization, much like what happens when X-rays or electron beams are absorbed by matter. These secondary electrons have energies of a few to tens of eV and travel lens of nanometers inside photoresist before initiating the desired chemical reaction. A contributing factor for this rather large distance is the fact that polymers have significant amounts of free volume. In a recent actual EUV print test, it was found that 30 nm spaces could not be resolved, even though the optical resolution and the photoresist.

The response of matter to EUV radiation can be captured in the following equation:

EUV photon energy = 92 eV = Electron binding energy + kinetic energy of the emitted photoelectron

where the electron binding energy is typically 79 eV for organic materials and 4-5 eV for metals. The photoelectron subsequently causes the emission of secondary electrons through the process of impact ionization. Sometimes, an Auger transition is also possible, resulting in the emission of two electrons with the absorption of a single photon.

Strictly speaking, photoelectrons, Auger electrons and secondary electrons are all accompanies by positively charged holes in order to preserve charge neutrality. An electron-hole pair is often referred to as an exciton. For highly energetic electrons, the electron-hole separation can be quite large and the binding energy is correspondingly low, but at lower energy, the electron and hole can be closer to each other. As the name implies, an exciton is an excited state; only when it disappears as the electron and hole recombine, can stable chemical reaction products form.

EUV photoresist images often require resist thicknesses roughly equal to the pitch. This is not only due to EUV absorption causing less light to reach the bottom of the resist but also to forward scattering from the secondary electrons (similar to low-energy electron beam lithography).

EUV Damage

Like other forms of ionizing radiation, EUV and EUV-generated electrons are a likely source of device damage. Damage may result from oxide desorption or trapped charge following ionization. Damage may also occur through indefinite positive charging by the Malter effect. If free electrons cannot return to neutralize the met positive charge, positive ion desorption is the only way to restore neutrality. However, desorption essentially means the photoresist is degraded during exposure, and furthermore, the desorbed atoms contaminate the optics.

EUV Defects

EUVL faces specific defect issues analogues to those being encountered by immersion lithography. Whereas the immersion-specific defects are due to unoptimized contact between the water and the photoresist, EUV-related defects are attributed to the inherently ionizing energy of EUV radiation. The first issues is positive charging, due to ejection of photoelectrons freed from the top resist surface by the EUV radiation. This could lead to electrostatic discharge or particle contamination as well as the device damage mentioned above. A second issue is contamination deposition on the resist from ambient or outgassed hydrocarbons, which results from EUV-or electron-driven reaction. A third issue is etching of the resist by oxygen, argon or other ambient gases, which have been dissociated by the EUV radiation or the electrons generated by EUV. Ambient gases in the lithography chamber may be used for purging and contamination reduction. These gases are ionized by EUV radiation, leading to plasma generation in the vicinity of exposed surfaces, resulting in damage to the multilayer optics and inadvertent exposure of the sample.

EUV Development

EUV has been the subject of ongoing research and development by many groups. It is unlikely for the technology to be used even at the 22 nm node in 2011, given that it is still under development in key such as light source, photoresists, and defect inspection. The difficulties faced stem from the dramatically higher energy of the EUV photon (92 eV for EUV light vs. 6.4 eV for 193 nm light), which underlines the difficulty of damage-free generation and control of EUV light and confining the energy adsorption within materials. It is also fundamentally impossible for EUV with low resolution enhancement and the energy adsorption within materials. It is also fundamentally impossible for EUV with low resolution enhancement and single patterning to compete with the larger depth of focus from the more established approach of using the 193 nm wavelength with strong resolution enhancement and double patterning.

ION BEAM LITHOGRAPHY

In analogy with the use of electrons, ions can also be applied for lithography. They too can be generated as well as accelerated by the electrode arrangements, filtered according to their energy and locally guided by electronic optics. Ion beams are utilized in nanolithography:

— to fabricate structure in ion beam sensitive resists
— for local implantation
— for structure transfer from structured resist layers in functional layer by ion beam etching
— to structure directly by local etching using a focused ion beam

Comparable to parallel light, parallel ion beams can be applied for proximity lithography. This process is characterized by only a small distance between the mask and substrate.

Ion Beam lithography is used on the one hand for projection exposure with masks but on the other hand it is also comparable to the electron beam lithography for direct exposure. In the case of the direct exposure with a finely focused ion beam, the higher particle mass in comparison to the electron mass causes a decrease of the required ion dose for resist exposure of about a factor 10–100 in relation to the electron dose. Thus, the writing rate can clearly be increased. However, whether sufficiently high writing rates can be obtained for the exposure of 300 mm wafers in the manufacturing is unlikely so far.

A high resolution requires a constant distance between the mask and substrate which means highly parallel arrangement, and a low beam divergence. Because ions beams are virtually not diffraction limited, greater apertures can also allow for a high resolution; combination of masks are applicable due to the highly parallel beam (Fig. 4.29(a). Small opening can be realized through the application of similar but slightly shifted masks (Fig. 4.29(a)). Owing to high resolution the ion beam lithography is of interest.

It consists of a vacuum container with an ion source, Fig. 4.29(b) an ion optical system an alignment system and table for the mask and substrates. Such systems exhibit a much higher productivity for structures in the nanometer range than serial writing electron beam systems. A exposure rate of 12 cm^2 s^{-1} is achieved with a resolution of 100 nm.

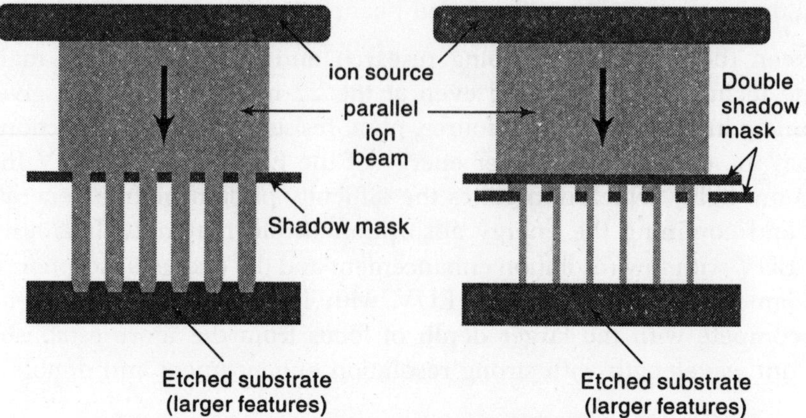

(a) Reduction of line width in ion beam lithographic etching by shadow masks using mask combinations

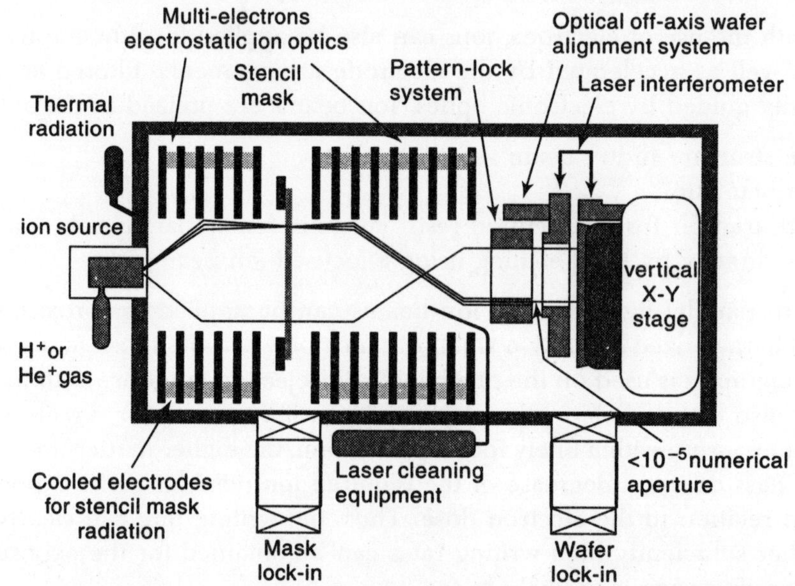

(b) Setup of a device for ion projection exposure with a stencil mask

Fig. 4.29

For the lithographic generation of structures in resists or the direct writing of small structures, lithography with a focused ion beam (focused ion beam lithography–FIBL; see below) and beam diameters in the lower nanometer range can be achieved. The interaction of energetic ions with the target material is more favorable than in the case of electron beams, because ion beams have only a reduced tendency to induce secondary electrons or X-ray. Therefore, the excitation pear is reduced, and the resulting proximity effect minimized. A beam of energetic ions can be focused and used for local implantation of foreign atoms without the need for masks. Therefore, the atoms to be implanted have to be used as the ion beam.

Ion beam lithography, using an expanded beam with a diameter of approximately a square centimeter can be applied in the projection procedure. Here, the mask is laid out as stencil mask for image projection so that the ions between the absorbers reach the photoresist layer on the substrate without dispersion.

A thin silicon diaphragm, which is etched from a single-crystal silicon wafer, Fig. 4.30 serves as mask material. The required mechanical stability of the mask is achieved by a back-up ring at the edge of the disk. However, the thermal stress of the mask structure leads to an uncontrolled distortion of the structural pattern due to the absorption of the ions, which is reduced by a small dose rate.

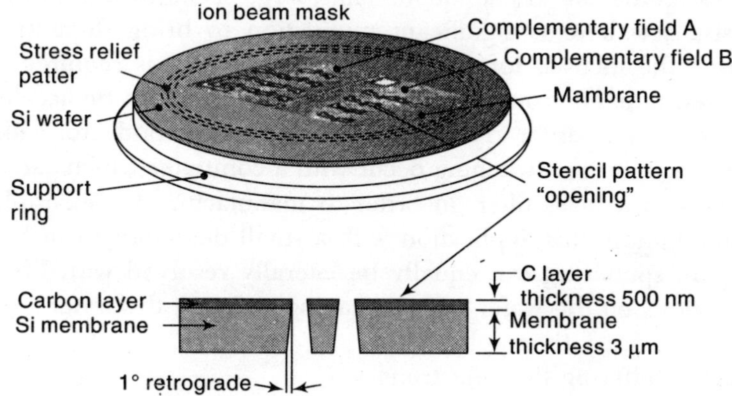

Fig. 4.30 *Mask structure in the ion beam lithography.*

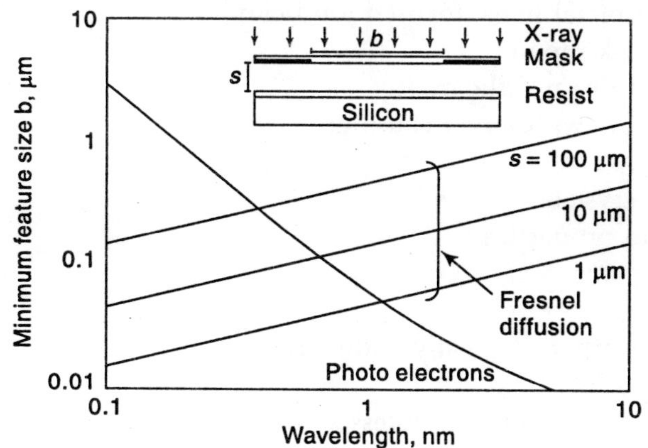

Fig. 4.31 *Resolution limit as a function of wavelength for X-ray lithography.*

Ion implantation plays a central role in the modern semiconductor technology. The Focused Ion Beam (FIB) toll cut away (mill) material from a defined area with dimensions typically in square microns or deposit material onto it. The accelerated particles with high surface (more than about 100 particle/surface atom) causes an erosion effect at the surface. With an implantation surface dose of about 1 particle/surface atom amorphization of the solid state occurs which is used, in wet chemical etching. There is a high interest in doping semiconductors within the dose range of about 1/100 particle/surface atom. This is implemented with diffusion methods of deposited dopant layers, which run at relatively high substrate temperatures. The ion implantation replaces this diffusion technique, since the doses can be substantially controlled more exactly via beam current and exposure time, the depth distribution is more defined and usually no substrate heating is necessary.

However, ions are always implanted on large areas in the resist, by the photolithography. The process which can be performed are numerous and critical to impurities: spinning of the resist, backing, mask positioning, exposing, developing rinsing, drying as well as removal (stripping) of the resist after the implantation. Thus, large sections of the implantation dose are wasted in the resist, so, there is significant motivation to bring dopants masklessly into a semiconductor. Here the focused ion beam implantation (FIB) is required: if it is possible to extract ions of desired dopants from a microscopic source they can be accelerated and focused in a particle-optical system, in order to intentionally dope a semiconductor with lateral resolution. The beam deflection and in/out switching occur with a computer which can select different ion types even by means of a mass filter, in order to manufacture for example, complementary doping profiles. In addition, this application with a small dose range can be extended to high dose: amorphizing an sputtering can equally be laterally resolved with FIB, which enable an analysis of cross sections on the wafer without having to break the wafer.

Advantages of Ion Beam Lithography

1. Less logn range scattering than electrons
2. Ion beams stay near initial trajectory
   → no doese adjustment for different patterns or substrates
3. Can directly write metal lines (focused ion beam)
   → suitable for mask repair

Disadvantages of Ion Beam Lithography

1. Ions interact strongly with target causing:
   (a) Ion mixing
   (b) Amorphizing crystals
   (c) Altered optical properties
   (d) Implanted dopants
   (e) Sputter etching
2. Ions are highly absorbed (typically within 10 nm)
   → Stencil type masks
   → The center of a ring falls out unless sub-ersoution supports are used

Focused Ion Beam

The focused ion beam technology (FIB) is based on possible point-like ion sources, which are referred to as emitters. These can be operated cryogenically and then yield elements like hydrogen, helium, nitrogen, and oxygen which are only present in the gas phase at room temperature and pressures below one atmosphere. By introducing gases or an organic gas compound, the FIB can selectively etch one material much faster than surrounding materials, or deposit a metal or oxide. The FIB is used for such tasks as site-specific cross-sectioning for interfacial microstructure studies, preferential removal or certain metals or oxide, semiconductor device editing or modifications, site-specific TEM sample preparation, and gain imaging. But, since the substrate to be implanted is typically grounded and hence the ion source must be of high-volume, cryogenic sources are relatively complex in construction and operation. However a He-ion beam, which has excellent point source characteristics, could be delivered. The life time of this source is only a few hours which is too short for technical application.

Heavier elements such as metals and all others which can bond with metal alloys are used for focused ion beam from the so-called sources of liquid-metal ion sources (LMIS). Their filling consists of only a single chemical element.

The extraction of focused ion beams from LMIS takes place a few millimeters away from a high-melting container of filament, in or on which a drop of the alloy is prevent and held via capillary forces. An equally high-melting needle (usually W) rises from the drop, which manifests a point often sharpened by electrolytic etching and must be moistened by the alloy. The extraction aperture which has a diameter of a few millimeters (typically 3 mm) and is often negatively charged with a high voltage of 4–9 kV is a few millimeters away from the needle. A liquid metal cone (Taylor cone) is formed at the point by electrostatic forces, whose radius of curvature of a few nanometers lies substantially below that of the solid metal point. At this point, ions are formed and extracted by the electrostatic point effect (excessive local field), which is still favored for the heating of the emitter (which is necessary for the melting of the alloy).

The LMIS is filled with gallium. Milling is achieved by accelerating concentrated gallium ions to a specific site, which etches off any exposed material, leaving a very clean hole of surface. This metal 302 K and therefore requires no heating. Heating to about 873 K for 10 seconds is necessary for moistening and remoistening of the metal needle in intervals of 10 to 100 hours.

The life span of a LMIS depends crucially on the steam pressure of the ingredients at the working temperature and on vacuum conditions: the lower the steam pressure and the vacuum pressure is, the higher is that the life time. With well pumped systems a vacuum pressure of about 10^{-9} mbar can be established on the LMIS, with which a Ga-LMIS achieves a life time of typically 10^3 h without constant heating with an emission current 10^{-5} A.

Structure of a FIB

FIB devices resemble scanning electron microscope columns, but the high voltage is reverse biased and all deflection units are electrostatic but not magnetic. Thus it is taken into account that a magnetic field would sort according to impulses. This causes (often inevitable) difficulties in double images and other focusing errors by isotope mixture of the ion source.

The structure of a FIB column is shown in Fig. 4.32a, the total structure with scanning electron microscope in Fig. 4.33b. Usually the ion source lies on the positive acceleration potential (from some 10 to some 100 kV) and the target (sample or wafer) is grounded. The focusing elements are the so-called single lenses which are composed of disk packages that direct the ion beam via drillings of approximately 3 mm large situated as exactly as possible in the column axle in which the electron beam is carried. The disks lie alternately on a high

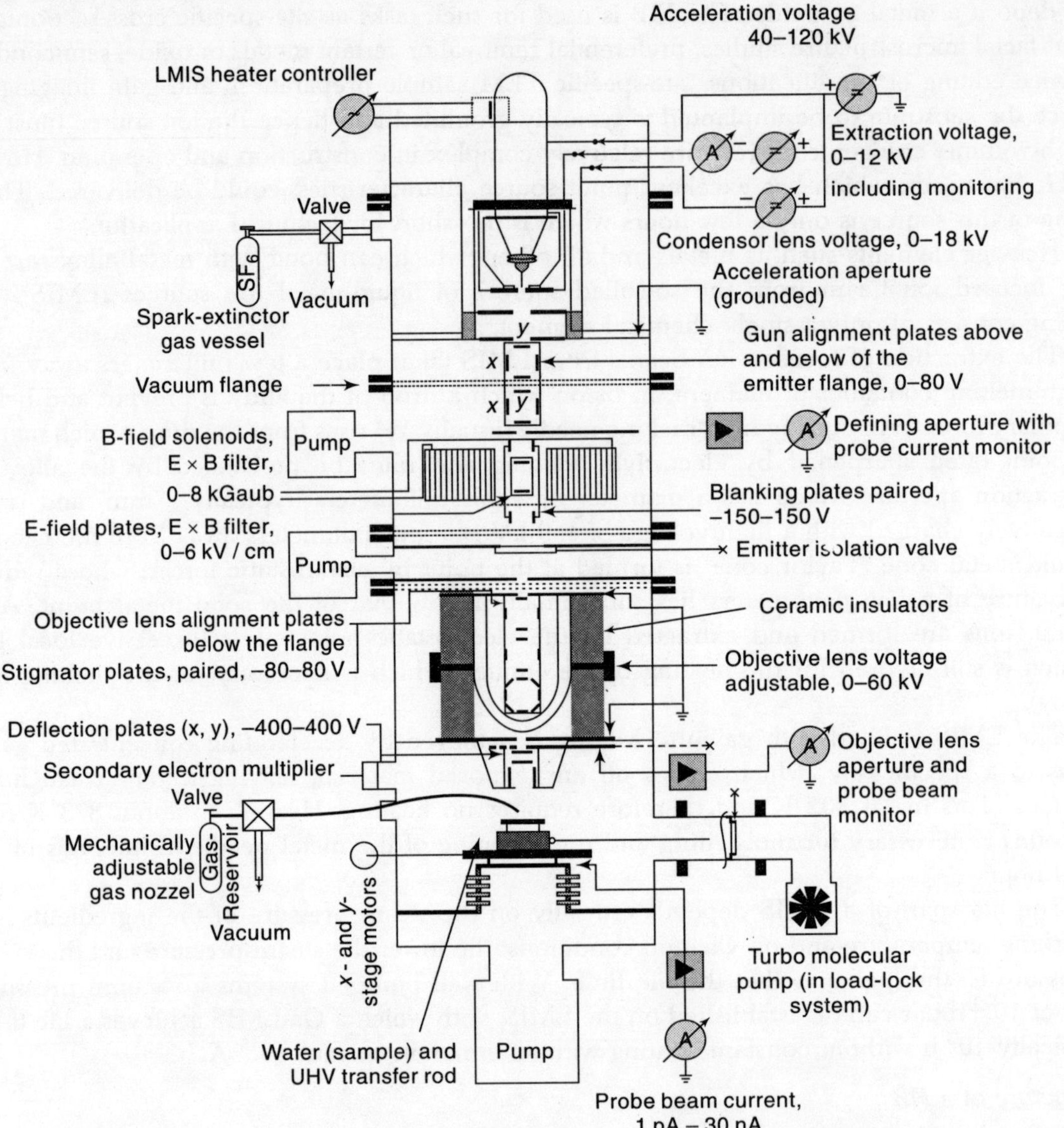

Fig. 4.32 *Functional setup of a FIB column.*

voltage which corresponds to about half of the acceleration voltage and the earth potential, through which focusing-defocusing is effectuated. However, the total effect is focusing which can simply be realized by the curved lines of the electric field in the surrounding of the holes, where the field is strongly inhomogeneous. With a single lens, there are two ways for electrical switching.

Now, the easiest way is to tap the focusing voltage from the positive accelerating voltage simply by a voltage divider which requires only one high voltage tank (which in general is like

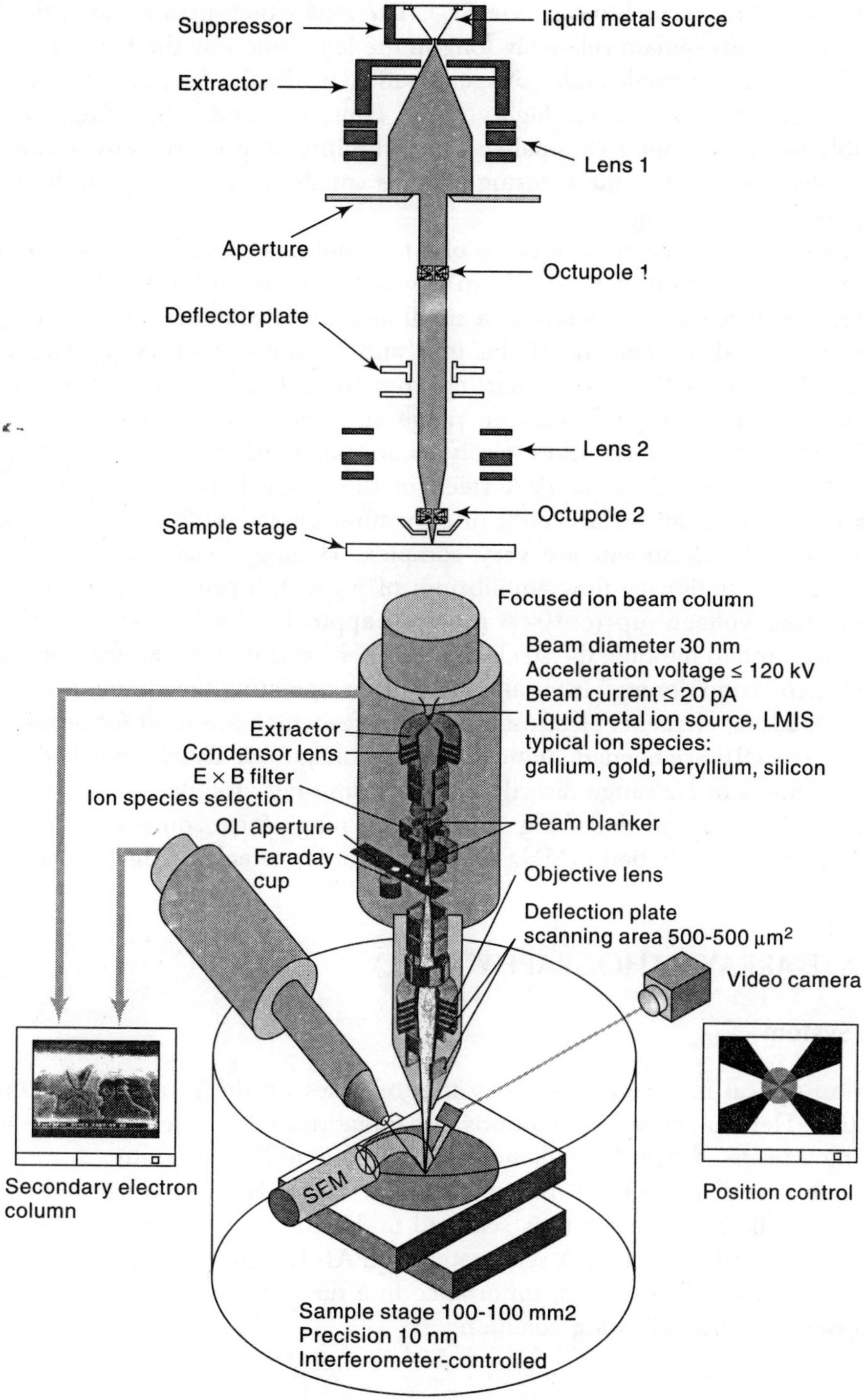

Fig. 4.33 ***REM-FIB system.***

the column isolated by quenching gas via SF_6). The ions which pass through the lens are thus delayed. Thus, the ions remain relatively long in the lens, whereby the focusing effect becomes stronger and a relatively small high voltage is sufficient. In the "accelerating mode", the lens package is occupied by a negatively high voltage. Thus, a second high voltage tank is required. The attainable focus is about 10% smaller. Since the ions stop in the lens during a short time when in "accelerating mode", the focusing effect is smaller and the magnitude of the required high voltage is about 10% higher.

A single lens operates at a disk distance of a few millimeters and a voltages of about 50 kV with yield strengths of about 2×10^5 V/cm^{-1}, which is close to the breakdown field strength (vacuum-pressure-dependent). Therefore, a small increase of the focusing voltage can lead to a strong rise in the leakage current of the lens and becomes a problem particularly by the associated timely focusing fluctuations starting from about 1 μA. Such leakage currents become relevant within the medium-high vacuum range and within the ultrahigh vacuum range. In high vacuum the rest gas works more easily as an extinguisher. They arise particularly from micro particles on polished plates (VA steel) of the single lens which causes electron field emission due to their small surface radii of curvature. Disassembling, postpolishing, cleaning, and assembling with adjustment are very complex. Healing (*conditioning*) by "nitridation" is more simple: in the stationary flow equilibrium of N_2 with a pressure of some 10^{-4} mbar, an increased focusing voltage (up to 100% more) is applied, which leads to a bluish luminous plasma discharge and to nitrating the steel surface. These nitrides have an extraordinary dielectric stability and such "conditioning" is usually enough for operating the system for several months.

Beside the focusing elements electrostatic pair plates which are used for adjusting, dimming, and deflecting the DIB are needed. Static adjusting voltages are usually blocked like alternating voltage by RC filters in Hz range directly at the column near the pair plates in order to obtain high stability. Dimming and deflecting voltages can range form some 10 to some 100 V and must be available as a wide-band (MHz to GHz) in order to achieve high dose accuracies and writing rates.

ZONE-PLATE-ARRAY LITHOGRAPHY (ZPAL)

The ZPAL System

ZPAL is an optical maskless technology that operates on the principle of diffraction rather than refraction. Here an, array of thousands of nanofabricated Fresnel zone plate lenses is used, each focusing a beam of light (or radiation-beam lets) onto the substrate.

A computer-controlled array of micromechanical spatial-High modulators turns the light to each lens on or off as the substrate is scanned under the array, thereby printing the desired pattern in a "dot-matrix" fashion. A schematic of ZPAL is shown in Figure 4.34.

Lithographic resolution is one of importance in a direct-write system such as ZPAL, and is often expressed with the following equation:

$$W_{\min} = k_1 \frac{\lambda}{NA}$$

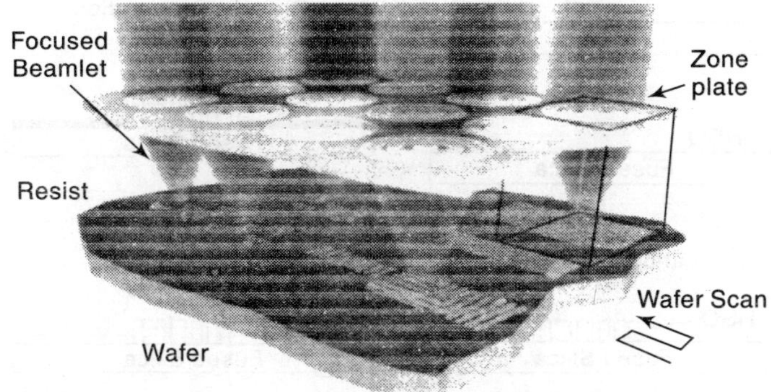

Fig. 4.34 *Zone-Plate-Array Lithography (ZPAL).*

Where W_{min} represents the minimum linewidth, k_1 is a proportionality factor that takes into account variable such as resist properties, environmental conditions, etc., λ is the wavelength of the radiation reaching the substrate, and NA is the numerical aperture.

The ZPAL prototype has already achieved high resolution patterning (W_{min} = 135 nm) corresponding to: $k_1 = 0.29$, $\lambda = 400$ nm and NA = 0.85.

Fabrication of Zone-Plate Arrays by NanoImprint Lithography (NIL)

The use of nanoimprint lithography is made for the fabrication of large arrays of high-numerical-aperture diffractive-lens arrays for zone-plate-array lithography (ZPAL). The Nanoimprint Lithography (NIL) system in the NSL is capable of 5 nm alignment. By using a template consisting of a small array of zone plates with this system a much larger array can be created.

The template is fabricated by scanning-electron-beam lithography (SEBL). The patterns are defined in HSQ. Self-aligned apertures are then created by chrome evaporation, followed by a Fulton-Dolan process. These steps are illustrated in Figure 4.35.

The temperature is then pressed into a NIL liquid deposited on top of a glass substrate. Shining UV light through the template cures the liquid, and thus, "fixes" the zone plate patterns in the cured polymer. The template is stepped, and the process is repeated until the required size of the array is obtained. The alignment capability is important to ensure that all the zone plates in the array are aligned to each other.

The Immersion Zone-Plate Array Lithography

To further improve the resolution, use of immersion lithography is made, which, scales λ down by a factor of n (where n is the refractive index of the medium between the zone-plate array and the substrate), thus producing a smaller diffraction-limited spot. Figure 4.36 shows simulated point-spread function (PSFs) of zone plates operating at $\lambda = 400$ nm under immersion and nonimmersion conditions. The immersion zone plate produces a smaller diffraction-limited spot.

Initial experiments on immersion Zone-Plate-Array Lithography (iZPAL), the gap between the zone-plate array and the substrate is filled with de-ionized water ($n = 1.34$ at $\lambda = 400$ nm).

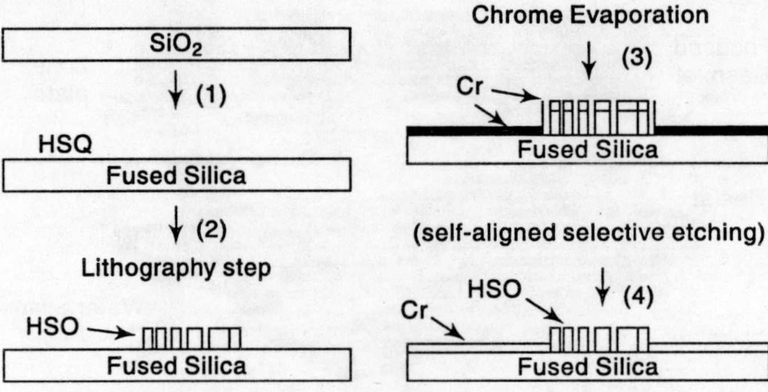

Fig. 4.35 *Fabrication of the NIL template.*

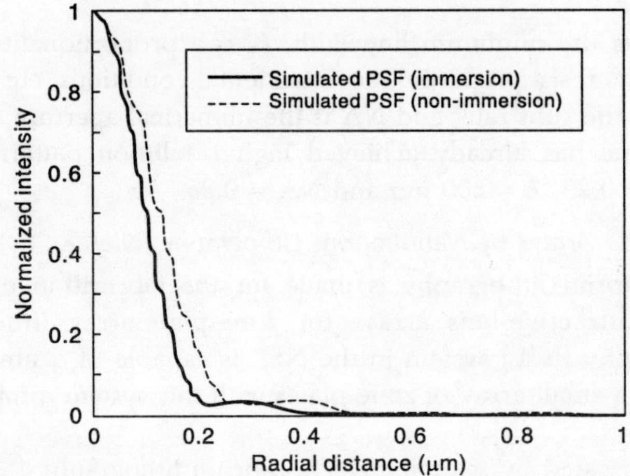

Fig. 4.36 *Simulated PSFs of NA = 0.85 zone plate operating at l = 40 nm under immersion and non-immersion conditions. The simulations were generated using a Finite-Difference-Time-Domain algorithm.*

A 26% decrease in the minimum feature size (W_{min} = 1000 nm given the values for k_1, λ, and NA listed above its expected. IZPAL setup is depicted in Figure 4.37.

Zone plates for the non-immersion scheme are fabricated by a robust planar process involving HSQ resist on a fused silica substrate. The proper functionality of a zone plate is dependent upon introducing a π-phase-shift between adjacent zones. The phase shift becomes difficult to achieve when the refractive indices of HSQ and the immersion liquid (D1 water) are almost identical ($n_{HSQ} = 1.39$, $n_{DIwater} = 1.34$ at $\Lambda = 400$ nm).

So, a new process in which the zones are patterned in Si-rich silicon nitride (nSiN = 2.41 at $\lambda = 40$ nm). The immersion zone plates are first patterned on PMMA USING e-beam lithography. The pattern is then transferred into a Nickel (Ni) layer through a litoff process, which is then used as an etch mask for reactive-ion-etching into the underlying Sin layer. Finally a chrome (Cr) absorbance layer is created, which serves to block out an light not incident on the zone plates. Scanning-electron micrographs of a completed immersion zone plate are shown in Figure 4.38.

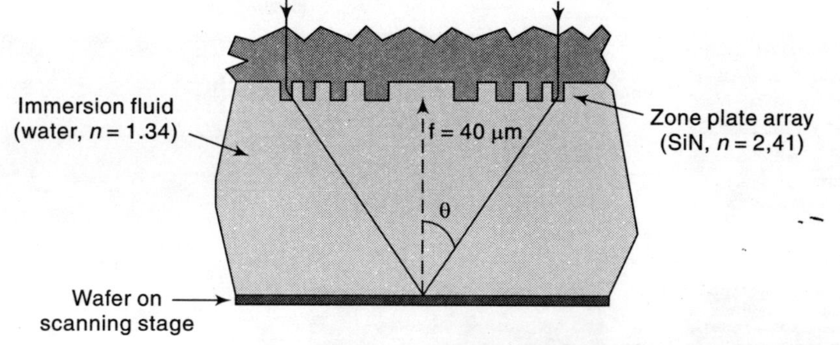

Fig. 4.37 *Immersion lithography for ZPAL.*

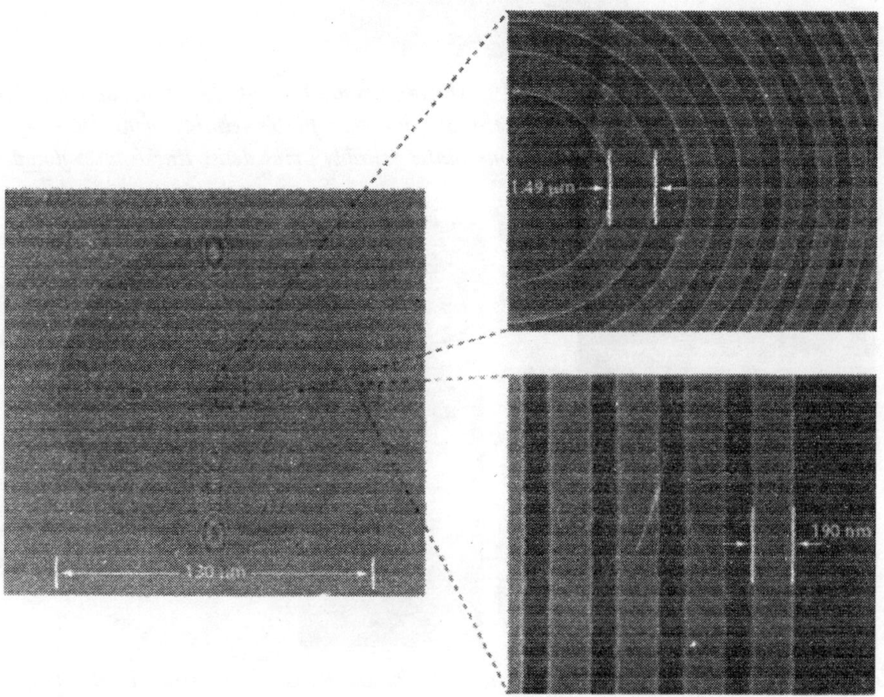

Fig. 4.38 *SEMs of an NA = 0.85 immersion zone plate, designed for l = 40 nm.*

The immersion zone-plate array is then integrated into ZPAL prototype system to test its lithographic performance. Dense lines/spaces are printed using both immersion and non-immersion ZPAL, and compared. These dense line/spaces have linewidths as low as 115 nm using immersion ZPAL (compared to 135 nm using non-immersion ZPAL) Fig. 4.39. Also at the same resolution, the immersion scheme prints are pattern patterned as compared to the non-immersion scheme.

SEMs of dense nested-L's and dense arrays of contact holes produced using immersion ZPAL are shown in Figure 4.40.

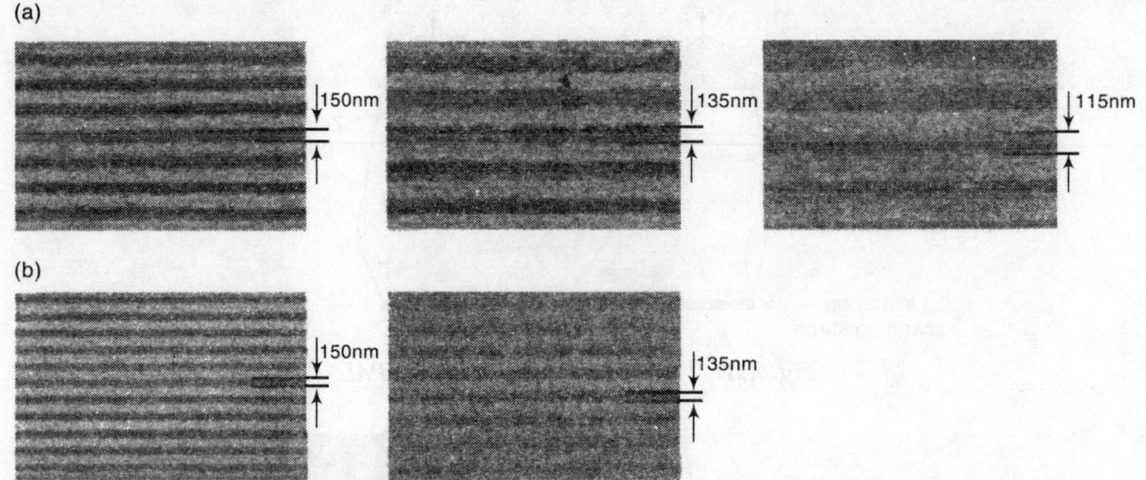

Fig. 4.39 *Scanning electron micrographs (SEM's) showing resolution of the immersion and non-immersion ZPAL systems at l = 400 nm. (a) The immersion zone plates reliably print dense lines/spaces down to 115 nm. (b) The non-immersion zone plates reliably print dense lines/spaces down to 135 nm.*

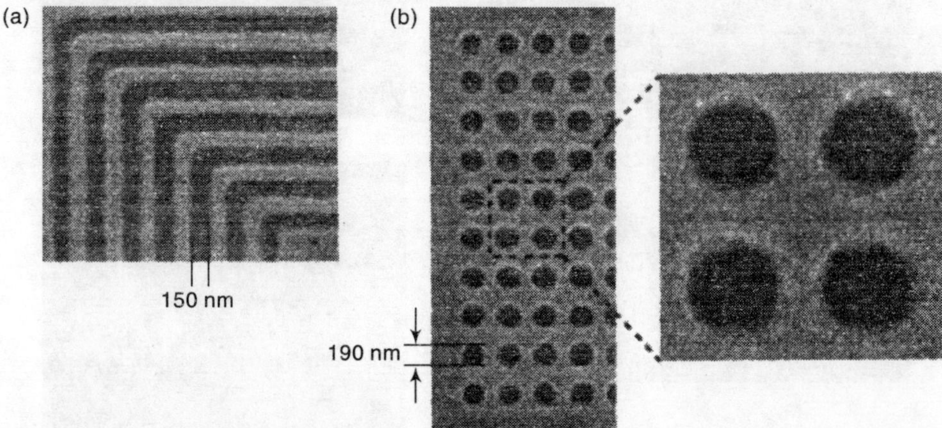

Fig. 4.40 *SEMS of (a) nested-L's and (b) arrays of contact holes, demonstrating the performance of the immersion ZPAL system.*

In addition of the resolution improvement, the immersion scheme improves ZPAL's exposure latitude and depth-of-focus, as is evidenced by the exposure window plots in Figure 4.41. The exposure window is quantitatively defined as the range of exposure dose and defocus values which produce linewidths that vary less than ±5% from the nominal linewidth (critical dimension). At focus, the immersion scheme possesses and exposure latitude of ~9.5% compared to ~5.9% for the non-immersion scheme. Furthermore, the exposure window for the immersion system remains fairly constant for defocus values upto ~200 nm compared to a more rapid decrease in the exposure window for non-immersion system.

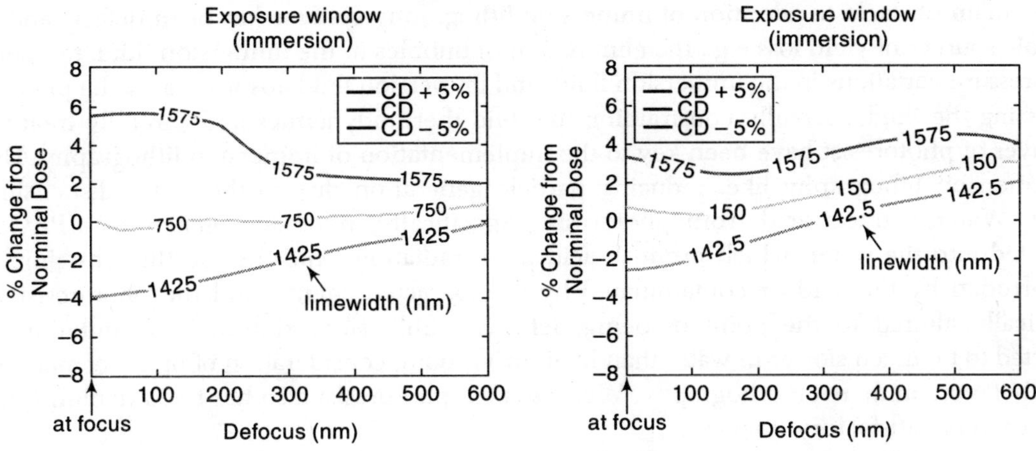

Fig. 4.41 *A comparison of the exposure windows for the immersion and non-immersion ZPAL systems. The area between the CD + 5% and Cd-5% curves defines the exposure window, which is clearly large for the immersion system.*

IMMERSION LITHOGRAPHY

Immersion lithography is a photolithography resolution enhancement technique that replaces the usual air gap between the final lens and the wafer surface with a liquid medium that has a refractive index greater than one. The resolution is reduced by a factor equal to the refractive index of the liquid. Current immersion lithography tools use highly purified water for this liquid, achieving feature sizes below 37 nanometers (Fig. 4.42).

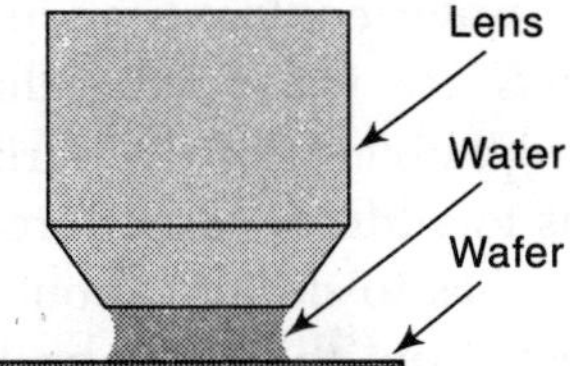

Fig. 4.42 *In immersion lithography, light travels down through a system of lenses and then a pool of water before reaching the photoresist on top of the wafer.*

The lenses in the highest resolution "dry" photolithography scanners bend light rays through a range of angles (numerical aperture) spanning nearly 90 degrees. As it is impossible to increase resolution by further bending, a "last-minute" wavelength reduction is possible by inserting an immersion medium between the lens and the wafer. The wavelength is reduced a factor equal to the refractive index of the medium. For example, for water at 193 nm wavelength, the index is 1.44.

The resolution enhancement from immersion lithography is therefore about 30–40% depending on materials used. The depth of focus, or tolerance in wafer topography flatness, is also 40–70% better (proportional to the refractive index of the imaging medium considered) than a corresponding "dry" tool at the same resolution.

Immersion lithography has the ability to extend resolution and depth of focus, between 65 nm and 45 nm nodes. This can also be achieved by double patterning on dry lithography tools. Since dry optical resolution is limited, the ability to control smaller feature sizes is more difficult.

The main obstacle to adoption of immersion lithography systems has been defects and other possible sources of yield loss e.g., the elimination of bubbles in the immersion fluid, temperature and pressure variations in the immersion fluid, and immersion fluid absorption by the photoresist. Degassing the fluid, carefully constraining the fluid thermodynamics and carefully treating the top layer of photoresist have been key to the implementation of immersion lithography. Defects in immersion lithography like. reducing particle generation due to the water dispensing are found. Water extracts acid from photoresist. Specifically, photoacid generators (PAGs) are extracted into the water, which produce acid upon radiation exposure. So the lens should not be corroded by the acid or contaminated by the extracted agents, and the photoresist is not chemically altered to the point of being defective. Still, since diffusion of contaminants is expected to be much slower in water than in air or vacuum, consideration of optics contamination actually favors immersion lithography. Water-soaked photoresist also has been demonstrated to produce very satisfactory images.

The above defect concerns have liked to consideration of using a topcoat layer directly on top of the photoresist. This topcoat would serve as a barrier for chemical diffusion between the liquid medium and the photoresist. In addition, the interface between the liquid and the topcoat would be optimized for watermark reduction. At the same time, defects from topcoat use should be avoided.

As a scanning speeds typically approach 50 mm/s for high-volume manufacturing, the actual resist-water contact time in any given exposures area is minimal. Hence the main concerns for defects are water left behind (watermarks) and loss of resist-water adhesion (air gap). The hydrophobicity of the surface and the water delivery/removal method are therefore the key areas to address. Other areas where defects may be enhanced are at the wafer edge, where the water has to do an "about-face" (reverse motion). It is important for the water not to pick up defects from the wafer backside.

Generally, implementation into manufacturing is only considered when defect yield reach a mature level, e.g., comparable to dry lithography levels.

Enhancements necessary to extend the technology beyond the 32 nm node are currently being investigated. Such enhancements include the use if higher refractive-index materials in the final lens, immersion fluid, and photoresist, in order to improve the resolution with single patterning. Currently, the most promising high-index lens material is luteium aluminum garnet, with a refractive index of 2.14. High-index immersion fluids are approaching refractive index values of 1.7. These new developments allow the optical resolution to approach ~30 nm. However, it is expected that at some point below 40 nm, current photoresists will limit further scaling. Polarization effects due to high angles of interference in the photoresist also have to be considered as features approach 40 nm. Hence, new photoresists will need to be developed for sub-40 nm applications.

Double patterning has received interest recently since it can potentially increase the half-pitch resolution by a factor of 2. This could allow the use of immersion lithography tools beyond the 32 nm node, potentially to the 16 nm mode. While double patterning improves pitch resolution, it still cannot dodge the need for new photoresist for improved single patterning resolution.

IMPRINT LITHOGRAPHY

Imprint lithography is a straightforward and reliable means of replicating patterns with feature sizes down to 20 nm, and potentially down to the single-molecule level. The imprint template consists of a thick, fused-silica blank coated with 100 nm of hydrogen silsesquioxane (HSQ), patterned by scanning-electron-beam lithography (SEBL). The patterned template is treated with alkyltrichlorosilane, which reduces the surface free energy of the template, and ensures that the template can be removed clearly after imprint. A planarization layer material is spun on a silicon wafer, followed by nanoliter droplets of a low-viscosity etch barrier fluid. The templates is leveled and brought into proximity with the substrate. Upon contact, the etch barrier fluid distributes itself by capillary action throughout the cavities in the template pattern. Low pressure is applied to the template, slightly compressing the etch barrier fluid, and allowing a three-point force measurement around the periphery of the template to ensure proper leveling. The gap between the template and substrate is maintained at ~100 nm, which allows freedom of movement of the template parallel to the substrate. Alignment is performed to the ~1-nm level using Fig. 4.43(a). Interferometric-Sparial-Phase Imaging (ISPI) technology.

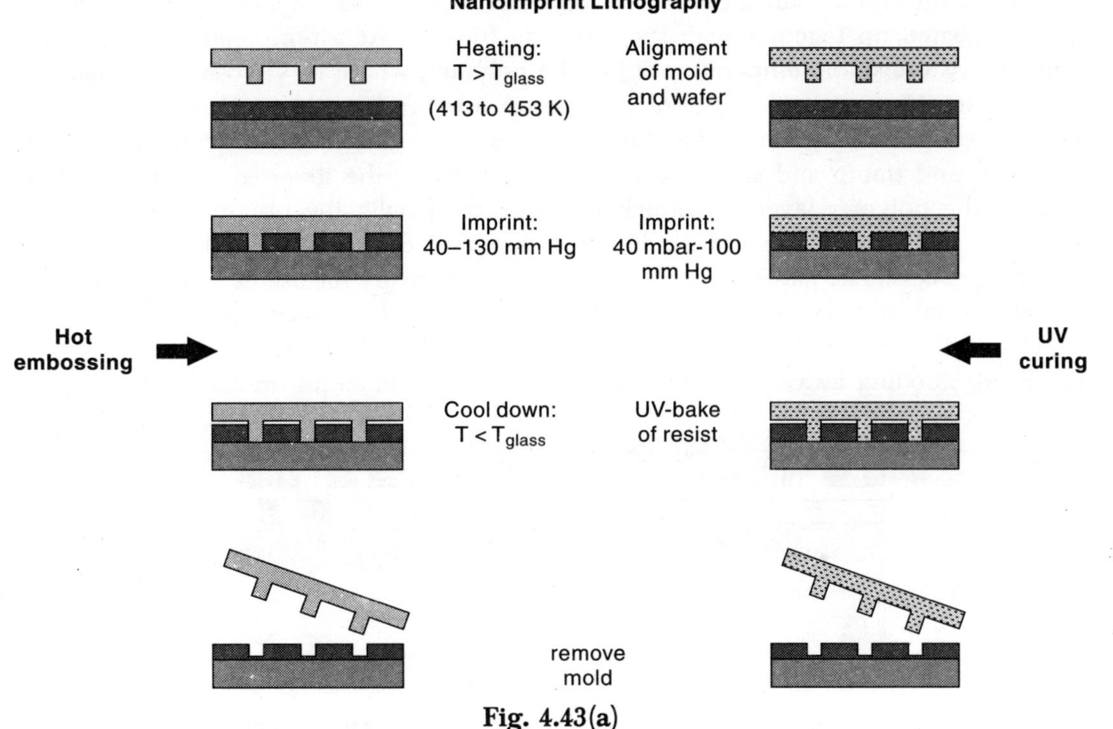

Fig. 4.43(a)

A brief exposure of the etch barrier fluid to ultraviolet light initiates crosslinking of nanometers in the fluid, after which the template is removed. The pattern is then transferred into the substrate with a bi-layer reactive-ion etch. The system can imprint on wafers up to 200 nm diameter, in field sizes up to 25 × 25 mm. Template leveling is controlled at three points to ~1 nm.

NANOIMPRINTING

Nanoimprinting stands for methods like embossing, printing, and modeling where definition of lateral pattern of a surface layer is mechanically performed. These techniques are characterized by the fact that a template (master, stamp) carrying the envisaged nanopattern is replicated on a thin surface layer on the substrate.

Some of these techniques are well known for patterning in the micrometer and sub-millimeter range typical for micromechanical devices and MEMS (micro electro mechanical system). The novel feature of nanoimprint is its application for nanometer patterning as well as its use as a lithography technique. Nanoimprinting means patterning of a thin layer, usually an organic material (polymer, resist) on top of a substrate. The patterned layer serves as a mask or enables mask definition for a subsequent etching step where the pattern is transferred to the substrate itself. Therefore in nanoimprint the template plays the role of the photomask in a conventional lithography process. This technique has been proposed first by and the term nanoimprint lithography (NIL) has become a generic term for all mechanical nano-patterning concepts. The process itself is a hot embossing process (*hot embossing lithography*, HEL). Its principle is explained in Fig. 4.44. A substrate, in most caves a silicon wafer, covered with a thin layer of thermoplastic polymer is heated up together with the template (stamp). At a temperature above the glass transition (glass transition temperature T_g) of the polymer, where its viscosity is sufficiently low to conform to the template, substrate and stamp are brought into contact and pressure is applied. As soon as the polymer has filled the stamp relief the stack is cooled to below the polymer's T_g and stamp and sample are separated, leaving the inverted pattern of the stamp frozen into the polymer layer as a thickness contrast. Finally the polymer remaining in the gaps, the residual layer, is removed in an anisotropic dry etch step, thus completing the lithography process. The patterned polymer layer is now ready for use as etch mask for pattern transfer to the substrate or as a mask for lift-off the patterned polymer layer is evaporated with metal. Due to the inhibited coverage of steep walls in an evaporation process the polymer can be dissolved, flooding away the metal on top of it. The remaining metal corresponds to the

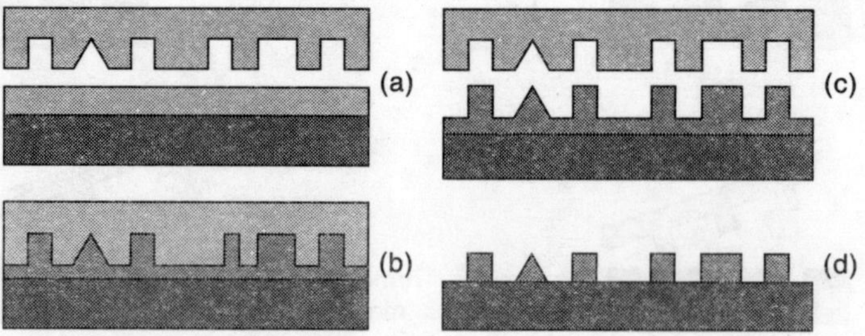

Fig. 4.44 *Principle of hot embossing lithography (HEL). Patterned stamp (template) and sample (substrate with spincoated polymer layer) are heated to process temperature (a) and brought into contact. As soon as the polymer has conformed to the stamp relief under pressure (b) the stack is cooled down. Separation of stamp and sample (c) is done below the glass temperature T_g, where the thickness contrast is frozen in the polymer layer. The residual layer remaining is removed in an anisotropic dry etch process (d) for opening of the mask windows.*

stamp pattern and may be directly used as an electrical wiring. Alternatively it may again serve as an etching mask for patterning of the substrate. Due to the superior mask selectivity of metals compared to a polymer this is the preferred technique for pattern transfer in the several nanometer range.

Stamps for nanoimprint made from silicon or from silicon with a patterned oxide or polysilicon layer on top are fabricated by conventional Si patterning technology. Stamps with nanometer patterns require complex lithography techniques and represent a major cost factor when large patterned areas are involved. For stamps with simple line patterns alternative techniques like holographic lithography or sidewall lithography (spacer technique) may be applied for stamp fabrication. In addition, low cost working stamps have successfully been prepared by hot embossing into a curable polymer layer on Si.

Nanoimprint is a parallel process, where the whole stamp area is replicated within one embossing step. Stamp areas of 4 diameter and more have been tested, where anti-sticking layers on the stamp improve large area imprint quality when densely patterned stamps are involved. Often self organised layers of the fluorinated siloxanes are utilized as anti-sticking layers, which can be applied from the liquid phase or gas phase. The latter is preferred in the case of fine patterns as it overcomes wetting problems in narrow trenches. Overall process times lie in the range of minutes due to the heating/cooling cycle and also due to the fact, that the polymer needs time to flow. Thus the process time strongly depends on the required maximum temperature (and therefore on the polymer's T_g and its molecular weight) and on stamp layout (pattern size and pattern density) hot embossing lithograhy demonstrates the ability to prepare patterns of a size of 6 nm.

Hot embossing. For hot embossing equipment available promises reproducible processing for areas of up to 6 inch diameter. Adjustment is done externally (optical alignment, clamping and transport to the imprint system), and applications on patterning with only one lithography level (patterned media for storage, sensors, nanofluidics, nano- and microsystems, biochemical systems) are also done. Based on production tools for microelectronics a step and stamp procedure is investigated and a combination with conventional optical lithography (mix and match) is tested for larger pattern sizes.

CONTACT LITHOGRAPHY

Contact lithography, also known as contact printing, is a form of photolithography whereby the image to be printed is obtained by illumination of a photomask in direct with a substrate coated an imaging photoresist layer.

The first integrated circuits had features of 200 micrometres which were printed using contact lithography. This technique was then substituted by proximity printing, where a gasp is introduced between the photomask and the substrate. Proximity printing had poorer resolution than contact printing (due to the gap allowing more diffraction to occur) but generated far less defects. The resolution was sufficient for down to 2 micrometre production. Then step-and-repeat projection system appeared which had a wide acceptance due to the reduction of the mask image.

Contact lithography is used in applications requiring thick photoresist and/or double sided alignment and exposure example: in advanced 3D packaging, optical devices, and micro-electromechanical systems (MEMS). In addition, the contact platform is the same as used in imprint processes.

Recently, two developments in contact lithography have made its comeback in semiconductor lithography. First, surface plasmon resonance enhancements including the use of silver films as lenses have been demonstrated to give resolution of less than 50 nm using wavelengths of 365 and 436 nm. Second, nanoimprint lithography is used for the semiconductor sector (e.g., hard-drive, biotechnology) and is also used for sub-45 nm semiconductor lithography, driving defect reduction practices and uniformity improvement for masks in contact with the substrate. Step-and-flash imprint lithography (SFIL), a popular form of nanoimprint lithography which involves UV curing of the imprint film, uses the same setup as contact lithography.

Operating Principle

Upon existing the photomask-photoresist interface, the image-forming light is subject to near-field diffraction as it propagates through the photoresist. Diffraction causes the image to lose contrast with increasing depth into the photoresist. This is explained by the rapid decay of the highest-order evanescent waves with increasing distance from the photomask-photoresist interface. The effect is partly mitigated by using thinner photorests. Contract enhancements based on plasmon resonances and lensing films have been disclosed. The chief advantage of control lithography is the elimination of the need for complex projection optics between object and image. The resolution limit in optical systems originates from the finite size of the final imaging lens and its distance from the image plane. More specifically, the projection optics can only capture a limited spatial frequencies spectrum from the object (photomask). Contact printing has no such resolution limit but is sensititve to the presence of defects on the mask or on the substrate.

Types of Contact Masks

There are several types of contact lithography masks.

The standard binary intensity amplitude mask defines dark and light areas where light is blocked or transmitted, respectively. The dark areas are patterned films consisting of chromium or other metal.

The light coupling mask has a corrugated dielectric surface. Each protrusion acts as a localized waveguide. Light is transmitted primarily through the protrusions as a result of this localized guiding effect. Since less contact area is needed, there is less potential for defects.

A hybrid nanoimprint-contact mask utilized both contact imaging and mechanical imprinting, and has been proposed to optimize imaging of both large and small features simultaneously by eliminating imprint residual layer issues.

Contact masks have traditionally been fairly large (>100 mm), but it is possible that alignment tolerances may require smaller mask sizes to allow stepping between exposures.

As in nanoimprint lithography, the mask needs to have roughly the same feature size as the desired image. Contact masks can be formed directly from other contact masks, or by direct writing (e.g., electron beam lithography).

As, thinner photoresist help in improving image contrast So reflections from the layer underlying the photoresist have to be taken into account when absorption and evanescent wave decay are reduced.

The resolution of contact lithography surpasses $\lambda/20$ periodicity.

The pitch resolution of contact lithography is enhanced by multiple exposures generating feature images between previously exposed features. This is suitable for nested array features, as in memory layouts.

Surface plasmons are collective oscillations of free electrons confined to metal surfaces. They couple strongly to light, forming surface plasmons polaritons. Such excitations effectively behave as waves with very short wavelength (approaching the x-ray regime). By exciting such oscillations under the right conditions, multiple feature appear in between a pair of grooves in the contact mask. The resolution achievable by surface plasmons polariton standing waves on a thin metallic film is <10 nm with a wavelength in the 380–390 nm range using a <20 nm silver film. In addition, deep narrow slits in metallic transmission gratings allows resonances to amplify light passing through the slits.

A layer of metal film, acts as a 'perfect lens' for amplifying the evanescent waves, resulting in enhanced image contrast. This requires tuning the permittivity to have a negative real part, e.g., silver at 436 nm wavelength. The use of such a lens allows imaging to be achieved with a wide tolerance of distance between mask and photoresist, while achieving extreme resolution enhancement by use of surface plasmon interference, e.g., a half-pitch of 25 nm with 436 nm wavelength. The perfect lens effect is only effective for certain conditions, but allows a resolution roughly equal to the layer thickness. Hence a sub-10 nm resolution is feasible with this approach.

The use of surface plasmon interference gives an edge over other lithography techniques, as the number of mask features are much less than the number of features in the desired image, making the mask easier to fabricate and inspect. While silver is the most commonly used metal for demonstrating surface plasmons for lithography, aluminum has also been used at 365 nm wavelength.

The resolution enhancement techniques allows 10 nm features to be contemplated. The most fundamental limitation appears to be photoresist roughness, which becomes predominant for shorter sub-wavelength periods where only the Zeroth diffraction order is expected to propagate. All the pattern details are conveyed by the evanescent waves, which decay more rapidly for finer resolution. So, the photoresist's inherent roughness becomes more significant than the pattern.

Defects

Defects are detrimental to contact lithography on surface contact, in two respects. First, a hard defect can widen the gap between the mask and the substrate. This can easily cause images based on evanescent waves or surface plasmon interference to disappear. Second, smaller, softer defects attached to the metal surface of the mask may not disturb the gap but can still alter the evanescent wave distribution or destroy the surface plasmon interference condition.

Oxidation of the metal surface also destroys plasmon resonance conditions (as the oxide surface is not a metal).

SOFT LITHOGRAPHY

Soft Lithography refers to a set of methods, for fabricating or replicating structures, using "elastomeric stamp, molds, and conformable photomasks" (see Fig. 4.45).

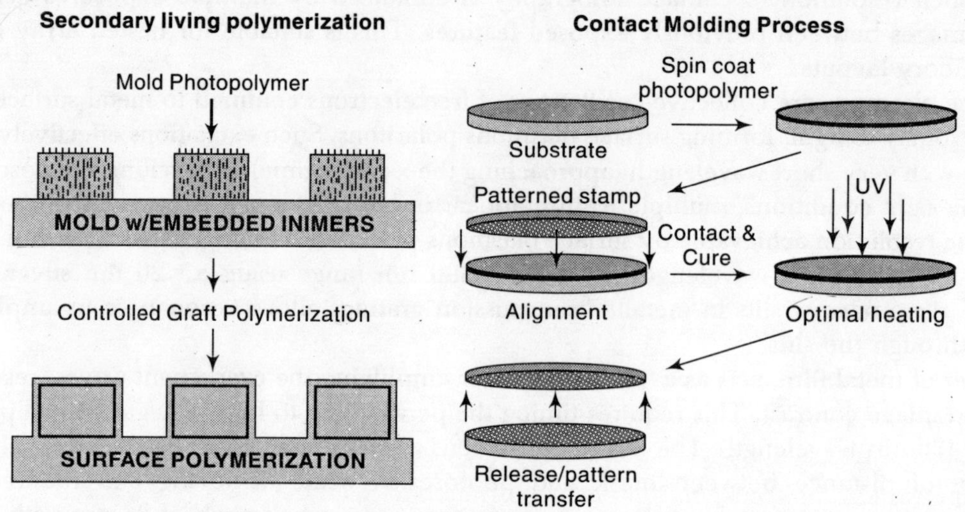

Fig. 4.45

It is called "soft" because it uses elastomeric materials. Soft lithography is used to construct features measured on the nanometer scale.

For top-down approaches like lithography or imprint molding, the problem is making reproducible features smaller than 50 nanometers (nm). For bottom-up approaches like self-assembly, the challenge is patterning large areas with exact positioning; and this is achieved by combining a molding step with a subsequent polymerization, i.e by including a molecule with both an initiator and a nanometer fragment for use in subsequent polymerization steps. Since some of the initial pattern's functionality remains exposed on the surface of the molded pattern, a second polymer can be directly synthesized onto the mold via living free-radical polymerization. Feature sizes between parts of the pattern can be reduced from a spacing 100 nm down to 20 nm.

Soft lithography includes the technologies of Micro Contact Printing (μCP), replica molding (REM), microtransfer molding (μTM), micromolding in capillaries (MIMIC) and solvent-assisted micromolding (SAMIM)) Platerning by etching at the nanoscale (PENs) one of the soft lithography procedures, which is as follows:

1. The steps of any of four favorite micro- or nano-scale lithography procedures (photolithography, EBL, etc.) are followed, to etch a desired pattern onto a substrate (usually silicon)
2. Next, the stamp is created by pouring a degassed resin overtop of the etched wafer. Common resins include PDMS and Fluorosilicone.
3. Removing the cured resin from the substrate, a stamp contoured pattern is acquired.

4. The stamp is then "inked" by placing it, (pattern-up), in a bath of inking solution (for example, ODT in ethanol) for a short period of time (Figure 4.46(a). The ink molecules will fall and adhere tc the surface of the stamp (Figure 4.46(b)) creating a single-molecule layer of the ink of the stamp.

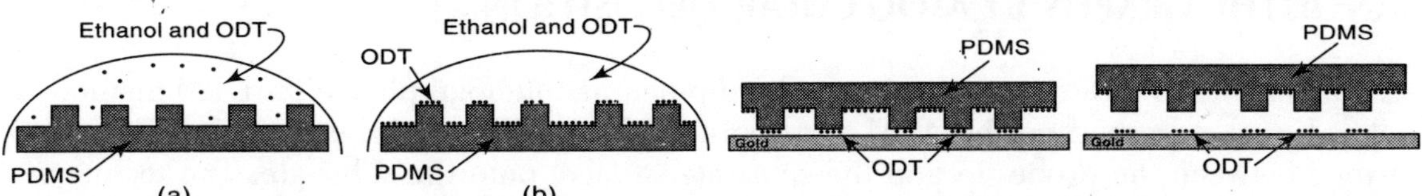

Fig. 4.46 *ODT from the solution settles down onto the PDMS stamp. Stamp now has ODT attached to it which acts as the ink.*

Fig. 4.47 *The PDMS stamp with the ODT is placed on the gold substrate. When the stamp is removed, the ODT in contact with the gold stays stuck to the gold. Thus the pattern from the stamp is transferred to the gold via the ODT "ink".*

5. The inked stamp is then pressed on the substrate and removed, leaving the desired single-molecule thick pattern on the substrate (Figure 4.47)
6. Steps 4 and 5 are repeated for each substrate on which the pattern is desired

Advantages

Soft lithography has some unique advantages over other forms of lithography (such as photolithography and electron beam lithography). They include the following

- Lower cost than traditional photolithography in mass production
- Well-suited for applications in biotechnology
- Well-suited for applications in plastic electronics
- Well-suited for applications involving large or nonplanar (nonflat) surface
- More pattern-transferring methods than traditional lithography techniques (more "ink" options)
- Does not need a photo-reactive surface to crate a nanostructure

DIP PEN NANOLITHOGRAPHY (DPN)

Dip Pen Nanolithography (DPN) is a scanning probe lithography technique where an atomic force microscope tip is used to transfer molecules to a surface via a solvent meniscus. The technique allows surface patterning on scales of under 100 nanometers. DPN is the nanotechnology analog of the dip pen (also called the quill pen), where the top of an atomic force microscope cantilevers acts as a "pen" which is coated with a chemical compound or mixture acting as an "ink," and put in contact with a substrate, the "proper".

DPN enables direct deposition of nanoscale materials onto a substrate in a flexible manner. The vehicle for deposition can include pyramidal scanning probe microscope tips, hollow tips, and even tips on thermally actuated cantilevers. Applications of this technology currently range through chemistry, materials science, and the life sciences, and include such work as ultra high density biological nanoarrays, additive photomask repair, and brand protection for pharmaceuticals.

NANOLITHOGRAPHY BY MOLECULAR DEPOSITION

Another way to create a nanopattern is by dip-pen nanolithography. With DPM method, ink molecules are adsorbed on the AFM and transferred by diffusion through the liquid meniscus (formed between the probe tip and the substrate surface) onto the substrate. The technique is illustrated in figure 4.48. The liquid can be water on another suitable solvent in which a material (the "pigment") is dissolved. The ink could also be in liquid form and require no solvent. Inks include organic compounds such as alkane thiols, dendrimers, polymers, or large biomolecules such as antibodies, proteins, or DNA.

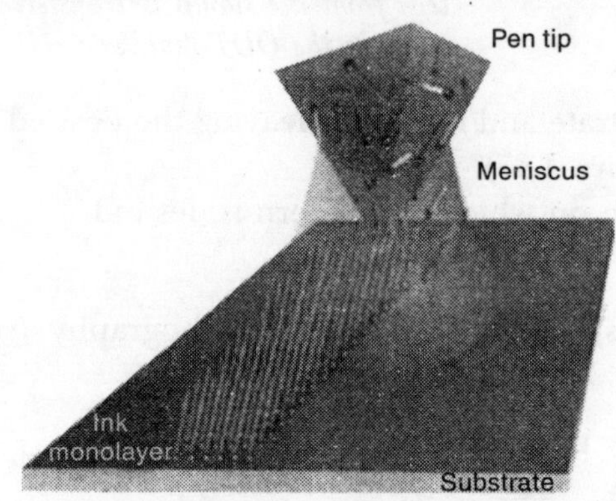

Fig. 4.48 *Illustration of the AFM probe being used to deposit molecules on a sample's surface. As the probe moves across the surface, molecules move down the probe and attach to the surface.*

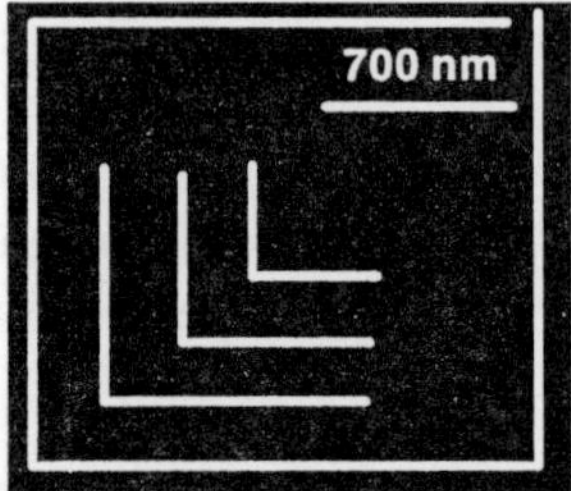

Fig. 4.49 *Lateral force microscope image of a pattern written with the DPM process. This pattern was written by an AFM that does not have XY calibrations sensors. As a result, at the right upper corner of the image two of the lines did not connect as intended.*

Figure 4.49 shows patterns formed by the DPN process a gold substrate using open loop control of the AFM tip. Note that the box drawn around the LLL pattern is not closed–the open loop control does not allow accurate location of the endpoint.

Advanced DPN experiments involve the overlay of patterned ink layers with the precise deposition of molecules.

AFM Lithography Instrumentation

Any atomic force microscope can be used for creating nanometersized patterns on a surface. However, the quality and complexity of the patterns depends on specific scanning probe hardware and software. The method of patterning is determined by the types of probes, substrates, and the specific software performance capabilities used to drive the scanning probe tool.

Hardware: The most critical hardware feature that is required is an X-Y calibration system. Because the piezoelectric ceramics used in AFM have unwanted characteristics such as creep and hysteresis, calibration sensors are necessary to guide the motion of the probe. Without the calibration sensors, the probe moves in an unpredictable motion, and it is hard to write complex patterns. As an example, in figure 4.49 the AFM do not have calibration sensors, and the lines at the upper right of the pattern is do not connect as intended. Conversely, Figure 4.50 is an illustration of a complex pattern drawn with the DPN process. Because the instrumentation has X-Y calibration sensors, a higher quality pattern is possible.

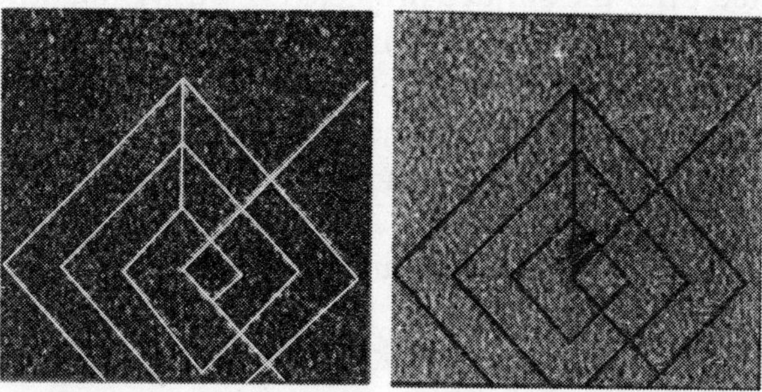

Fig. 4.50 *Lateral Force Microscope (LFM) images of patterns created using DPN. With X-Y calibration sensors the lines at the edges of the boxes intersect. Line widths vary from 60 nm–100 nm.*

Software: Software is used for defining the pattern that will be drawn with an AFM and then for drawing the pattern on a materials surface. Figure 4.51 shows a control window that may be used for creating patterns with an AFM. The pattern may be drawn on a section of the screen as a combination of dots and lines or it may be imported from a bit-map file. The software allows writing of the patterns by applying a specific force or a specified voltage. Also, the rate of scanning can be specified.

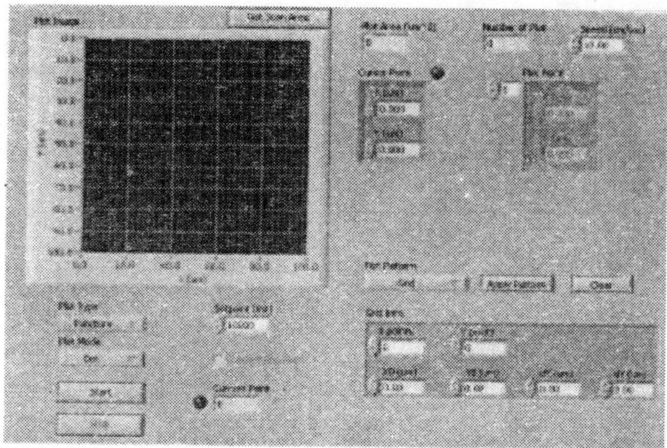

Fig. 4.51 *A typical control window for modifying surfaces with an AFM. Each of the button has a drop down menu that facilities selection of parameters.*

ATOMIC BEAM LITHOGRAPHY

Instead of beams of energetic ions, atomic beams can be used to induce chemical changes in resists or to the structure directly. A disadvantage is the lack of a possibility for focusing and guiding by electrical or magnetic fields due to the missing charge.

However, atomic beams can be focused by intensive light. Standing waves of laser light act as an array of convex lenses and focus the atomic beam. So gratings with a period of half a wavelength can be realized. For an additive structure generation by neutral atomic beams, the atoms are directly projected onto the surface-projected intensity maxima of a UV radiation. So line gratings of aluminum with a width of 70 nm and a distance of 155 nm have been obtained. Interference patterns that cross allow the generation of two dimensional grating structures.

Apart from directly structured deposition, nanotechnical masks techniques are also accessibly by atomic beams. A proximity radiation through a micro grating succeeded in the local removal of a molecular monolayer of helium atoms, and octadecylsiloxane layers on silicon were structured by neutral argon ions in a similar manner.

MULTILAYERS RESISTS

Multilayers resist processes are applied when a single resist layer cannot meet the requirements for the resist. One problem is the sharpness of the contours of projections into the resist layers of certain thickness, when the thickness cannot be minimized due to the relief on the substratre or projection required against etching processes. Very high numerical apertures are applied for the projective optics to enable high resolution, so that the depth of focus is significantly reduced. Even in the sub-micrometer range the height of the structure often exceeds the depth of focus, so that a high-resolution projection through out the resist is hampered. Thus either a planarizing step is integrated into the technology or resist has to fulfill the planarizing function for the lithographic process. (See Fig. 4.52)

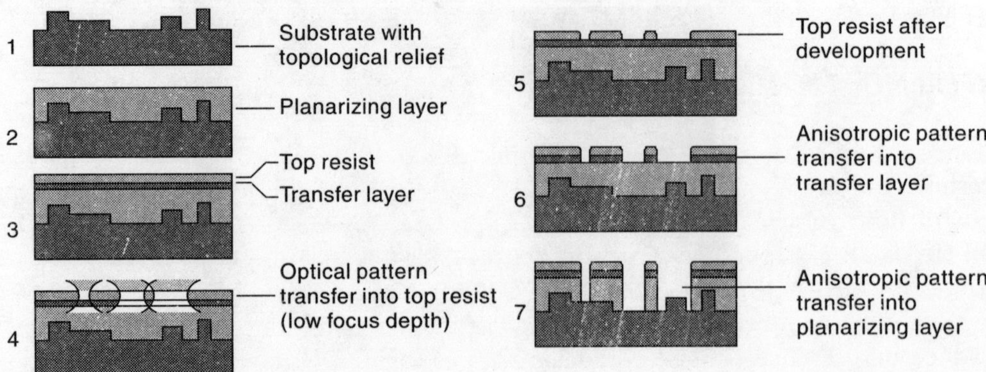

Fig. 4.52 *High-resolution photolithography by multilayer resist technology. Planarization of the topological relief ("bottom layer"), projection into a small layer volume with low focus depth in a plane, thin light-sensitive ("top resist") layer and pattern transfer using an etch-selective layer ("transfer layer")*

Multilayer resists provide a separation of functions: the actual collection of the optical image is achieved by a thin layer that is on top of a thicker layer, which both levels out the topological relief and protects the structural elements. Multilayer resist technology requires a precise and anisotropic pattern transfer from the primary structure of the thin upper layer into this planarizing subplayer, which is usually realized by dry etching. Although silicon-containing or other doted organic resist layer were developed to make them more resistant to an oxygen plasma, the precision and anisotropy of such two-layer resist systems are not sufficient for high quality lithography.

Therefore, triple-layers systems with an additional thin layer as the transfer masks are applied in most cases (Fig. 4.45). The thin intermediate layer ("transfer layer") is precisely structured under medium selectivity from the projected pattern in the upper layer. A material is chosen that allows for a highly selective etching of the planarizing layer compared with the transfer layer. Thereby the transfer layer is preserved in the subsequent structure transfer process. Hence precise resist structures with high aspect ratios are realized, which are also applicable for demanding etch processes down to the nanometer range.

GRATING

A grating is a simple constituting periodic lines and spaces. The grating structure is defined in a thin layer of light-sensitive polymer called photoresist–"resist" because of its resistance to chemical or plasma etching, which is often used to subsequently transfer the pattern into an underlying substrate. The spatial period of the grating is defined as the distance in which the line-space pattern repeats itself. *Nanoruler* is designed to pattern grating of high accuracy to detecting workpiece motion in precision tools) using a method known as optical encoding, with an accuracy that is some 10–100X better than laser interferometers. The Nanoruler utilizes a patterning method called *scanning-beam-interference lithography* (SBIL), that is capable of rapidly patterning large grating (>300 mm diameter) in only a few minutes. Such super-accurate gratings serve as optical encoder plates. Another important application for the Nanoruler is the patterning of nano-accurate gratings necessary for locking an electron beam using technique called spatial phase locked electron beam lithography (SPLEBL).

INTERFERENCE LITHOGRAPHY

Interference lithography (sometimes called holographic lithography) is new technique capable of patterning gratings with periods down to 150 nanometers and with distortions that are hundreds of times smaller than traditional techniques. Patterning on substrates well over one meter in size is also be possible.

Interference lithography is a widespread tool for making diffraction gratings. The principle of IL is straightforward. Figure 4.53 illustrates the concept. The coherent beams of light interfere and produce interference fringes–parallel planes of high and low light intensity. Imagine two expanding circular water waves, excited by the dropping of two stones in to a quiet pond. Where the two waves intersect, a stable interference pattern, called a standing wave, appears. These fringes are recorded in photoresist, which after wet development, become granting lines.

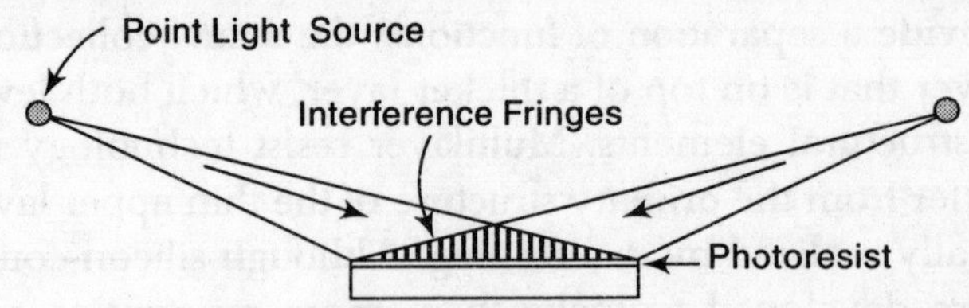

Fig. 4.53 *A schematic of the MIT interference lithography system.*

Chemical etching and other processes can then be used to transfer the granting structure permanently into a substrate.

The new interference lithography (IL) technique overcomes problems of granting fabrication methods such as mechanical ruling, interference lithography any scanning electron-beam lithography. By crossing and interfering two ultraviolet (UV) laser beams, periodic interference fringes result which are captured and imaged in a photosensitive coating.

The new IL technique causes narrow beams to interfere, creating a small granting "image". Large gratings are written by scanning a substrate under the image with a high-precision air bearing stage. This new technique is known as *scanning beam interference lithography* (SBIL) and it is the first-ever SBIL tool, the "Nanoruler," because it "rules" gratings by scanning the substrate, similar to the time-honoured "ruling" technique of scratching a diamond used as length scales. The SBIL method can be thought of as a hybrid of the older ruling and holography techniques, but with the added virtue of sophisticated electronic feedback that ensures long-range fidelity.

The IL process is extremely fast compared to ruling since al the grooves are formed simultaneously in a single exposure. The exposure times is typically below one minute. This significantly relaxes the requirements on environmental stability. The process is static, i.e., there are no mechanical moving parts, and the long coherence length of the laser determines the spatial coherence of the granting. Spectral defects called "ghosts," which are prominent in ruled gratings, are absent in interference gratings. The size and/or the period of a mechanically ruled granting is governed by diamond wear but, IL is capable of patterning meter-sized gratings with deep-sub-micron periods, about laser's wavelength, coherence length and power.

The early interference gratings did not have high diffraction efficiency, because of an apparent lack of control in shaping the groove profile. However, with the maturing of the fabrication technology, a large number of techniques can not be leveraged to properly manipulate the grating profile in order to yield the desired efficiency. For example, metallic reflection gratings have been produced through IL and that have a diffraction efficiency exceeding 95%. In other words, nearly all of the incoming light energy is concentrated in one diffracted beam.

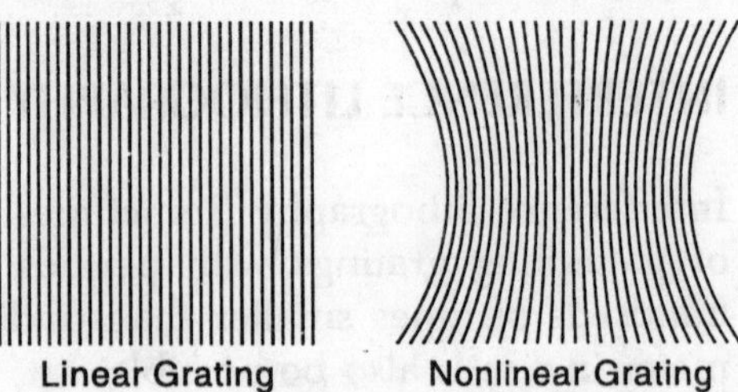

Fig. 4.54 *A linear grating vs. a nanolinear grating (schematic only). The grooves on the linear grating form perfect parallel straight lines, whereas those on the linear grating are a family of curves.*

Where IL fails to shine is in the area of producing large linear gratings with uniform diffraction efficiency. Figure 4.54 graphically depicts the difference between a linear and a nonlinear grating. The only way that traditional single-exposure IL can generate linear gratings is if the two interfering beams are plane waves.

Scanning Beam Interference Lithography (SBIL)

To produce gratings that are not only large but also perfectly linear and uniform, scanning beam interference lithography was invented. The Nanoruler was designed to pattern gratings with a precision of better than 1 nm across a 300 mm substrate.

SBIL is a hybrid of mechanical ruling and traditional interference lithography. Figure 4.55 demonstrates the SBIL concept in its skeleton. Two narrow UV laser beams, roughly 2 mm in diameter, interfere and produce a small patch of interference fringes. The fringes expose a small spot in the photoresist coating on a substrate. The substrate is affixed to a stage. As the stage moves along the fringe direction a strip of resist is thus exposed. This is called a scan. In order to build up a large grating, the stage must move so as to stitch together many scans. Because of its motion, and nanometer-accurate gratings, the SBIL tool has been named as the Nanoruler.

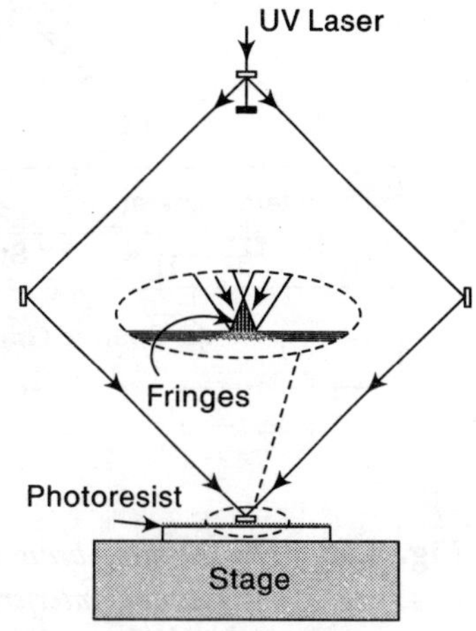

Fig. 4.55 *The SBIL concept.*

SBIL has advantages over ruling an IL. The small beams used in SBIL provide a major benefits in ease of obtaining small wavefront distortions and thereby highly linear fringes within the interference spot. By scanning the fringes, nonlinear distortions along the scan direction are averaged out. Overlapping adjacent scans leads to further and more significant smoothing. Instead of a single diamond tip, SBIL with thousand of "tips" in parallel, improves the system throughout, example, to create a grating 300 × 300 mm in size with a period of 400 nm, (assuming its tip stays intact!), has to run continuously for 50 days, whereas the Nanoruler can finish the job in around 20 minutes.

The scanning-electron and atomic-force microscopes are essential tools for the nanotechnology era.

Scanning electron-beam lithography (SEBL) is another popular tool for fabricating gratings. This method is also very slow and expensive, although not nearly as slow as mechanical ruling. In addition, gratings written by SEBL suffer form large distortions and errors due to beam deflection jitter and drift.

The concept of SBIL is to combine the sub-1 nm displacement-measuring capability of laser interferometry to control a high-performance air-bearing stage, with the interference of narrow coherent beams, to produce coherent, large-area, linear gratings and grids Fig. 4.56.

To produce gratings with sub-nm distortion over areas many tens of centimeters in diameter. SBIL require sophisticated environmental controls to mitigate the effects of disturbances such as acoustics, vibration, and air turbulence, and variations of temperature, pressure, and humidity. The system, also features real-time measurement and control of optical phase using heterodyne fringe detection, acousto-optic modulator phase locking and a high-speed digital signal processor (DSP) controller (see Fig. 4.57).

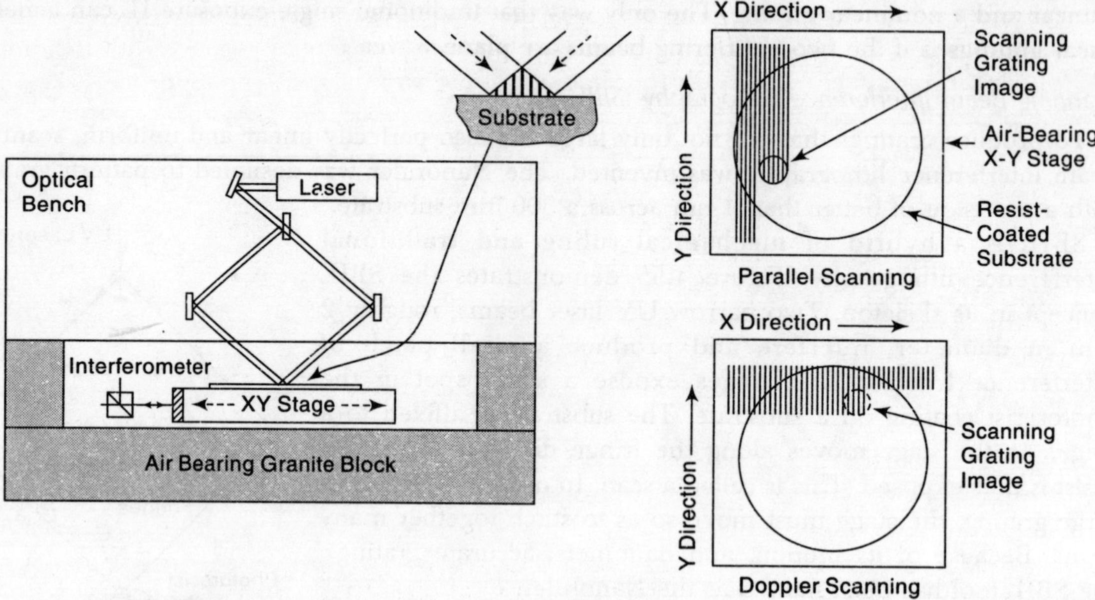

Fig. 4.56 *The scanning-beam-interference lithography (SBIL) system. A pair of narrow, low-distortion beams overlap and interfere at the substrate, producing a small grating "image." The substrate is moved under the beams, writing a large area grating. Tightly overlapped scans ensure a uniform dose.*

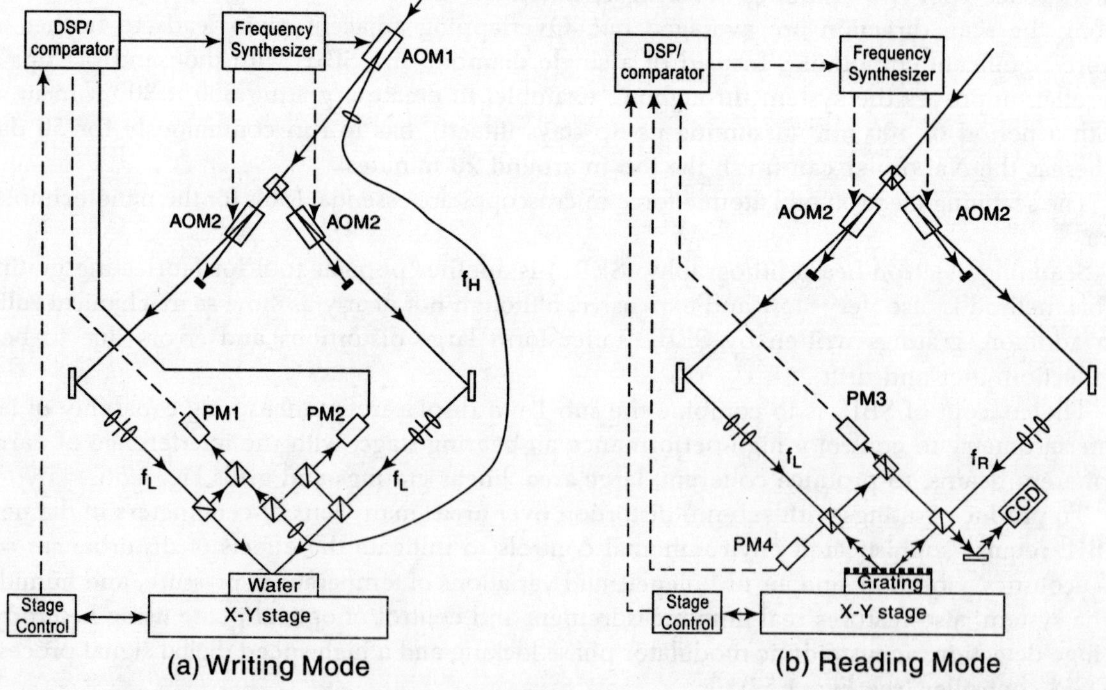

Fig. 4.57 *SBIL acousto-optic (AO) modulator phase locking system. Both writing and reading modes are depicted.*

An important feature of SBIL is the ability to both write and read gratings with nanometer control of grating phase. (~2 nm 3σ repeatability of the writing and reading process. Fig. 4.58 shows a photograph of a 300 nm-diameter wafer patterned with the Nanoruler.

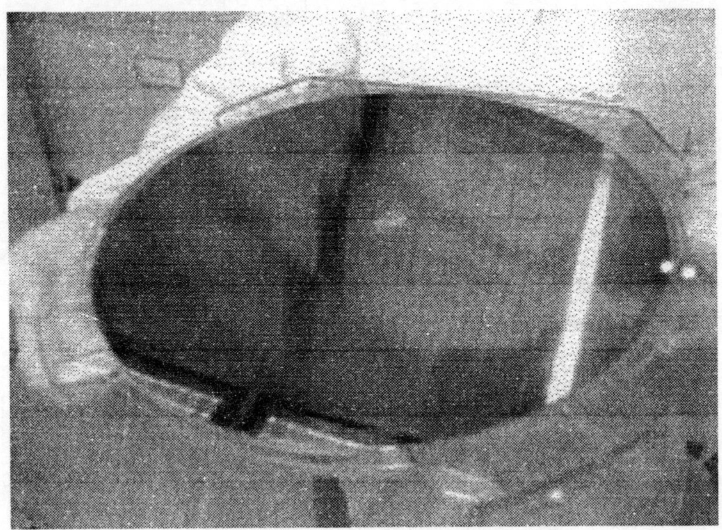

Fig. 4.58 *A 300 mm-diameter silicon wager patterned with a 400 nm-period grating by the Nanoruler. The grating is diffracting light from the overhead fluorescent bulbs.*

Interference Lithography for Patterning Variable-period grating

Scanning-beam interference lithography (SBIL) patterns large-area, linear, low-phase-distortion gratings with a pair of small diameter (millimeter size) phase-locked laser beams. With these a new system that generalizes the concept of phase-locked scanning beams for patterning continuously varying (chirped or quasi-periodic) patterns. These structure can subsequently be used to fabricate chirped X-ray reflection gratings for astronomical imaging applications, chirped fiber Bragg gratings for time delay or spectral filtering applications, and/or diffractive optical elements.

Figure 4.59 shows the experimental diagram of the variable-period scanning-beam-interference lithography (VP-SBIL) system. For controlling the grating period and orientation, the system employs dual-axis picomotor-driven gimbal mirrors to produce symmetric deflections of a pair of interfering beams around the optical axis without translation. Two objective plano-convex lenses (f# = 4.25, 2.12) are used in a 4-f optical configuration. Such a lens system allows the conjugate points of beam deflection (on mirrors) to overlap at the focal plane of the second objective lens. The spot size of image overlap is reduced to half the beam diameter as the ratio of focal lengths $f_1/f_1 = 0.5$. This relaxes the maximum period variation ($\Delta\Lambda$) constraint over the image diameter (D) that requires $\Delta\Lambda/\Lambda << \Lambda/D$ where Λ is the grating period.

To attain phase stability during grating patterning, homodyne fringe locking is adopted using an imaging detector, analog fringe-locker and a piezo-actuated mirror in closed-loop. In the present experimental configuration, two-axis beam rotation can generate any fringe orientation.

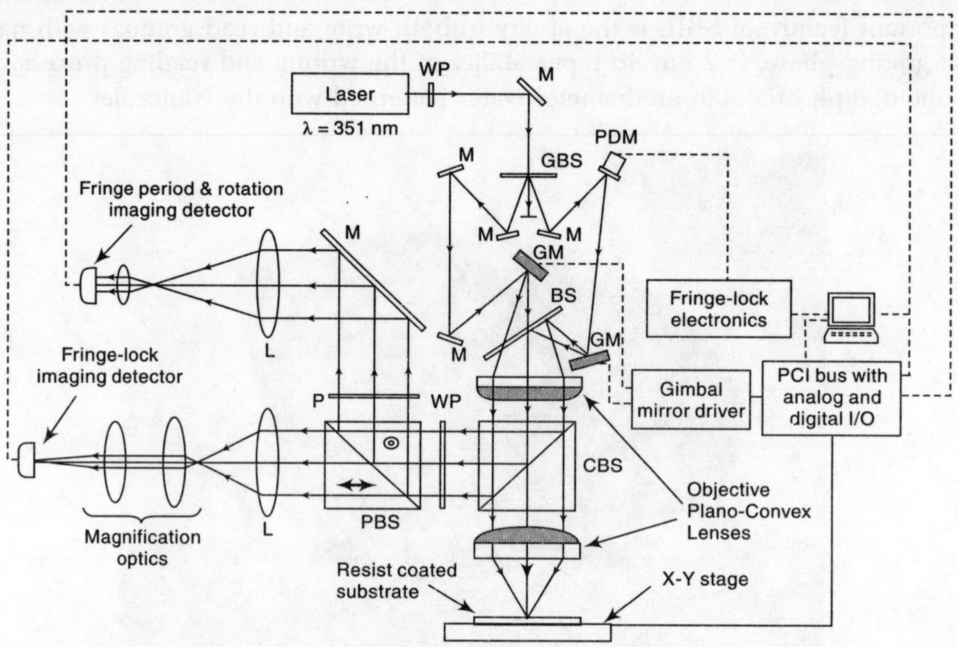

Fig. 4.59 *Experimental diagram of variable-period scanningbeam-interference lithography system. M: mirror, L: lens, P: polarizer, GM: gimbalmirror, WP: wave plate, BD: beam splitter, PDM: piezo-actuated mirror, GBS: grating BS, CBS: cubic BS, PBS: polarizing BS.*

However, variation in grating period (23 μm to 1000 μm) is limited by the range of deflection produced by the gimbal mirrors (± 10°) and by the numerical aperture (NA) of the lens system. Using position-sensitive detectors with an appropriate imaging and Fourier lens configuration, closed-loop beam steering is implemented to vary the grating period and orientation in a predetermined fashion. Typical requirements for x-ray reflection grating fabrication are $\Lambda_{ave} \sim$ 2 μm and chirp factor $\Delta\Lambda/\Lambda \sim 5\%$.

Fig. 4.60 shows two grating images of period 2.0 μm and 4.0 μm obtained on a static substrate by changing the angle between the beams using the picometer-controlled gimbal mirrors. Line uniformity in the image and indicates minimal fringe distortion over the entire beam overlap. The picometers can be constantly driven to write large-area gratings with continuously varying period and orientation on a substrate mounted to a precision X-Y stage. The piezo-actuated picomotors (which produce displacement jitter and exhibit low bandwidth operation) will be subsequently replaced by voice coil-actuated fast steering mirrors.

Nanofabricated Metal Transmission Grating

Wide-grid metal transmission gratings have many useful and novel optical properties that are enhanced *when the grating period and/or slit dimensions approach or go below the wavelength of light.* These benefits are particularly attractive in the UV and x-ray bands where the performance of conventional optics is poor. This benefit generally requires control of the grating period and achieving feature dimensions in the nanometer to picometer range, by *boosting grating transmission efficiency, reducing* the defect levels, and improving the control of grating feature geometry.

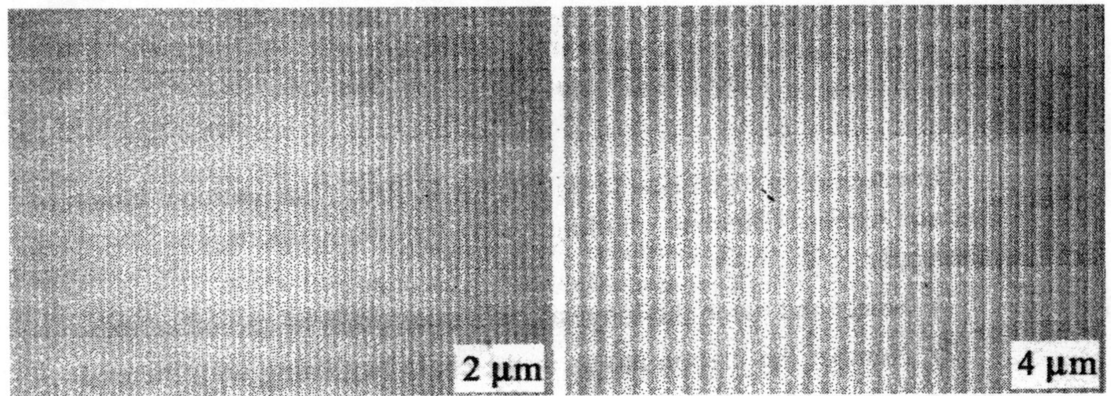

Fig. 4.60 *Optical micrographs of gratings written by VP-SBIL with period (a) 2.0 m and (b) 4.0 mm.*

Metal transmission gratings are generally fabricated with electroplated gold and supported by submicron-thick polyimide membranes or coarse meshes of electroplated gold or nickel. The thin and fragile grating members requires a carefully engineered coarse support mesh and metal frame to withstand the rigors of rocket launch and space environment. Transmission grating period down to 100 nm and sizes up to 30 × 30 mm have been fabricated. Grating patterning is performed by interference lithography (IL) using a variety of tri-level resists schemes, followed by reactive-ion etching and metal electroplating. Advanced IL tools in the world for high-yield fabrication of transmission gratings are also being developed.

Also, High-dispersion X-ray and extreme ultraviolet (EUV) transmission gratings are fabricated eg: *Solar EUV Monitor* (SEM) on the *Solar and Heliosperic Observatory* (SOHO) mission, the *Chandra* x-ray telescope, and the *Geostrationary Operational Environmental Satellites* (GOES N-Q) missions. The *Chandra* telescope provides high-resolution imaging and spectroscopy of X-ray-emitting astrophysical objects, with unprecedented power and clarity.

A scanning-electron micrograph of a 200 nm-period gold grating form *Chandra* mission is shown in Fig. 4.61. This granting is used in the High Energy Transmission Granting Spectrometer (HETGS) which provides high-resolution X-ray spectroscopy in the λ = 0.1-1.4 nm band. Period control of 40 picometers is required to meet telescope resolution requirements. Figure 4.62 is an example of an X-ray spectra obtained by *Chandra* using gratings.

Transmission gratings are used to block the intense Hydrogen Lyman Alpha (λ = 121.6 nm) deep-UV radiation obtained on IMAGE (Magnetospheric Imaging Medium Class Explorer).

Figure 4.63 depicts a 200 nm period atom nanofilter grating with 45 nm wide slots, designed to block deep-UV radiation. Slot widths need to be controlled to within a few nanometers for optimal UV blocking.

Grazing-incidence X-ray reflection gratings are an important component of advanced high-resolution spectrometers and other X-ray optics. These have traditionally been fabricated by diamond scribing with a ruling engine or by interference lithography followed by ion etching. These methods result in gratings which suffer from a number of deficiencies, including high surface roughness and poor groove profile control, leading to poor diffraction efficiency and large amounts of scattered light.

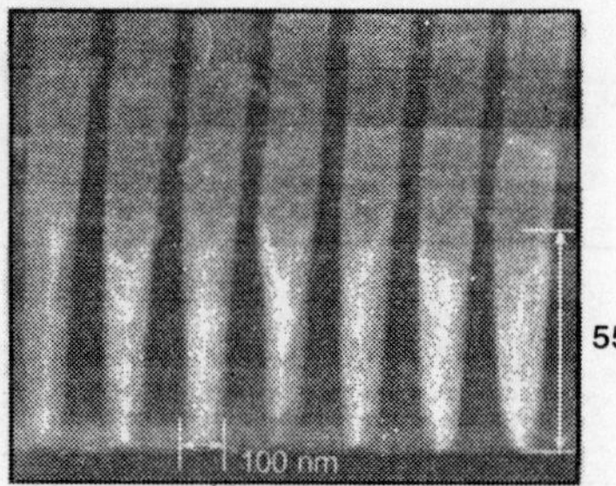

Fig. 4.61 *Scanning-electron micrograph of a 200 nm-period gold x-ray transmission grating used in the HETGS instruments on the Chandra Observatory, cleaved to show the grating line sidewals. The HETGS provides high-resolution x-ray spectroscopy in the λ = 0.1–1.4 nm band. The gold bars are 100 nm wide, or approximately 400 gold atoms.*

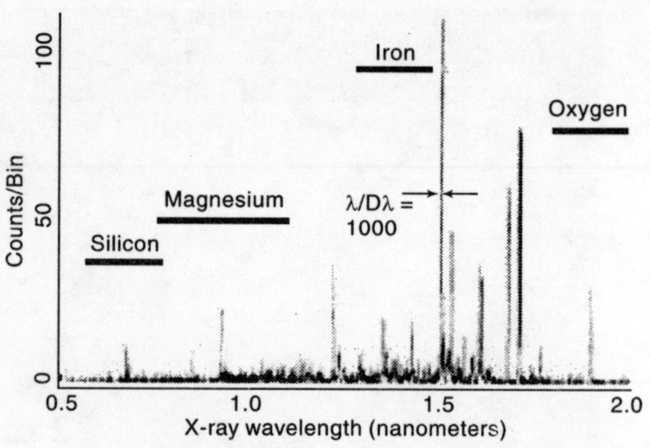

Fig. 4.62 *X-ray spectra of the binary star Capella obtained via gold diffraction grating on the Chandra x-ray astronomy satellite.*

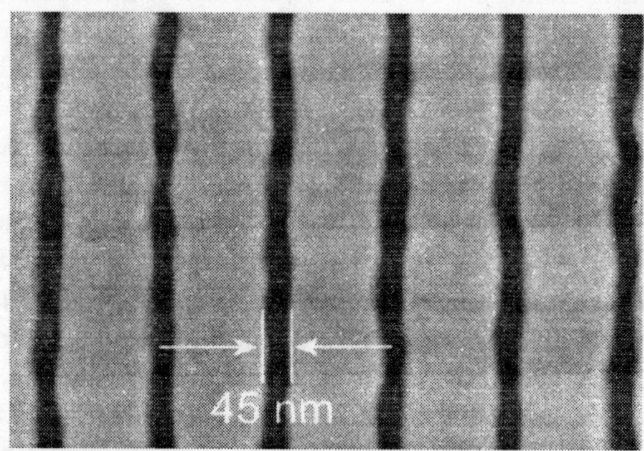

Fig. 4.63 *Scanning-electron micrograph of a deep-UV blocking grating used in atom telescope on the NASA IMAGE and TWINS missions. The grating blocks deep-UV radiation while passing energetic neutral atoms. Due to the narrow slow width of 45 nm and the large slot depth (~500 nm), the UV transmission is extremely low (~10^{-6} at λ = 121.6, while decreasing the transmitted atomic flux by only a factor of 10.*

Methods for fabricating blazed x-ray reflection gratings which utilize special (111) silicon wafers, cut 0.5-10 degrees of the (111) plane are developed. Silicon anisotropic etching solutions, such as potassium hydroxide (KOH), etch the (111) planes very slow compared to other crystallographic directions, resulting in the desired super-smooth groove surface. Similar off-cut (111) silicon substrates to fabricate blazed diffraction gratings, but utilized a second KOH etch step that compromised the grating facet flatness and is unsuitable for small grazing angle X-ray diffraction.

Gratings are patterned using interference lithography with the λ = 351.1 nm wavelength, and transferred into substrates using tri-level resist processing, reactive-ion etching (RIE), and silicon-nitride masking during KOH etch. The narrow (~100 nm) ridge of silicon which supports the nitride mask is removed using a novel chromium lift-off step followed by a CF_4 RIE. The result is extremely smoothed sawtooh patterns, which, after applying a thin evaporative coating of Cr/Au, are suitable for X-ray reflection (see Fig. 4.64).

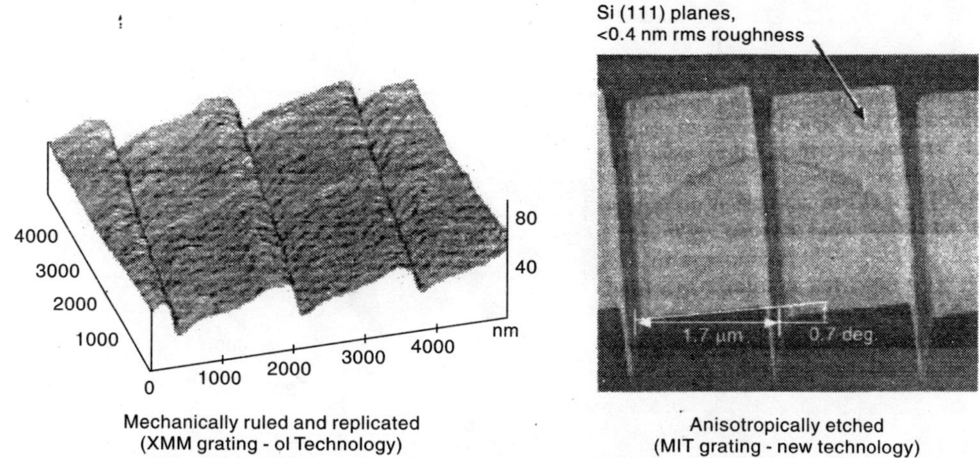

Fig. 4.64 *(a) An AFM image of a traditional mechanically-ruled and replicated x-ray reflection grating, (b) An AFM image of a blazed x-ray reflection grating fabricated by anisotropic etching of special off-cut (111) silicon wafers. Note the improvement of groove surface flatness and smoothness, leading to significantly improved performance.*

Presently, UV and thermal nanoimprint lithography for gratings is developed which demonstrate the extremely high fidelity of the nanoimprint replication process, where added surface roughness is < 0.2 nm (see Fig. 4.65).

Grating-based Devices

Optical telecommunications have widespread application in the field of Bragg gratings. A Bragg grating is formed by creating a periodic corrugation of refractive-index modulation in an optical wavelength. Such a structure behaves as a wavelength selective filer, reflecting a narrow band of wavelengths while transmitting all other wavelengths.

A combination of several different types of lithography to generate Bragg-grating devices is used. In interference lithography, two coherent laser beams are crossed, generating a standing-wave interference pattern. This standing-wave pattern is used to expose photoresist, yielding a coherent submicron-period grating. This grating can be used directly as the device grating or it can serve as a precision reference for subsequent electron-beam lithography steps. For devices that require long Bragg gratings with engineered phase shifts or variations in grating strength, we use a technique called spatially-phase-locked e-beam lithography (SPLEBL), which combines the long-range spatial coherence of interference lithography with the flexibility of scanning e-beam lithography. Gratings written by standard e-beam lithography contain inherent pattern-placement errors that limit device performance. SPLEBL reference the interference-generated grating during the e-beam exposure to minimize these placement errors.

Figures 4.66 and 4.67 show a variety of grating-based devices.

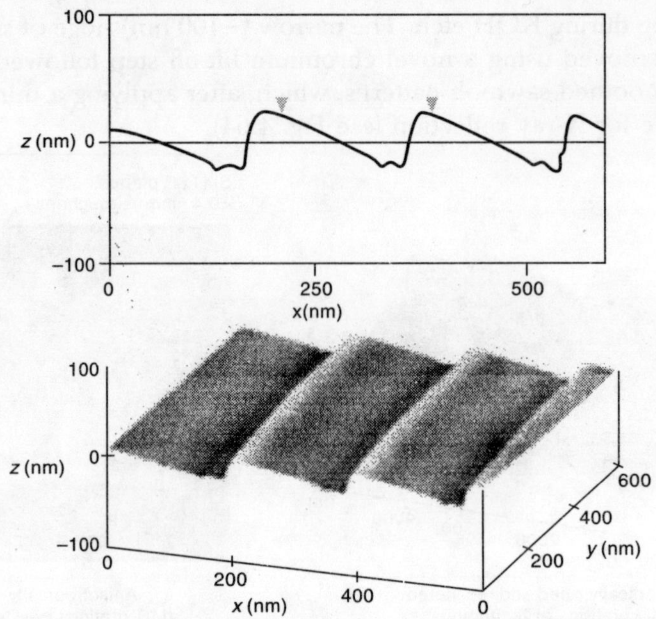

Fig. 4.65 *AFM image of 200 nm-period thermal-nano-imprint grating with 7° blaze angle. The groove surfaces are extremely smooth with a RMS surface roughness of < 0.2 nm.*

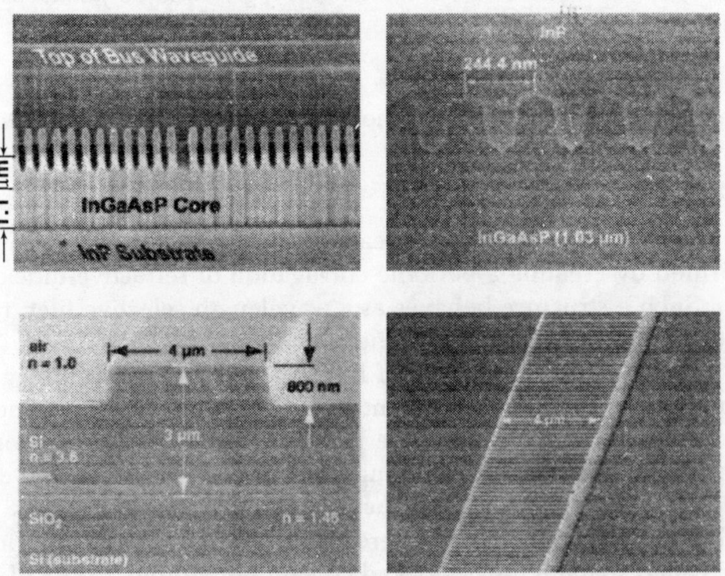

Fig. 4.66 *Extremes of Bragg-grating filters in two different materials systems. (a) Scanning-electron micrograph depicting a quarter-wave-shifted, 244.4 nm period Bragg grating etched into the top surface of an InGaAsP waveguide, and the subsequent lnP overgrowth. (b) Silicon-on-insulator (SOI) ride waveguide cross-section and SOI waveguide with Bragg grating in the top.*

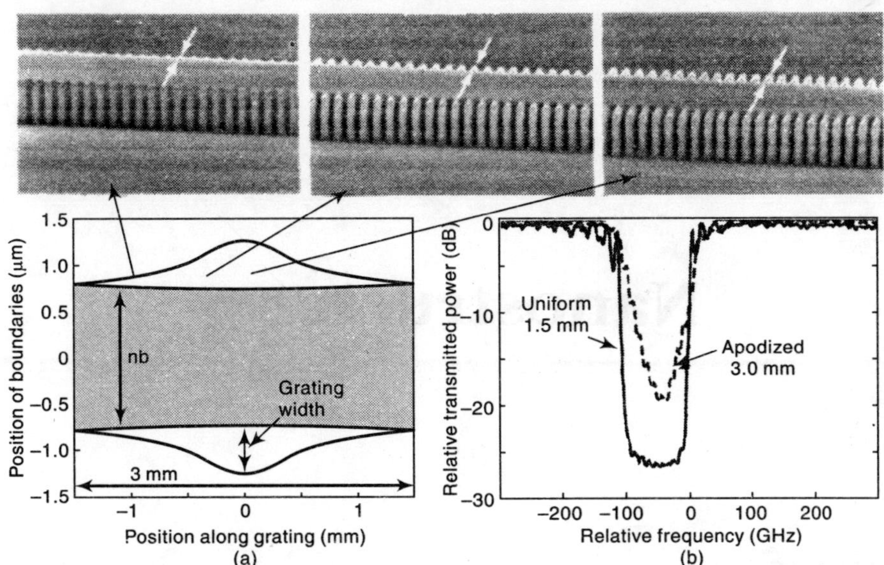

Fig. 4.67 *(a) Apodized Bragg gratings fabricated in the sidewalls of a SOI ridge waveguide. (b) Transmission spectra for the TE-mode of uniform and adopized waveguide-grating devices. The reduction of side-lobe levels for the apodized devices is readily apparent.*

5

Nanostructures

NANOMATERIALS

Nanomaterials is the study of how materials behave when their dimensions are reduced to the nanoscale. Materials characterised by structures whose dimensions are of the order of 10^{-9} meters or nanometer are called nanomaterials. Nanomaterials possess properties which are intermediate between those of the atoms which constitute a material and the bulk material. It is possible to precisely tune material properties simply by controlling their growth by adding atoms and ensuring the resulting nanostructures posses exactly the required number of atoms example, the electronic and magnetic properties of a wide range of materials. It will be also possible to undertake atom-by-atom construction of complex nanostructures which possess defined properties.

Materials reduced to the nanoscale can suddenly show very different properties compared to what they exhibit of a macroscale enabling unique applications. For instance, opaque substance become transparent (copper); inert materials become catalysts (platinum); stable materials turn combustible (aluminum); solid turn into liquids at room temperature (gold); insulators become conductors (silicon).

A unique aspect of nanotechnology is the vastly increased ratio of surface area to volume present in many nanosccale materials which opens new possibilities in surface-based science, such as catalysis. A number of physical phenomena become noticeably pronounced as the size of the system decreases. These include statistical mechanical effects, as well as quantum mechanical effects, for example the "quantum size effect" where the electronic properties of solid are altered with great reductions in particle size. This effect does not come into play by going them macro to micro dimensions. However, it becomes dominant when the nanometer size range is reached.

Nanosize powder particles (a few nanometres in diameter, also called nanoparticles) are potentially important in ceramics, powder metallurgy, the achievement of uniform nanoporosity and similar applications. The strong tendency of small particles to form clumps ("agglomerates") is a serious technological problem that impedes such applications. However, a few dispersants such as ammonium citrate (aqueous) and imidazoline or oleyl alcohol (nanoqueous) are promising additives for deagglomeration.

Nanocrystals Fundamentals

The electronic and optical properties of molecules and crystals are determined by the constituent atoms, their interactions (bonding) and their spatial arrangement. The evolution from single atom to crystal is a consequence of the increasing strength of the interactions and of the number of constituents. Two main states of the matters can be individuated: clusters (few atoms) and crystals (many atoms).

In clusters, as the cluster size increased, its properties can be described in terms of particle size and shape but not with the number of atoms and their spatial configuration. When clusters assume nanometer dimensions, they are named nanosystems or quantum systems. When nanosystems are ordered domains of atoms, the word *nanocrystals* or nanocrystallites is widely used.

Low dimensional systems may be classified according to their dimensions as quantum wells (two-dimensional confinement), quantum wires (one-dimensional confinement) and quantum dots (quasi zero-dimensional confinement). Quantum wells and quantum wires posses a translational symmetry in one or two dimensions, and a statistically large number of electronic excitations can be created.

Each nanocrystal has inhomogeneously broadened absorption and emission spectra due to distribution of sized, shape fluctuations, environment imhomogeneities and other features. Hence, the study the optical properties of nanocrystal forms a strongly interdisciplinary field. The most studied nanocrystal systems are semiconductors. The materials considered are II-VI (CDSe, CdTe, CdS), III-V (GaAs, InAs), I-VII (CuCl, CuBr, AgBr) compounds and group IV (Si and Ge) semiconductors.

The interest for the optoelectronic properties of Si rose after the observation of photoluminescence at room temperature in porous Si (PS).

In comparison to PS, Si nanocrystals embedded in amorphous silica (a-SiO_2) are better candidates for optoelectronic nanophotonic applications, because of their robustness and stability and their better compatibility with the mainstream Complementary Metal Oxide Semiconductor (CMOS) technology. Si nanocrystal can be produced by different technique either directly from a gas phase or indirectly by recrystallization within a matrix: ion implantation of Si in SiO_2, laser ablation gas evaporation, sputter deposition and Si/SiO_2 multilayers formation (Figure 5.1). The common feature of recrystallization techniques are the deposition or formation of sub-stoichiometric silica films, with a large excess of Si, followed by a high temperature annealing. The annealing causes a partial phase separation between the two constituent phases, Si and SiO_2, with the formation of small Si nanocrystals. The size and density of the Si nanocrystals can be controlled by the deposition and the annealing parameters. The degree of phase separation depends on the annealing temperature and time. A fine control of size distribution has been achieved by annealing of amorphous Si/SiO_2 superlattices, also an almost monodispersed size distribution has been demonstrated.

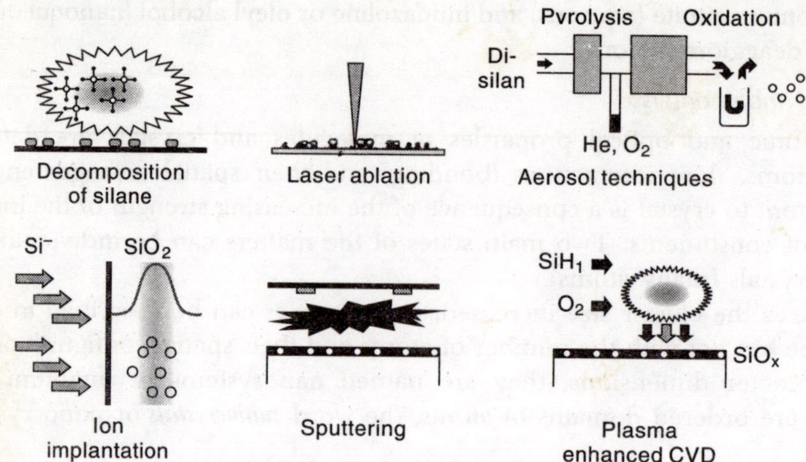

Fig. 5.1 *Sin nanocrystals fabrication techniques. The three lower techniques produce nanocrystals in a SiO_2 matrix.*

Nanocrystalline Materials

Included here are ceramics, metals, and metal oxide nanoparticles. These are a class of materials with a nanometer-sized microstructure. These materials are assembled from nanometer-sized building blocks, mostly crystallites. The building blocks differ in their atomic structure, crystallographic orientation, or chemical composition. In the crystallites, incoherent interfaces are formed between them, depending on the atomic structure, the crystallographic orientation, and the chemical composition of adjacent crystallites. These materials are microstructurally heterogeneous, consist of crystallites and are in the regions between adjacent grain boundaries. It is this heterogeneous structure on a nanometer scale that is crucial for many of their properties and distinguishes them from glasses, gels, etc. that are microstructurally homogeneous.

Grain boundaries make up a major portion of the materials at nanoscale, and strongly affect properties and processing. The properties of nanoscale materials deviate from those of single crystals (or coarsegrained polycrystals) and glasses with the same average chemical composition. This deviation results from the reduced size and dimensionality of the nanometer-sized crystallites as well as from the numerous interfaces between adjacent crystallites.

Nanocrystallites of bulk inorganic solids exhibit size dependent properties, such as lower melting points, higher energy gaps, and nonthermodynamic structures. In comparison to macro-scale powders, increased ductility is observed in nanopowders if metal alloys. In addition, quantum effects from boundary values become significant leading to quantum dots lasers.

One of the primary applications of metals in chemistry is their use as heterogeneous catalysts in a variety of reactions. In general, heterogeneous catalyst activity is surface dependent. Due to their vastly increased surface area over macro-scale materials, nanometals and oxides are ultra-high activity catalysts. They are also used as desirable starting materials for a variety of reactions, especially solid-state routes. Nanometals and oxides are also widely used in the formation of nanocomposites. Aside from their synthetic utility, they have many useful and unique magnetic electric, and optical properties.

Processing Methods

Some processes of manufacturing nanocrystaline materials are given below:

Starting phase	*Techniques*	*Nature of product*
Vapour	Inert gas condensation	3D
	Physical vapour deposition-evaporation and sputtering	1D
	Plasma processing	
	Chemical vapour deposition	3D
	Chemical reactions	3D
Liquid	Rapid solidification	3D
	Electrodeposition	1D, 3D
	Chemical reactions	
Solid	Mechanical attrition	3D
	Devitrification	3D
	Spark erosion	3D
	Sliding wear	3D

I. *Plasma processing:* A non-transferred arc thermal plasma reactor is used to vaporize coarse metal powders. A supersonic nozzle is used to quench the powder, and the powders are collected in a filter system that can be sealed and dismantled in a glove box. A drawback of the process is oxidation during fine powder handling.

II. *Combustion process:* The combustion process utilizes oxidizers and fuels. In controlled environments, high temperature is generated by the exothermic redox reactions between decomposition products of the oxidizer and the fuel. Owing to the fast heating and cooling, there is nucleation of crystallites but little growth, resulting in nanocrystalline ceramics. Some of the process features are as below.

(a) It is a versatile process leasing to the synthesis of single phase, solid solutions, and composites as well as complex compound oxide phases.
(b) It uses cheap raw materials.
(c) It is a scalable and high production rate process.
(d) The products are of high purity and loosely agglomerated, resulting in high sinterability.

III. *Mechanical alloying:* A high-energy ball milling process is used to produce nanostructured powders.

IV. *Supercritical fluid:* The production of nanostructured particles or their films depends upon the substantial change in solubility of a solute in a solvent, at the supercritical point (temperature and pressure) of the solvent.

V. *Sol–gel method:* Processing technique that can be scaled up to mass production are as follows.

(a) Powder synthesis techniques–Magnetic nanocrystals are synthesised as free-standing powders or nanoencapsulates that are compacted to form a bulk alloy with nanocrystalline grains. An example is carbon as synthesis which is the primary method used in nanoencapsulated, nanotubes and endohedral fullerenes. This technique allow for the production of metal or metal carbides nanoparticles with protective coatings.

(b) Rapid solidification processing–Amorphous alloys to be produced by a variety of rapid solidification processing routes with cooling rates > 10^4 K/s for eutecric alloys.

(c) Solidification processing of bulk amorphous alloys–Bulk amorphous alloys are formed by more conventional solidification routes at slower cooling rates.

(d) Crystalline of amorphous precursors–Bulk alloys with nanocrystalline grains can also be produced by solid state (crystallisation) of an amorphous precursor.

Some of the methods are given below in details.

VI. *Crystallization from the amorphous phase:* One of the popular methods use din producing nanocrystalline materials is by first making at amorphous phase of appropriate composition. This amorphous phase is then subjected to certain elevated temperature, above the primary crystallization temperature, in controlled time. Doing so yields nanocrystalline grains with amorphous intergranular material. By and large, the crystallization process is done by annealing in an inert gas furnace. However, there are few exception: Nanocrystalline Finemet of ≈ 15 nm nanocrystallite size has been produced by laser annealing of the amorphous precursor. This technique offers some flexibility as compared to conventional furnace-annealing techniques. The use of flash annealing of a meltspun. Finemet precursor has also been reported. Very fine nanocrystalline grain size achieved by this process has resulted in enhanced softy magnetic properties.

VII. *Rapid solidification techniques:* High enough cooling rated during solidification can be achieved when two important requirements are satisfied. The solidification rate, *R* during ideal cooling is related to the section thickness, *z* (in mm) through the relation: $R = 10^4 z^{-2}$, suggesting that the solidification rate increases by two orders of magnetic for a decrease in section thickness by one order of magnitude. Hence, it is imperative that the molten metal be delivered in a stream which is thin enough in at least one dimension and has high surface areas to volume ratio to allow rapid heat removal. Second to rapidly remove heat, contact area between the melt and the cooling medium needs to be maximized by rapidly increasing the liquid alloy surface area. This could be achieved by spreading the melt as thin layer on a substrate, or by physically disintegrating the melt into small droplets by atomisation. It is obvious that the molten metal can be delivered in the form of either droplet, cylindrical stream or ribbon stream and the melt stream can be cooled by gas, liquid, or solid.

Rapid solidification methods could be classified into three broad categories as follows.

(a) *Spray methods:* The melt is fragmented into droplets prior to quenching. Powder is produced by this method. The mechanism of atomisation and the means of cooling can be different in different techniques. Some atomisation methods are by gas, water, ultrasonic, rotating, soluble gas, electrohydrodynamic, spark erosion, and twin roll means.

(b) *Chill methods:* The continuity of the melt is preserved up to and during quenching. This is achieved by bringing the melt into contact with a chilling substrate. Various methods is involve injecting the melt into a die cavity, forming the melt into a thin selection by forgoing between a hammer or piston and an anvil, extruding the melt on to a chilled surface or extraction of the melt by contact with a rotating disc. Some chill methods are by using dies, piston and anvil, twin roller quenching, free fight melt spinning, chill block melt spinning, planar flow casting, and melt overflow.

(c) *Surface methods:* The material acts as a heat skin–rapid melting and solidification take place at a limited depth at the surface. This method has also been referred to as self-quenching, laser annealing, laser glazing etc. An extremely high power density of laser or electron beam, typically of the order of 10^6 W/cm^2 is concentrated on the alloy surface. There are two ways of performing tis surface treatment. One is where the surface is rapidly scanned in a direction perpendicular to the beam, while other is where the material is not moving. Cooling rates of 10^6–10^8 K/s have been reported.

As an example, consider the production of the amorphous soft magnetic materials precursor, which can be in bulk form, patterned magnetic arrays or thin substrate form. This amorphous cobalt based alloy stripes are generally manufactured by the liquid quenching method which resorts to the single roll technique. The extremely small thickness of less than 4.8 μm resulting from the substrate method notably enhances soft magnetic properties such as permeability and core loss in the high frequency range. Additionally, magnetic cores as electromagnetic apparatus can be produced from thin Co-based amorphous alloy strips. Once of the technical challenges faces is that of making the layers thinner than 5 μm. These thin strips contain relatively numerous pinholes because they entrain bubbles during the reduction of plate thickness and, therefore, pose problems of practicability as well as adaptability for high frequencies.

Iron-based alloys are also of considerable interest. Production of thin soft magnetic alloy strips of Co-based alloys, comprises the following steps: (a) ejecting molten alloy through a nozzle onto the surface of a rotating cooling member, and (b) rapidly quenching the alloy thereby producing thin amorphous alloy. It is also possible to produce thin Co-based amorphous alloy strips of thickness less than 4.8 μm.

The Co soft magnetic material is used in noise filters, saturable reactors, miniature inductance elements for abating spike noise, mains transformers, choke coils, zero-phase current transformers, magnetic heads etc., i.e., devices which are expected to exhibit high levels of permeability at high frequencies.

The method for the production of thin soft Fe-based magnetic alloy strips consists of rapid quenching of the molten alloy and heat-treating the quenched alloy strip at a temperature exceeding the crystallization temperature of the alloy. Specifically, materials of high permeability are effectively used in various magnetic parts such as current sensors in zero-phase current transformers and noise filters. In the case of a noise filter, for example, a switching power source is widely used as establishing powder source for electronic equipment devices. In the switching power source, adoption of a measure for the abatement of noise constitutes an important task. High-frequency noise having as switching frequency as its basic frequency and noise in the MHz range issuing from as load such as, for example, the logic circuit of a personal computer poses a challenge.

VIII. *Sol-gel synthesis as a processing method:* Sol-gel synthesis is widely used due to some advantages such as those listed below.

(i) Produces materials (both metals and ceramics) at low temperatures (around 573 k *vis-à-vis* about 1773 k for conventional techniques).
(ii) Able to produce in large a quantities (commercially viable) relatively cheaply.
(iii) Can synthesize many materials.
(iv) Co-synthesize two or more materials simultaneously.
(v) Coat one or more materials onto other materials (metal or ceramic particulates, and three-dimensional objects).
(vi) Produce extremely homogeneous alloys and composites.
(vii) Synthesize ultra-high purity (99.9999%) materials.
(viii) Tailor the composition very accurately right from the early stages of the process, because the synthesis is actually carried out at atomic level.

The sol-gel process is able to control the microstructure of the final products accurately and thus control the physical, mechanical, and chemical properties of the final products.

IX. *Powder injection moulding*

(a) Powder injection moulding is the combination of metal injection moulding (MIM) and ceramic injection moulding (MIM) (Lall 1992). Powder moulding is a productive and widely used technique shaping plastics. However, there are several properties of metal and ceramics, such as high strength, high stiffness, high operating temperature, which have strong advantages over polymer. Besides, metals and ceramics exhibit electrical, magnetic and thermal properties not possible with polymer. Thus, powder injection moulding (PIM) has been used for the production of metallic and ceramic products, exploiting benefits that have been proven over the decades in plastic injection moulding.

Power injection moulding process consists of four major steps, namely feedstock, injection moulding, debinding and densification as shown in figure 5.2. In the first stage, a feedstock is compounded from the powder and binder system. The second stage requires the aid of an injection machine to produce the "green" moulds. Removal of the binder yields the third stage. Sintering and finishing which is the final stage in PIM gives the final product.

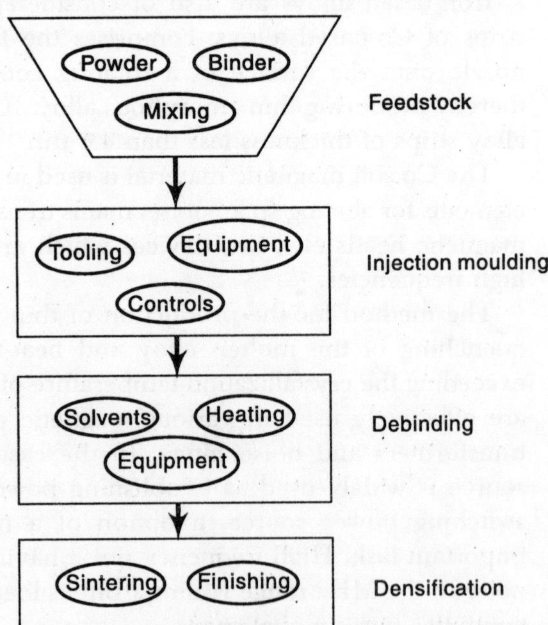

Fig. 5.2 *Flow chart for the PIM process showing the four major process steps.*

As a technology, PIM has been around for some time. There are five factors that impact on the selection of PIM for any application as shown in Figure 5.3. There are a large selection of materials, complex

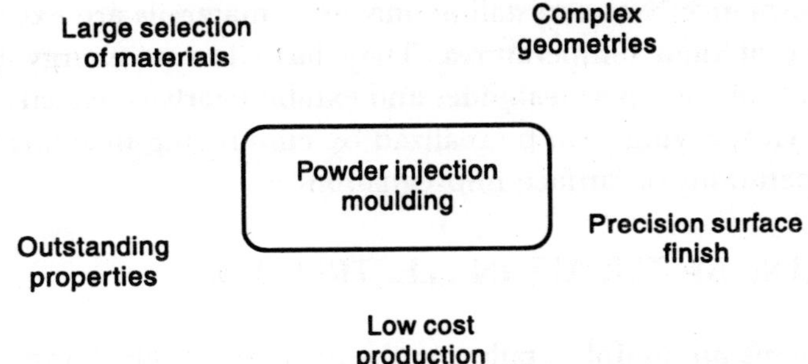

Fig. 5.3 *Five important factors for the selection of PIM.*

geometries, surface finish, outstanding properties and relatively low cost production. Using PIM to make products with magnetic properties is one of the key applications of PIM. The next section will give a brief description of powder injection moulding of conventional magnetic materials.

(b) *Powder injection moulding of conventional magnetic materials*

A large number of amorphous alloys have been prepared by various rapid solidification techniques. Among amorphous alloy systems, iron- and cobalt based amorphous alloys have been used in application of magnetic materials. Powders for rapid solidification are produced mainly by inert gas atomization and centrifugal atomization. Both processes produce spherical particles in the size range 20 to 100 μm in diameter by employing cooling rates up to 10^6 K/s. However, for nanostructured magnetic materials, the particle size should range form 1–100 nm.

Conventional magnetic materials used in powder injection moulding can be amorphous. Amorphous magnetic materials, which impose less obstacles to movement of magnetic domain boundaries and have low hysteresis losses, and high permeability, have been developed for various electronic devices such as sensors and motors.

Although power injection moulding of conventional magnetic materials can produce complex shape, good surface finish and good dimensional control, because of the amorphous phase structure, it cannot obtain other advantages which nanostructured magnetic materials possess, as will be highlighted in the next section. For example, sensors made by nanocrystalline materials can exhibit improved sensitivity compared to sensors made form amorphous materials. Advantages and applications of powder injection moulding of nanostructured magnetic materials will also be presented in the following section.

(c) *Powder injection moulding of nanostructured magnetic materials*

By the use of powder injection moulding technique with nanostructured magnetic materials, the following advantages can be obtained:

(a) High shape complexity: PIM can produce more complex parts than either investment casting or traditional press and sinter techniques.
(b) High density: PIM parts produced with out feedstock will have density levels between 97.5 and 99.5%. These levels typically exceed traditional press and sinter techniques, and are comparable to investment casting.

(c) High performance: Nanocrystalline magnetic materials are exceptionally strong, hard, and ductile at high temperatures. They have lowest energy losses (narrowest B/N hysteresis loop), high permeabilities and exhibit nearly or exactly zero magnetostriction.
(d) Low cost: Cost savings can be realized by eliminating machine operations, secondary coining operations or surface impregnation.

NANOCRYSTALLINE MATERIALS IN ELECTRONICS

Nanostructured systems are useful in tailoring the magnetic, optical and electronic properties of materials. It is obvious that nanostructured materials will have an increasing impact on electronics, since the drive for increasing smaller dimension in electronics translates into the requirement for high functionality, increased memory density, and higher speed. The hope of utilizing effects obtained with an additional tenfold reduction in dimension drives the study of nanoscale devices based on new operating principles. Advantageous manipulation of the properties of materials can be made by systematic nanoscale variation of composition.

Quantum effect devices or single electron devices are of great potential utility for future electronic circuits. For example, one class of devices operates on the interference effect typical of wave phenomena in modifying electron transport. These devices tend to be analogs of microwave or optical devices such as waveguides, modulators, or interferometers. All of these devices rely on the elimination of scattering in order to preserve the phase of the electron wave. The requires combination of highly ordered dimensions. Below are some examples of practical applications:

Next Generation Computer Chips

With convergences as one of the main drivers in the microelectronics, industry, smaller and better performance electronic devices are undoubtedly requires. A significant reduction in size of a microprocessor means a considerable increase in the operating speed. However, researchers are faced with several technological impediments to these advancements, namely, lack of the ultrafine precursors to manufacture these components; poor dissipation of the tremendous amount of heat generated by these microprocessors due to faster speeds; and poor reliability (short mean time to failure). Ultra-high purity, nanocrystalline starting materials can help the industry to overcome these barriers with better thermal conductivity, and longer lasting, durable interconnections.

Phosphors for High Definition TV

Better resolution of television screens could be achieved by reducing the size of the phosphors, which make up the pixels. Nanocrystalline zinc selenide, zinc sulphide, carmium sulphide, and lead telluride synthesized by the sol-gel technique are noteworthy candidates for improving the resolution of monitors. It was envisioned that through the use of monophosphors, the cost of these displays could be made much more affordable so as to enable them to be purchased by an average household.

Low Cost Flat-panel Displays

A huge market in the portable computer industry is represented by flat-panel displays. Japan has become the market leader primarily because of its research and development efforts on the materials for such displays. Nanocrystalline phosphors again play a major part in enhancing the resolution of these display devices. In addition, much higher brightness and contrast could be expected from flat panel displays constructed out of nanomaterials due to their enhanced electrical and magnetic properties.

Large Electrochromic Display Devices

There is a class of materials in which an optical adsorption band can be introduced, or an existing band can be altered by the passage of current through the materials, or by the application of electric field. An electrochromic device utilizes such materials. Nanocrystalline materials such as tungsten oxide gel are used in very large electrochromic display devices. The reaction governing electrochromism, which basically is a reversible coloration process under the influence of an electric field, is the double-injection of ions and electrons, which combine with the nanocrystalline tungstic acid to form a tungsten bronze.

These devices could be used in public billboards and ticker boards to convey information. They are quite similar to LCD displays, and they change colour when a voltage is applied. When the voltage polarity is reversed, the colour is bleached. The resolution, brightness, and contrast of these devices are greatly dependent on the tungstic acid gel grain size, and this is where nanotechnology comes into play.

NANOSTRUCTURES

Nano-structures are materials with size features on the order several nanometers. These could be: quasi-2 dimensional (e.g. quantum wells), quasi-1 dimensional (e.g. nanowires), quasi-O-dimensionl (e.g. quantum dots), porous materials with nanosized pores (e.g. aerogels).

Their small size makes them interesting for optical, electrical chemical, mechanical, thermal, and magnetic properties. These materials are flexible as their properties depend strongly on their size.

The properties of nanostructured materials are controlled by size effects; changes of the dimensionality of the system, changes of the atomic structure, and alloying of components (e.g. elements) that are not miscible in the solid and/or the molten state.

Also, the volume of an object decreases as the third power of its linear dimensions, but surface area decreases as its second power. For example the power of a drill (or any other machine) is proportional to the volume, while the friction of the drill's bearings and gears is proportional to their surface area. For a normal-sized drill, the power of the device is enough to handily overcome any friction. However, scaling its length down by a factor of 1000, for example, decreases its power by 1000^3 (a factor of a billion) while reducing the friction by only 1000^2 (a factor of "only" a million). Proportionally it has 1000 times less power per unit friction than the origional drill. if the origional friction-to-power ratio was, say, 1%, that implies the smaller drill will have 10 times as much friction as power. The drill is useless.

This is why, while super-miniature electronic integrated circuits can be made to function ,the same technology cannot be used to make functional mechanical devices in miniature; the friction overtakes the available power at such small scales. So while one may see microphotograph of delicately etched silicon gears, such devices are curiosities with limited real world application, for example in moving mirrors and shutters.

Surface tension increases in the same way, causing very small objects to tend to stick together. This makes any kind of "micro factory" impractical: even if robotic arms and hands are scaled down, anything they pick up will be impossible to put down. Thus, molecular evolution has resulted in working cilia, flagella, muscle fibers, and rotary motors in aqueous environments, all on the nanoscale. These machines, however, exploit the increase of the frictional forces at the micro or nanoscale.

Classification of Nanostructures

As indicated above, a reduction in the spatial dimension, or confinement of particles of quasiparticles in a particular crystallographic direction within a structure leads to changes in physical properties of the system. Hence, the classification of nanostructured materials and system depends on the number of dimensions which lie within the nanometre range, as shown in Figure 5.4:

(a) systems confined in three dimensions
(b) systems confined in two dimensions
(c) systems confined in one dimension.

Nanoparticles are also quasi-crystalline, i.e. the atoms are packed together in an icosahedral arrangement and show non-crystalline symmetry characteristics. Such quasi-crystals are stable at the nanometre or, at most, the micrometre scale.

Nanoparticles may be present within another medium, such as nanometer sized precipitates in a surrounding matrix material. These nanoprecipitates have a specific morphology (e.g., spherical, needle-shaped or plate-shaped) and possess certain crystalographic orientation relationships with the atomic arrangement of the matrix depending on the nature of the interface which leads to coherency strains in the particle and the matrix, example self-assembled semiconductor quantum dots. The connectivity of such nanometre-sized regions as, in the case of a nanoporous material, is also important.

Systems confined in one dimension, or quasi-2D systems, include discs or platelets, ultrathin films on a surface and multilayered materials; the films themselves could be amorphous, single-crystalline or nanocrystalline.

System confined in two dimensions, or quasi-1D systems, include nanowires, nanoroads, nanofilaments and nanotubes: again these could either be amorphous, single crystalline or polycrystalline (with nanometre-sized grains). The term 'nanoropes' is often employed to describe bundles of nanowires or nanotubes.

Nanoparticles and nanopores exhibit three-dimensional confinement. In semiconductor terminology such systems are called quasi-zero dimensional, as the structure does not permit free particle motion in any dimension.

Table 5.1 gives examples of nonostructured systems which fall into each of the three categories described above. It can be argued that self-assembled monolayers and multilayered Langmuir-Blodgett films represent a special case in that they represent a quasi-2D system with a further nanodimensional scale within the surface film caused by the molecular self-organization.

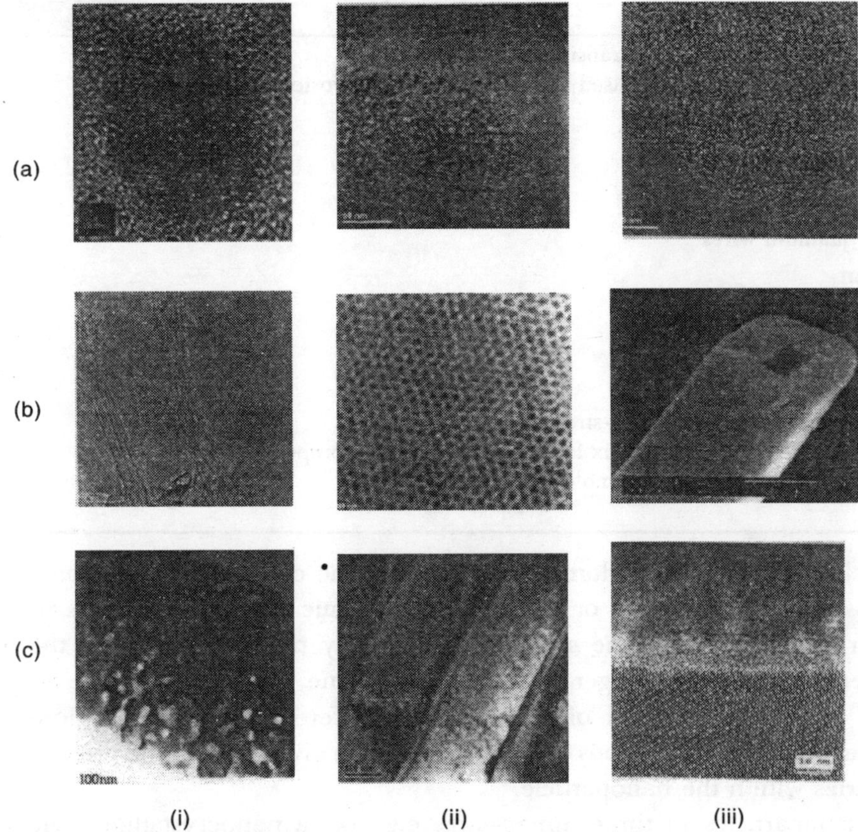

Fig. 5.4 *Calassification of nanostructures*

(a) Nanoparticles and nanopores (three dimensions): (i) high-resolution TEM image of magnetic iron oxide nanoparticle, (ii) TEM image of ferritin nanoparticles in a liver biopsy specimen, and (iii) high-resolution TEM image of nanoporosity in an activated carbon.

(b) Nanotubes and nanofilaments (two dimension): (i) TEM image of single-walled carbon nanotubes prepared by CD, (ii) TEM image of ordered block copolymer film, and (iii) SEM image of silica nanotube formed via templating on a tartaric acid crystal.

(c) Nanolayers and nanofilms (one dimension): (i) TEM image of ferroelectric thin film on an electrode, (ii) TEM image of cementite (carbide) layers in a carbon steel, and (iii) high-resolution TEM image of glassy grain boundary film in an alumina polycrystal.

Table 5.1 *Example of Dimensionality Systems*

1D confinement
Nanolaminated or compositionally modulated materials
Grain boundary films
Clay plates
Semiconductor quantum well and superlattices
Magnetic multilayers and spin valve structures
Langmuir-Blodgett films

(*Contd.*)

(*Contd.*)
Silicon inversion layers in field effect transistors
Surface-engineered materials for increased wear resistance or corrosion resistance
2D confinement
Carbon nanotubes and nanofilaments
Metal and magnetic nanowires
Oxide and carbide nanorods
Semiconductor quantum wires
3D confinement
Fullerenes
Colloidal particles
Nanoporous silicon
Activated carbons
Nitride and carbide precipitates in high-strength low-alloy steels
Semiconductor particles in a glass matrix for non-linear optical components
Semiconductor quantum dots (self-assembled and colloidal)
Quasi-crystals

Nanoparticles may have a random arrangement of the constituent atoms or molecules (e.g., an amorphous or glassy material) or the individual atomic or molecular units may be ordered into a regular, periodic crystalline structure which may not necessarily be the same as that which is observed in a much larger system. If crystalline, each nanoparticle may be either a single crystal of itself composed of a number of different crystalline regions or grains of differing crystallographic orientations (i.e., polycrystalline) giving rise to the presence of associated grain boundaries within the nanoparticle.

Consider nanoparticles in three dimensions, e.g., i.e. a nanocrystalline solid consisting of nanometre-sized crystalline gains each in a specific crystallographic orientation. As the grain size d of the solid decreases the proportion of atoms located at or near grain boundaries, relative to those within the interior of a crystalline grain, scales as $1/d$. This has important implications for properties in ultrafine-grained materials which will be principally controlled by interfacial properties rather than those of the bulk.

Many properties are continuously modified as a function of system size. Often these are extrinsic properties, such as resistance, which depend on the exact size and shape of the specimen. Other properties depend critically on the microstructure of the material; for example, the Hal-Petch equation for yield strength, σ, of a material as a function of average grain size (d) is given by

$$\sigma = k\langle d\rangle^{-1/2} + \sigma_0 \tag{1}$$

where k and σ_0 are constants. Intrinsic materials properties, such as resistivity, should be independent of specimen size, even if many of the intrinsic properties of matter at the nanoscale are not predictable from those observed at larger scales. This is because new phenomena like, quantum size confinement (leading to changes in electronic structure) the presence of wave-like transport process, and predominance of interfacial effects do emerge.

Synthesis, Characterization and Processing of Nanostructured Materials

Nanostructures materials can be synthesized by a number of techniques, such as inert gas condensation, plasma processing, physical and chemical vapor deposition, electrodeposition, mechanical alloying, rapid solidification, sol-gel, micro-emulsion, spark erosion, and severe plastic deformation. Inert gas condensation, mechanical allying, and electrodeposition are among the methods that are used.

Gas-phase Synthesis

Gas-phase synthesis is a well-known technique for the production of en extensive variety of nanostructured materials example; reinforcing fillers, rheology control additives and pigments.

Although a number of gas-phase synthesis processes exists, they all have in common the fundamental aspects if particle formation mechanisms that occur once the product species is generated.

The product quality and application characteristics of nanostructured materials depend on the size distribution, morphology and state of aggregation, i.e. the size and number of primary particles defining the degree of aggregation. In gas phase reactors the final product characteristics are determined by fluid mechanisms and particle dynamics within a few milliseconds at the early stages of the synthesis process.

Three major formation mechanisms dominate particle formation (Figure 5.5). The chemical reaction of the precursor leads to the formation of product monomers (clusters) by nucleation or direct inception and to the growth of particles via the reaction of precursor molecules on the surface of newly formed particles which is called surface growth.

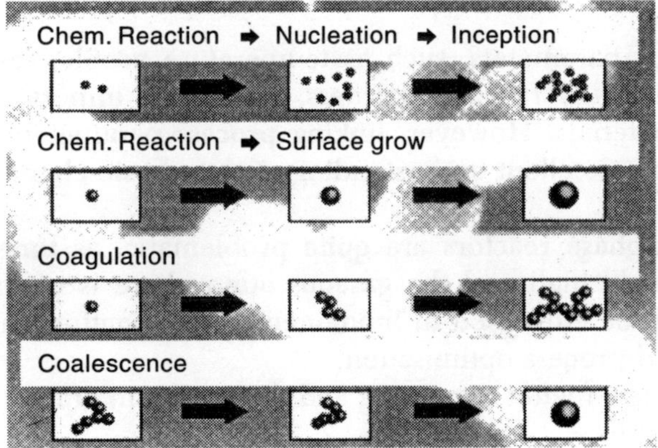

Fig. 5.5 *Particle formation mechanisms*

Coagulation is an intrinsic mechanism which inevitably occurs at high particle concentrations and therefore in all industrial aerosol processes. Particles, dispersed in a fluid move randomly, and, along their trajectories, they collide with each other. Resulting in coagulation due to adhesive forces.

Finally, coalescence and fusion are sufficiently fast in the high temperature zones of the reactor to effect a reduction in the level of aggregation or even or formation of spherical particles, due to sintering processes.

Flame Reactor

Gas-phase reactions are characterized by very high temperatures and extremely short residence times. The manufacturing process and conditions determine the size and morphology of the product and hence their suitability in given fields of application.

Flame reactors are one of the most common reactor designs for the production of high-purity nanostructured materials in large quantities, especially for the production of silica, titania and alumina. Powders, liquids and vapours can be used as precursors.

Due to the high energy density in the flame, the precursor concentration can also be quite high, with flame temperatures reaching 1273–2673 k.

The residence time in the highest temperature region is very short, usually between 10 and 100 ms. This hot zone is crucial for the formation of the primary particles, which in the process range form only a few nm up to 500 nm. Beyond this zone, only the size and the morphology of the aggregates can be influenced.

The specific surface areas of these materials is up to $400m^2\ g^{-1}$.

The shape of the flame is influenced by the type of fuel and air inlet used. If the gases are pre-mixed, the flame reaction takes place right at the burner mouth, creating a very short, homogeneous flame. With diffusion flames, the fuel and the air or oxygen are fed separately to the burner mouth. The reactants have to diffuse together before they combust, which creates a significantly longer flame.

Computer Simulation

The control of reactor parameters, such as temperature profile, residence time in the hot zone and concentration of the reaction partners, is of great importance in producing tailor-made nanostructured materials. However, linking process parameters measurably to product characteristics requires an excellent understanding of the physio-chemical fundamentals of gas phase synthesis.

Measurements in gas-phase reactors are quite problematic, as time scales are very small, temperatures are extremely high and the gaseous atmosphere is often aggressive. Therefore, computer simulation can be employed in fundamental investigation and may provide a better roadmap for product and process optimisation.

Processes with particulate matter increasing and decreasing in amount and size over time can be described mathematically.

Gas-phase reactors usually produce non-homogeneous profiles for temperature velocities, concentrations of species etc. Therefore, when simulating complex reactor geometries, a technique has to be coupled with computational fluid dynamics (CFD). Depending on the information needed, different algorithms ranging from as basic but fast synthesis mechanisms and their effect on the product to a CFD coupled technique to emulate real production reactors.

For example, particle formation in a pre-mixed flame reactor the average time-temperature of the particles' trajectories is approximated. The adiabatic flame temperature is assumed to drop exponentially. The precursor mole fraction entering the reactor is set to 0.01 and reacts at an extremely small time-scale (<1 ms).

Therefore, the inception of particles occurs almost instantaneously, driving up the number of particles in the system. Consequently, and intense coagulation process sets in and the aggregate volume grows continuously, forming either larger spherical particles or agglomerates. Coalescene induces the primary particles diameter to grow in regions of high temperatures.

Downstream, or at larger residence times, the temperature drops and causes the coalescence to cease until coagulation dominates the formation process (t > 20 ms). As Figure 5.6 shows, the evolution of the primary particle size levels out at this point. However, the number of primaries within the agglomerates continues to grow, due to coagulation.

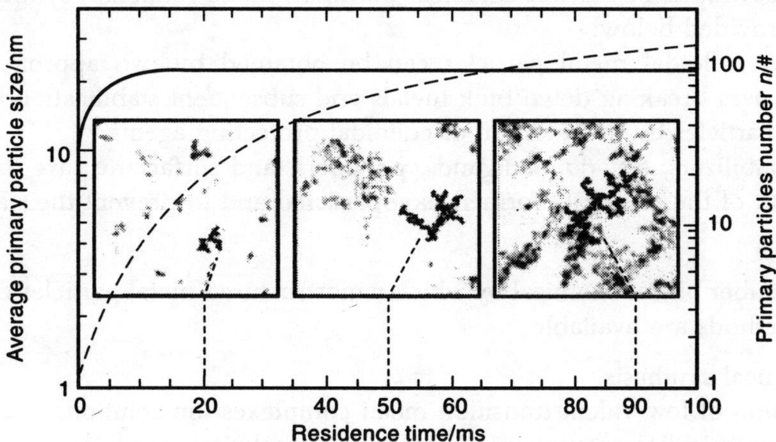

Fig. 5.6 *Comparison of simulation results and experiment*

The calculated particles size evolution is compared to experimental data obtained by thermophoretic sampling, which makes possible direct measurement in the flame at various distances (residence times) from the burner mouth.

Figure 5.6 also shows TEM-picture sections of particles collected form the flame after residence times of 20, 50 and 90 milliseconds and the corresponding average aggregates obtained from simulation. With increasing residence time, the advancing degree of aggregation can clearly been seen, but there is hardly any change in primary size. The results from experiment and simulation are in excellent conformity, validating the simulation technique.

Several products based on flame synthesis are:

(i) nanostructured zinc oxide, which can be sued as a UV filter in sunscreen lotions or transparent coatings,

(ii) extremely fine particles of ceria (cerium dioxide), which are ideally suited to polishing integrated circuit wafers in the chemical mechanical planarisation (CMP) process, due to their high abrasivity.

(iii) Indium tin oxide (ITO) can equip polymer surfaces with transparent, antistatic properties, and can also be used for IR absorption.

WET CHEMICAL SYNTHESIS OF NANOPARTICLES

Nanomaterials have affected considerable interest due to their peculiar mechanical, electronic, optical, magnetic as well as the thermodynamic properties that differ significantly from those materials either in the bulk or the single molecule. The unique properties can be attributed to their special structures particles between 1 and 50 nm sizes that are prevented from agglomerating by protecting shells are regarded as nanostructured metal colloids. The term nanoparticle is used here in the sense of covering a wide range of nano structured materials, including colloidal materials, nanoclusters, nanorods etc. In the case of colloidal particles, depending on the nature of the protecting shell, they can beredispersed in water ("hydrosole") or organic solvents ("organosols"). A synopsis of the wet chemical transition metal reduction synthesis of colloidsl nanoparticle is provided below:

Nanostructured colloidal metal particles can be obtained by two approached. The "top down" approach uses breaking down bulk metals and subsequent stabilization of the resulting nanosized metal particles by the addition of colloidal protecting agents.[1,2]

A variety of stabilizers, e.g. donor ligands, polymers, and surfactants, are used in order to control the growth of the primarily formed nanoparticled and to prevent their agglomeration.

Reduction Methods

There are a number of wet chemical by which nanostructured metal particles can be oriented. The following methods are available:

1. Electrochemical synthesis,
2. Decomposition of low valent transition metal complexes ion solution,
3. Preparation in micelles, reverse micelles and encapsulation, and
4. Transition metal salt reduction method.

Of these methods, transition metal salt reduction for the preparation of protected nanocolloids is given below:

Metal Salt Reduction

The chemical reduction of transition and salts in the presence of stabilizing agents to generate zero valent metal colloids in aqueous or organic media was achieved by Faraday. The reliable method has become the most common synthetic tool in this field. The first reproducible was the preparation of gold metal colloids (e.g., for 20 nm) by reduction of $[AuCl_4^-]$ with sodium citrate. He also proposed a mechanism for the stepwise formation of nanoclusters based on nucleation, growth, and agglomeration, which in essence is still valid. From the data of thermodynamic and kinetic experiments gained by modern analytical technique, this model has been refined as illustrated in Fig. 5.7.

The metal salt is reduced to give zerovalent metal atoms in the embryonic stage of the nucleation, which collides in solution with other metal ions, metal atoms or clusters to form an irreversible "seed" of stable metal nuclei. The diameter of this "seed" nuclei can be controlled to well below 1 nm depending on the difference in redox potentials of the metal salt and the reducing agent used and the strength of he metal-metal bonds.

PROPERTIES OF NANOSTRUCTURES

Structural Properties

The increase in surface area and surface free energy with decreasing particle size leads to changes in interatomic spacings. For Cu metallic clusters the interatomic spacing is observed to decrease with decreasing cluster size, due to strain induced by the internal pressure, arising from the small radius of curvature in the nanoparticle. But, for semiconductors and metal oxides interatomic spacing increase with decreasing particle size.

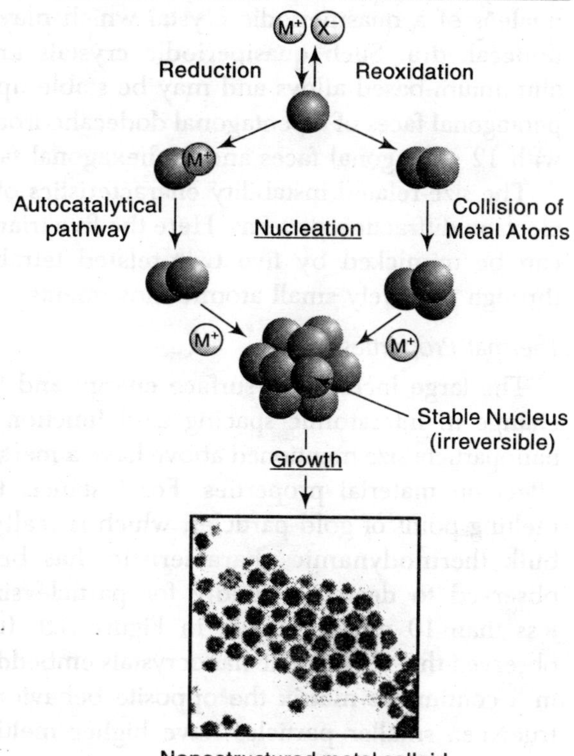

Fig. 5.7 *Formation of nanostructured metal colloids via the "salt reduction" method.*

The apparent stability of metastable structures in small nanoparticles and clusters, due to their bulk atomic arrangement is lost. Metallic nanoparticles, such as gold, are known to adopt polyhedral shapes such as cubo-octahedra, multiply twinned icosahedra and multiply twinned decahedra (Figure 5.8). These nanoparticles are regarded as multiply twinned crystalline particles (MTPs) in which the shapes can be understood in terms of the surface energies of various crystallographic planes, the growth rates along various crystallographic directions and the energy required for the formation of defects such as twin boundaries. However, there is compelling evidence that such particles are not crystals but are **QUASIPERIODIC CRYSTALS** or crystalloids. These icosahedral and decahedral quasicrystals from the basis for under growth of the nanocluster, up until a size where they will switch into more regular crystalline packing arrangements.

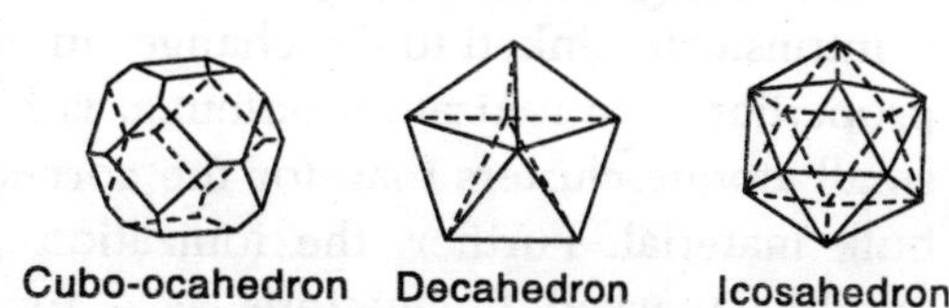

Fig. 5.8 *Geometrical shapes of cubo-octahedral particles and multiply twinned decahedral and icosahedral particles.*

Now, in the cubic close-packed and hexagonal close-packed structures, exhibited by many metals, each atom is coordinated by 12 neighbouring atoms. All of the coordinating atoms are in contact, although not evenly distributed around the central atom. However, there is an alternative arrangement in which each coordinating atom is situated at the apex of an icosahedron and in contact only with the central atom. If however, the central atom is allowed to reduce in diameter by 10 %, the coordinating atoms come into contact and the body now, has the shape and symmetry of a regular icosahedron with point group symmetry 235, indicating the presence of 30 two fold, 20 threefold and 12 fivefold axes of symmetry. This geometry represents the nucleus of a quasiperiodic crystal which may grow in the forms of icosahedra or pentagonal dodecahedra. Such quasiperiodic crystals are known to exist in an increasing number of aluminium-based alloys and may be stable up to microcrystalline C_{20} (dodecahedrene with 12 pentagonal faces of a pentagonal dodecahedron, but unstable) and C_{60} (the well-known buckyball with 12 pentagonal faces and 20 hexagonal faces of a truncated icosahedron).

The size-related instability characteristics of quasiperiodic crystals are distinguished by their electron diffraction patterns. Here the five triangular faces of the fivefold symmetric icosahedron can be mimicked by five twin-related tetrahedra (with a close-packed crystalline structure) through relatively small atomic movements.

Thermal Properties

The large increase in surface energy and the change in interatomic spacing as a function of nanoparticle size mentioned above have a marked effect on material properties. For instance, the melting point of gold particles, which is really a bulk thermodynamic characteristic, has been observed to decrease rapidly for particle sizes less than 10 nm, as shown in Figure 5.9. It is observed that for metallic nanocrystals embedded in a continuous matrix the opposite behavior is true; i.e., smaller particles have higher melting points.

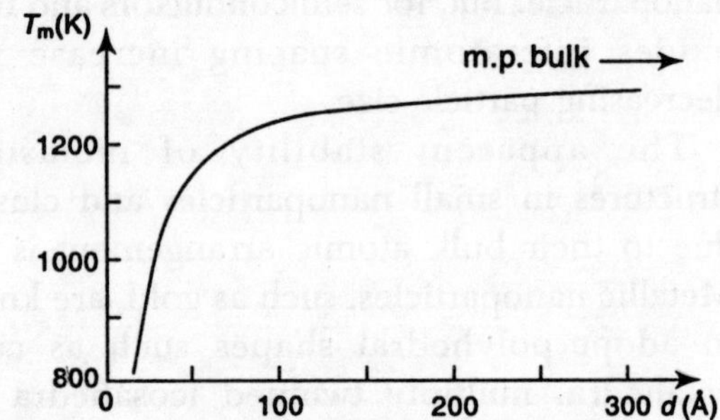

Fig. 5.9 *Schematic diagram of the variation in melting point of gold nanoparticles as a function of particle size.*

Chemical Properties

The change in structure as a function of particle is intrinsically linked to the changes in electronic properties. The ionization potential is higher for small atomic clusters than for the corresponding bulk material. Further, the ionization potential exhibits marked fluctuations as a function of cluster size. Such effects appear to be linked to chemical reactivity, such as the reaction of Fe_n clusters with hydrogen gas (Figure 5.10).

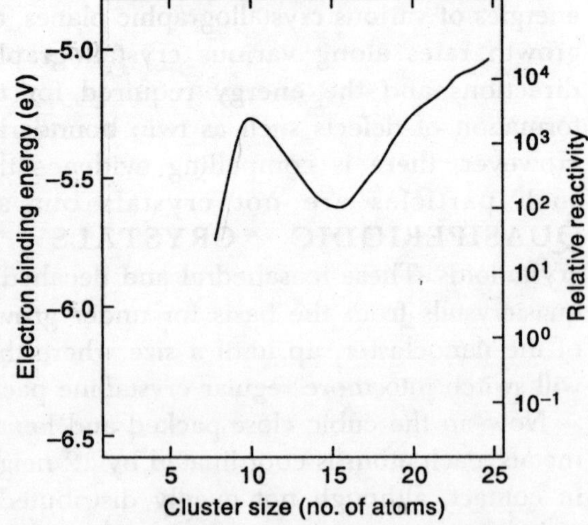

Fig. 5.10 *Schematic diagram of the dependence of the electron binding energy and relative chemical reactivity of iron clusters to hydrogen gas as a function of cluster size.*

As mentioned earlier Nanoscale structures such as nanoparticles and nanolayers have very high surface area to volume ratios and have potentially different crystallographic structures so it leads to a radical alteration in chemical reactivity. Catalysis using finely divided nanoscale systems increase the rate, selectivity and efficiency of chemical reactions such as combustion or synthesis whilst simultaneously significantly reducing waste and pollution. Gold nanoparticles small than 5 nm in diameter are known to adopt icosahedral structures rather than the normal face centred cubic arrangement. This structural change is accompanied by an extraordinary increase in catalytic activity. Also, nanoscale catalytic supports with controlled pore size, select the products and reactants of chemical reactions based on their physical size and transport to and from internal reaction sites within the nanoporous structure. Additionally, nanoparticles often exhibit new chemistries as distinct from their larger particulate counterparts; for example, many new medicines are insoluble in water when in the form of micron-sized particles but will dissolve easily when in a nanostructured form.

Mechanical Properties

Many mechanical properties, such as toughness, are highly dependent on the ease of formation or the presence of defects within a material. As the system size decreases, the ability to support such defects becomes increasingly more difficult and mechanical properties will be altered accordingly. Novel nanostructures, which are very different from bulk structures in terms of the atomic structural arrangement, will obviously show very different mechanical properties. For example, single and multi-walled carbon nanotubes show high mechanical strengths and high elastic limits that lead to considerable mechanical flexibility and reversible deformation.

As the structural scale reduced to the nanometer range, for example, in nanolayered composites, a different scale dependence from the usual Hall-Petch relationship for yield strength often becomes apparent with large increases in strength. In addition, the high interface to volume ratio of nanostructured materials appears to enhance processes such as plasticity, ductility and strain to failure. Many nanostructured metals and ceramics are observed to be superplastic, i.e. they are able to undergo extensive deformation without necking of fracture. This arises from grain boundary diffusion and sliding, which is significant in a fine-grained material.

Magnetic Properties

Magnetic nanoparticles are used in ferrofluids, colour imaging, bioprocessing, refrigeration as well as high storage density magnetic memory media. The large surface area to volume ratio results in a substantial proportion of atoms having a different magnetic coupling with neighbouring atoms, leading to differing magnetic properties. Figure 5.11 shows the magnetic moments of nickel nanoparticles as a function of cluster size.

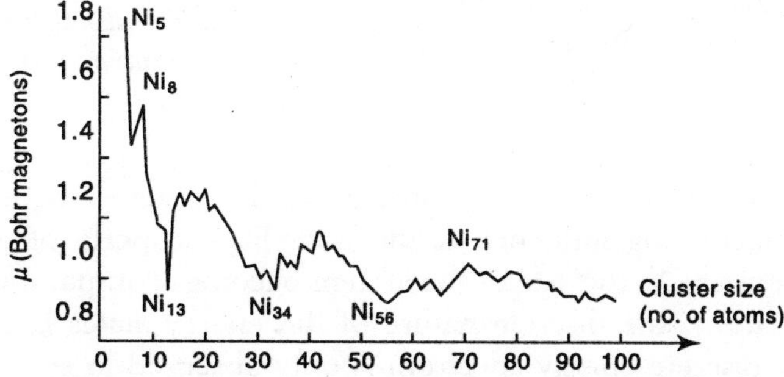

Fig. 5.11 *Diagram for the variation in magnetic moments of clusters vs. cluster size.*

Whilst bulk ferromagnetic materials usually from multiple magnetic domains, small magnetic nanoparticles often consist of only one domain and exhibit a phenomenon known as superparamagnetism. In this case the overall magnetic coercivity is then lowered: the magnetizations of the various particles are randomly distributed due to thermal fluctuations and only become aligned in the presence of an applied magnetic field.

Giant magnetoresistance (GMR) is a phenomenon observed in nanoscale multilayers consisting of a strong ferromagnet (e.g., Fe, Co) and a weaker magnetic or non-magnetic buffer (e.g., Cr, Cu); it is usually employed in data storage and sensing. In the absence of a magnetic field the spins in alternating layers are oppositely aligned through antiferromagnetic coupling, which gives maximum scattering from the interlayer interface and hence a high resistance parallel to the layers. In an oriented external magnetic field the spins align with each other and this decreases scattering at the interface and hence resistance of the device.

Optical Properties

In small nanoclusters the effect of reduced dimensionally on electronic structure, effects the energies of the highest occupied molecular orbital (HOMO), the valence band, and the lowest unoccupied molecular orbital (LUMO), the conduction band. Optical emission and absorption depend on transitions between these states; semiconductors and metals, show large changes in optical properties, such as colour, as a function of particle size. Colloidal solutions of gold nanoparticles have a deep red colour which becomes progressively more yellow as the particle size increases; indeed gold colloids have been used as a pigment for stained glass since the seventeenth century. Figure 5.12 shows optical absorption spectra for colloidal gold nanoparticles of varying sizes. Semiconductor nanocrystals in the from of quantum dots show similar size-dependent behavior in the frequency and intensity of light emission as well as modified non-linear optical properties and enhanced gain for certain emission energies or wavelengths. Other properties which may be affected by reduced dimensionality include photocatalysis, photoconductivity, photoemission and electroluminescence.

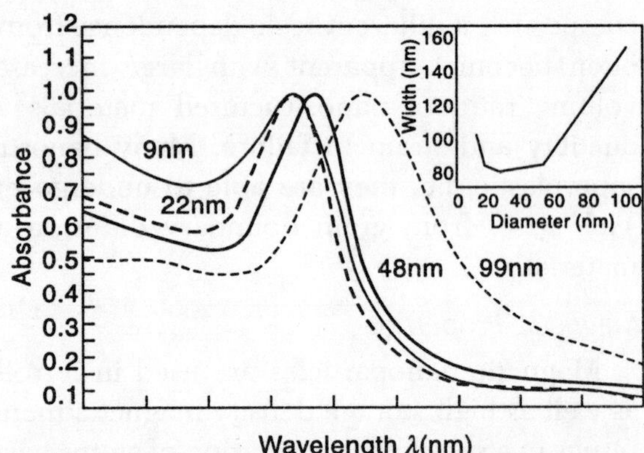

Fig. 5.12 *Size dependence of the optical absorbtion wavelength for gold nanoparticles and (inset) the corresponding value of the full width at half maximum (FWHM) of the absorption peak.*

Electronic Properties

The changes which occur in electronic properties as the system length scale is reduced are related mainly to the increasing influence of the wave-like property of the electrons and the scarcity of scattering centres. As the size of the system becomes comparable with the de Broglie wavelength of the electrons, the discrete nature of the energy states becomes apparent once again, although a fully discrete energy spectrum is only observed in systems that are confined in all three dimensions. In certain cases, conducting materials become insulators below a critical length size, as the energy bands cease to overlap. Owing to their intrinsic wave-like nature, electrons can tunnel quantum mechanically between two closely adjacent nanostructures, and if a voltage is applied between two nanostructures which aligns the discrete energy levels in the DOS, *resonant tunnelling* occurs, which abruptly increases the tunnelling current.

In macroscopic systems, electronic transport is determined primarily by scattering with phonons, impurities, or other carriers or by scattering at rough interfaces. The path of each electron resembles a random walk, and transport is said to be *diffusive.* When the system dimensions are smaller than the electron mean free path for inelastic scattering, electrons can travel through the system with randomization of the phase of their wavefunctions. This gives rise to additional localization phenomena which are specifically related to phase interference. If the system is sufficiently small so that all scattering centres can be eliminated completely, and if the sample boundaries are smooth so that boundary reflections are purely specular, then electron transport becomes purely *ballistic,* with the sample acting as a waveguide for the electron wavefunction.

Conduction in highly confined structures, such a quantum dots, is very sensitive to the presence of other charge carriers and hence the charge state of the dot. These coulomb *blockade* effects result in conduction processes involving single electrons and as a result they require only a small amount of energy to operate a switch, transistor or memory element.

All these phenomena are utilised to produce radically different types of components for electronic, optoelectronic and information processing, such as resonant tunnelling transistors and single-electron transistors.

BULK BEHAVIOR OF NANOSTRUCTURED MATERIALS

Bulk nanostructured materials (BNMs) are defined as bulk solids with nanoscale or partly nanoscale microstructures. This category of nanostructured materials has a relatively recent focus due to new discoveries of unique properties of some nanoscale materials.

Classification

The classification of bulk nanostructured materials is dependent on the number of dimensions which lie in the nanometer range. Thus depending on which dimensions fall within this range, they may be classified as: (a) zero-dimensional (0D); e.g., nanopores and nanoparticles; (b) one-dimensional (1D); e.g., layered or laminate structures; (c) two-dimensional (2D); e.g., filamentary structures where the length is substantially greater than the cross-sectional dimensions, and finally (d) three dimensional (3D); e.g., structures typically composed by consolidated *equiaxed* (all three dimensional equally nanoscale) monocrystalline. The classification system is useful in describing the type of structure to be manufactured. Figure 5.13 illustrates all four types and, of these four types, 3D and 1D are of must importance.

Early, refined microstructures, for example, small grain sizes, provided attractive properties such a increased strength and toughness in structural materials. A classic example of this is age hardening of aluminum alloys. Maximum hardness is observed with a mixture of grains with the dimensions of 10-100 nm. Therefor, the important microstructural feature of age hardened aluminum alloys is nanoscale. There are a number of other examples of nanoscale microstructures providing optimized properties. For example, the critical current density J_c of commercial superconducting Nb_3Sn in controlled by grain size and is inversely proportional to grain size, with grain size of 50–80 nm providing high values of J_c.

There are a number of bulk properties that change when the microstructure is nanoscale, example ferromagnetic materials with nanoscale microstrucutres have potential applications as soft magnetic materials and permanent magnet materials, information storage, magnetoresistance spin valves and magnetic nanocomposite refrigerants.

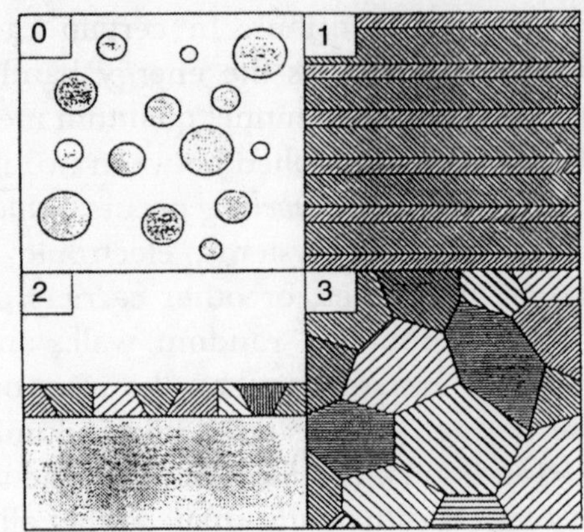

Fig. 5.13 *Schematic illustration of nanomaterials showing fine structural scale indifferent numbers of dimensions. These may be zero-dimensioned; thin deposited layers showing fine scale in one dimension; surface with two-dimensional nanostructures; and three dimensional nanostructures.*

Fabrication by Electrodeposition

The formation of nanoparticles and bulk nanostructured materials (BNMs) by electrodeposition; i.e., the deposition of a conductive material from a plating solution by the application of electrical current, is achieved using either two separate electrolytes or, from one electrolyte, by exercising appropriate control of the deposition parameters. These processes are applied to the synthesis of pure metals, alloys and composites. They have few size and shape limitations. The electrodeposition of nanostructured materials is possible, if a large number of nuclei form on the surface of the electrode, and their growth is restricted by the use of high current density for a short duration or the use of organic additives as growth inhibitors or the control of bath temperature, and manipulation of bath pH.

Pulsed electrodeposition (PED) differs from direct current (DC) electroplating techniques in that the peak current density is very high (1-2 A cm^{-2}), much larger than in DC electroplating, but that the duration of this high current density is very short. The deposition rate in the PED is governed by a number of controllable parameters, including the pulse duration, the time between pulses, and the pulse current. Cations are depleted from the electrolyte adjacent to the electrode during the application of the pulse and their levels are restored by the diffusive transport of material to this depleted region between pulses. The size of nuclei is dependent on both pulse duration and current density, whilst a balance is struck between the replacement of cations close to the electrode by diffusion and the minimisation of grain coarsening when determining an appropriate interpulse time.

Addition of suitable organic compounds to the electrolyte leads to the inhibition of grain growth. These adsorb on the electrodeposit, in a reversible way, and block active sites for growth, such as surface steps, hence leading to a reduction in the growth rate of crystallites.

They also act to inhibit surface diffusion of adatoms, so that nucleation becomes the preferred method of electrolyte depletion. The crystallite size depends in this instance on the concentration of inhibitor in the electrolyte.

At high temperature, diffusion at the surface is much enhanced, thereby resulting in a much higher rate of grain growth. Inhibiting molecules also decrease in efficacy because the increased rate of desorption from growth sites (due to higher temperature) will increase the number of active attachment sites. But by reducing the bath temperature, it will contribute to the inhibition of growth. Changes in electrolyte temperature will change the width of the grain size distribution and the average grain size.

Control of the redox potential of the species by addition of a complex is particularly important if alloys with components of widely differing redox potentials are to be electrodeposited. If this equilibrium of the complexes depends additionally on the pH of the solution, a defined redox potential and therefore a defined alloy composition may be obtained by variation of the pH value.

PROPERTIES OF BULK NANOMATERIALS

For application of bulk nanocrystalline materials is necessary to investigate in detail their hardness,

1. *Mechanical Properties*

1. Hardness: For the mechanical properties of nanocrystalline materials, special characterises the resistance of materials to plastic deformation during indentation of a harder body into it. Diamond is used generally as an indentor material.

The hardness of the material is determined by yield limit σ_y. The grain size has strong effect on microhardness; this effect has been studied extensively on metal, alloys and ceramics with a grain size D greater than 1 μm. According to the Hall-Petch law.

$$\sigma_o = \sigma_i + k_y D^{-1/2}\,, \tag{1}$$

where σ_o = yield stress σ_i is the frictional stress opposing dislocation motion and k_y is a constant. At a temperature, $T/T_{\text{melt}} < 0.4\text{–}0.5$ (T_{melt} is the melting point), hardness H (Vickers microhardness) is linked with the yield limit σ_i by empirical relationship $H/\sigma_i \approx 3$. This leads to the following size dependence of hardness

$$H \approx H_i + kD^{-1/2} \tag{2}$$

Where H and k are constants.

Deformation is carried out by diffusion sliding, then at a low temperature T/T_{melt}, the deformation rate $\varepsilon \equiv d\varepsilon/dt$ is

$$\frac{d\varepsilon}{dt} = B\sigma\Omega\delta d_{\text{dif}}/k_B T D^3, \tag{3}$$

where B is proportionality coefficient; σ is applied stress, Ω is the atomic volume, δ is the thickness of the grain boundary, d_{dit} is the coefficient of grain boundary diffusion.

Equations (1) to (3) show that a decrease in the grain size should result in a strong change in the mechanical properties. In particular, equations (1) and (2) predict hardening of the material with a decrease in the grain size *D*. At the same time, equation (3) shows that at the nanometer size of grain, diffusion sliding becomes important even at room temperature, greatly increasing the deformation rate. Thus, the effect of the grain size on the strength properties of a nanocrystalline material is ambiguous and depends on the ratio between the change of the yield limit and the deformation rate. In addition, it is important to take into account the possible increase of the grain boundary diffusion coefficient d_{dif} with a decrease in the grain size.

At 300 K, the microhardness of nanocrystalline materials is 2–7 times higher than *H* of coarse-grained materials.

The grain size of Cu and Pd is varied by annealing. The results show that a decrease in the grain size of coarse-grained copper from 25×10^3 to 5×10^3 nm is accompanied by an increase in microhardness. Microhardness of nanocrystalline copper (n–Cu) *D* ~ 16nm was is ~2.5 times figure 5.14 higher than that of copper with a grain size of 5×10^3 nm, but with a decrease in the grain size of n-Cu from 16 to 8 nm *H* decreased by ~25%. The decrease of *H* was also found in the case of a decrease in the grain size of n-Pd from 13 to 7 nm.

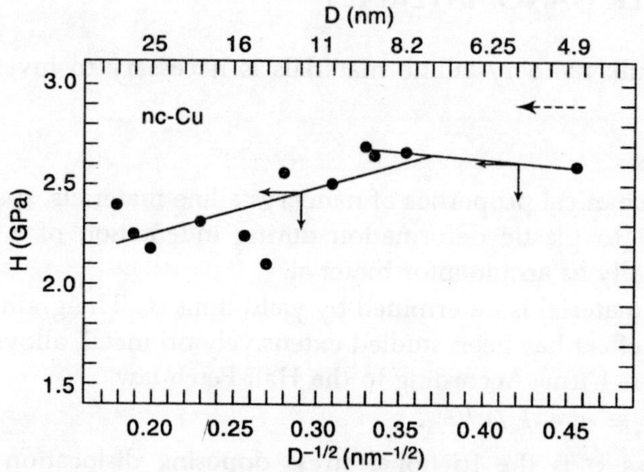

Fig. 5.14 *Dependence of the microhardness H of bulk nanocrystalline copper nc-Cu on* $D^{-1/2}$ *(D is the grain size): the growth of grains of nc-Cu was achieved by increasing annealing time at 423 K (dashed arrow indicates the direction of increase of annealing time).*

There is a decrease in the *H* value of nanocrystalline alloys Ni–P, TiA1Nb, Tia1, $NbA1_3$, with a decrease in the grain size from 60–100 to 6–10. This effect is determined by the increase in the contribution of diffusion mobility in the deformation with decreasing grain size *D*.

If microhardness H is measured on a single specimen, that is successively heated to produce ever-increasing grain sizes, then dependence *H* (*D*) is different. In this case, an increase in grain size *D* up to some critical value leads to a rise in *H*; the following increase in *D* is accompanied by decreasing *H* and corresponds to the Hall-Petch law.

The results of (Ag, Cu, Pd, Se, Fe and Ni) Several independently produced specimens of a nanocrystalline metals having grains of different sizes, the hardness of these specimens increases with a decrease of *D* to 4–6 nm; the dependence of *H* on $D^{-1/2}$ is governed by the Hall-Petch law (3). The hardness of the intermetallic compounds and alloys in which the grain size D is varied by annealing depends in a more complicated manner on D; initially an increase in D leads to an increase in H and then decreases H. In other words, in the case of grain growth as a result of annealing, the Hall-Petch law (decrease of H with increasing D) is fulfilled only at $D>$ 12–20nm. The hardness, strength, and deformation behavior of nanocrystalline materials is unique and not yet well understood. As grain size is reduced through the nanoscale regime (<100nm), hardness typically increases with decreasing grain size and can be factors of 2 to 7 times harder for pure nanocrystalline metals (10 nm grain size) than for large-grained (> 1 μm) metals. Deformation is not operable in nanocrystalline materials at the smallest grain sizes (< 30 nm), because at these grain sizes, mobile dislocations are unlikely to occur.

2. Elastic Properties

Measurements of the elastic constants on nanocrystalline materials prepared by the inert gas condensation method gave values, for Young's Modulus, E, that were significantly lower than values for conventional grain size materials. The main reason of that is the presence of extrinsic defects-pores and cracks, in nanocrystalline materials compacted form powders. Now, work on porosity-free materials has supported these and it is now believed that the intrinsic elastic modules of nanostructured materials are essentially the same as those for conventional grain size materials until the grain size becomes very small, e.g., < 5 nm, such that the number of atoms associated with the grain boundaries and triple junctions becomes very large.

3. Ductility and Toughness

It is well known that grain size has a strong effect on the ductility and toughness of conventional grain size (>1 μm) materials. For example, the ductile/brittle transition temperature in mild steel can be lowered about 313K by reducing the grain size by a factor of 5. On a very basic level, mechanical failure, which limits ductility, is an interplay or competition between dislocations and cracks. Nucleation and propagation of cracks can be used as the explanation for the fracture stress dependence on grain size. Grain size refinement can make crack propagation more difficult and therefore, in conventional grain size materials, increase the apparent fracture toughness. However, the large increases in yield stress (hardness) observed in nanocrystalline materials suggest that fracture stress can be lower than yield stress and therefore result in reduced ductility. The results of ductility measurements on nanocrystalline materials: metal are mixed and are sensitive to flaws and porosity, surface finish, and method of testing (e.g., tension or compression testing). In tension, for grain sizes <30 nm, essentially brittle behavior has been observed for pure nanocrystalline metals that exhibit significant ductility when the grain size is conventional.

In some metals, Cu for example, ductile behavior is observed in compression, along with yield strengths about twice those observed in tension. While it is likely that the flaws and porosity present in many nanocrystalline materials samples, seriously affect the results of mechanical tests and may be partly responsible for the asymmetry of results in terms of shear banding (see below) may also be important. The above behavior is presumably due to the inability of usual dislocation generation and motion to occur at these smallest nanocrystalline materials grain size.

4. Superplastic Behavior

Superplasticity is the capability of some polycrystalline materials to exhibit very large tensile deformations without necking or fracture. Typically, elongations of 100% tp>1000% are considered the defining features of this phenomenon. As grain size is decreased it is found that the temperature T_{sp} is lowered at which superelasticity occurs, and the strain rate for its occurrence is increase the reason of that is that grain-boundary sliding is an important deformation mechanism during superplasticity. Typically, $T_{sp} > 0.5\ T_m$ (T_m is the melting temperature of the material). The nanocrystalline materials are found to be superplastic at a temperature about 673–773 K below that attained for microcrystalline materials. This corresponds to $T_{sp} \approx 0.36\ T_m$. Sample of nanocrystalline nickel, nanocrystalline aluminum alloy, and nanocrystalline nickel aluminide before and after deformation, their T_{sp}, strain rate, and $(T_m - T_{sp})/T_m$ ratio are shown in Figure 5.15.

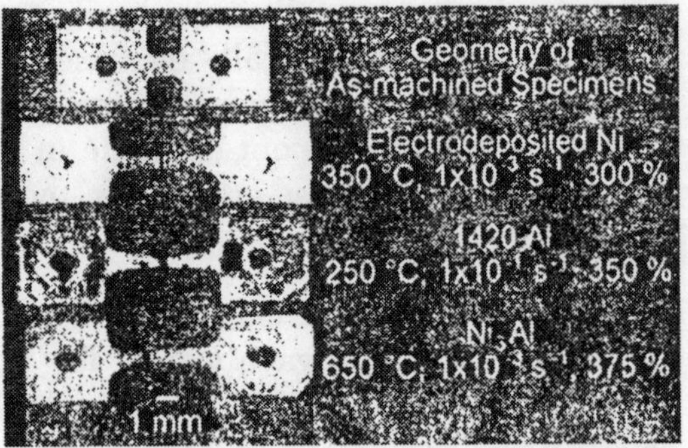

Fig. 5.15 *Miniature tensile specimens, shown in the as machined geometry and after deformation.*

Reasons of the Unique Mechanical Properties of Nanocrystalline Materials

The most important deformation mechanism during superplasitcity is grain-boundary sliding. The large volume fraction of grain boundaries in nanostructured materials provides low-temperature of high-strain-ratesuperplasticity.

While there are still only limited data on the mechanical behavior especially tensile properties-of nanocrystalline materials, some generalizations may be made regarding the deformation mechanisms. A crystal is never perfect; a variety of imperfections like dislocations can mar the ordering. Crystals dislocation, is a line defect that runs through the length of the crystal. Another type, called an edge dislocation, is shown in Figure 5.16. This defect occurs when there is a missing row of atoms. In the figure the crystal arrangement is perfect on the top and bottom. The defect is the row of atoms missing from region b. This mistake runs in a line that is perpendicular to the page and places a strain on region a.

Fig. 5.16 *Crystalline structure defect.*

Dislocation are formed when a crystal is grown, and great care must be taken to produce a crystal free of them. Dislocations are stable and will exist for years.

It is likely that for the large end of the nanoscale grain sizes, about 50–100 nm, dislocation activity dominates for temperatures lower than $0.5\ T_m$. As grain size decreases, dislocation activity apparently decreases. The essential lack of dislocations at grain sizes below 50 m is presumably the result of the image force that act on dislocations near surfaces or interfaces. The lack of dislocations in small, confined spaces such as single-crystal whiskers has been well known for many years. Creation of new dislocations is also made difficult as the grain size reaches the lower end of the nanoscale (<10nm). Stresses needed to activate dislocation sources, such as the Frank Read source, are inversely proportional to the distance between dislocation pinning points. Since nanoscale grains will limit the distance between such pinning points, the stresses to activate dislocation sources can reach the theoretical shear stress of a dislocation-free crystal at the smallest grain sizes (~2 nm). Thus, at the smallest grain sizes we may have new phenomena controlling deformation behavior. It has been suggested that phenomena may involve grain boundary sliding and/or grain rotation accompanied by short-range diffusion-assisted healing events.

FABRICATION OF NANOSTRUCTURES

Nanostructures can be made in numerous ways; a board classification divides methods into either those which build form the bottom up, atom by atom, or those which construct from the top down using processes that involve the removal or reformation of atoms to create the desired structure. The two approaches are schematically represented in Figure 5.17.

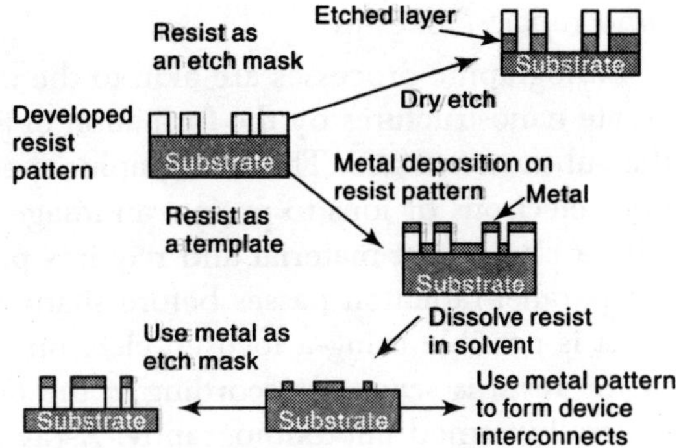

Fig. 5.17 *Top-down and bottom-up processes and their relationship to biological processes and structures.*

In the bottom-up approach, atoms, molecules and even nanoparticles themselves can be used as the building blocks for the creation of complex controlling nanostructures. By controlling the size of the building blocks i.e. their surface, and internal chemistry, their organization and assembly, it is possible to engineer properties and functionalities of the overall nanostructured solid or system. These processes are complex chemical syntheses. On the other hand, top-down approaches are inherently simpler and rely either on the removal or division of bulk material, or on the miniaturization of bulk fabrication processes to produce the desired structure with the appropriate properties. When controlled, both top-down and bottom-up methods may be viewed as a essentially different forms of microstructural engineering. As shown in Figure 4.7, biological processes are essentially intermediated between top-down and bottom-up processes, however, in reality they constitute complex bottom-up processes. A brief overview of some of the more common fabrication methods for nanostructures are discussed below:

Milling

One nanofabrication process is high-energy ball milling, also known as mechanical alloying. As shown in Figure 5.18, coarse-grained materials (usually metals but also more recently ceramics and polymers) in the form of powders are crushed mechanically in rotating drums by hard steel or tungsten carbide balls, usually under controlled atmospheric conditions to prevent unwanted reaction such as oxidation. This deformation causes large reduction in grain size (via the formation and organization of grain boundaries within the powder particles). Different components can be mechanically alloyed together by cold welding to produce nanostructured alloys. A nanometre dispersion of one phase in another also achieved. Microstructures and phases produced in this way are thermodynamically metastable.

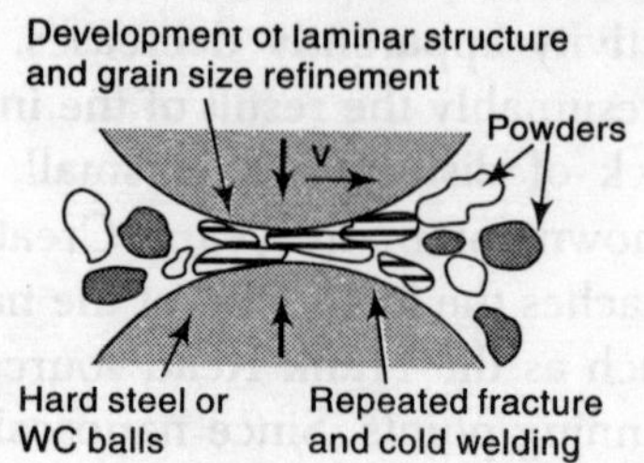

Fig. 5.18 **The mechanical alloying process.**

Generally any form of mechanical deformation under shear conditions and high strain rates leads to the formation of nanostructures, because energy is being continuously pumped into crystalline structure to create lattice defects. The severe plastic deformation that occurs during machining, cold rolling or sliding wear has also been reported to form nanostructured material.

Lithographic Processes

Lithographic processes are akin to the emulsion-based photographic process and are used to create nanostructures by the formation of a pattern on a substrate via the creation of a resist on the substrate surface. The lithographic method uses either visible or ultraviolet (UV) light, X-rays, electrons or ions to project an image containing the desired pattern onto a surface coated with a photoresist material and requires prior fabrication of an absorbing mask through which the parallel radiation passes before shadowing onto the photoresist. Also, direct writing of the resist is possible using a focused electron, ion or possibly X-ray beam. Here the focused beam or the resist is scanned according to the desired pattern design. These different techniques are generally termed photolithography, X-ray lithography, electron beam lithography or ion beam lithography depending on the radiation employed (see Chapter 3).

The resist material, a polymer, metal halide or metal oxide, is chemically changed during irradiation, thus altering the solubility or composition of the exposed resist. In a positive resist, irradiated areas are dissolved as they are more soluble than unexposed material, whilst in a negative resist irradiated areas are insoluble. As shown in Figure 5.19, the resist is then used as template, or as an etch-mask, for subsequent deposition or etching of, the underlying substrate.

The pattern transfer process is divided into solution-based wet chemical etching or dry etching procedures in a reactive plasma doping using ion implantation techniques, or thin film deposition. Dry etching is used for high resolution pattern transfer directly or indirect. The reactive ion species which combine with the elements in the substrate material are removed into a vacuum system.

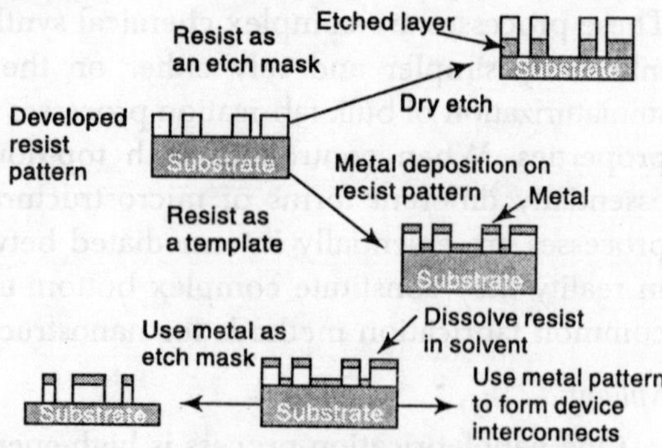

Fig. 5.19 *Representation of various types of photolithographic process.*

Lithographic techniques are used in the semiconductor processing industry for the fabrication on integrated circuits, optoelectronic components, displays and electronic packaging. The uniformity and reproducibility of the fabrication process, and the time required to pattern a given area of a planar device are done in microns per hour. For the production of integrated circuits and microelectromechanical systems (MEMS), the resolution of the lithographic technique is also done. Fundamentally, the wavelength of the radiation used in the lithographic process determines the detail in the resist and hence the final planar nanostructure. The resolution ranges from a few hundred nanometres for optical techniques to tens of nanometres for electron beam techniques. Phenomenologically, throughput and resolution of lithographic techniques broadly follow a power-law dependence; the resolution is approximately equal to $23A^{0.2}$, where A is the areal throughput. Optical, UV and X-ray lithography are fast. Whereas electron and ion beam methods provide nanometre resolution, but are, slower as they involve exposure while using scanned focused probes controlled by computer-aided design (CAD) software. Parallel projection electron beam systems (such as the Scalpel method) and X-ray proximity printing methods will give high-resolution.

Principles of Fabrication

Fabrication removes a deposited material from certain locations. Lsayer deposition is used with reduced thicknesses. High lateral resolution and a high precision methods with respect to positioning are required for fabrication. These requirements exclude standard optical projection methods as applied in photolithography. A reduction of exposure wavelength is necessary. Beam probe techniques are suited for the direct fabrication or transfer at the nanoscale.

The process of removal of functional material depend on its thickness and the subtractive patterning of a molecular monolayer or a functional group on a plane solid surface. Nanometer structures are possible even with small aspect ratios. Wet chemical and plasma chemical methods of subtractive, patterning and isotropic gas phase processes are suitable for fabrication. The ultrathin functional layer is covered by a resist layer that is structured by a lithographic process. This pattern is transferred into the ultrathin layer, prior to removal of the mask.

The subtractive patterning of ultrathin layers is based on a series of individual reactions (that remove the layer in the apertures of the mask and transfer the material into the gas phase). The amount of material and the required reaction rate differs, because process times are in the second to minute range and the removal of ultrathin layers requires only a low rate of removal. The removal of ultrathin layers sometimes occurs in one single step, e.g; for leavage of a small group of atoms or the local removal of molecules from a molecular monolayer (Fig. 5.20A). Mask-guided processes of surface chemistry are not limited to the removal of molecules or atoms. The binding of new atoms or group of atoms to the surface is also possible (Fig. 5.20B). Because the new material is added, such a process is an additive pattern transfer. In the case of ultrathin layer i.e. monomolecular layers, there is no clear dividing line between subtractive and additive processes.

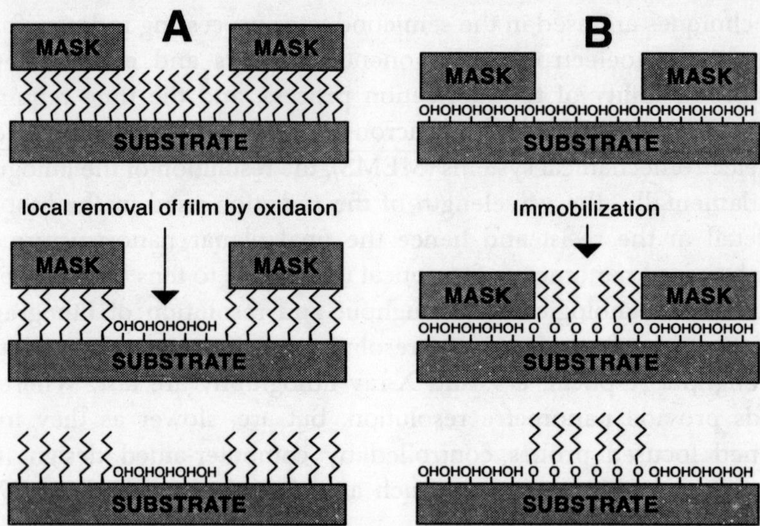

Fig. 5.20 *Fabrication of lateral microstructured molecular monolayers by a lithographic process. A, Local oxidation of alkylated surface in the window of mask. B, Local alkylation of an OH-group rich surface in the windows of a mask.*

Such additive process can be applied in nano technology for structures consisting of several layers of atoms or molecules. As regards the subtractive methods, the steepness of the mask edges has to be sufficient in the case of thick masks, to facilitate structures in the nanometer range. An example is galvanic fabrication, which is also applicable on the nanometer scale.

Lift-off Process

The lifting is a specific process (Fig. 5.21) similar to the additive process (a mask is generate). The functional layer is deposited on top of the mask layer, so that the functional materials is only in direct contract with the support in the mask apertures after removal of the mask layer,

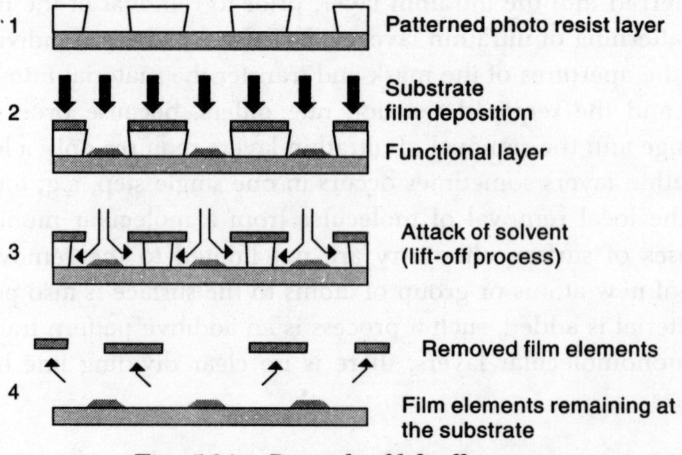

Fig. 5.21 *Principle of lift-off process.*

all parts of the functional layer on top of the mask material and removed ("lifted"). From the standpoint of deposition, lifting is more a subtractive than an additive process, due to the whole substrate surface being covered by the functional material.

A great advantage of lifting is the independence of the process from the chemical solubility of the material that represents its etching behavior. The applied etching bath is independent of the functional material, because it only has to remove the masking material lifting has the potential to fabricate very small structures with high precision. The quality of the mask edge and the deposition of functional material inside the aperture of the mask material determine the edge quality of the final structure in the functional layer. A change in the dimensions of the structure and bent or round edges, is observed in etching processes pre-serve the dimensions makes lifting an ideal process for nanostructures for successful lifting, the mask layer, covered by the functional layer has to be removed in a certain time frame. So a sufficient number and density of holes in the functional layer are needed to ensure the access of the solvent onto the makes layer material. This requirement is met for small lateral dimensions of the structures and for densely packed structures and when the functional material does not cover the edges of the mask layer (Fig. 5.22 (a) and (b).

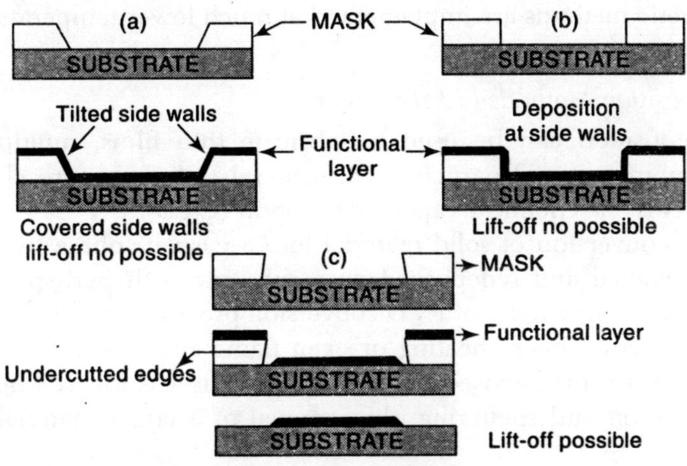

Fig. 5.22 *Limits of lift-off process.*

Lifting can be also applied in the case of nanoporous functional materials, so that the solvent and the resulting product can penetrate this layer. In this process intermediate structures of layer elements are observed, which are free of the substrate but connected laterally. Continuous nanopores are also observed in ultrathin inorganic layer, especially after evaporation and sputtering. The density of pores decreases sharply with thicker layers.

Lifting processes are not applicable for compact layers with thicknesses greater than the thickness of the mask layer.

Lifting processes are assisted by underetched mask flanks (Fig. 5.22(c)). Also the direction of impact of the deposited atoms or molecules influences subsequent lifting. Through isotropic distribution, mask edges are usually covered, thereby slowing down the access of the solvent.

Anisotropic deposition yield shadow effects, especially for underetched edges. For thin layers, the tendency to form nanopores and the low ratio of functional to mask layer thickness makes lifting interesting. Thus it has a greater importance for nanotechnology.

Bottom-up Processes

Bottom-up processes effectively encompass chemical synthesis and/or the highly controlled deposition and growth of materials. Chemical synthesis is carried out in either the solid, liquid or gaseous state. Solid-state synthesis usually involves an iterative procedure of bringing solid reaction precursors into intimate contact by mixing and grinding and then promoting atomic diffusion process via heat treatment at high temperatures to form a reaction product with a relatively large grain size unless grain growth inhibitor are present. Consequently, true nanoscale systems are difficult to obtain via solid-state routes, part from perhaps the controlled growth of second phase nanoscale precipitates within a primary matrix such as in precipitation-strengthened steels, aluminium alloys or other nanocomposite systems.

The focus below is primarily on liquid and gas phase fabrications. Diffusion of matter in the liquid and gas phases is typically many orders of magnitude greater than in the solid phase; therefore these synthetic methods are implemented at much lower temperatures, thus inhibiting unwanted grain growth.

1. Vapour Phase Deposition Methods of Fabrication

Vapour phase deposition can be used to fabricate thin films, multiplayers, nanotubes, nanofilaments or nanometre-sized particles. The general techniques are classified as physical vapour deposition (PVD) or chemical vapour deposition (CVD).

PVD involves the conversion of solid material into a gaseous phase by physical processes; this material is then cooled and redeposited on a substrate with perhaps some modification, such as reaction with a gas. Examples of PVD conversion processes includes thermal evaporation (such as resistive or electron beam heating or even flame synthesis), laser ablation or pulsed laser deposition (where a short nanosecond pulse from a laser is focused onto the surface of a bulk target), spark erosion and sputtering (the removal of a target material by bombardment with atoms or ions).

One example of PVD technique is vapour phase through a jet into a low-pressure ambient background to produce supersaturation of the vapour. Flow rates can approach supersonic speeds and the process can lead to the nucleation of extremely small clusters, ranging from a few atoms upwards, which can be analysed using mass spectrometry. Although the quantities of such clusters are low, these can be of great use of studying the physics of small, usually metallic, clusters, which often are composed of magic numbers of atoms due to their very stable geometric and electronic configurations. A schematic diagram of such a system is shown in Figure 5.23.

Another PVD process is direct gas phase condensation. Here a material, often a metal, is evaporated from a temperature-controlled crucible into a low-pressure, inert gas environment – ultra-high vacuum (UHV) inert gas condensation. The metal vapour cools through collisions with the inert gas species, becomes supersaturated and then nucleates homogeneously; the particle size is usually in the range 1–100 nm and can be controlled by varying the inert gas

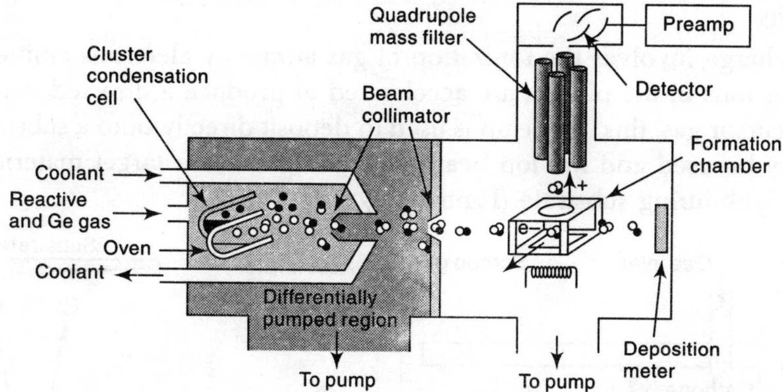

Fig. 5.23 *Diagram of cluster formation via vapour phase expansion from an oven source.*

pressure. Particles can be collected on a cold finger cooled by liquid nitrogen, scraped off and compacted to produce a dense nanomaterial Figure 5.24 shows the experimental apparatus.

CVD involves the reaction or thermal decomposition of gas phase species at elevated temperatures (typically 773–1273k) and subsequent deposition onto a substrate. A simple example is aerosol spray pyrolysis involving aqueous metal salts which are sprayed as a fine mist, dried and then passed into a hot flow tube where pyrolysis converts the salts to the final products. Since materials are mixed in solution, homogeneous mixing at the atomic level is possible and pyrolysis at relatively low temperature provides particles in the size range 5–500nm.

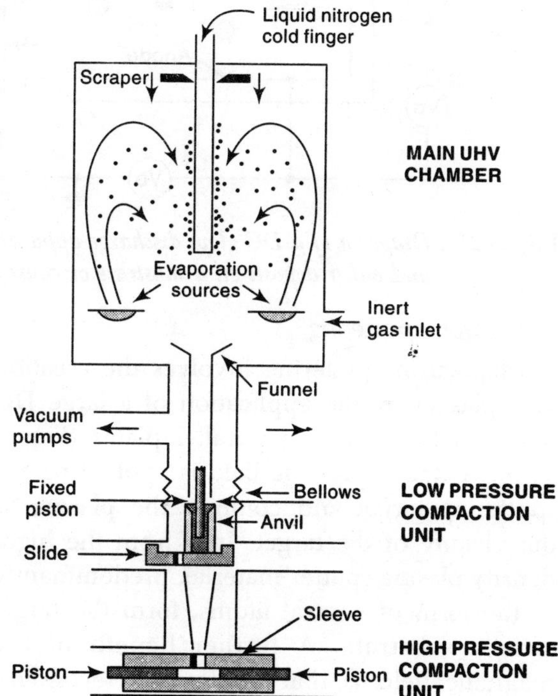

Fig. 5.24 *Schematic diagram of an inert gas condensation apparatus.*

Several CVD processes employ catalysts to enhance the rates of certain chemical reactions. When the catalyst is in the form of a nanometre-sized dispersion of particles, it provides a templating effect; for example, in the production of carbon nanotubes using the decomposition of ethyne or ethene with hydrogen. Fe-, Co- or Ni-based catalysts are employed. The size and size distribution of the catalyst particles also determines the internal diameter of the nanotube.

2. Plasma-assisted Deposition Methods of Fabrication

The use of plasmas (i.e., ionized gases) during vapour deposition allows access to substantially different chemical and physical processes. There are several different types of plasma deposition reactor for plasma-assisted PVD.

DC glow discharge

DC glow discharge involves the ionization of gas atoms by electrons emitted from a heated filament. The gas ions in the plasma are accelerated to produce a directed ion beam. If the gas is a reactive precursor gas, this ion beam is used to deposit directly onto a substrate; alternatively an inert gas may be used and the ion beam allowed to strike a target material which sputters atoms onto a neighbouring substrate (Figure 5.25).

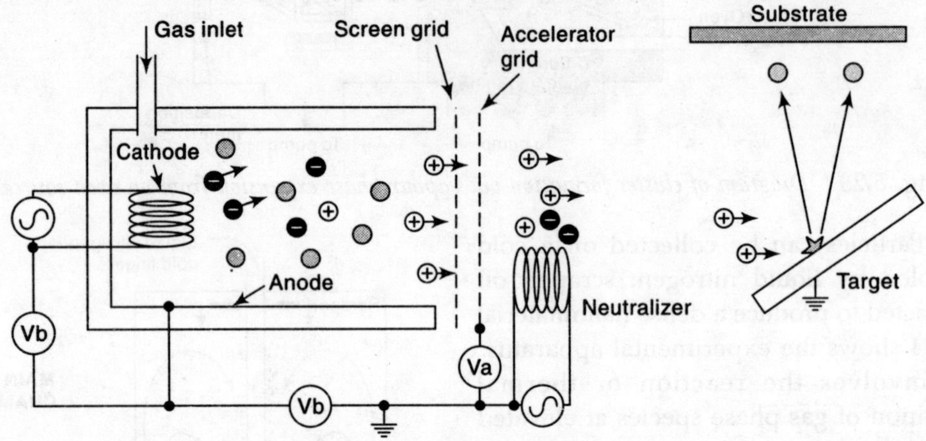

Fig. 5.25 ***Diagram of a DC glow discharge apparatus in which gas atoms are ionized by an electron filament and either deposit on a substrate or cause sputtering of a target.***

Magnetron sputtering

Magnetron sputtering involves the creation of a plasma by the application of a large DC potential between two parallel plates (Figure 5.26). A static magnetic field is applied near a sputtering target and confines the plasma to the vicinity of the target. Ions from the high-density plasma sputter material, predominantly in the form of neutral atoms, form the target onto a substrate. A further benefit of the magnetic field is that it prevents secondary electrons produced by the target from impinging on the substrate and causing heating or damage. The deposition rates produced by magnetrons are high enough (~1μm/min) to be industrially viable; multiple targets can be rotated so as to produce a multilayered coating on the substrate.

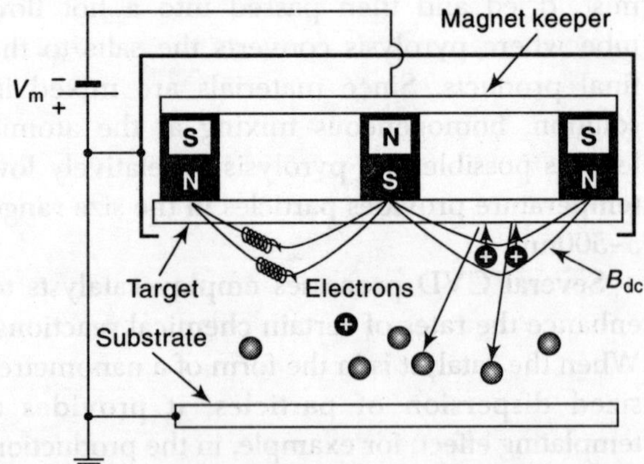

Fig. 5.26 ***Diagram of a magnetron sputtering apparatus that uses a magnetic field to contain ionized gas molecules close to a sputtering target, giving large deposition rates.***

Vacuum arc deposition

Vacuum arc deposition involves the initiation of an arc by contacting a cathode, constructed from the target material, with an igniter attached to an anode; this generates a low-voltage, high-current arc, which is self-sustaining. The arc ejects predominantly ions and large, micrometre-sized droplets from a small area on the cathode. The ions in the arc are accelerated towards a substrate and deflected using a magnetic field if desired; any large particles being filtered out before deposition (Figure 5.27). A vacuum cathodic arc can operate without a background gas under high-vacuum conditions, however, reactive deposition can be achieved by introducing a background gas such as nitrogen. The high ion energy produces dense films even at low substrate temperatures and consequently arc technology is commonly used for the deposition of hard coatings. As well as DC sources, plasmas can also be produced using radio (MHz) frequency and microwave (GHz) frequency power; these methods have the advantage of being able to provide higher current densities and higher deposition rates.

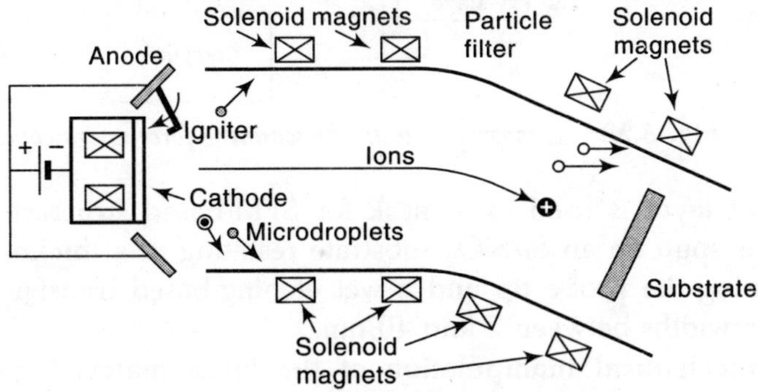

Fig. 5.27 *A vacuum arc deposition apparatus.*

In the plasma-assisted PVD processes described above, all vapour phase species originate from a solid target. Instead, plasma-enhanced CVD (PECVD) employs gas phase precursors that are dissociated to form molecular fragments which condense to from thin films or nanoparticles. The dissociation temperatures required for PECVD tend to be much lower than for conventional CVD processes due to the high energy of the plasma, and this may be importance for deposition on sensitive substrates such as semiconductor and polymers.

NANOFABRICATION BY SCANNING PROBE TECHNIQUES

1. By Scanning Force Probes

Scanning probe techniques, such as scanning force or scanning tunneling microscopy, are important ultramicroscopic methods for fabrication of nanostructures due to their ease of application and extremely high resolution.

Scanning force microscopy (SFM) generates a signal by interaction of the scanning tip with the object. The force between the tip and substrate is used for visualization of the substrate topography, and for the induce movement of weakly bound particles on the surface or for scratching into the upper substrate layer.

Structure generation takes place in analogy with imaging: the probe tip and substrate approach towards each other in the z-direction, while the distance or interaction force is regulated usually based on a light pointer (Fig. 5.28). Movement of the substrate in the *x*-and *y*-directions results in structures that are generated at the contact point of the tip with the surface.

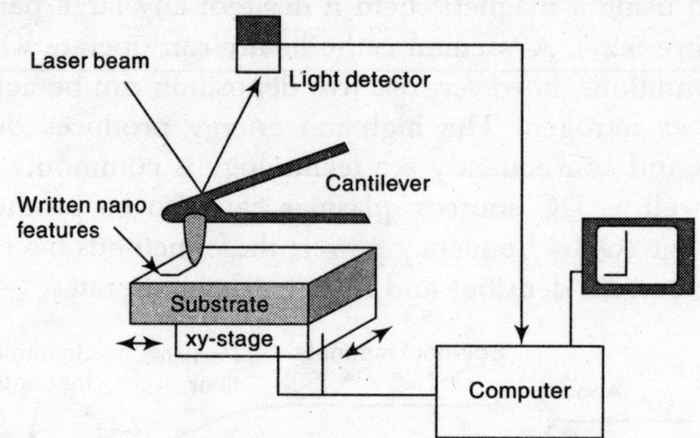

Fig. 5.28 *Lithography with the scanning force microscope.*

An ultrathin resist layer is used as a mask for SFM-based structure generation. A highly diluted photoresist is spun on an Si/SiO_2 substrate resulting in a thickness of 5.4 nm, prior to writing a pattern using the probe tip and a wet etching-based transfer of this pattern in the substrate, yields line widths between 2 and 40 nm.

In addition to mechanical manipulation of the layer material, individual cluster and nanoparticles are also addressed by SFM operations. Thus, 50 nm gold nanoparticles are pushed, by directed movement of the scanning tip, into a gap between two microstructured electrodes. This is arrangement exhibits single electron tunneling effects in a subsequent electrical characterization. Metal nanoparticles, nanorods and nanotubes are examples of scanning probe manipulations.

SFM is also able to dissect very small individual objects adsorbed onto surfaces. So a single DNA-plasmid could be cut into pieces down to dimensions of 20 nm.

The family of SFM-based structure generation also includes dip-pen nanolithography (DPN). Here molecules are transferred from the scanning tip to the surface during the scanning process.

Scanning force microscopy is also used for thermally based shaping of materials. Thus, the advantage of local heat transfer over small dimensions is combined with the small probe diameter. High resolution is realized for systems of a deformable material with fairly low heat transfer of a substrate that is rather heat conductive ,e.g., a thin PMMA layer on an Si wafer. The energy of a heated tip in contact with the PMMA layer is transferred and leads to a local softening of the material, so that the geometry of the tip is transferred into the surface topography of the resist layer. The adjacent regions of the resist are not affected, because the heat is efficiently carried off by the Si substrate, so that a steep lateral temperature gradient is realized. The pixel distances achieved of 40 nm in a PMMA layer of 40 nm thickness result in pixel densities with an interesting potential for future memory applications in the 400 Gbit cm^{-2} range.

Electrical Structure Generation by SFM

Electrically induced structure generation utilizes an electrical current to change the local surface state of the material. Often metals or semiconductors are transformed into their oxides, using the surface humidity layer or oxygen as a reaction partner.

Nanostructure generation by anodic oxidation with an SFM tip prefers materials like Si, Al and several transition metals, with a tendency towards stable oxide surface layers. The substrate is used as anode. Voltages in the range between 2 and 30 V are used to overcome the oxidation potential of the substrate. The voltage drops between the substrate and tip as a result of resistive surface layers on the scanning tip.

Conductive SFM is applied to form thin layers of Ti into line electrode structures of 75 nm width in isolated surroundings. Structures in InAs/AlGaSb are fabricated at 3–8 V. The properties of certain metals for the excellent local oxidation by SFM are utilized for SFM mask processes. Thus sub-micrometer structures in Si have been generated by an Al mask that was patterned by the local oxidation of thin Al mask that was patterned by the local oxidation of a thin Al mask layer prior to anisotropic pattern transfer by an RIE process, using a combination of laser and SFM lithography.

Also thin dielectric layers are accessible for conductive SFM structure generation. Therefor, the local current transforms the material into a state that is selectively removable. So dielectric layers are usable as resist material for nanolithography, when the underlying material is electrically conductive and used as the anode. A 100 nm thick spin-on glass layer (polysiloxane) has been structured at 70–100 V into linewidth of 100 nm. Also organosilane monolayers are used as high-resolution resists in SFM tip induced local oxidation. Trimethylsilyl layers yielded resolutions down to 20 nm and maximum writing speeds of up to 5 mm s^{-1}.

Pure organic layers are also applicable. Even a monolayer can act as a resist, when the subsequent process has a high selectivity. Trenches of 100 nm width are fabricated in Si monocrystals using a multi-layer resist process by local anodic oxidation of an octadecylsilyl monolayer using an SFM tip, and a subsequent transfer of this pattern into a thin SiO_2 layer, which acts as a transfer mask for an anisotropic wet etching of the Si substrate by a TMAH (tetra-methylammonium hydroxide) solution.

Scanning probe lithography is seriously hampered by the low writing sped. The speed can be increased by electronic control of the current during the scan and by the application of a parallel arrangement of several tips.

2. By Scanning Tunneling Microscope (STM)

STM is not only a microscopic method for highest (e.g., atomic) resolution, but also a tool for manipulation of very small structures, which had already been applied before the introduction of the SFM.

The manipulation of single atoms and small molecules with a scanning tunneling probe is achieved by tuning the interaction between the tip and the atom/molecule on the hand and between the atom/molecule and the surface on the other. Preferential treatment of one or the other interaction is achieved by adjustment of the STM potential in combination with the distance between the tip and surface. Thus single Cu atoms are moved or CO molecules are picked up and transferred by the STM tip on a crystalline Cu surface (Fig. 5.29). Also, Si atoms are moved on an Si(001) surface and arranged into lines, resulting in trench structures of 3 nm width and 2–3 nm depth.

Experiments at 4 K show the feasibility of manipulating individual atoms and arranging them into structures. On crystalline Ni, adsorbed Xe atoms are assembled by the STM tip into an IBM logo in an area of 5 nm × 16 nm. The information density of such structures is about 1 bit nm^{-2}, which corresponds to 100 Tbit cm^{-2}. However, the deposition and manipulation of individual atoms and small molecules, and the generation of structures with atomic resolution is limited. Therefore, the patterns created by individual noble gas atoms on crystalline surfaces at low temperatures represent ultimate limits of nanostructure technology, as they do not exhibit direct relevance for the fabrication of nanotechnical devices.

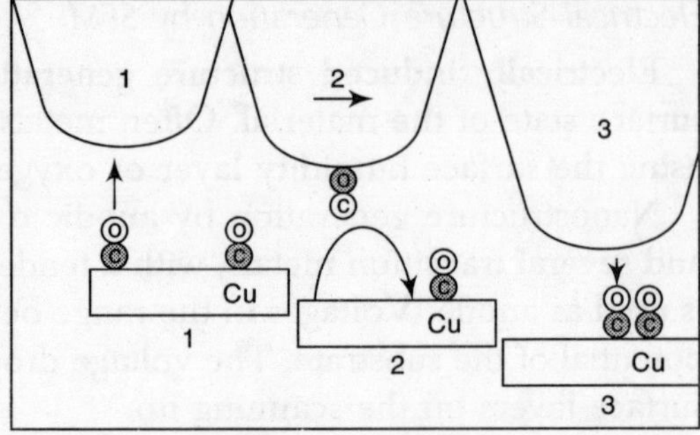

Fig. 5.29 *Scanning probe-based manipulation of single molecules on solid substrates; transfer of CO onto a Cu substrate.*

In (Fig. 5.30). the scanning tip positioning is realized by either force sensing (SFM) or is based on the tunneling current (STM). The latter uses the tunneling current for both distance control and local electrochemical transformation of the surface.

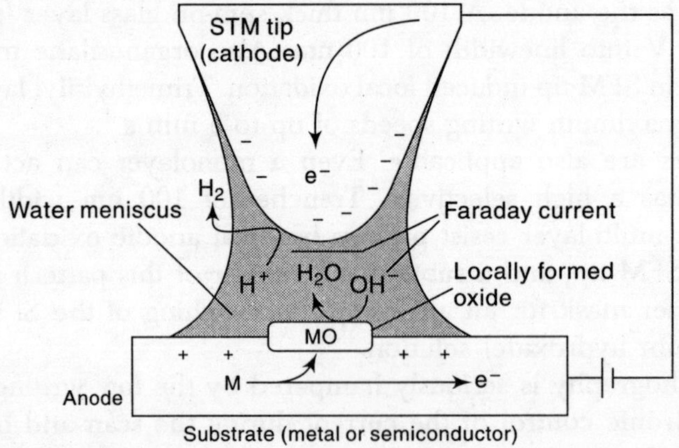

Fig. 5.30 *Electrochemical processes during the anodic manipulation of metals and semiconductors with the scanning tunneling microscope.*

Electrical Structure Generation by STM

Nanofabrication by STM depends on parameters such as material, potential, current, continuous or pulsed mode, besides increase in temperature, activation of surface diffusion for adsorbed particles, electromigration, field ionization and electrochemical reactions.

STM allows the modification of a mono molecular or even mono atomic adsorbate layers on the substrates. These processes are particularly efficient when the adsorbates really transfer into the gas phase. So special Si-H layers on monocrystalline Si are accessible for ultimate lithography. At negative tip bias, a passivating H layer is removed under UHV conditions from a monocrystalline Si surface. In this way, areas of 20 nm squares as well as line gratings with a period of only 3 mn are achieved (1 nm lines with 2 nm distances). These adsorbate-free areas oxidize in the presence of oxygen and create an oxide layer, so that the structures can be incubnated with ammonia. Nanostructured H layers on Si surfaces are also used to inhibit surface selective reactive deposition processes. Thus a CVD deposition of Al layers from a dimethylaluminum hydride occurs on H-free surfaces only, and STM-written structures are transformed into alumina structures with line widths below 3 nm. STM-generated areas with low H concentration in H-rich and less reactive surroundings can be applied for the selective deposition of organic molecules in nanostructures.

STM is also able to structure easily oxidizable metals and semiconductors in an anodic process. Lines of 100 nm width are structured in 30–50 nm thick chromium layers (Fig. 5.31). A simple electrochemical model includes a thin aqueous adsorbate layer as the electrolyte, the tip as the cathode and the substrate as the anode. Anodic oxidation of silicon yields SiO_x lines and meanders of high quality with line below 100 nm.

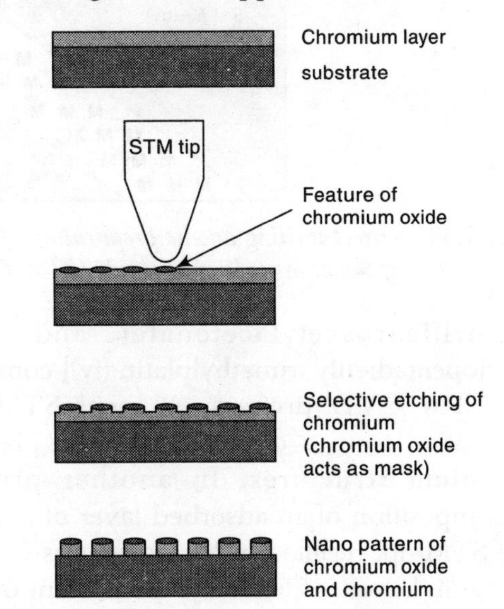

Fig. 5.31 *Farbrication of chromium nanostructures by a combination of local STM oxidation and a subsequent selective etching process.*

The tunneling tip also allows for a potential-controlled manipulation of individual atoms and small clusters, e.g., the extraction, transfer and deposition of Si on a substrate surface. Extremely plane surfaces are required for reproducible manipulations of atoms and molecules. This requirement is met by crystalline surface. Such as monocrystalline Cu.

Electrical pulses allow the transfer of groups of atoms from the probe tip onto the substrate surface (Fig. 5.32). Larger clusters of about 100 nm diameter are created in the case of tungsten or aluminium tips. Line gratings with a periodicity of 100 nm are fabricated by an equidistant arrangement of deposited Al clusters. Voltage pulse-induced material transfer from a gold tunneling tip creates gold islands of 10–40 nm diameter. A W-tip writes lines of only 0.8 nm width on a Ge monocrystalline surface.

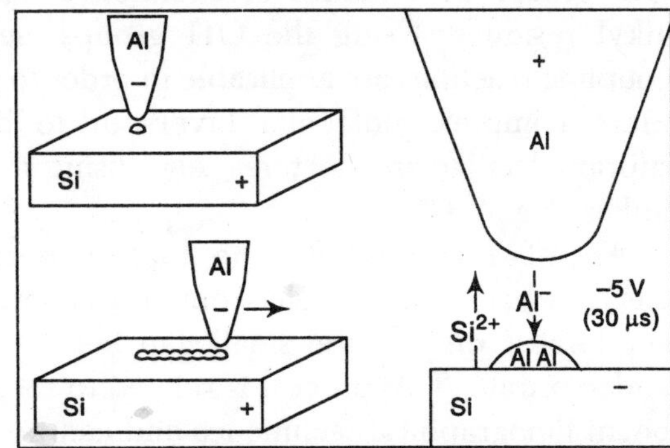

Fig. 5.32 *Preparation of aluminium nanostructures by transfer from a scanning probe tip using an electrical pulse.*

An additive writing process of metal structures is made from the supply of cleavable gases. To create noble metal structures, an electrically induced local decomposition of organometallic gases are also used. (Fig. 5.33). Dots of only 30 nm diameter are made with gold [dimethyl-gold

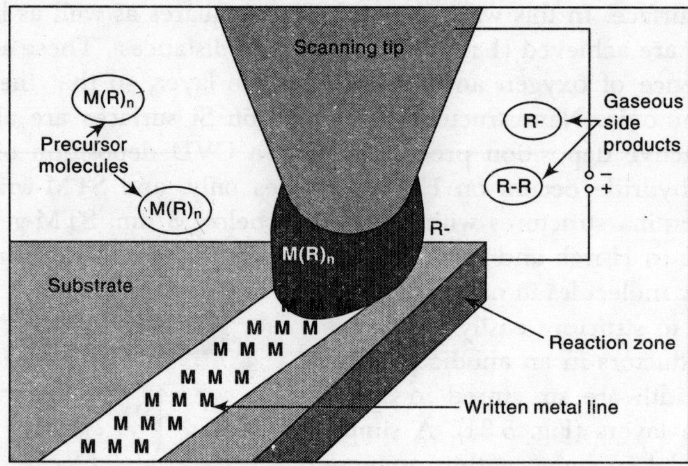

Fig. 5.33 *Direct-writing anodic preparation of metal nanostructures by scanning probe-induced decomposition of metal organic compounds (STM-CVD).*

(iii)-trifluoroacetylacetonate] and platinum [cyclopentadienly-trimethylplatin (iv)] compounds in an STM CVD process. A local STM-induced deposition from a dimethylcadmium gives 20 nm cadmium structures. In another process the decomposition of an adsorbed layer of ferrocene by the STM tip, resulted in line structures with a width of 3 nm. Layers of trimethyl aluminium on graphite substrates yielded tip-induced structures of 20 nm width, sometimes even holes of 2.7 nm diameter.

Local tip based oxidation of monomolecular alkyl layers results in areas with high densities of reactive OH groups in a matrix of chemically fairly inert alkyl residues. Using the OH groups, secondary coupling reactions are applicable in order to generate more complex molecular layers or to build up supramolecular architectures, e.g., using a siloxane linker (Fig. 5.34).

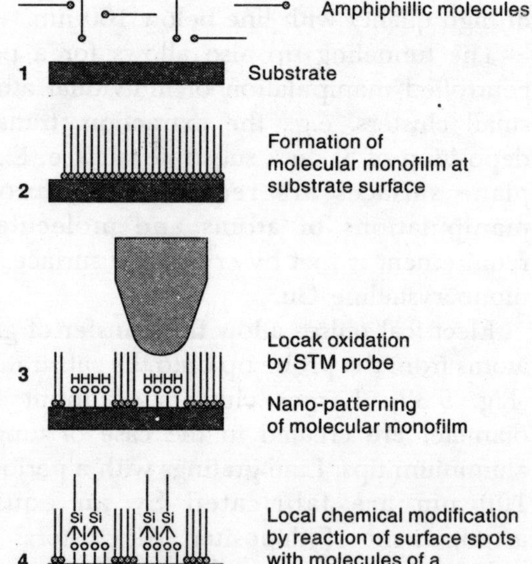

Fig. 5.34 *Preparation of a lateral nanostructured molecular monolayer by local oxidation using a scanning probe tip and subsequent local surface modification.*

An STM is used for the local generation of electron beams as the electrons are generated and accelerated only in the small gap between the tip and substrate. This process is related to the electron beam lithographical techniques and can be used for the exposition of typical electron beam resists with structure widths of about 15 nm being achieved. Direct structure generation with a low energy electron beam (LEEB) uses a beam extracted from the STM tip (cathode) with electron energies of 30–160 eV and beam currents of 1 nA–1 μA. At substrate temperatures of 873–973k oxide flms on Si substrates are decomposed, yielding structure dimensions of below 50 nm. In addition to the ultrathin layers of standard electron beam resists. Langmuir-Blodgett-films (LB-films) are also applicable as resist layers. Structures written with STM in LB-films are transferred into aluminium by wet etching.

By soft lithography techniques (Figure 5.35) pattern a resist by physically deforming (or embossing) the resist shape with a mould or stamp, which is coated with a chemical that reacts with the resist at the edges (of the stamp). These techniques produce patterned structures in the range 10 nm and above. One of the main limitations on resolution arises from plastic flow of the polymeric materials involved. Master moulds are fabricated using lithographic techniques, micromachining or naturally occurring surface relief on the substrate materials. Once the master mould is formed, a low molecular weight pre-polymer is poured and then cured to yield a polydimethylsiloxane (PDMS) elastomeric 'stamp' which is peeled off the master. Thus, on the bottom of the PDMS stamp are an inverted replica of those on the master. They are inked with a molecule of interest across their surfaces and placed against a substrate, transferring molecules in a pattern that reflects the features of the stamp. Elastomeric PDMS stamps are used as conformal contact masks for phase-shifting photolithography, facilitating the fabrication of structures as small as 50 nm in photoresist. Channels formed in PDMS stamps are used to mould three-dimensional structures, a process known as micromoulding in capillaries (MIMIC).

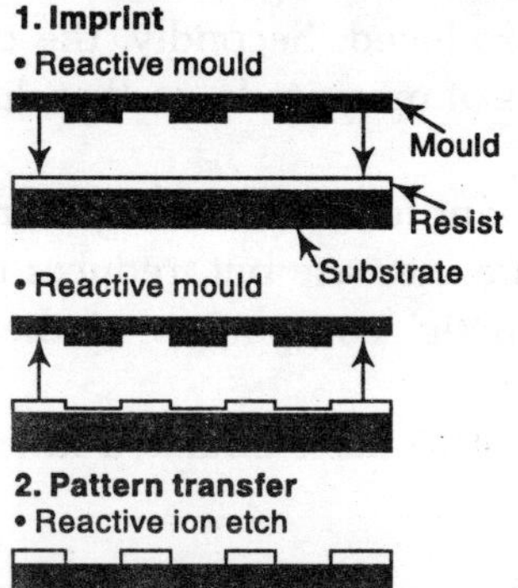

Fig. 5.35 *Schematic representation of the soft lithographic process which patterns a resist using a stamp followed, if required, by subsequent etching.*

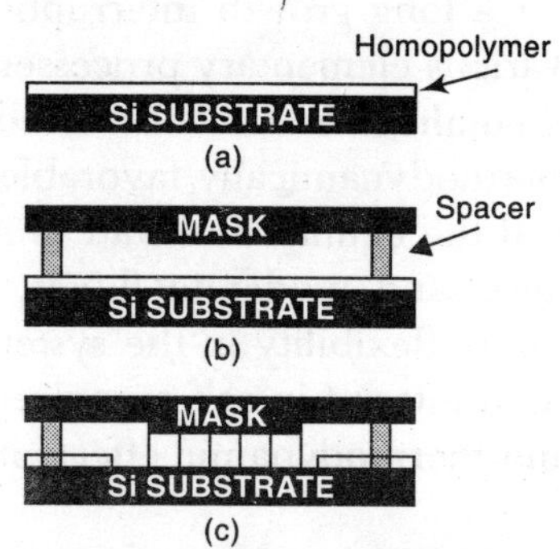

Fig. 5.36 *Schematic representation of the self-assembly process of a surface polymer film lithographically induced by a mask.*

Another variation of this technique is lithographically induced self-assembly (Figure 5.36), in which a mask is used to induce and control, via electrostatic and hydrodynamic instabilities, the formation of periodic supramolecular pillar arrays on a thin polymer melt initially deposited flat onto a substrate. Microcontact printing of self-assembled monolayers (SAMs) is achieved by transfer of a SAM 'ink' onto a substrate surface from a patterned 'stamp', possessing elastomeric properties, to achieve intimate contact with the surface.

GROWTH AND CHARACTERIZATION TECHNIQUES

To build nanostructures one has to deal with individual atoms, major success in nanofabrication has been achieved by epitaxial growth via self-organization phenomenant surface. In modern epitaxial growth techniques like molecular beam epitax (MBE) or metalorganic chemical vapor deposition (MOCVD), it is a prerequisite in semiconductor nanotechnology. Different ways of controlling chemical composition, point defect concentration, growth rate, etc., plays a significant role in optimizing growth parameters to obtain high-quality structures with planar or periodically-modulated interfaces. A range of materials are used in MBE growth and two approaches are used to spontaneous formation of ordered nanostructure on crystal surfaces. (1) Thermodynamic approach and (2) Kinetic approach.

Thermodynamic approacheses are generally based on the assumptions, firstly, that a real system is approximated as a closed system, and secondly, that all kinetic processes are sufficiently fast to drive the system into its equilibrium (lowest free energy) state. Kinetic approaches, assume that approximation by a closed system is inappropriate, or that the equilibrium state cannot be reached on the time scale. of real experiments. Then all the main properties of the system, such as concentrations and types of point defects, geometrical size and shape of nanoislands, etc., are defined by particular kinetic pathways. This is, however, Firstly, a realistic kinetic model which leads to the same result as thermodynamics, if the growth rate is extrapolated to zero, or a long growth interruption or annealing is introduced. Secondly, the characteristic rates of various elementary processes differ by a few orders of magnitude, so that different types of partial equilibrium are considered.

The thermodynamically favorable state is highly important for constructing a proper kinetic pathway. If the equilibrium state is reached the growth rate increase but, reduces the substrate temperature, and moderate fluxes, i.e., enhance the 'kinetic' component of the process, to achieve more flexibility in the system.

Increased interest in self-organized growth of nanostrucutres has stimulated interest between kinetic and thermodynamic effects in MBE growth.

MOLECULAR BEAM EPITAXY (MBE)

MBE Apparatus

Molecular beam epitaxy (MBE) is a technique for the epitaxial growth of materials. It operates via the chemical interaction of one or several molecular or atomic beams of different intensities and compositions, which occurs on the surface of a heated single-crystalline substrate. A schematic representation of the MBE apparatus is shown in Fig. 5.37.

Source materials are place in evaporation cells composed of a crucible, whose shape and dimensions ensure that required angular distribution of atoms or molecules in the beam, a resistive heater, and thermal screens. The angular distribution of the beam and the distances between the sources and the substrate determine the homogeneity of the parameters of epilayers and heterostructures grown by this technique. A manipulator with a substrate holder is used to ensure the required position of the substrate relative to the cells and to heat it to the required

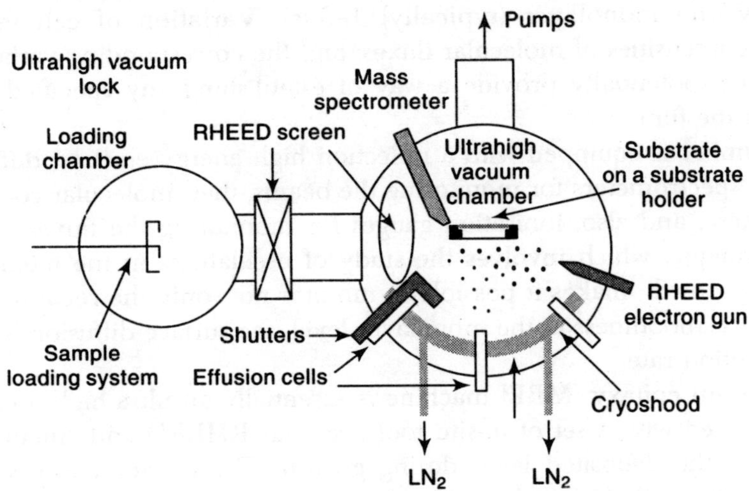

Fig. 5.37 *Molecular beam epitaxy apparatus.*

temperature. The homogeneity of the resulting films is often improved by rotating the substrate.

For typical distances between the sources and the substrate, the molecular beam condition is ensured (i.e., the free path of the particle is larger than the geometrical size of the chamber), if the total pressure does not exceed 10^{-4} mm. However, all MBE systems maintain an ultrahigh vacuum ($\approx 10^{-11}$ mm) and operation is usually oil-free because the substrate has to be atomically clean before growth. A low level of doping and control over the properties of the grown materials and structures is obtained if uncontrolled fluxes of atoms reaching the substrate surface are weak. An ultrahigh vacuum is essential to have an equilibrium between the rate of gas evolution and the rate of pumping. Thus, crucible materials with the lowest possible rate of gas evolution must be used. The usual crucible material for MBE with III–V compounds is boron nitride, which combines a low rate of gas evolution with a weak chemical activity right up to temperatures of the order of 1773k. Also, it is important to ensure cryogenic screening around the substrate so that it minimizes stray fluxes of atoms and molecules from the walls of the chamber, and from the heated components of the apparatus. Not only the total pressure, but also the partial composition of the atmosphere is monitored and, if necessary, altered. By exposing the substrate to a hydrogen flux. A hydrogen flux can be used in order to reduce the partial pressures of the most active components of the background atmosphere and hinder the incorporation of undesirable impurities. Only ultra-pure materials can be used as source materials in MBEs.

One advantage of MBE technology is the intrinsic feasibility of controlling profile of the composition and doping of a growing structure at the monolayer level. This feasibility is ensured by a molecular beam regime during growth that excludes any interaction between atoms or molecules in the beam and between different beams, in combination with a relatively low growth rate. An abrupt change in the composition and /or degree and nature of doping are achieved by opening or closing the relevant fluxes using the shutters with which each cell is equipped. The operation time of a shutter (<0.1 s) is usually considerably less than the time needed to grow one monolayer (typically) 1–5 s). Variation of cell temperatures and, consequently, the intensities of molecular fluxes, and the corresponding variation, if necessary, of the growth rate, potentially provide a way of establishing any specified composition and doping profile in the film.

The MBE chamber is equipped with a reflection high energy electron diffraction (RHEED) system and mass spectrometers for monitoring the beams, their molecular composition, and the residual atmosphere, and also, ionisation gauges for monitoring the fluxes. A modification of the RHEED technique, which involves the study of oscillations in the intensity of diffraction reflections during growth, makes it possible to monitor not only the reconstruction of the film surface, but also its smoothness at the monolayer level, the surface diffusion length of migrating atoms, and deposition rate.

A molecular beam epitaxy (MBE) machine is essentially an ultra-high-precision, ultra-clean evaporator, combined with a set of in-situ tools, such as RHEED and Augur spectroscopy for characterization of the deposited layer during growth. The reactor consists of an ultra-high-vacuum chamber (typically better than ~5×10^{-14} atmospheric pressure) of approximately 1.5 m diameter (Figure 5.38)

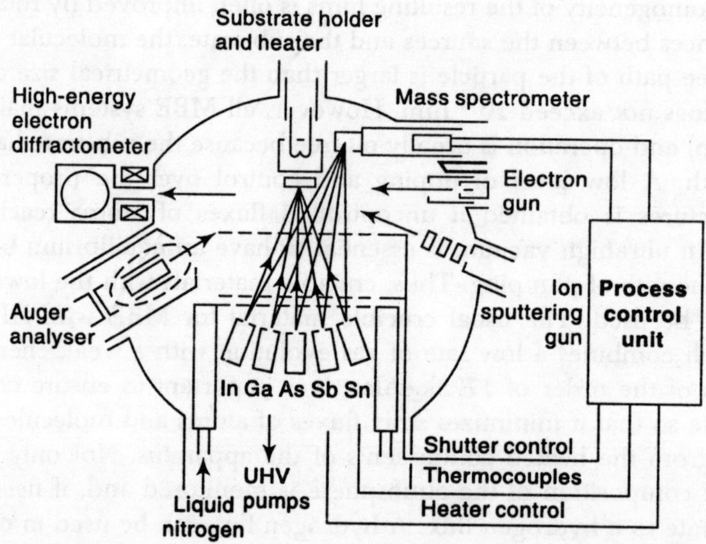

Fig. 5.38 *Schematic diagram of a molecular beam epitaxy thin film deposition system.*

MBE is growth technique in which epitaxial, single atomic layer (~0.2–0.3nm) are grown on a heated substrate under UHV conditions, using either atomic or molecular beams evaporated from effusion sources with openings directed toward a heated substrate usually consisting of a thin (~0.5nm) wafer cut from a bulk single crystal. The sources can be ether solid or gaseous and an MBE machine will typically have an array of multiple sources, which can be shuttered to allow layered, alternating heterostructures to be produced. Semiconductor quantum wells, superlattices and quantum wires and metallic or magnetic multilayers for spin valve structure are deposited using this technique.

Standard MBE uses elements in a very pure from as solid contained within a number of Knudsen cells. In operation the cells are heated to the temperature at which the elements evaporate, producing beams of atoms which leave the cells. The beams intersect at the substrate and deposit the appropriate semiconductor, atomic layer by atomic layer. The substrate is rotated to ensure even growth over its surface. By operating mechanical shutters in front of the cells, it is possible to control which semiconductor or metal is deposited. For example, opening the Ga and As cell shutters results in the growth of GaAs. Shutting the Ga cell and opening the Al cell switches the growth to AlAs. As the shutters can be switched rapidly, in comparison to the rate at which material is deposited, it is possible to grow very thin layers exhibiting very sharp interfaces. Other effusion cells contain elements required for doping, and it is possible of monitor the growth by observing the electron diffraction pattern produced by the surface.

MOVPE

Another, main epitaxial technique is metalorganic vapour phase epitaxy (MOVPE), also known as metalorganic CVD (MOCVD). In this technique the required elements are transported as components of gaseous compounds such as metal alkyls and non-metal hydrides (e.g., trimethylgallium $(CH_3)_3Ga$, arsine AsH_3) TO A suitable chamber where they flow over the surface of a heated substrate. These compounds break down and react so as to deposit the relevant semiconductor, with the remaining waste gases being removed from the chamber MOVPE is used extensively for the production of compound semiconductor (e.g., III–V) thin films, such as the production of GaAs and AlGaAs layers using trimethyl gallium (TMG):

$$Ga(CH_3)_3 + AsH_3(g) \rightarrow GaAs(s) + 3CH_4(g),$$

$$xAl(CH_3)_3(g) + (1-x)Ga(CH_3)_3(g) + AsH_3(g)$$
$$\rightarrow Al_xGa_{1-x}As(s) + 3CH_4(g).$$

Valves in the gas allow gas switching, thereby enabling layered structures to be deposited. However, since rapid switching of a gas is more difficult than for an atomic beam, and because of residual gas adsorption on the walls of the feed pipework, the interfaces between adjacent layers grown by the gas source epitaxy methods are not as abrupt as those attainable with solid MBE. For the same reason, solid source MBE is capable of growing thinner layers, and the slow growth rates involved (of order 1 μm per hour) generally give rise to higher precision in layer thicknesses compared to MOVPE and MOMBE. High growth rate of MOVPE is suited for production, especially for semiconductor laser where thick cladding or waveguiding layers must be grown in a reasonable timescale, in addition to the thin quantum well layers. MOVPE is also easier to scale up to larger systems, permitting simultaneous growth on multiple substrates. Gas source MBE, including metal organic MOMBE, operates with speeds and precisions which are intermediate between those of solid-source MBE and MOVPE. All the metalorganic epitaxy techniques have safety implications as the gases used are highly toxic.

The aim of molecular beam epitaxy is to produce a perfect single-crystal structure whose morphology and lattice spacing exactly match those of the substrate material. This requires the equilibrium lattice constant of all the different epitaxial layers to be closely matched to that of the substrate, other wise dislocations will form combinations of layers, thus, a suitable substrate cannot be identified, in which case in initial buffer layer is deposited of another materials.

MBE growth is generally far from equilibrium and various growth modes can be identified as shown in Figure 5.39. The prevalence of each mode depends on the exact substrate temperature and deposition rate.

- 1D step propagation growth occurs at high temperature and/or low growth rates. It occurs on vicinal, stepped surfaces when the surface diffusion of the growing species is pronounced. Deposited materials diffuses to a step and is incorporated into the laterally growing terrace. The growth mode produces very flat and smooth interfaces.
- 2D nucleation and growth occurs at intermediate temperatures and/or growth rates and involves the initial nucleation of islands, which grow and coalesce to form a network of interconnected islands separated by holes. These holes are finally filled and a flat surface is formed. The process then repeated in a cyclical fashion.
- 3D rough growth generally occurs at low temperatures and high deposition rates. Islands nucleate on existing islands and extensive faceting occurs, producing very rough interfaces.

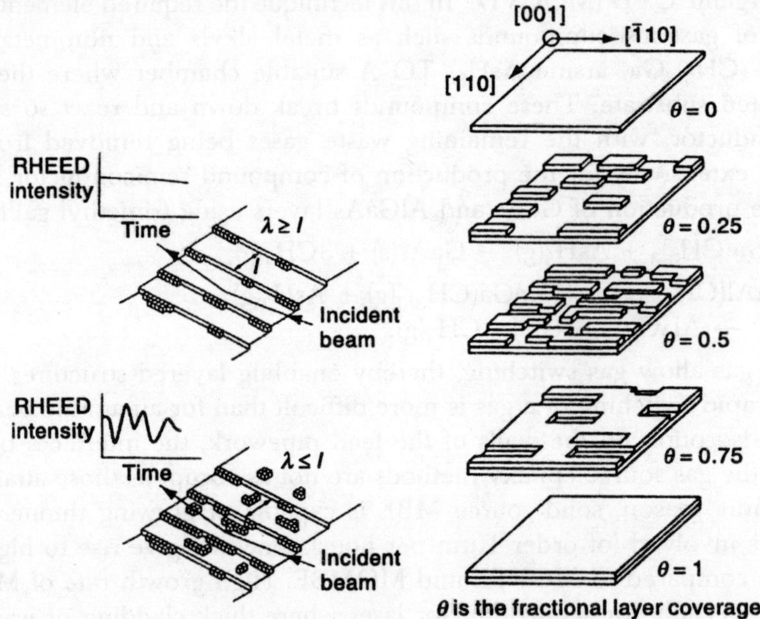

Fig. 5.39 *MBE growth modes: (left) 1D step propagation (and (right) 2D nucleation and growth*

Because the dopants in both MBE and MOVPE systems exist as additional switchable sources, they may be added in known amounts and distributed over known depth concentration profiles.

Both MBE and MOVPE are very successful in growing quantum well structures with a sensitive control of layer thickness. However, it is much more difficult to achieve fie control of lateral dimensions in devices for the production of either quantum wires or quantum dots. One technique for achieving lateral control is by the use of patterning. Patterning can be carried out before or after epitaxial growth (or both) using lithography or etching. In addition, under certain growth conditions, quantum wires or dots may form spontaneously during the epitaxial growth of strained materials.

Liquid Phase Methods

A variant of many of the PVD processes are thermal spraying techniques in which a spray of molten or semi-molten solid particles generated by either an electrical thermal source (e.g., plasma spraying or by chemical combustion (e.g., flame spraying or high-velocity oxygen fuel spraying) are deposited onto a substrate and undergo rapid solidification. This is extensively used to produce nanocrystalline coatings from nanocrystalline powder, wired or rod feedstocks previously fabricated by the mechanical milling or precipitation routes.

More generally, liquid phase chemical synthesis involves the reaction of a solution of precursor chemicals in either an aqueous or non-aqueous solvent; these precursors react and/or naturally self-assemble to form a solution and ultimately results in nucleation of the product either homogeneously (in solution) or heterogeneously (on external species such as vessel walls or impurities). Initial nuclei then grown into nanometre-sized particles or architectures according to both thermodynamic and kinetic factors. The nature, size and morphological shape of the precipitated structure can often be controlled by parameters such as temperature, pH, reactant concentration and time. A multicomponent product may require careful control of co-precipitation conditions in order to achieve a chemically homogenous final product.

Colloidal Methods

Colloidal methods rely on the precipitation of nanometre-sized particles with a continuous fluid solvent matrix to form a colloidal sol. Generally a finely dispersed system is in a high free energy state, as work has essentially been done to break up the solid, which this is equivalent to the free energy required to produce the increased surface area. The colloidal material therefore tends to aggregate due to attractive van der Waals forces and lower its energy. The presence and magnitude of an energy barrier to agglomeration depends on the balance of attractive and repulsive forces between the particles, as shown in Figure 5.40. The energy of a colloidal

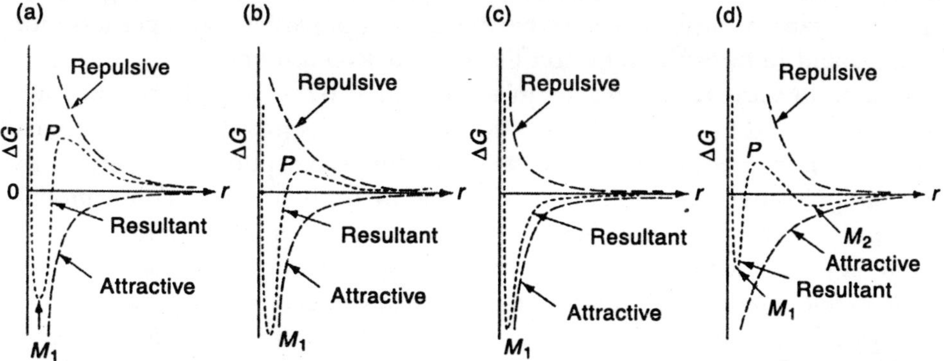

Fig. 5.40 *Possible energy balance between attractive and repulsive interparticulate force as a function of interparticle separation. In (a) there exists a large energy barrier (P) to strong aggregation of particles (minimum M_1); this is smaller in (b) and absent in (c) owing to reduced long-range repulsive forces in (d) can lead to weak flocculation of particles (minimum M_2).*

solution which is available for aggregation is K_BT. Agglomeration (the formation of strong compact aggregates of nanoparticles) and also flocculation (the formation of a loose network of particles) is prevented by increasing the repulsive energy i.e. potential energy, which is achieved by electrostatic or steric stabilization.

Surface charges on colloid particles are easily produced by ionization of basic or acidic groups as the pH is varied; example, $TiCl_4$ hydrolyses in the presence of a base and form a colloidal solution of TiO_2. Here surface OH^- group on TiO_2 clusters act as an electrostatic colloid stabilizer produced by the attachment of a capping layer to the particle surface. Additional chemicals are added to the colloidal solution that binds to the cluster surface and blocks vacant coordination sites, to prevent further growth. These additives are polymeric surfactants or stabilizers that attach electrostatically to the surface, or anionic capping agents which covalently bind to the cluster. A further possibility is to precipitable a material within a volume of space defined by a micelle or membrane which acts a barrier to further growth.

Subsequent processing of colloids involve a number of approaches, including additional colloidal precipitation on particle surfaces to produce a core-shell nanoparticle structure, deposition on substrates to produce quantum dots, self-assembly on substrates as ordered 2D and even 3D arrays, and finally embedding in other media to form a nanocomposite. Colloidal methods are relatively simple and have been extensively used for the production of metal and semiconductor nanocrystals using optimized reactions and reaction conditions. One problem in many colloidal methods is that colloid solution particles increase in their size as a function of time.

FABRICATION BY SOL-GEL METHODS

Materials with a combination of the excellent thermodynamics of composites with the rapid kinetic of monomolecular materials are used to decrease the particle size of the constituent phases in composite energetic materials. Thus the surface area of the materials increase i.e. there is increased contact area between fuel and oxidizer, Hence a much larger fraction of the composite reacts instantaneously. The effective downsizing of particles also permits more intimate mixing of phases and reduces the diffusion distances between them.

That a decrease in particle size results in a qualitative increase in burn rates in solid-solid mixtures of oxidizers and fuels. An increase in contact area is possible with nanomaterials and lead to energetic nanocomposites with new and sometimes exceptional properties. By reducing the diffusion distances the mass and thermal transport problems that conventional energetic composites suffer from are mitigated. Using this approach the energy release rates of nanocomposite energetic materials may approach those of monomolecular materials.

Pyrotechnic nanocomposites of MoO_3 and Al burn at extremely rapid rates (> 100 m/s). The initiation, energy release, and safety properties of energetic materials are dependent on particle size and morphology. Therefore energetic nanocomposites will have very interesting properties.

There exist numerous physical and chemical synthesis and proceeding routes (e.g., vapor condensation methods, sol-gel micellular, sonochemical, sonication, and ball milling) to prepare nanomaterials of both oxidizers and fuels.

Sol-gel methods involve a set of chemical reactions which irreversibly convert a homogeneous solution of molecular reactant precursors (a sol) into an infinite molecular weight three-dimensional polymer (a get) forming an elastic solid filling the same volume as the solution. Typically this involves a hydrolysis reaction followed by condensation polymerization, for example.

$$\text{tetraethylorthosilicate (TEOS) } (Si(OC_2H_5)_4)$$
$$+ \text{ ethanol}(C_2H_5OH) \xrightarrow{\text{react and dry}} SiO_2 + \text{other product}$$

Basic Process

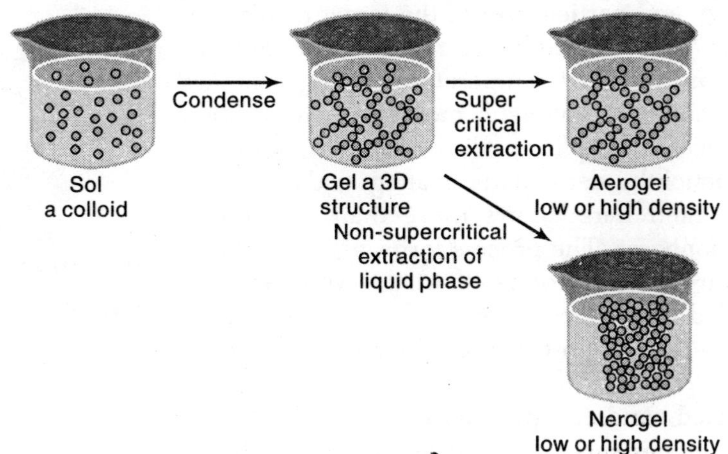

Fig. 5.41 *The basic sol-gel process.*

Molecular precursors are induced to undergo hydrolysis and condensation in solution to form a sol.

Mixtures of precursors are used to produce binary or ternary systems, each molecular precursor having its own reaction rate that is dependent on the conditions (e.g., pH, concentration, solvent and temperature). The polymer gel so formed is a 3D skeleton surrounding interconnected pores Fig. 5.41, and this can be dried and shrunk to form a rigid solid form. In a single precursor component system, the final material can be designed to have interconnected nanoscale porosity and hence a high surface area. For example, nanoporous silica can have a variable percentage porosity that depends on the precursor and the solvent employed.

Also, precursors containing nanometre-sized particles used are aqueous colloidal sols. The transformation to a gel is most often achieved by changing the pH or the concentration of solution composed of interconnected nanometre particles surrounding pores that are generally bigger than in the corresponding material derived from molecular precursor.

The main benefits of sol-gel processing are the high purity and uniform nanostructure achievable at low temperatures. The method is often used to produce metal-oxide nanomaterials. Further processing usually involves forming the gel using a number of techniques (see below) followed by gentle drying to remove the solvent (this often lead to shrinkage so care is needed to prevent cracks forming during this process). Gel can be cast and moulded to form a microporous and dried to produce a monolithic bulk material (e.g., a xerogel or an aerogel) used to form filters and membranes. They are spin coated or dipped to produce thin (typically 50–500 nm) films on substrates. These films are used for electronic thin film devices, for wear, chemical or oxidation protection, as well as for their optical properties (e.g., anti-reflection). Thus, fibers can be drawn from the gel; e.g., silica fibres for light transmission.

In general, aerogels have lower densities, higher porosities, and higher surface areas than xerogels, but both materials contain a architecture of interconnected particle sand pores with nanoscopic dimensions.

Sol-gel approach to energetic materials offer the possibility to precisely control the composition, density, morphology and particle size of the target material at the nanometer scale. An added feature is near or ambient temperature gelation and low temperature drying schemes prevent degradation of the energetic molecules. In addition the numerous processing options allow the fabrication of powders, porous monoliths, dense parts, coatings and thin films. The method presented here allows more than the simple mixture of two nanophases. The process forms a porous three-dimensional structures network of nanoparticles whose cavities can be partially filled with the complementary oxidizer or fuel. A schematic of the sol-gel nanostructures energetic materials is shown in Figure 5.42.

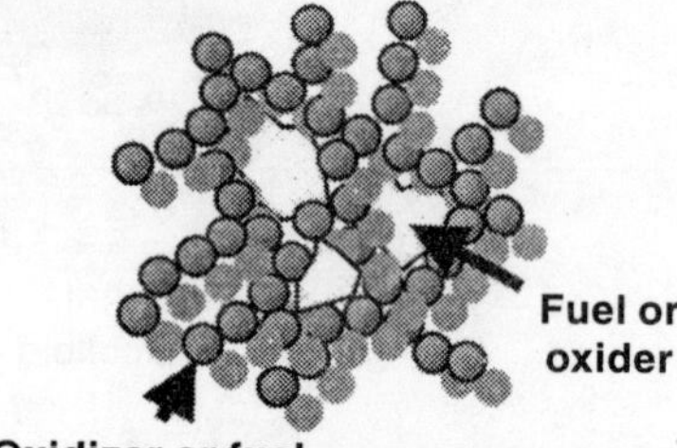

Fig. 5.42 *Idealized representation of sol-gel nanostructures energetic material.*

The interconnected nanoscale porosity in the dried gel is filled via incorporation of a second material using techniques such as liquid infiltration or chemical reaction. These materials are known as nanocomposites. Diphasic gels use the initial gel for the precipitation of a second phase by sol-gel routes. Also, an organic material, incorporated as a monomer within an inorganic gel host; is polymerized to from a hybrid material. If a dense rather than a nanoporous material is desired, dying is followed by sintering at higher temperature. The high surface area leads to rapid densification, which can be accompanied by significant grain growth if temperature are too high.

The nansostructuring of the composit performs two unique functions relative to simple mixing of nanopowders. First, it provides vary intimate mixing of the phases as the sol-gel phase is formed *in situ*, in a colloid of the added particulate phase, and it grows around the suspended particles. This methodology promotes very intimate mixing of oxidizer and fuel. Second, it provides a three-dimensional nanostructure where the two phases are interlocked and will resist separation upon manipulation and have different properties relative to mixtures of nanopowders.

The method is very versatile as the composition of the sol-gel nanostructured material can be readily varied to give the resulting material varied properties. For example the skeletal sol-gel matrix in Figure 5.42 can be made up of oxidizer (e.g., Fe_2O_3, WO_3, SnO_2) or fuel (e.g., hydrocarbon resin or carbon) and the imbedded particulate phase may be fuel (Al) or oxidizer (e.g., NH_4ClO_4). In the metal oxide/Al systems the primary output would be heat and light whereas the organic sol-gel nanostructured material would generate gas upon reaction. The compositional versatility of this approach may lead to applications in a wide range of areas. The unique compositional and processing properties of sol-gel nanostructured energetics make it and attractive technology for new applications of energetic materials. For example, the method may be amenable to creating energetic coatings on wafers or on Microelectromechanical (MEMS) devices.

The experimental

Below is given an experimental procedure followed by conclusion of its result for the fabrication of sol-gel nanomaterials.

Preparation of $Fe_xO_y/Al(s)$ pryotechnic nanocomposites. To prepare a $Fe_2O_3/Al(s)$ pyrotechnic nanocomposite a Fe_2O_3 sol is generated and carefully, monitorsed. Ultra fine grain aluminum is suspended in ethanol and sonicated for 15 minutes. Near the gelation point of the Fe_2O_3 sol, the UFG Al is added and the composite mixture geled soon after. Samples prepared by this method are subjected to the above drying conditions for processing to aerogels or to xerogels.

Safety characterization. Small scale testing of energetic materials and other compounds is done to determine their sensitivity to various stimuli including thermal degradation, friction impact and static spark. Tests such as Thermal Analysis, Drop Hammer (Impact Sensitivity), Friction and Spark are performed.

The sensitivity these materials toward electrostatic discharge is measured on a electrostatic discharge tester. Samples are loaded into Teflon washers and covered with a 1 mm thick tape. The sensitivity is defined as the highest energy setting at which 10 consecutive "no-go" results are obtained using a 10 kV potential. The frictional sensitivity of the sol-gel nanostructured energetic materials is evaluated using a high friction sensitivity tester. The tester employs a fixed porcelain pin and a movable poecelain plate that executes a reciprocating motion. Weight affixed to a torsion arm allows for a variation in applied force between 0.5 and 36.0 kg. The relative measure of the frictional sensitivity of a material is based upon the largest pin load at which is less than two ignitions (events) occur in ten trials.

"Drop-Hammer Machine" is used to determine the impact sensitivity of these materials relative to the primary calibrants.

Preparation of resorcinol/formaldehyde/ammonium perchlorate energetic nanocomposites. A porous organic solid matrix is prepared by the polycondensation of resorcinol with formaldehyde (RF) using Na_2CO_3 as a catalyst. Subsequent crystallization of an oxidizer, ammonium perchlorate (AP), within the pores of the gel matrix, completes the synthesis. The oxidizer is first dissolved in water, and then added to the reacting RF sol prior to gelation. Gelation of the RF sol occurrs first. Crystallization of the AP is then incudes by exchanging the liquid residing in the pores for a solvent in which the oxidizer is insoluble. The final step of removing the pore fluid is done by slow evaporation resulting in a dense solid.

Processing of iron (III) oxide gels. The wet gels are covered and allowed to age for at least 24 hours under ambient conditions. After that they are immersed in a both of absolute ethanol where they are washed for ~1 week. Aerogel samples are processed in supercritical point drier. The solvent liquid in the wet gel pores is exchanged for $CO_2(l)$ for 2-3 days, after which the temperature of the vessel is ramped up to ~318 k, while maintaining a pressure of ~100 bars. The vessel is then depressurized at a rate of about 7 bars per hour. (CAUTION: Take special care when doing supercritical drying of energetic composites. Use only CO_2 (low temperature ~318 k) supercritical extraction vessel.

Physical characterization. Surface area analysis is performed using an surface are analyzer. Samples of aerogel (0.1 – 0.2 g) are heated to 473 k under vacuum (10^{-5} mm) for at least 24 hours to remove to remove all adsorbed species. Nitrogen adsorption is taken at five relative pressures from 0.05 to 0.20 at 77 K, to calculate the surface area high-resolution transmission electron microscopy (HRTEM) is performed at 300 keV using zero loss energy filtering to remove inelastic scattering. The images are taken under bright field conditions and slightly defocused to increase contrast. The images are also recorded on a 2k × 2k CCD camera Powdered samples are dispersed in methanol and drops of the mixture are deposited onto a carbon coated Cu grid using a pipette. Scanning electron microscopy (SEM) is carried out. Typical accelerating voltages used for aerogel samples ranged from 1.8-6 kV and depended on sample conductivity. No sample preparation (i.e., coating with conductive layer of Au) is performed on the samples.

Conclusions

Sol-gel methods are developed to prepare high surface area nanostructures iron (III) oxide materials. These materials are made with high surface area (~250-350 m^2/g^{-1}), either crystalline or amorphous, and with reasonable purities using a simple one step chemical method. In addition, nanometer-sized iron (III) oxide nanostructures are prepared with two distinctly different particle morphologies. Figure 5.43 shows SEM images of two different morphological types, one with needle-like morphology and the other with particulate morphology. Thus, sol-gel nanostructured materials with these two types of iron (III) oxide nanoparticles have very ignition, safety, and energy release properties.

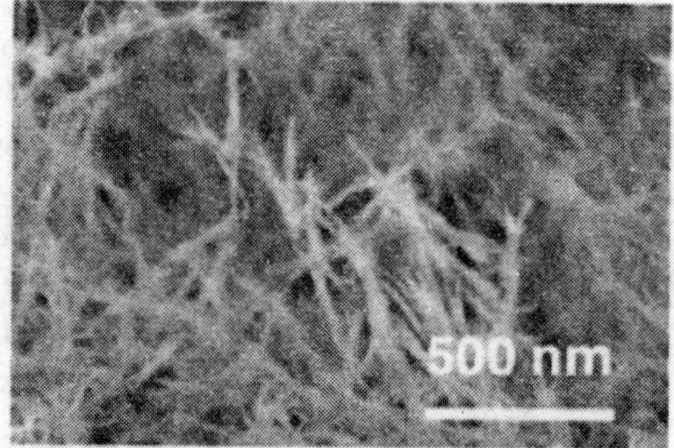

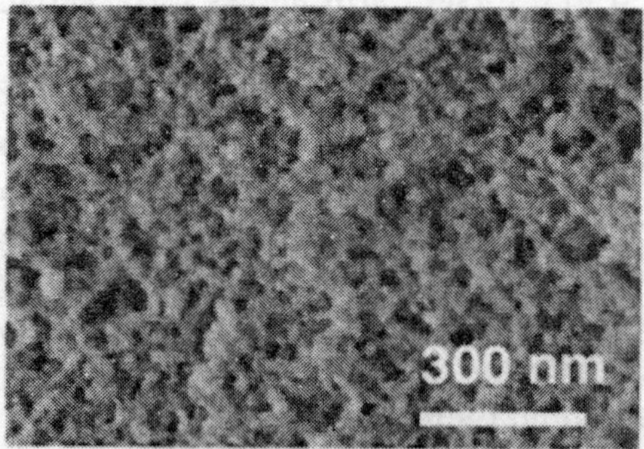

Fig. 5.43 *Scanning electron micrograph images of two iron (III) oxide sol-gel nanomaterials with markedly different particle morphologies.*

The intimacy of mixing between oxidizer and fuel is an important factor in the behavior of these materials. One problem, nanomaterials suffer from, is the tendency for agglomeration into larger aggregates. In such a case the properties of the resulting composite is influenced by the size of the aggregate and not the size of the individual nanoparticles, thus, defeating the potential benefit of the nanomaterial. Even though nanosized components are used, there is no guarantee that the sol-gel composite will have such mixing. To characterize this degree of mixing nanostructured composite material is analysed using transmission electron mixroscopoy (TEM) a Fe_2O_3/UFG Al material that is stoichiometric is used Electron microscopy results for one such composite is shown in Figure 5.44.

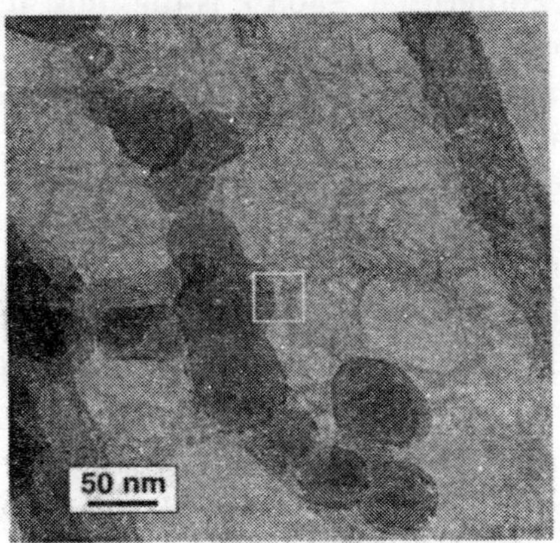

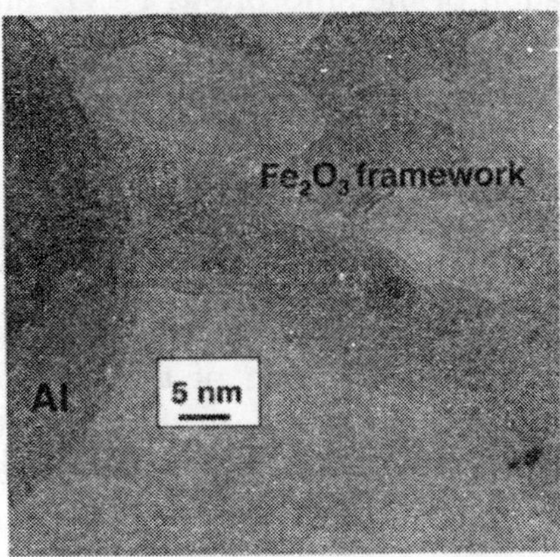

Fig. 5.44 *Transmission electron micrograph of sol-gel Fe_2O_3/UFG Al nanocomposite.*

In this figure the image on the right is a close up of the portion of the image on the left in the white box. In the images, the fibrous network is the iron (III) oxide phase and the spheres are the UFG SAL. The exquisite mixing between the constituent phases as well as the nanostructured nature of the composite is observed. Here added particulate (UFG Al) phase is incorporated into the porous network of the skeletal iron (III) oxide phase. The demonstrated fine degree of mixing as well as the three-dimensional nanostructure of the composite gives these materials very important properties.

Some of the thermal properties of these materials are investigated and contrasted with those of conventional sized (μm-sized) thermites. Figure 5.45 contains the differential thermal analysis (DTA) traces for one energetic composites and one sol-gel nanostructured energetic. It is clear

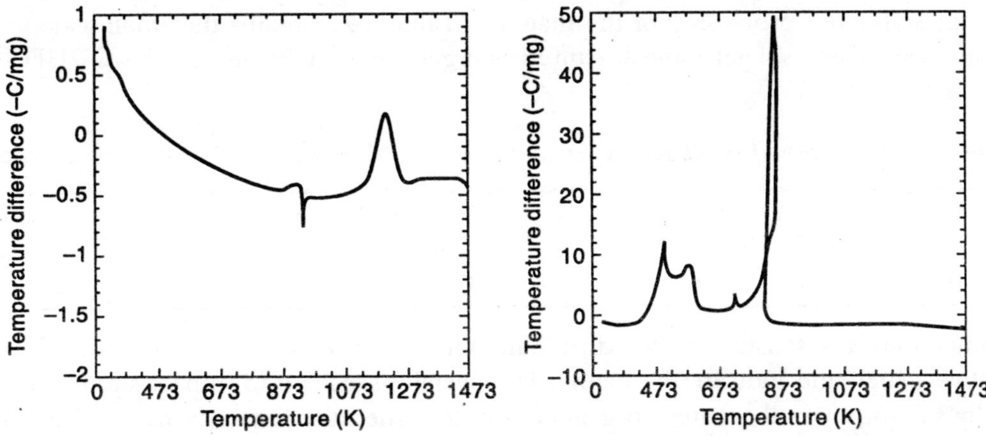

Fig. 5.45 *Differential thermal analysis traces of two thermite materials. The sample on the left is micron Fe_3O_3/Al (2 μm) and the one of the right is sol-gel nanostructured xerogel Fe_2O_3/UFG Al.*

from these DTA traces that the thermal behavior of these materials is quite different. In the sol-gel xerogel Fe_2O_3/UFG Al nanocomposite there are thermal events at ~563, ~333, and ~863 k. It is found that the two-lower temperature events are related to a phase transition and crystallization of the amorphous Fe_2O_3 phase. The exotherm at ~863 k is the most interesting as it corresponds to the thermite reaction. This exotherm is very narrow and sharp, possibly indicating a very rapid reaction. Another pint to be made her is that the thermite reaction takes place at a temperature markedly below the melt phase of bulk Al (T_m = 933 k).

Alternatively, the DTA trace of the dry powder mix of Fe_2O_3/Al micron-sized powder differs from that of the nanostructured material. There are no low temperatures in this DTA, as the oxidizer in this mixture is crystalline Fe_2O_3 and not amorphous hydrous sol-gel iron (III) oxide. The main features of the trace occur at ~933 K and 1188 K. The 933 K endotherm corresponds to the melting of the micron-sized Al and the ~1188 K exotherm is from the thermite reaction. In this case the thermite reaction takes place after the melting of AL. The ignition point of the traditional thermite material is ~598 K higher than that for the nanocomposite. This peak is also much broader than that seen in the nanocomposite DTA and likely indicates a less rapid reaction. Although qualitative, the DTA analyses indicate that the thermal behaviour of the sol-gel Fe_2O_3/UFG Al and the conventional Fe_2O_3 composite are quite different. Itis observed that the nanostructured energetics burn are much easier to ignite and burn much moire rapidly than the thermite composites.

Burn rates of nanostructures energetic materials are measured. Loose powders of nanostructured aerogel and xerogel Fe_2O_3/UFG Al are placed in an open-ended tube (confined geometry) and ignited with a CO_2 laser. A high-speed camera is used to capture the ignition and the propagation of the flame front. Burn velocities for selected samples of these materials are 895 m/s for the aerogel material and 320 m/s for the xerogel materials. These values are much higher than those for materials.

Small scale testing of energetic materials is done to determine their sensitivity to various stimuli including thermal degradation, friction impact and static spark. These characteristics are crucial as they often determine the necessary procedures employed in the synthesis, handling, storage, life-style, and processing of the material. Table 5.2 contains the small-scale safety test results for two of our sol-gel nanostructured energetic materials: aerogel Fe_2O_3/UFG Al and xerogel Fe_2O_3/UFG Al.

Table 5.2 *Summary of small-scale safety data for two sol-gel nanostructured energetic materials.*

Material description	*DH_{50} (cm)*	*Spark*	*B.A.M. Friction*	*DSC Onset (K)*
aerogel Fe_2O_3/UFGAl	89	1/10 @ 0.03 J	0/10 @ 36 kg	463
xerogel Fe_2O_3/UFGAl	101	0/10 @ 1 J	0/10 @ 36 kg	506

Both materials are sensitive to the drop hammer test. However, the values for the materials are relatively high and indicate the neither is especially sensitive to impact stimuli. This is not especially significant as thermites, in general are not especially impact sensitive Neither of the two is friction sensitive and only the aerogel sample is space sensitive. Please not that it is spark sensitive at a test energy of only 0.03 J whereas, the other material is not sensitive at 1 J. Hence, that when handling the aerogel Fe_2O_3/UFG Al material procedures and conditions that minimize static electricity build up should be rigorously employed.

Aerogel composites are more sensitive to ignition than their xerogel counter parts. In fact, aerogel composites made are spark sensitivity in small-scale safety experiments and have significantly lower drop hammer (DH_{50}) values (~100 cm for xerogels and ~90 cm for aerogel materials). More sensitive materials can be an advantage for certain applications but a disadvantage for safety concerns. The reason for the different in reactivity between the two materials has to do with their different thermal conductivities. The thermal conductivity of an aerogel is much lower than that of a xerogel of the same material. Therefore, an aerogel composite will have a more difficult time dissipating a thermal stress than a xerogel composite. This results in the more rapid formation of "hot spots" in the aerogel material when it is thermally stressed by a spark or drop hammer. Once one of the hot spots reaches the ignition point the reaction is self propagating and the entire composite is ignited.

Sol-gel methodology offers other advantages of safety and stability in energetic material processing. For example, ambient temperature gelation and low temperature drying schemes prevent degradation, and the water like viscosity of the sol before gelation, allows easy casting to near net shapes, which is preferred over the alternative, hazardous machining.

Inorganic Sol-gel Nano composites

Several metal oxides undergo the thermite reaction. Sol-gel methods are used and utilize for known ones to prepare solgel nanostructure senergetic materials with different compositions. The compositional versatility of the sol-gel method is one of its greatest value.

A large number of inorganic metal oxidizers of particular interest to energetic materials applications are readily prepared using sol-gel methods. Figure 4.46 contains TEM images of two such oxides MnO_2 and SnO_2.

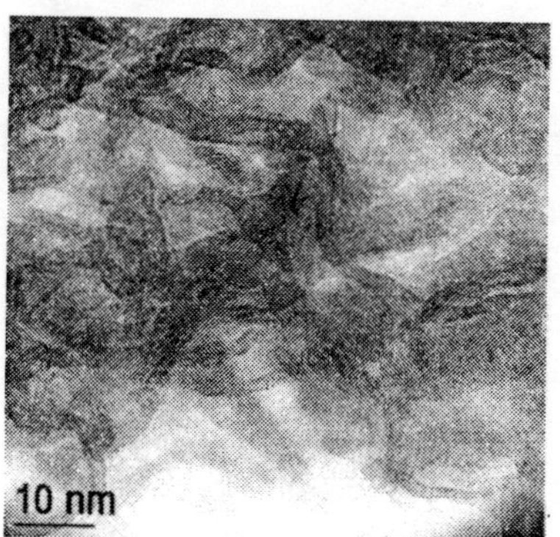

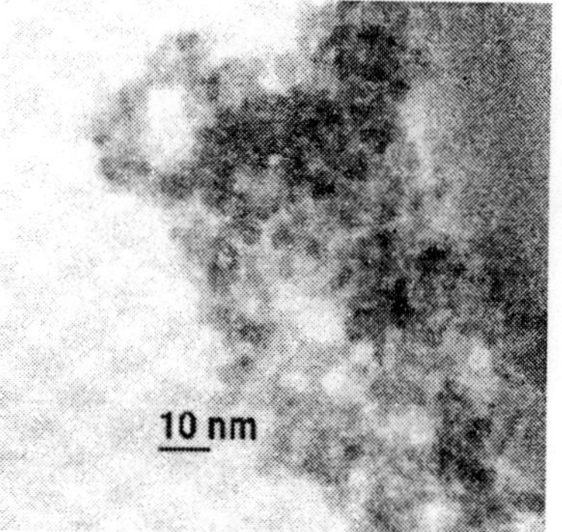

Fig. 5.46 *TEM image of sol-gel nanostructured MnO_2 on then left and SnO_2 on the right.*

The images in Figure 5.46 clearly indicate that the materials are made up of a homogeneous distribution of interconnected particles with extremely small primary dimensions (~5-10 nm) with high surface areas. Both of these oxides undergo strongly exothermic thermite reactions with aluminum. Like the iron (III) oxide materials these solids can be prepared in a straightforward manner in a variety of safe and non-toxic solvents. It is also clear the materials have significant porosities.

Organic Sol-gel Nanocomposites

In these types of composites organic sol-gel methods are used to produce a skeletal matrix of hydrocarbon fuel with oxidizer added in particulate form to fill the void spaces. This nanostructured material is unique from the thermite type materials, as it generates significant amounts of gas upon decomposition. In this particular example, a solid skeleton of fuel based on resorcinolformaldehyde has nanocrystalline ammonium perchlorate, the oxidizer, trapped within the pores. At optimum stoichiometry it has approximately the energy density of HMX. Using a sol-gel procedure to make aerogels, a porous organic solid matrix is first prepared by the polycondensation of *resorcinol with formaldehyde* (RF). Subsequent crystallization of an oxidizer, *ammonium perchlorate (AP)*, within the pores of the gel matrix, completes the synthesis, followed by atmospheric drying of the wet composite to form a dense xerogel material. But atmospheric drying leads to the growth of large crystals of AP. Therefore, freeze drying of the wet composite material, to form cryogels, inhibits crystal growth.

To determine whether the material is energetic differential scanning calorimetry (DSC) is performed. DSC plots for a RF-AP nanocomposite and neat ammonium perchlorate are compared in Figure 5.47. The trace for the RF-AP nanocomposite on the right Fig. 5.47(b) showed an exotherm at about 523 K, indicating that it is indeed energetic. Pure AP Fig. 5.47(a) has significantly less exothermicity and integrated decomposition enthalpy in the absence of the fuel gel skeleton (RF). It appears that the RF/AP composite undergoes decomposition via different pathway that pure AP.

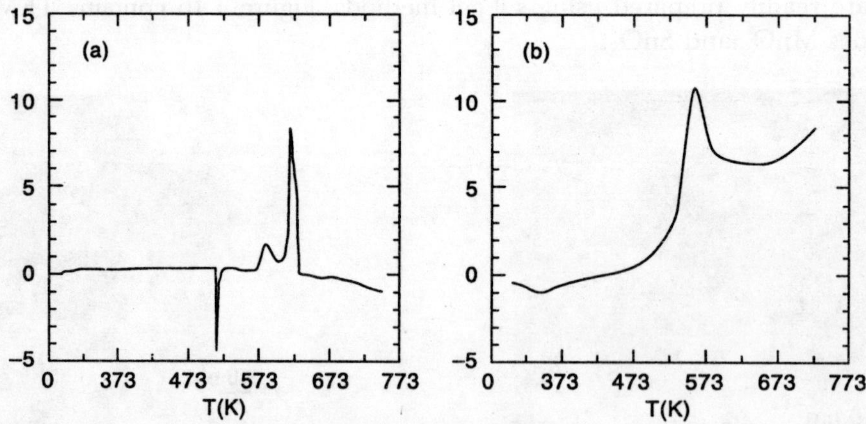

Fig. 5.47 *Differential scanning calorimetry scans for neat AP on the left and an RF/AP nanostructured energetic material on the right. The value on the y-axes is Heat flow (units: W/g)*

Electrodeposition

Electrodeposition involves inducing chemical reactions in an aqueous electrolyte solution via use of an applied voltage; as either anodic or cathodic processes. In addition to deposition of metallic thin films, nanostructured materials include the deposition of metal oxides and chalcogenides. The electrodeposition process obviously requires an electrically conducting substrate.

In an *anodic* process, a metal anode is electrochemically oxidized in the presence of other ions in solution, which then react together and deposit on the anode. For example:

$$Cd \rightarrow Cd^{2+} + 2e^-,$$
$$Cd^{2+} + Te^{2-} \rightarrow CdTe.$$

Meanwhile in a *cathodic* process, components are deposited onto the cathode from solution precursor,. In the case of metals this process is known as electroplating:

$$Cd^{2+} + 2e^- \rightarrow Cd,$$
$$HTeO_2^+ + 4e^- + 3H^+ \rightarrow Te + 2H_2O,$$
$$Cd^{2+} + Te^{2-} \rightarrow CdTe$$

If a metal is oxidized and then undergoes hydrolysis or the local pH at the electrode is changed by liberating hydrogen gas electrochemically and producing OH^- ions, then metal oxide materials may be deposited.

Electrodeposition is performed at low temperatures to minimize interdiffusion when a multilayered thin film material is being prepared. The film thickness is controlled by monitoring the amount of the charge delivered, whereas the deposition rate is followed by the variation of the current with time. The composition is controlled by the magnitude of the applied potential, which can be used to deposit non-equilibrium phases. Pulsing or cycling the applied current or potential in a solution containing a mixture of precursors allows the production of a multilayered material. The potential during the pulse determines the species deposited whilst the thickness of individual layers is determined by the charge passed. The substrate is transferred periodically from one electrolytic cell to another. The final films range in thickness from a few nanometres to tens of microns and are deposited onto large specimen areas of complex shape.

Electrodeposition is also performed within a nanoporous membrane which serve to act as a template for growth; for example, anodized aluminium has cylindrical nonporous of uniform dimensions and electrodeposition within this membrane produces nanocylinders. Deposition on planar substrates can also limit nanocrystal growth and produce ordered arrays; if the growth is epitaxial then any strain due to lattice mismatch between the nanocrystal and the substrate can be growth-limiting. Furthermore, it is possible to modify the surface of substrates (e.g., by STM or AFM) to produce arrays of defects which can act as nucleation sites for the electrodeposition of nanocrystals.

NANOSTRUCTURES FROM FULLERENES

The application of fullerene-based building blocks such as single nanoparticle, clusters, agglomerates, nanotubes, and nanolayers, largely depends on the combination of "nanostructured materials" (NSM). These are materials with a structure of zero to three dimensions on a length scale less than 100 nm. They are nanospheres, nanowires, and nanolattices.

While chemical methods have been employed to fabricate nanoparticles, nanospheres, nanolattices, and nanowires via different chemical modifications of fullerenes, physical methods have shown to be advantageous for the fabrication of nanotubes via template processes and high temperature methods. Among the three-dimensional nanoscale building blocks, the nanoparticles with higher symmetry are easily synthesized by chemical modification, whereas nanowires with low symmetry are synthesized by chemical modification only with difficulties. As a result, there are many more reports on the fabrication of fullerene-based nanoparticles than one fullerene-based nanowires. There aer no fullerene-based nanotubes fabricated by chemical methods. Though nanotubes have been successfully prepared via template processes, to remove the template materials completely from the nanotubes formed is still a difficult process. As for the high temperature methods, the nanotubes formed cannot be considered as fullerene-based nanostructures because over 1073 K fullerenes are completely decomposed. Therefore, it is advisable in this case to consider fullerenes not as building blocks, but as a carbon source. Fullerene-based nanostructures are attractive because their particular shape introduces curved interfaces and because large numbers of functional groups can be readily introduced into a single cage of fullerenes. Their unique structural and electron features have led to phase behavior in the application of organic solar cells. The fullerene-based nanostructure interface has been systematically controlled by the structure, solvent etc.

In comparison with zero-dimensional, fullerene-based nanostructures (the so-called quantum dots or nanoparticles), 1D nanostructures (including wires, rods, belts, and tubes) provide better model systems to investigate the dependence of electronic transport, optical, and mechanical properties on the size confinement and dimensionality. Nanowires, as interconnects and active components in fabricating nanosclae electronic and photonic devices.

Buckminsterfullerene (C_{60}) exhibits electronic, conducting, and magnetic properties, and has enormous applications such as electronic and optical materials, super conductors, building blocks for new compounds and materials, optical limiters and light emitters

C_{60} Diameter

It is known that the diameter of C_{60} is approximately 0.7 nm, measured from the nucleus of one carbon to the nucleus of the carbon on the opposite side. The mean hydrodynamic radius of C_{60} calculated from the Stokes-Einstein equation is $R_h = 0.76 \pm 0.15$ nm, which is larger that one half of the nearest-neighbor distance (r) of C_{60}, i.e., $r/2 = 0.50$ nm, as determined by X-ray on individual crystals. It is also larger than the radius of 0.56 ± 0.07 nm for the C_{60} molecule in a Langmuir film. The single crystal X-ray structure of benzene-solvated C_{60} ($C_{60} \cdot 4\ C_6H_6$) indicates that three benzene rings are oriented parallel to the C_{60} molecular surface. The distances from the center of the solvated benzene rings to the center of C_{60} are 0.674, 0.677, and 0.681 nm.

The molecular nanostructure and morphology of C_{60} are shown in Figure 5.48. In toulene solution or in benzene solution, the cohesive energy of C_{60} is 129.8 kJ/mol^{-1}.

FABRICATION

The ability to convert the Buckminsterfullerenes into new materials and compounds has found great interest in many areas. To date, the common methods for the fabrication of

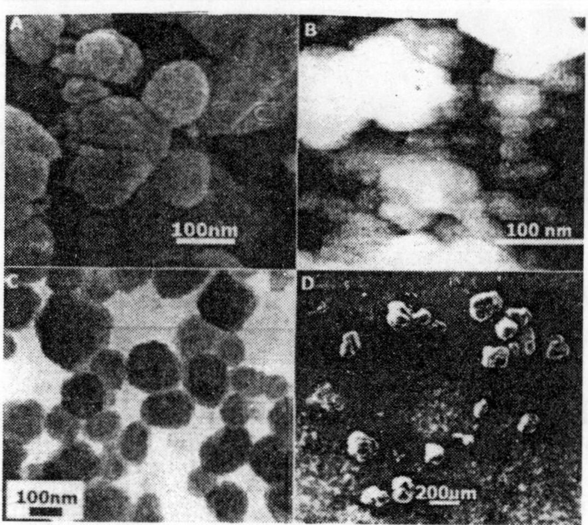

Fig. 5.48 *Scanning electron microscopic images (SEM) of (A) $C_{60} > 99.5\%$ purity), (B) AFM, (C) transmission electron microscopic image (TEM), (D) SEM images of the C_{60} aggregates.*

nanoscale building blocks include the template method arc discharge method, and vapor-liquid-solid method. However, due to many special features of the fullerene family, the fabrication technologies for fullerene-based nano-scale building blocks are very different.

Several techniques have been developed for the production of fullerene nanoparticles, nanowires, and carbon nanotubes.

1. *Nanoparticles and Nanospheres*

(a) Spray-drying technique: to form fullerene-rhodium nanocomposites with the fullerenes C_{60} and C_{70} in toulene. Here the nanocomposite particles are formed by thermal decomposition in a nitrogen atmosphere at a processing temperature of 823 k, which is sufficiently low to avoid stronger particle growth of C_{60} and rhodium, but high enough for thermal decomposition. The pure fullerene particles obtained by this process show lattice fringes corresponding to the prominent *d* spacing of 0.2 nm in C_{60}. The 50/50 wt% fullerene/Rh composite shows 3–5 nm fullerene-Rh crystallites which is consistent with the value of 4 nm, calculated from the X-ray diffraction peak broadening.

(b) Vapour Condensation: Here, fullerenes are vaporized in a continuous flow reactor at a temperature of 673–773 k using nitrogen as the carrier gas. The particles obtained are spherical in shape and have narrow size distribution with an average size of 30 to 40 nm. The formation of nanosphere fullerene particles is also via the spray-drying/pyrolysis and vapor condensation routes: Solid C_{60}/C_{70} is dispersed in a toulene/acetonitrile mixture and the formation of nanoparticles with an average diameter in the colloidal stage is detected by TEM immediately after the components are mixed, see Fig. 5.49. The average center-to-center distance between the fullerene molecules is roughly estimated to be about 1 nm. Therefore, it is evident that the fullerene colloid particles are formed as a result of crystallization process. Further, it is shown that smaller particles are formed when the crystalline fullerene particles are isolated from a concentrated solution. The negative

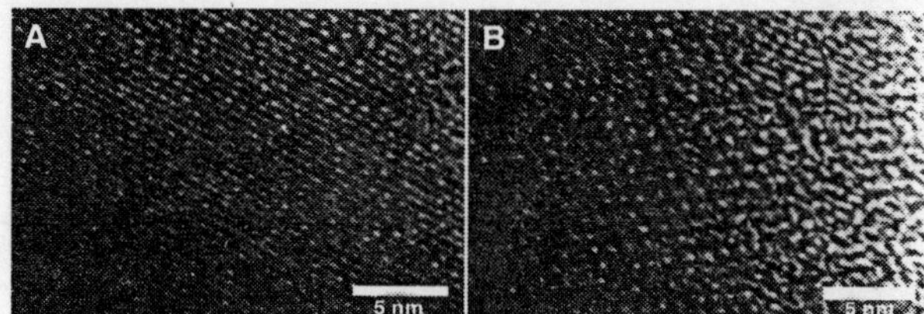

Fig. 5.49 *TEM pictures of a fullerene particle formed in a toulene/acetonitrile mixture. Panels A and B show different portions of the same particle. The toulene solution contained a mixture of 50% C_{60} and 50% C_{70}. The total fullerene concentration in the dispersion was 0.01 g/dm^3.*

values obtained for the electrophoretic mobility of the particles suggests that the fullerene dispersions in polar organic solvents are stabilized by repulsive electrostatic interactions.

The possible mechanism for the formation of fullerene nanoparticles is that initially, the instantaneous small emulsion droplets of toulene are formed immediately after mixing with acetonitrile. The droplets exist for a every short period of time because toulene is completely miscible with acetonitrile and its molecules begin to diffuse into the dispersion medium. When the concentration of fullerene inside the toulene droplets reaches the saturation point, the crystallization process quickly takes place and fullerene micro-crystals are formed. Each emulsion droplet leads to the formation of one colloidal particle. If the size of the toulene droplets is uniform and independent of the fullerene concentration, the droplets do not coalesce during their short existence.

This method is probably the most suitable one because it allows the preparation of the nanoparticles and the control of their size and shape. However, the separation of the nanoparticles from the micro-emulsion solution is very difficult.

Other processes for the production of fullerene nanostructures of 2–30 nm include their information in the gaseous state, e.g., in the flames of benzene, acetylene, or ethylene, when premixed with oxygen and an inert diluent.

Nanowires

An ordered porous alumina template with narrow pore size distribution is used to fabricate nanotubes from C_{60}. For the process, the template is repeatedly dipped into a toulene solution of C_{60} and the solvent allowed to dry between dipping. After a number of "dip-and-dry" cycles, the templates are kept at 773 K for 5 h in an argøn atmosphere and then cooled to room temperature. Figure 4.46 shows the as–grown nanotubes formed with the toluene solution C_{60} by the dip-and-dip dry method.

Figure 5.50A shows that the template is filled with C_{60} nanotubes of metallic sheen and smooth surfaces. Figure 5.50B depicts an array of aligned C_{60} nanotubes, which have smooth and clean surfaces. Their diameters are 220–310 nm and their length 60 μm, which corresponds to the diameter of the pores of membrane and the thickness of the template used. After selectively dissolving the template, the remaining nanotube array still exhibits its hexagonal long-range order. The tip of a C_{60} nanotube formed in an alumina template is uncovered by etching the alumina substrate with aqueous sodium hydroxide, it yielded tubes with a

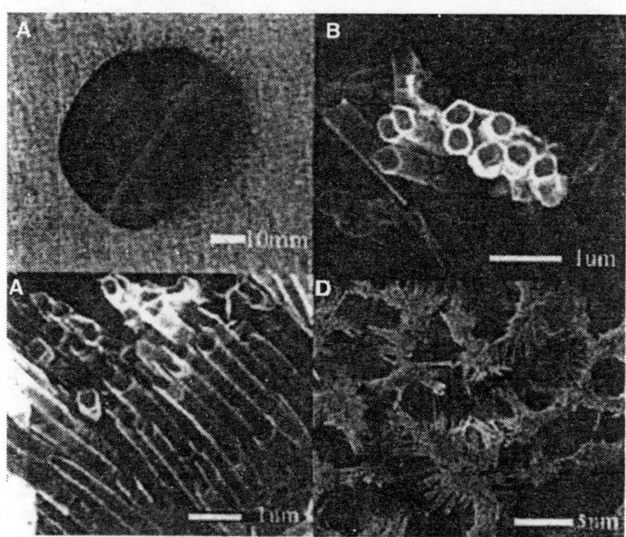

Fig. 5.50 *Photograph of the alumina template with a diameter of 2.5 cm (A) SEM images of C_{60} nanotubes. (B) Top of C_{60} nanotubes after removal of the alumina template. (C) Ordered array of C_0 nanotubes after complete removal of the template. (D) Large area of aligned C_{60} nanotubes.*

monodisperse diameter of 30 nm (Fig. 5.50C). The nanotubes grown from the membrane exhibit an ordered arrangement, which is perpendicular to the substrate (Fig. 5.50D).

The comparative analysis of the IR spectra of C_{60} and the C_{60} nanotubes provides the evidence of the formation of new bonds (Fig. 5.51). However, the results of X-ray microanalysis do not indicate unambiguously that these nanotubes are composed only of carbon (Fig. 5.51). Other elements existing in these nanotubes could stem from the alumina template.

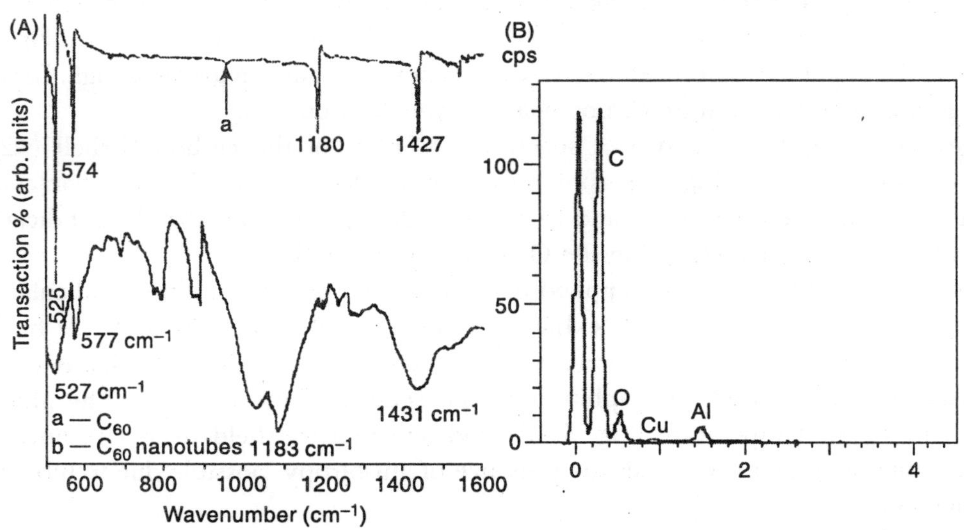

Fig. 5.51 *(A) FTIR spectra (KBr) of C_{60} and C_{60} nanotubes after the complete removal of the template. (B) Energy-depressive X-ray microanalysis of a single C_{60} nanotube.*

When the ferrocene/C_{60} mixtures are heated under Ar, the formation of aligned, carbon-Fe-filled nanotubes is observed (Fig. 4.52).

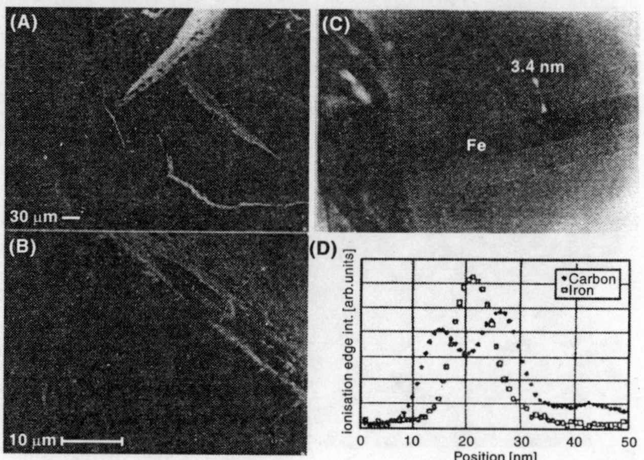

Fig. 5.52 *SEM images of the flake-like particles (films) consisting of aligned Fe-filled nanotubes at lower (A) and higher (B) magnification. The material is grown by pyrrolyzing ferrocene/C_{60} mixtures (e.g., 1:1) at 1173–1323 K in an argon flow using a conventional two-stage furnace. (C) HRTEM image of two crossed Fe-filled nanotubes exhibiting different diameters and high degrees of graphization (interlayer spacing ~0.34 nm); (D) Concentration profiles of C and Fe across the diameter of an Fe-filled tube derived form EELS line scans (See white arrow).*

The SEM images of the products (Figs. 5.52A and 5.52B) show that the external diameters and length of the as-prepared carbon nanotubes are 20–70 nm and ~40 nm, respectively. HRTEM, nanoscale elemental mapping, and high-resolution electron energy loss spectroscopic (HREELS) of the material show that ~40–60% of these tubes are filled with Fe (see Fig. 5.52D). Also, the degree of graphitization exhibited by the nanotube walls is extremely high. In addition, the average number of concentric carbon-coated layers is about 25.

The near-edge electron energy loss spectrum (NEELS) of the carbon K-shell (<284 eV) confirms that the material is highly graphitic with the L-edge (<708 eV) being characteristic of metallic Fe. It is observed that the C and Fe concentration profiles anticorrelate, indicating that pure Fe is indeed encapsulated within the C layers [Fig. 5.52D].

Plots of the magnetization versus applied field (measured parallel and perpendicular to field for an individual "carpet;" Fig. 5.53A) exhibit saturation at fields of ~14000 Oe. At high fields, the hysteresis loop also exhibits a decrease in magnetization (Fig. 5.53A). This is due to the diamagnetic influence caused by graphitic material coating of the Fe wires. The plots of the coercive field versus temperature (Fig. 5.53B) indicate that the field decreases linearly with increasing temperature, which reveals the presence of anisotropy, possible due to tub geometry and orientation.

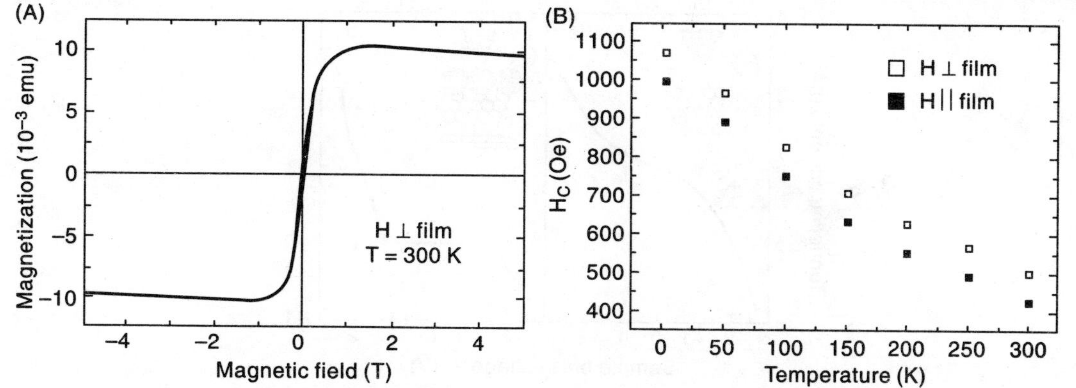

Fig. 5.53 *(A) Magnetization vs applied field curves, measured parallel and perpendicular to the applied field on individual carpets, exhibiting saturation fields of ~14000 Oe. By increasing the applied field, the hysteresis curve also undergoes a decrease in magnetization; (B) The coercive fields at different temperatures, which decrease in a linear fashion with increasing temperature and reveal traces of anisotropy. When the field is applied parallel and perpendicular to the axis of the nanowires, the demagnetization factor is zero and close to 2p, respectively.*

The small size, anisotropy, and the single isolated domain nature of the encapsulated Fe crystals, is responsible for this enhanced coercivity and Fe nanostructures do not degrade and they are thermetically sealed within the graphitic tubes. This kind of nanowire arrangement finds applications in the fabrication of high-density magnetic recording devices as well as fine particle magnets in magnetic inks and toners in xerography.

APPLICATION OF FULLERENE-BASED NANOSTRUCTURES

1. In Molecular Devices

The negative differential resistance (DNR) effect, which is characterized by the phenomenon of decreasing current with increasing voltage for the current–voltage (*I–V*) curve, is the essential property of electronic devices and has wide applications in fast switches, oscillators, and frequency-locking circuits. It has been shown experimentally and theoretically that an NDR on the atomic-scale is possible in tunneling structures with a scanning tunneling microscope (STP) tip.

The NDR molecular device consists of two C_{60} molecules, of which one is physically adsorbed on the hexanethiol self-assembled monolayer, and the other is adsorbed on the Pt–Ir tip. Such controllable tunneling structures and the associated known electronic states ensure the stability and reproducibility of the NDR device.

The attachment of a C_{60} to the tip is a reversible process. When applying a voltage pulse of about 6 V, the adsorbed C_{60} on the tip will desorb. (Fig. 5.54), (a), (b), (c), and (d) correspond to the tunneling structures of the base Pt–Ir tip over thiols, bare Pt–Ir tip over C_{60}, C_{60} tip over thiols, and C_{60} tip over C_{60}, respectively.

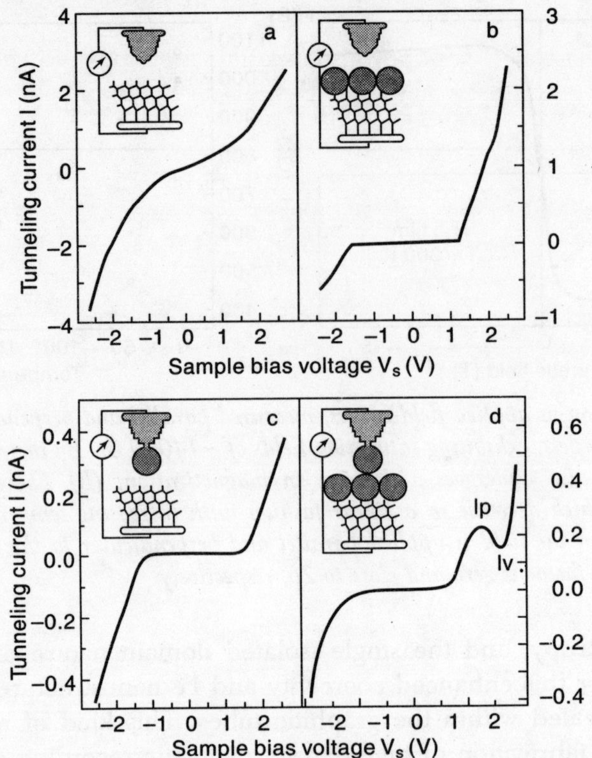

Fig. 5.54 *I–V curves obtained from four different tunneling structures, as shown in the insets. (a), (b), (c) and (d) correspond to bare Pt–Ir tip over thiols, bare Pt–Ir tip over C_{60}, C_{60} tip over thiols, and C_{60} tip over C_{60}, respectively.*

Figure 5.54(a) shows displays the I–V curve of the bare Pt–Ir tip over the thiols. This curve shows metallic behaviour, which is similar to the typical STS results and also confirms that no C_{60} is adsorbed on the tip. The I–V curves, dominated by the LDS of the C_{60} molecules, are very similar. In both cases, an energy gap of about 2 eV is observed, which is close to the energy gap between the highest occupied molecular orbital of the free C_{60} molecule (about 1.8 eV). The *I–V* curve in (c) also reproduces the main features of the STS of a C_{60}-adsorbed Pt–Rh tip on graphite surface [101]. This further confirms the adsorption of a C_{60} molecule on the tip.

When C_{60} molecules are simultaneously adsorbed on the tip and on the thiols, an NDR at the positive bias voltage is observed in the *I–V* curve [Fig. 5.54(d)]. The high symmetry of the free C_{60} molecule leads to the highly degenerated and largely separated energy levels. Thus, narrow-featured LDS is expected when C_{60} molecules are adsorbed on the Pt–Ir tip and the chemically inactive thiol SAM. The NDR phenomenon is due to the narrow.

Structure in the LDS of the C_{60} molecules on both of the electrodes. The positive result provides a motivation in exploring the potential practical NDR molecular device.

2. Application in Materials Science

Three main fields, given below, are emerging in the application of fullerene-based nanomaterials.

3. Applications in Porous Materials

Two types of fullerene-based nanoporous materials are prepared: (a) porous nanoparticles from fullerene derivatives directly obtained by chemical modification of fullerness, and (b) inorganic–organic nanocomposites obtained by incorporating fullerene molecules or their derivatives in an inorganic matrix.

(a) The first type comprises fullerene derivatives, in which the cages of fullerness are untouched and some organic groups such as ketone and hydroxyl groups are incorporated into the oxidized fullerens, example preparation via the oxidation of C_{60} by $KMnO_4$. The pore size of the resulting hollow oxidized fullerene nanospheres is centered at 8 nm and the BET surface area is found to be 222.63 $m^2\ g^{-1}$. Another example is the oxidization of fullerene with manganese dioxide in concentrated hydrochloric acid. The pore size distribution of the oxidized fullernesses is found to be 8.6 nm. The Langmuir surface area of the hollow nanosphere, calculated from the N_2 adsorption, was 222.18 m^2g^{-1}.

(b) The second type of fullerene-based nanoporous materials are zeolite and Aerogels. They are produced via the introduction of nanoparticle of fullerenes into porous materials.

(i) *Zeolites:* are composed of nanoporous materials with pores generally in the range 0.3–1.3 nm. The intriguing feature of using zeolites as a host matrix for fullerene nanoparticles is their pore dimensions that are very close to the one of C_{60} (0.7 nm diameter). Zeolites wit pores larger than 1 nm allow a fullerene diffusion into the pores and have enough space to accommodate it. C_{60} molecules will be confined in a monodimensional array within the nanoporous host material.

Interesting results are obtained with the incorporation of C_{60} into VPI-5 zeolites C_{60} is introduced into the VPOI-5 zeolite by impregnation with a benzene solution or from the gas phase.

A microstructure is formed consisting of alternating corner-sharing AlO_4 and PO_4 tetrahedrons, linked together to give a monodimensional 18-atom ring channel. Fullerenes incorporated into this structure should form a monodimensional array.

A comparative study of different thermal analysis thermogravimetric analysis (DTATGA) data of undoped and fullerene-doped VPI-5 gives an approximate stoichiometry of one C_{60} for 35 $AlPO_4$ units, which corresponds to a two-thirds filling level of the pores. The incorporation of a small fraction of C_{60} into zeolites results in a clear modification of fullerene-doped VPI-5. The fullerene-doped VPI-5 shows characteristics similar to those of crystalline C_{60} and might be blocked in structural defects. The X-ray diffraction studies of the C_{60} product indicate that the product obtained studies of the C_{60} product indicate that the product obtained was highly crystalline with no super lattice reflections. This material was stable in air without any phase transition and, therefore, C_{60} stabilizes the VPI-5 structure (Fig. 5.55).

After the incorporation of fullerene nanoparticles into the VPI-5 zeolites, remarkable changes in the fluorescene spectrum were observed. The main peak in the 1.7 eV region, due to the emission in crystalline powders, broadened, increased in intensity, and was shifted to the visible region, peaking around 2 eV. The photoluminescence spectra of doped and undoped VPI-5 with C_{60} nanoparticles are shown in Figure 5.56.

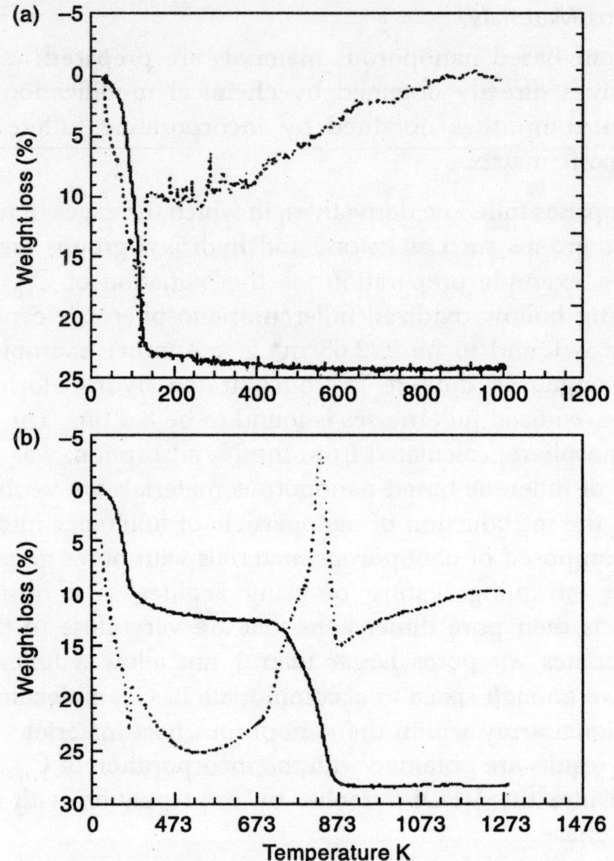

Fig. 5.55 *Thermogravimetric and differential thermal analysis profiles of: (a) VPI-5 zeolites, and (b) trapped VPI-5 sample 60 prepared from a benzene solution.*

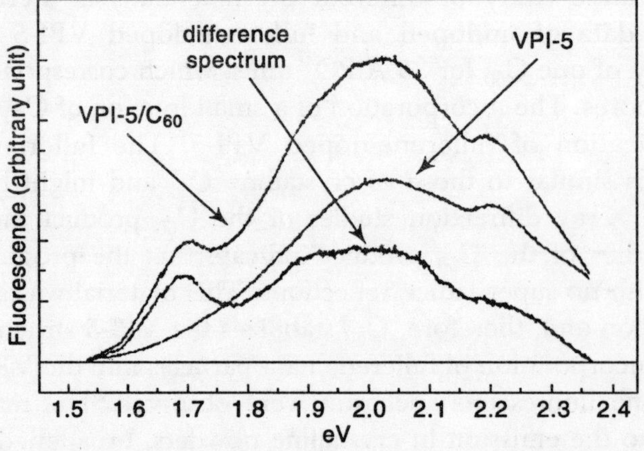

Fig. 5.56 *Photoluminescence spectra of: (a) VPI-5, (b) C confined in VPI-5, and (c) pure C powder. Excitation wavelength 514.5 nm.*

In the 830–540 nm interval, a set of five lifetimes such as 10–80 ps, 180 ps, 450 ps, 1.2 ns, and 4.6 ns, pertaining to the C_{60} nanoparticle-doped VPI-5 is observed (Fig. 5.56). After comparison with the fluroescence from the zeolite matrix and solid C_{60}, only the 1.2 ns lifetime is attributed to the emission from confined C_{60} molecules in the pores. This supports the incorporation of C_{60} in the zeolites, mainly as single molecules. A lifetime of 1.2 ns and the broadening of the spectrum, with respect to solid C_{60}, after the incorporation in VPI-5 is similarly observed in solution.

(ii) Aerogels: (a) A fullerene-doped silica aerogel is prepared by the sol–gel process followed by the supercritical drying technique Or (b) By introducing nanoparticles of C_{60} and a C_{60}/C_{70}, mixture in silica aerogels by dissolving in toulene and adding it to an acid-catalyzed sol for tetraethoxysilane (TEOS).

(a) Higher solution transmission electron microscopic (HRTEM) images together with electron diffraction revealed some interesting features of the fullerene-doped silica aerogel (Fig. 5.57). When the C_{60}/C_{70} nanoparticles are doped with the silica aerogel with a particle size in the range of 1 to 20 nm. The fullerenes and silica particles are not distinguishable. The electron diffraction pattern (EDP) shows a diffuse broad ring, which indicates that it is 'amorphous' over the whole

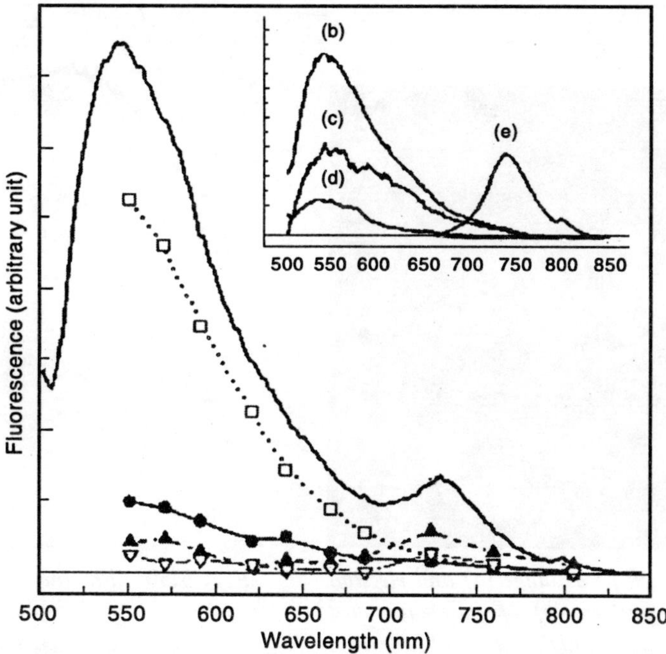

Fig. 5.57 *C60-VPI5 emission spectra recorded at an excitation wavelength of 496 nm: (a) before washing the sample with benzene and (b) after being abundantly rinsed with benzene; (cz) calculated difference spectrum of spectra (b) and (d), where (d) emission spectra of pure VPI-5; (e) spectrum of C60 powder recorded. Spectral distribution of lifetimes issued from glkobal analysis of C60-VPI-5 fluroescence decays (▲ = 180 ps; ▼ = 450 ps; • vs = 1.2 ns; □ = 4.6 ns). The lines between the points are drawn in order to visualize the spectra distribution of the components.*

specimen, just as that of undoped silica aerogel. Both of these results indicate that fullerene nanoparticles exist as small clusters or even individual nanoparticles inside the silica aerogel. When specimens are doped strongly with C_{60}/C_{70} (for example, .125 mol%, as shown in Fig. 5.58), a distinctly different morphology is obtained. The dark field images of the sample show the presence of separately distributed crystallized particles, corresponding to fullerene particles entrapped in disordered networks of the silica aerogel (such as the bright particle image in Fig. 5.58(a). The typical EDP of these crystalline particles is in accord with the (110)-spacing of the face-centered cubic C_{60} (d = 5.01 nm), which shows that the fullerene molecules tend to aggregate, forming crystallites under strong doping.

The fullerene-doped composites are found to be able to emit intense visible light (2.26 eV) under Ar^+ laser (488 nm) excitation at room temperature. The luminescent intensity increased with a low doping of fullerenes (0.05 mol%), and then decreasing with strong doping (2.5%) (Fig. 5.59).

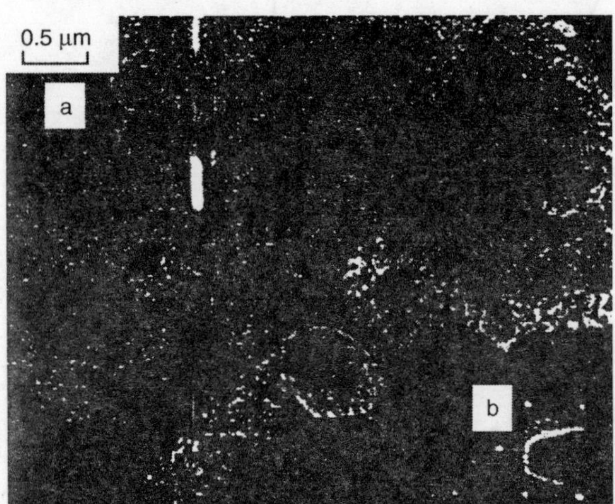

Fig. 5.58 *HRTEM dark field image (a) and electron diffraction pattern (b) of fullerene-doped silica aerogel with high doping of fullerenes (1.25 mol%), showing the fullerene crystallites inside the silica networks.*

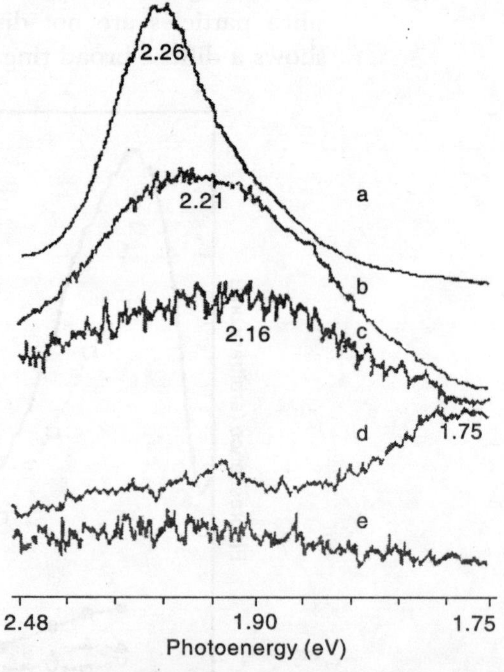

Fig. 5.59 *Photoluminescence spectra of fullerene-doped silica aerogel with an increasing doping concentration of fullerene; (a) 0.05 mol%, (b) 1.25 mol%, (c) 2.5 mol%, with the PL of pristine C_{60}/C_{70} (d), and silica aerogel (e) with an Ar^+ laser (488 nm) excitation.*

(b) The concentration of C_{60} in the silica gel is very low (0.001–0.5%) due to its low solubility in the toulene. After several days of aging, the gels are dried under supercritical conditions. The C_{60}/C_{70} mixture is introduced directly into the sol, which is sonicated to attain a homogeneous fullerene distribution (Fig. 5.60).

The dark parts of the sample indicate the presence of crystallized particles corresponding to fullerene particles (e.g., the particle labeled "a" in Fig. 5.60). The EDP of these crystalline particles is in accord with the (110)-spacing of the face-centered cubic C_m ($d = 0.5$ nm). Further, it is found that the size of particle and the micropores of the silica aerogel increase accordingly.

It is observed that the fullerne-doped sample (0.05 mol%) emits a visible light peak at 2.26 eV under an Ar^+ laser 9488 nm, 10 mW) excitation (Fig. 5.61(a). No photoluminescene of the as-prepared silica aerogel could be observed under the same experimental conditions (Fig. 5.61(b).

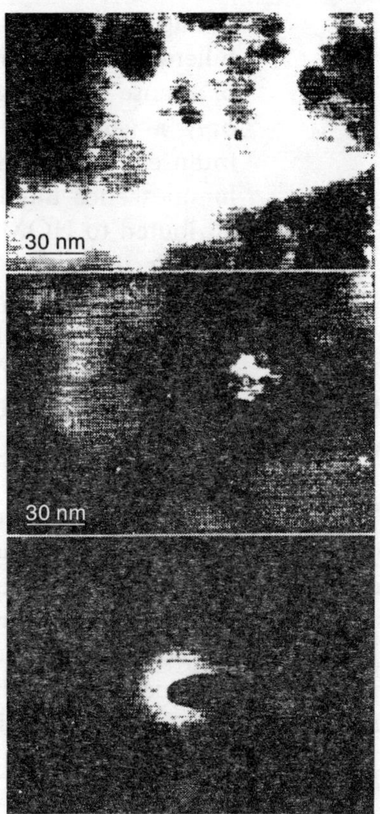

Fig. 5.60 *HRTEM bright-field image (top), dark-field image (middle), and EDP (bottom) of fullerene-doped silica aerogel with strong doping of fullerene (1.25%) showing the fullerene gram "a" inside the microscopes of silica aerogel.*

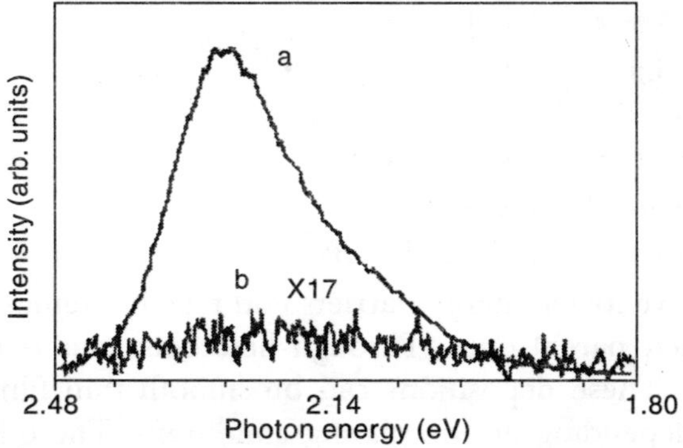

Fig. 5.61 *PL spectra of fullerene-doped silica aerogel (0.05 mol%) (a) and of the as-prepared pure silica aerogel (b) under Ar^+ (488 nm, 10 mW) excitation.*

The introduction of C_{60} or C_{60}/C_{70} mixtures into silica aerogels produce fullerene clusters that yield quantum confinement effects (QCE) and the emission of intense visible light. While under normal conditions fullerene powders exhibit only a very weak photo-luminescence (peaking around 1.7 eV), in aerogels, intense visible light emissions around 2.1 and 2.3 eV are observed. The luminescence intensity also increases and shifts toward blue. These effects are attributed to HOMO-LUMO transitions in C_{60}, which are usually forbidden by the selection rules. The small fullerene clusters formed during the incorporation into the aerogels yield a quantum dot effect that produces a blue shift and a larger quantum yield of photoluminescence.

CRYSTALLIZATION OF NANOSTRUCTURE BY PLASMA ENHANCED CHEMICAL VAPOR DEPOSITION (PECVD)

The plasma deposition technique is based on the strategy of assembly or etching small particles in the state of electrons, atoms, ions, and/or excited molecules. it provides a convenient, effective, vapor-solid or vapor-liquid-solid method for fabrication of nanomaterials. Organic products produced by PECVD are pinhole-free and highly cross-linked amorphous structures.

Below is PECVD which is for obtaining nanocrystalline structures without any catalysts and templates. Specific example of silicon nanocrystalline films made by PECVD is considered to illustrate its application in nanofabrication and crystallization of other nanocrystal structures.

Fabrication

PECVD is an effective way to nanofabricate crystalline inorganic and organic polymer materials. The crystal phase structure and morphology depend on the plasma conditions to a large extent.

In a plasma reactor there exists a lot of highly reactive species such as electrons, excited neutrals, free radicals, and ions.

In PECV, precursors feed into the plasma reactor and are partly ionized into active particles in the following ways:

excitation: $A + e^- \rightarrow A^* + e^-$

ionization: $A + e^- \rightarrow A + 2e^-$

dissociation: $A_2 + e^- \rightarrow 2A + e^-$

electron attachment: $A + e^- \rightarrow A^-$

photoemission: $A^* \rightarrow A + hv$

charge transfer: $A+ + B \rightarrow A + B^+$

These reactive species have lower energy barriers and reaction temperatures for physical and chemical reactions than their parent ones. Through heterogeneous reactions of these reactive species with solid surfaces. These depositions can be smooth thin films, rough powders, fine nanoclusters or oil drops depending on the plasma conditions. These heterogeneous reactions include adsorption, recombination, metastable deexcitation, sputtering, polymerization, or combinations of these processes.

The interaction of the active species with the substrate surface and the nanocrystal formation in PECVD is schematically illustrated in Figure 5.62.

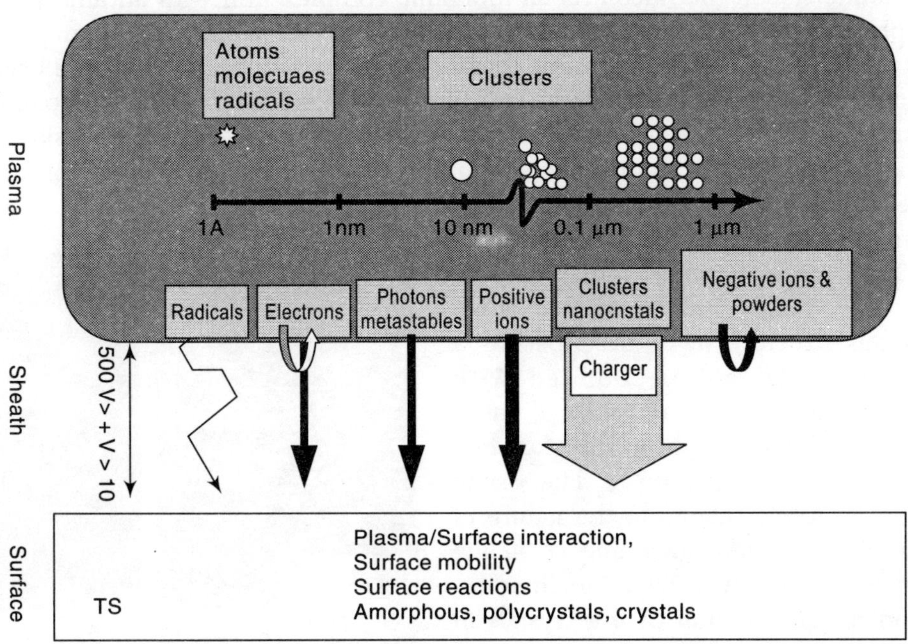

Fig. 5.62 *Nanocrystal formation process in PECVD.*

By control-line the process parameters, the deposition structure and properties can be controlledy. These parameters are power, total pressure, flow rate of monomer gases, substrate temperature, and discharge frequency, discharge time, electrode spacing, electrode materials, reactor geometry, and pre-coated catalysts. They are interactive in determining deposition property and deposition rates.

A typical plasma reactor is usually composed of a gas handling system (consisting of precursor gas supplies, mass flow controllers, vacuum systems, comprising pumps and pressure controllers), a plasma reactor, a power supply, and a safety device. The deposition process has the following steps: (1) the wafer with or without catalysts is loaded in the reactor and the system is pumped down to 10^{-5} mm or below to minimize impurities and water vapor; (2) the substrate holder is heated to the desired temperature; (3) the feedstock is admitted, and the flow rate and chamber pressure can be set to desired levels with the aid of a throttle valve; (4) the power is coupled to the plasma. At the end of the run, the heater, the power source, and the gas flow are turned off and the system is purged with argon flow to cool the reactor. The wafer is removed after the reactor cools to below 373k.

The power sources most often used in PECVD are radio frequency (RE, 13.56 MHz) power and microwave (MW, 2.54 GHz) power. Because of operating convenience and other advantages, dielectric barrier discharge plasma (DBD, called atmospheric PECVD here) and hot filament plasma (HFCVD) are also used in the fabrication of nanomaterials.

nc Structures

Nanocrystalline silicon (nc-silicon) is an inorganic complex film with an amorphous silicon continuous phase embedded with countless discrete silicon nanocrystallites ranging from one to several tens of nanometers. It is interesting that these silicon nanocrystallites have been avoided in the manufacture of microelectronic silicon film until significant visible photoluminescence (PL) at room temperature. This material has an efficiency for absorption for visible light, and since then a large number of studies have been devoted to ne-silicon as a promising structure for flat panel displays, solar cells, thin film transistors (TFTS), and sensors.

The atomic structure of nanocrystalline Si consists of Si ordered domains with diamond, cubic-fcc, or cubic-bc8 crystal structure fig. 5.63. The amorphous structure of Si described as a disordered lattice of atoms, bonded with tetrahedral coordination, and with a small distortion both in bond length and angle when compared to diamond crystalline Si. The visible light emission is closely related to the nature of the nanosize Si crystallites, including crystallite size, fraction of crystallinity, orientation, interface structure, and the distribution of the crystals in the amorphous phase. Controlling these structural factors, especially size and the fraction of crystallinity, is a key point to improving PL and other properties.

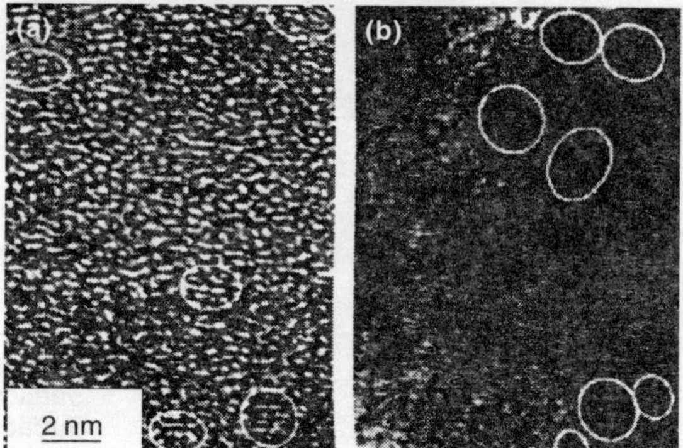

Fig. 5.63 *HRTEM images of the nc-Si film by PECVD: (a) fcc ordered domains and (b) bc8 ordered domains embedded in the amorphous matrix.*

Compared with the progress of the controllable synthesis of metal of inorganic nanostructures, the study of organic and polymeric nanomaterials is still in its infancy, although some of the organic and polymeric materials have already shown special properties different from other metal or inorganic nanomaterials. Polymer nanostructure are mostly investigated for two purposes. One is to nanofabricate conjugated polymer structures with higher crystallinity for semiconductor or conductor applications such as FET's, LED's, molecular wires, and molecular devices. Another is to get nanostructures for insulating interconnections between other conducting or semiconduction nanostructures such as BCP's, nanocable shells, AFM tip coating, insulation coatings for biological implants such as integrated circuits, bond pads wires, and neural probes. PTFE like polymers have many remarkable properties, including low dielectric constants, low coefficients of friction, high thermal stabilities, good solvent of chemical resistance, low permeability constant, low surface tension, and good bio-compatibility. This makes them the best candidates for the second kind of applications.

Properties of nc-Structures

Silicon nanocrystal are the most extensively studied nanomaterials after carbon nanotubes. Different morphologies of silicon nanocrystals have different properties and applications. Among the most interesting phenomena observed include light luminance for nanocrystalline silicon films, field emitting diode (FED) properties for silicon nanotubes or nanorods, and single electron tunneling in silicon single crystals. They have potential applications in nanoscale electronic devices, catalysts, optical and ultra sensitive chemical sensors, field effect transistor (EFT's), light emitting diodes (LED's) and photovoltaic devices.

1. The structure of ni-Si film, depends on important parameters such as temperature, precursor gases, via voltage of the electrode, and discharge modes.

   A simple way to improve the crystallinity of nc-Si films is by heat annealing, example at strong room temperature photoluminescence in the wavelength range 650-950 nm is observed in silicon oxide (SiO*x*) thin films prepared by PECVD and then annealed in a pure nitrogen atmosphere.

   The starting annealing temperature for obtaining PL depends on the PECVD conditions.

   Silicon-containing gases such as SiH_4, SiF_4, which can be diluted by H_2, Ar or some other gases are used as precursor gases depending on the application the films are normally amorphous SiH films heating SiH_4 gas above 773 k has promoted the crystallinity from the early stage of growth without increasing the surface roughness.

   As the temperature increases, the integrated intensity of the PL peaks increases up to 1523 k. An increase of Si concentration in SiO*x* films and a marked red shift of the luminescence peak is detected. The mean radius value increases and the distribution becomes wider.

   Through hydrogen or argon dilution of silane plasma, direct growth of nc-Si: H is completed without any heat treatment. For example, the SAED pattern of Si nanoparticles grown in pure SiH_4 shows diffuse rings assigned to an amorphous structure in Figure 5.64 (a). However, the nanoparticles obtained in 95% Ar-diluted plasma exhibit fcc ordered structure in Figure 5.64(b). Moreover, the growth rate of nanoparticles is about 100 times faster than that of pure SiH_4. Similar phenomena have been observed for hydrogen dilution of silane plasma. In this way, nc-Si film could be obtained directly at low temperature (room temperature to 523 °k) and no post-annealing treatment is needed.

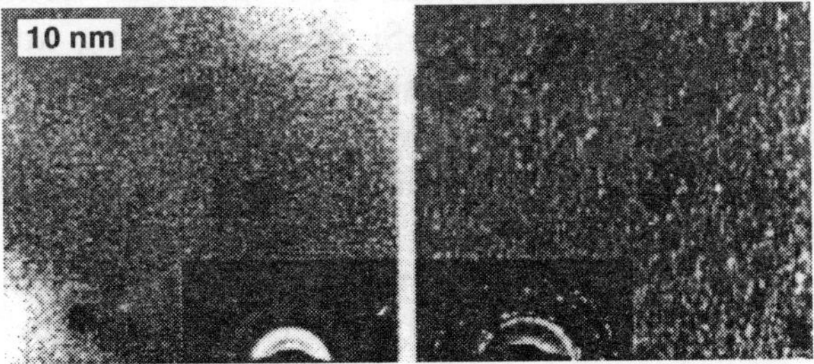

Fig. 5.64 *TEM images of Si nanoparticles obtained from (a) pure SiH_4 discharge and (b) high Ar-diluted SiH_4 discharge.*

(2) The ratio of silane/carrier gas has been found to be important in obtaining different shapes of nanocrystals and surface roughness. As shown in Figure 5.65, the nanocrystal amy change from elongated shape with a length of 30 nm to long grains with a length 0.7-1 nm, with a decrease of silane concentration.

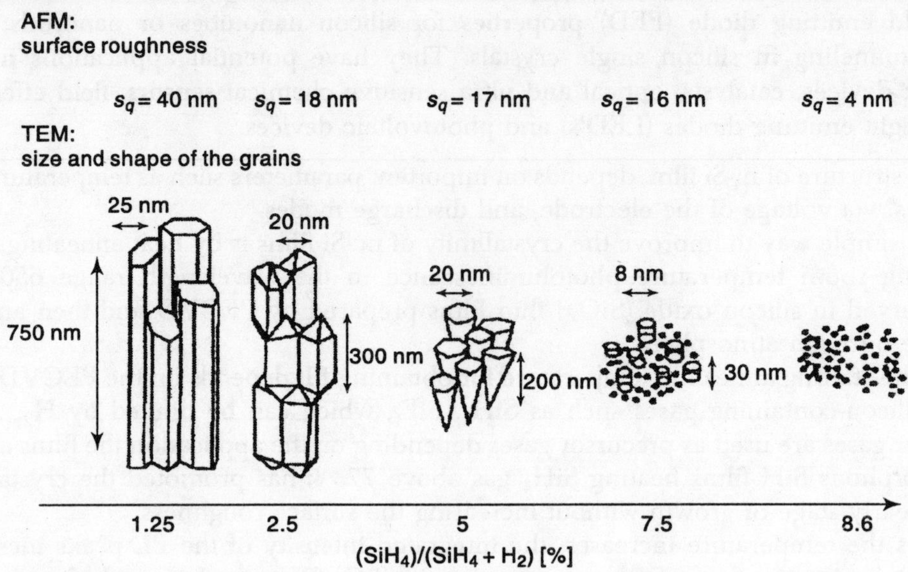

Fig. 5.65 *Schematic picture of the evolution of the microstructure with increasing concentration of silane in hydrogen.*

On the other hand, the crystallinity will be improved by changing the discharge mode. With ECR (electron cyclotron resonance) plasma, a weak magnetic field is superimposed perpendicular to the cylinder axis. Hydrogen dilution is not necessary for the deposition of nc-Si from silane plasma. Nc-Si: H film and high deposition rates of up to 8 Å/s in a pure silane could be obtained at substrate temperature below 473 k. This is attributed to a high degree of dissociation of SiH_4, under these conditions, which results in a high hydrogen flux.

SICN

Using a two-stage plasma PECVD process, SiCN nanorod arrays have been synthesized. First, a SiCN buffer layer is deposited on Si substrates using an ECR plasma. Then, the substrates are transferred to a microwave plasma for the growth of nanorods along a preferred orientation. Quasi-aligned single crystal SiCN nanorods with a six-side-rod-shape of about 1–1.5 mm in length and about 20–25 nm in diameter have been prepared. The growth of the SiCN nanorods is self-mediated without the addition of any metal catalysts. Characteristics current-voltage measurements indicate a low turn-on field of 10 V/mm. Field emission current density in excess of 4.5 mA/cm^2 has been observed at 36.7 V/mm Fig. 5.66. Moreover, they exhibit rather stable emission current under constant applied voltage.

For the mechanism of the nc-Si film growth, two models have been applied to explain the process of the nanocrystalline formation. One model emphasizes on-the-surface reaction. As shown in Figure 5.67, sufficient flux density of atomic hydrogen from hydrogen diluted silane plasma realized full-surface coverage by bonded hydrogen and produces local heating through hydrogen recombination reactions on the growth surface of the film. These two events occurring on the surface enhance the surface diffusion of film precursors (SiH_3). As a consequence, film

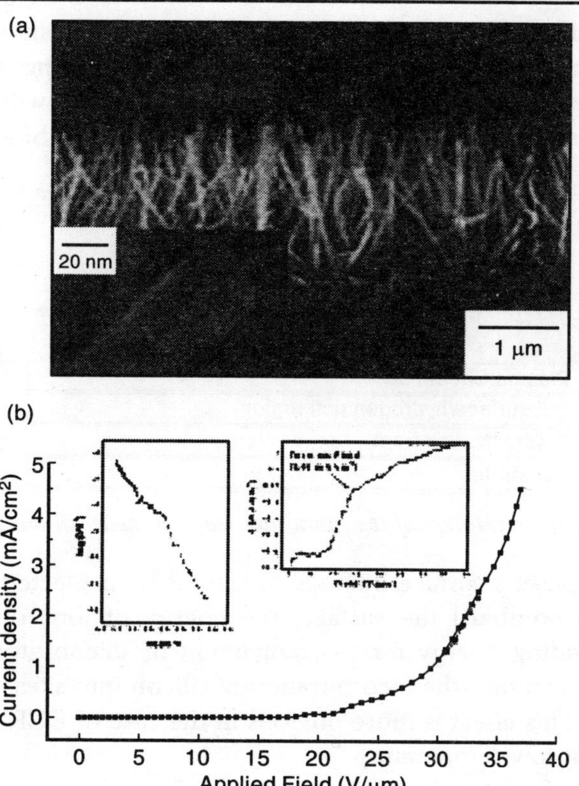

Fig. 5.66 *Morphology and property of SiCN nanorods: (a) Typical SEM image of the SiCN nanorods. (b) Field emission J–E curve and the corresponding F–N plot of the SiCN nanorods.*

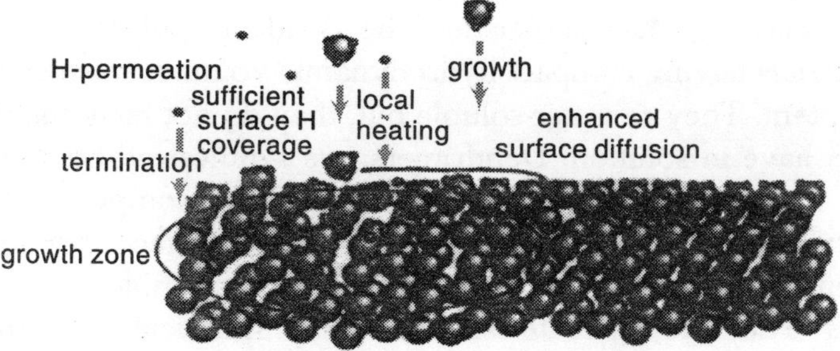

Fig. 5.67 *Schematic of the structure realization model in the formation of nc-Si film by PECVD.*

precursors adsorbed on the surface can find energetically stable sites, leading to the formation of atomically crystalline surface. The effects of hydrogen dilution on the nc-Si crystallinity could be partly explained by this model.

Another model details, the substrate reaction. Cross-linking reactions occur underneath the surface from a few nm to a few tens of nm. In this region, chemical annealing, a hydrogen-mediated reaction, promotes the rearrangement of the silicon network by breaking weak silicon–silicon bonds and by reforming stable crystalline bonds. As shown in Figure 5.68, many hydrogen atoms are permeating into the sub-surface region, giving rise to crystallization of amorphous networks through the formation of a flexible network with a sufficient amount of atomic hydrogen in the sub-surface region without any removal of Si atoms by etching.

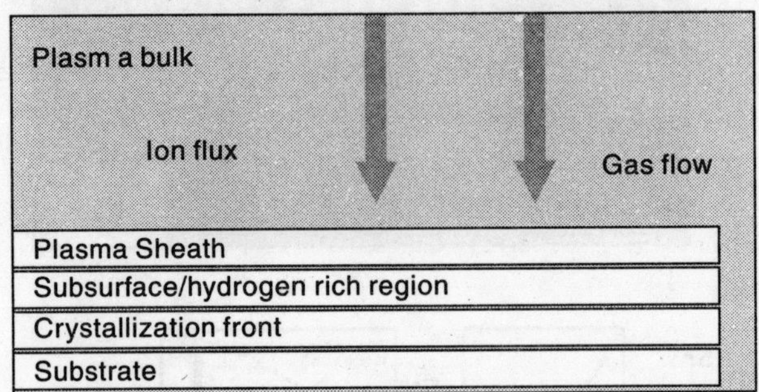

Fig. 5.68 *Schematic representation of the subsurface reaction model in nanocrystalline Si growth.*

In these two models, plasma-surface interaction cannot be neglected in the formation of nc-Si film. When the ions bombard the surface, the energy of ions dissipates by etching the amorphous silicon, providing energy for rearrangement of silicon atoms. The process results mobile silicon atoms. It promotes the incorporation of silicon into sites of minimum energy and forms stable crystallites. This effect is more obvious in the case of ECR plasma, so formation of ncSi can be completed at low temperature.

INORGANIC AND ORGANIC NANOPARTICLES

Dendrimers (Organic Nanoparticles)

A new structural class of macromolecules, the dendritic polymers, are nanometer sized, hyperbranched materials having compact hydrodynamic volumes in solution and high, surface, functional group content. They re water-soluble but, they do not have the thickening properties that many polymers have in solution. Dendrimers, are synthesized by step-wise convergent or divergent methods Dendrimers are defined by their three components: a central core, an interior dendritic structure (the branches), and an exterior surface (the end groups). Over 50 compositionality different families of these nanoscale macromolecules, with over 200 end-group modifications. They are characterized by nearly spherical structures, nanometer sizes, large numbers of reactive endgroup functionalities, shielded interior voids, and low systematic toxicity.

Materials engineering, industrial, pharmaceutical, and biomedical applications. Specifically, nanoscale catalysts, novel lithographic materials, rheology modifiers, and targeted drug delivery systems, MRI contrast agents, and bioadhesives represent some of the potential applications.

Fluorescent Semiconductor Nanocrystals

Fluorescent: Fluorescence corresponds to the fast emission of a photon (quantum of light) by a molecule/particle following the absorption of an excitation photon, higher energy. Fluorescence stops when the excitation is turned off, in contract to phosphorescence. Molecules/particles which possess this property are fluorophores. The probability of this process is termed the quantum yield (maximum value 1). Fluorescent semiconductor nanocrystals consists of two parts: the core and the shell fig. 5.69. The fluorescence of the crystal is given by the core size and its intension is improved by surrounding it with a shell of a higher bandgap semiconductor 5.70. Moreover, a shell of surfactant organic molecules (the ligand shell) passivates the nanocrystal surface and prevents aggregation.

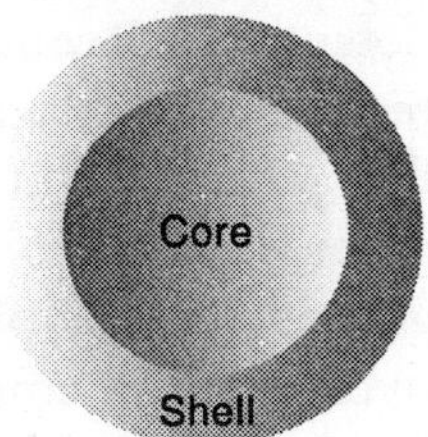

Fig. 5.69

Fluorescent semiconductor nanocrystals are nanometer sized pieces of crystal compromising from a few tens to several thousands or more of atoms. Due to a quantal effect, quantum confinement, their electronic and optical properties are size dependent., when shined with UV light they emit narrow spectrum fluorescence light whose "color" shifts towards shorter wavelengths as the size of the crystallites decreases.

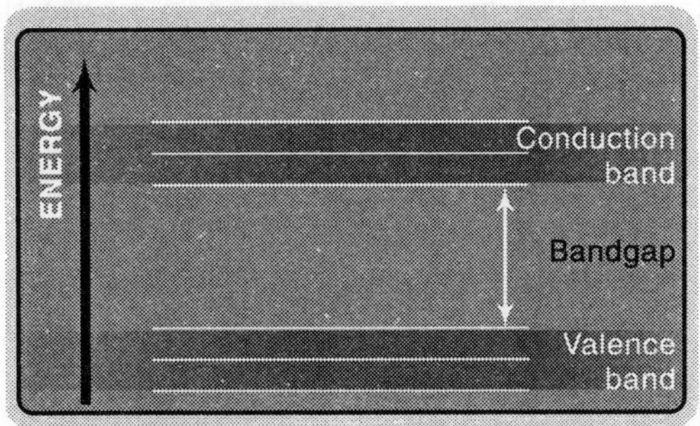

Fig. 5.70 *Energy diagram of the core nanocrystal. The bandgap is the energy difference between the lower edge of the conduction band and the upper edge of the valence band*

Application of nanocrystals is in the fields of biology and medicine. Labelling of live cells, in vivo imaging of growing, single particle tracking in cell trafficking, intracellular staining, as light absorbing or emitting centers in photovoltaic devices or electroluminescent diodes as tags in microbeads to prepare molecular barcodes.

Fluorescence Emission Principle

As in bulk semiconductor, a bandgap separates the valence and the conduction band. At room temperature, the conduction band is empty. When shining the nanocrystals with UV light, electrons are excited from the valence band to the conduction band.

A possible relaxation process (with a probability equal to the quantum yield) is the emission of a photon at the bandgap, the energy excess being dissipated as heat. With CdSe nanocrystals, the visible spectral range is between 500 and 650 nm. With other materials, the spectral range is from the near infrared (CdTe, InASs,...) to the near ultraviolet (ZnS, ZnSe,...). In this sense, semiconductor nanocrystals are inorganic flurophores by analogy with standard organic flurophors (rhodamin or fluroescein.)

For CdS nanocrystals covered with a proper shell, quantum yield above 0.5 at room temperature is obtained.

Polyhedral Silsesquioxanes (Inorganic-Organic Hybrid Nanoparticles)

Hybrid inorganic-organic composites are an emerging class of new materials. Materials are designed with the good physical properties of ceramics and the excellent choice of functional group chemical reactivity. Silicon containing organic polymers, polysilsesquioxanes, have potential replacement for silicon-based inorganics in the electronics, photonics, and other materials technologies. Hydrolytic condensation of trifunctional silanes yields polyhedral clusters having the generic formula $(RsiO_{1.5})_n$. Hence they are known by name silsesquioxanes. Each silicon atom is bound to an average of one and a half (sesqui) oxygen atoms and to one hydrocarbon group (ane). Typical functional groups alkoxy-or chlorosilanes, silanols, and silanolates are hydrolyzed or condensed. These inorganic-organic hybrids offer a unique of physical, chemical, and size dependent properties. These properties can not be realized from just ceramics of organic polymers alone. Silsesquioxanes are therefore depicted as bridging the property space between these two component classes of materials. Many of these silsesquioxnae hybrid materials also exhibit an enhancement in properties such as solubility, thermal and thermomechanical stability, mechanical toughness, optical transparency, gas permeability, dielectric constant, and fire retardancy etc.

Quantum Corrals

The closed—mostly circular—barrier of iron atoms on a copper surface causes interesting phenomena, revealing the wavelike behaviour of the electrons trapped within.

Quantum corrals of any shape can be made; among the most common are circular, rectangular, and elliptical structures. Depending on their size and shape, these fascinating nanostructures exhibit tantalizing phenomena such as quantum mirage.

In a quantum mirage the electronic signature of an impurity atom, placed inside the corral, is projected to another location inside the corral, while the impurity remains on the same position. This phenomena depends on the geometry of the corral, and on the nature of the substrate and the impurity.

Although quantum mirages are associated with an elliptical corral shape, theoretical investigations show that this phenomenon can also occur in other corral structures due its quantum mechanical nature.

An other interesting phenomena in quantum corrals is the effects of impurities on the 'electron sea'. The Kondo effect is an excellent example. It provides a useful means to explore the phenomenon of quantum mirage.

When placing a magnetic impurity on a specific spot inside an elliptical quantum corral of special sizes and shapes, Fig. 5.71 the phenomena of quantum mirage occurs. More precisely, when the Co atom is placed in one of the focal points of the elliptical carrol, the perturbations of the electron sea caused by the Kondo effect appears at the other focus, just like it would be mirrored.

Nanocomposites

Nanocomposites are materials with a nanoscale structure that improve the macroscopic properties of products. Typically, nanocomposites are clay, polymer or carbon, or a combination of these materials with nanoparticle building blocks.

Nanocomposites, materials with nanoscales separation of phases are of two types: multilayer structures and inorganic/organic composites. Multilayer structures are typically formed by gas phase deposition or from the self-assembly of monolayer. Inorganic/organic composites are formed by sol-gel techniques, bridging between clusters (as in silsequioxanes), or by coating

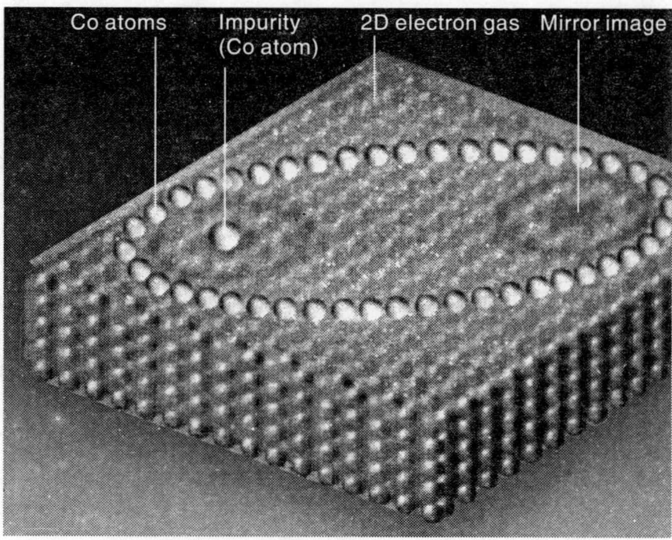

Fig. 5.71

nanoparticles, in polymer layers for example. Nanocomposites greatly enhance the properties of materials. For example, ppm level impurities results in the formation of nanoscale aluminide secondary phases in aluminum alloys, increasing their strength and corrosion resistance. Magnetic multilayered material are one of the most important aspects of nanocomposites as they have led to significant advances in storage media.

Nano-Intermediates

Nanostructured films, dispersions, high surface area materials, and supramolecular assemblies are the high utility intermediates to many products with improved properties such as solar cells and batteries, sensors, catalysts, coatings, and drug delivery system. They have been fabricated using various techniques.

Nanoparticles are obvious building blocks of nanosystems but, require special techniques such as self-assembly to properly align the nanoparticles. Recently air resistant, room temperature systems for nanotemplates with features as small as 67 nm are observed. Also, electron-beam systems are used to fabricate devices down to 40 nm.

NANOSTRUCTURED MAGNETIC MATERIALS

Magnetic materials are the key to the future of the storage industry. The recent development of thin film heads, the demand for higher density of information storage, and the emergence of completely new technologies call for entirely new types of magnetic materials with much higher magnetization and coercivity, H_c.

Given below is a brief introduction to the language and technical terms involved in the study of nanomagnetism.

Magnetostatics

In common usage, magnetic materials are those that exist in a state of permanent magnetization without the need to apply a field. For all permanent magnets there is a north pole (N) and a south pole (S), and lines of magnetic field pass from N to S. As a net properties of a magnet are some form of sum over a large number of atomic or electronic magnets (referred to as magnetic dipole moments), the magnetic materials can be separated into three categories.

Diamagnetism, paramagnetism and ferromagnetism. Diamagnetism is a fundamental property of all atoms (molecules), and the magnetization is very small and opposed to the applied magnetic field direction. Many materials exhibit paramagnetism, where a magnetization develops parallel to the applied magnetic field as the field is increased from zero, but again the strength of the magnetization is small. Ferromagnetism is the property of those materials which are intrinsically magnetically ordered and which develop spontaneous magnetization without the need to apply a field. The ordering mechanism is the quantum mechanical exchange interaction. A variation on ferrimagnetism is ferrimagnetism, where different atoms possess different moment strengths but there is still an ordered state below a certain critical temperature.

Magnetic Anisotropy

Consider a single crystal of a ferromagnet, and apply a magnetic field along different crystallographic directions, the magnetization varies with field depending on the chosen direction for the applied field (Figure 5.72). This phenomenon goes under the generic title of magnetic anisotropy. For Fe, as shown in Figure 5.72, ⟨100⟩ are the easy directions (low field required to saturate), ⟨110⟩ and ⟨111⟩ directions are hard (high field to saturate). This form of anisotropy is

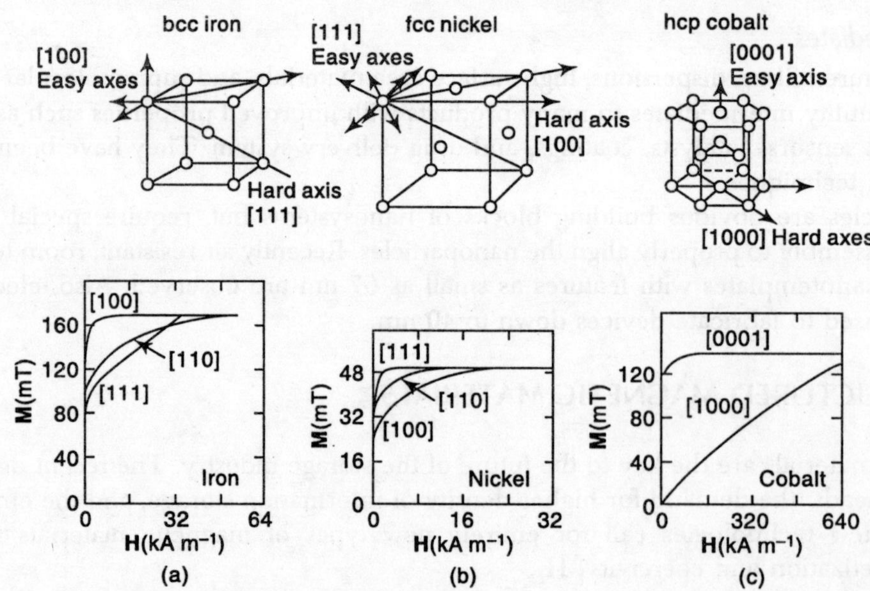

Fig. 5.72 *The crystal structure and magnetization loops For Fe, Ni and Co, demonstrating the anisotropic nature of the magnetization process.*

termed magnetocrystalline anisotropy and is closely related to the detailed electronic structure of the crystal.

Fe is a body-centered cubic solid; that is, the Fe atoms arrange on the cube corners and at each cube centre in a regular three dimensional pattern. The ⟨100⟩ directions are along a cube edge, the ⟨110⟩ directions along a face diagonal and the ⟨111⟩ directions along as cube body diagonal.

Domains

All ferromagnetic materials below the Curie temperature are composed of small-volume regions, called domains, in which there is a mutual alignment in the same direction of all magnetic dipole moments Fig. 5.73(a).

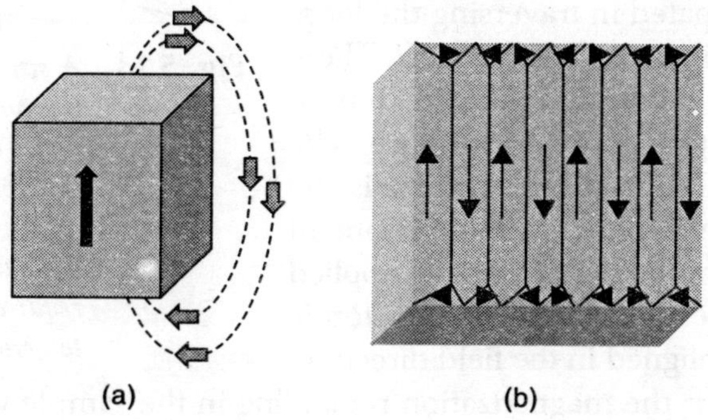

Fig. 5.73 *(a) A single-domain ferromagnetic particle, (b) a multidomain ferromagnetic particle with closure domains on the top and bottom surfaces.*

Within these domains, the spontaneous magnetization present is equal to the saturation magnetization of the material, and so the individual domains are fully magnetized at all times. In the absence of an applied field, there is no net magnetic moment or field generated by the material because the magnetization direction of each domain is randomly oriented. During magnetization of the material, domains whose magnetization directions have a component in the direction of the applied field will grow at the expense of those that do not. Once all the unfavourably oriented domains have been eliminated by domain wall movement, the magnetization direction of the single domain that remains will be rotated so as to be parallel to that of the applied field and saturation is thus achieved. As the magnetic field is reduced by reversal of field direction, the curve does not retrace its original path.

The Magnetization Process

The magnetic field strength **H** can be defined by

$$\mathbf{B} = \mu_0 \mathbf{H}, \tag{1}$$

where $\mu_0 = 4\pi \times 10^{-7}$ H m^{-1} is the permeability of free space. The magnetic induction B has the units of tesla (T). The flux $\Phi = BA$ can be defined, where A is a cross-sectional area. Flux has the units of weber (W).

If a ferromagnetic material is placed in a field **H**, Equation (1) becomes

$$\mathbf{B} = \mu_0(\mathbf{H} + \mathbf{M}) = \mu 0(\mathbf{H} + \chi\mathbf{H}) = \mu_r\mu_0\mathbf{H}, \quad (2)$$

where **M** is the magnetizations of the sample (the magnetic dipole moment per unit volume). Here, χ is the susceptibility of the magnetic material, and $\mu_r = 1 + \chi$ as the relative permeability of the material (both χ and μ are dimensionless). Take any ferromagnet and increase the field from zero to some peak value, decrease the field back through zero to an equal and opposite value and then return again to the original peak value, hysteresis loop (Figure 5.74) is formed. Energy is dissipated in traversing the loop. This energy loss is primarily manifested as heat. There are several key points on the loop. It is a key tool in the analysis of permanent magnet performance. The saturation magnetization $\mathbf{M}_s$ is the maximum is the maximum value that the magnetic dipole moment per unit volume can take in the direction of an applied magnetic field. In this state all the contributing atomic (electronic) moments are aligned in the field direction.

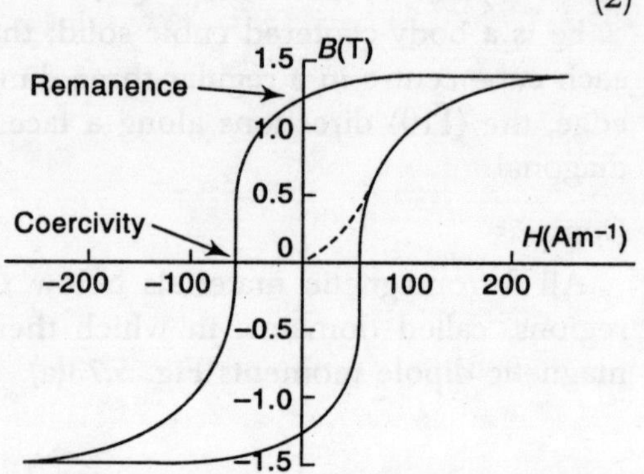

Fig. 5.74 *A schematic diagram of a magnetization (or hysteresis) loop. Remanence is the magnetization remaining when the field is reduced to zero from that required to saturate the sample. Coercivity is the field required to bring the magnetization to zero from remanence.*

The magnetization $\mathbf{M}_r$ is the magnetization remaining in the sample when the applied field is reduced to zero from the value which saturated the sample. The coercivity $\mathrm{H_c}$ is the field required to reduce the magnetization to zero from saturation (Figure 5.74), and is applied in the opposite direction to the original saturating field. If the magnetization is reduced to zero from less than its saturating value than $\mathbf{H}_c$ is termed the coercive field (always less than the coercivity). The ratio M_r/M_s and the area enclosed in the loop are important in determining the applicability of a given ferromagnet.

These loops are a graphical representation of the relationships between an applied magnetic field and the resulting induced magnetization within a material. The field is generated by the magnetized material (B_i)

When added to that of the applied field (H) is known as the normal induction (B_n) or simply B. Simply this induction has two components, it is defined as

$$\mathrm{B} = \mathrm{Bi} + \mathrm{H}$$

The loops the properties of magnetic materials as it is magnetized and demagnetized. When a magnetic field is applied to unmagnetized material, the intrinsic induction (Bi) is established within it, parallel to the applied field. Ig H is sufficiently strong, the magnet will become fully magnetized at the saturation flux density (Bsat). When the field is reduced to zero, the magnet returns to the residual value or remanence (Br), as long as the magnet is within a closed magnetic circuit. Some of these magnetic properties can be improved with nanostructured magnetic material.

Nanomagnetic Materials

A broad classification of nanomagnetic materials may be made as follows:

- *Particulate nanomagnets:* These are granular solids where one or more phases are magnetic; nongrained layers grown in columnar films; quasi-granular films made by heat-treating multilayers of immiscible solids.
- *Geometrical nanomagnets:* columnar films or nanowires; needleshaped particles in a matrix.
- *Layered nanomagnets:* multilayers with metallic and/or other nanometre-thick films on a supporting substrate.

1. Particulate Nanomagnets

There are two routes to produce these nanophase materials: (1) form an amorphous phase then heat-treat to develop the nanostructure; (2) cool more slowly so that the nanostructure is developed without further treatment. The alloy chemistry can be varied to go from the softest (lowest coercivity and anisotropy) to the hardest (highest coercivity and anisotropy) magnetic materials known:

$$\text{Soft} = \text{FeCuNbSiB} \Rightarrow \text{NdFeB} = \text{hard}.$$

It is the physical nanostructure and the resultant interplay between exchange and anisotropy energies that is significant here. In very small grains that are magnetically coupled, exchange forces cause the moments in neighbouring grains to lie parallel, overcoming the intrinsic easy-axis properties of the industrial grains. The coupling dominates over length scales shorter than the exchange length, L_{ex}, for given exchange stiffness, A. This length L_{ex} is of the order of the domain wall width. It is an expression for the length scale over which the direction of magnetization can vary:

$$L_{ex} = \sqrt{\frac{A}{K_1}} = 35 \text{ nm for } Fe_3Si.$$

For *permanent nanomagnets* the archetype comes from a nanocomposite of $Nd_2Fe_{14}B$ grains of size 20–30 nm in a paramagnetic Nd- and B-rich matrix. Figure 5.75 illustrates the unit cell of $Nd_2Fe_{14}B$ which has the *c*-axis as the uniaxial anisotropy axis. To be useful, a permanent magnet material requires a high remanence (M_r/M_s close to 1), a high coercive field and a high Curie temperature.

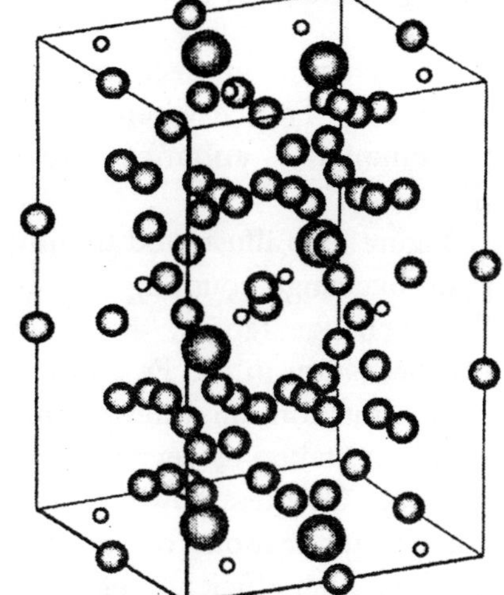

Fig. 5.75 *Tetragonal unit cell of $Nd_2Fe_{14}B$. The c axis (direction of uniaxial magnetic anisotropy) is vertical in this image.*

Nanocomposites are usually used to make working magnets. The nanocomposite is formed by taking rapidly solidified flake, or particles produced from flake, and then bonding with a filler (e.g., a polymer) or by forming diffusion bonds between particles or flakes by heat treatment under high pressure at elevated temperature. The route used to produce the final product significantly affect the overall magnetic properties, as illustrated by Figure 5.75.

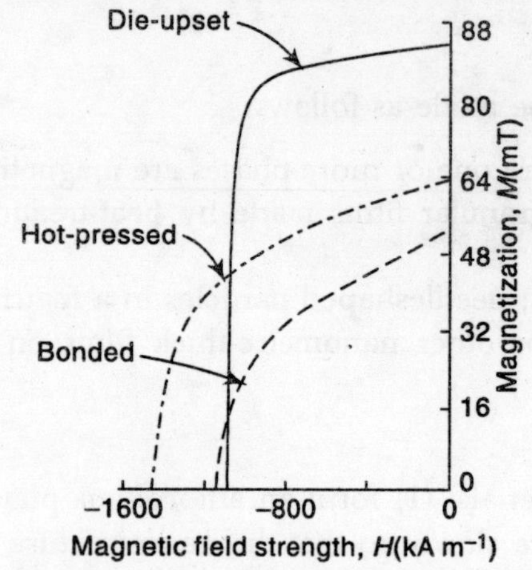

Fig. 5.76 *The effects of final magnetic production on hysteresis loop shape in the second quadrant for NdFeB magnets. The die-upset nanocomposite involves forcing liquid metal into the die to minimize porosity. A hot-pressed nanocomposite magnet is a powder of flake consolidated under high temperature and pressure and the bonded nanocomposite is also a powder or flake treated with a binder, such as an epoxy resin.*

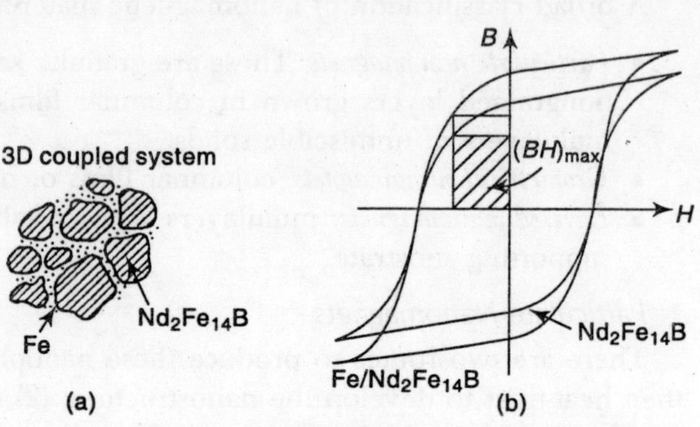

Fig. 5.77 *(a) A schematic diagram of the exchange-coupled alloy microstructure of $Nd_2Fe_{14}B$. (b) The corresponding hystreresis loop.*

In the nanocomposite material there can be exchange coupling between grains, especially if the alloy is Fe-rich when bcc-Fe nanoparticles accumulate between the NdFeB grains, leading to remanence enhance effects whereby the remanence can be increased significantly (Figure 5.77).

Figure 4.78 illustrated the importance of controlling the intergranular Fe thickness to ensure good exchange coupling between the $Nd_2Fe_{14}B$ grains. If the thickness of the Fe layer exceeds the exchange length, then coupling between NdfeB grains will be lost as the direction of magnetization in the Fe grain will rotate towards the intrinsic easy direction in the Fe crystal.

It is the continual development of alloy composition and control of physical nano-structure which has taken us very close to the theoretical limit for performance with the NdFeB family of magnets. Cordless power tools and miniature earphones on portable audio devices are just two current applications of such advanced nanomagnetic materials. Work is also in progress to produce high-efficiency electric motors for use in transportation. The energy product $(BH)_{max}$, is the parameter used to define the energy stored in an air gap in the magnet. This point is marked on Figure 5.77 for illustration. The highest value reported so far in an NdFeB type magnet is around 450 kJ m^{-3}.

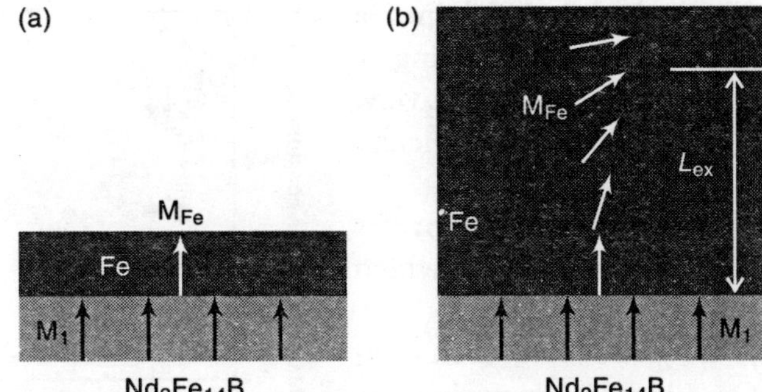

Fig. 5.78 *The effect of soft layer thickness on exchange coupling between grains in a permanent magnet nanocomposite. Lack arrows represent the magnetization direction in the $Nd_2Fe_{14}B$ phase and white arrows represent the magnetization direction in the Fe layer.*

2. Layered Nanomagnet Films (Thin Films/Multilayers)

Here a nanoparticulate magnetic film is grown on a suitable seeding layer. If the seed layer can be grown with a high crystallographic texture (close-packed direction) out of the plane, and with small grain size, then the functional overlayer can be both textured and nanostructural. Such films are used for recording media (Figure 5.79).

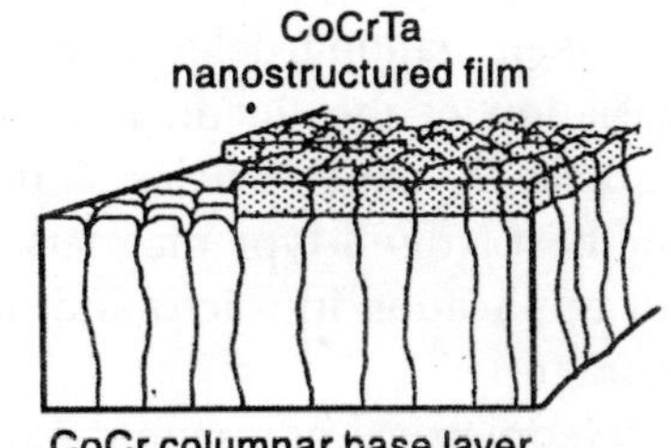

Fig. 5.79 *A schematic picture of a granular magnetic recording medium. In this example the crystallographic texture of the CoCr underlayer ensures that the functional CoCrTa layer has the necessary nanostructure.*

Modern recording media require small grain size to ensure low noise with many grains per bit. High coercivity is also essential so that the written information remains stable overtime. If the grains can be magnetically decoupled, using nonmagnetic grain boundaries, stability against thermal erasure of information may also be increased., The growth of a suitably doped nanomagnetic layer on a columnar seed layer ensures these criteria are met. The doped grains are about 30 nm thick.

Also in layered magnets, effects at surfaces and interfaces are observed; for example, there is growing evidence that in chemically homogeneous films with thickness less than 20 nm, the magnetic elastic coefficients are markedly different to the bulk. This is shown in Figure 5.80 for $Ni_{81}Fe_{19}$, where the composition is chosen for zero magnetostriction *in the bulk*. There are major implications for such services as spin values, where the soft magnetic layers are 'buried' in a stack of many layers. This is inevitably going to place the layers under strain.

The reduced symmetry at a surface of interface (atoms only coordinated over a hemisphere) can leads to changes in magnetic anisotropy. As films (or particles) decrease in dimension, surface effects dominate volume effects.

3. Geometrical Nanomagnets

Self-assembly is developing as a very important technique in this area. One route to the production of geometric nanomagnets comes from templating. Using lithographic techniques, a mask is formed which can be stamped into a polymer resist. Then using lift-off or etching, a pattern of magnetic islands remains. Alternatively, layered films may be directly lithographically patterned.

Block copolymer techniques, or porous alumina, are used to give a template in which nanowires are grown by electrodeposition.

If nanoparticles can be produced, then self-assembly by surfactant and ligand structuring can be used. This offers a number of advantages in that large areas must be covered by self-assembly, and the spacing between nanoparticles might be covered by self-assembly, and the spacing between nanoparticles might be controlled by variation of the ligand lengths. Self-assembled nanocomposite have been formed using FePt and Fe_3O_4 nanoparticles as precursors, giving an energy product of 160 kJ m^{-3}. This is one of the best NdFeB-type magnets. Thin film self-assembled permanent magnets have great potential for application in micro-and nanoelectromechanical systems (MEMS and NEMS) and also data storage.

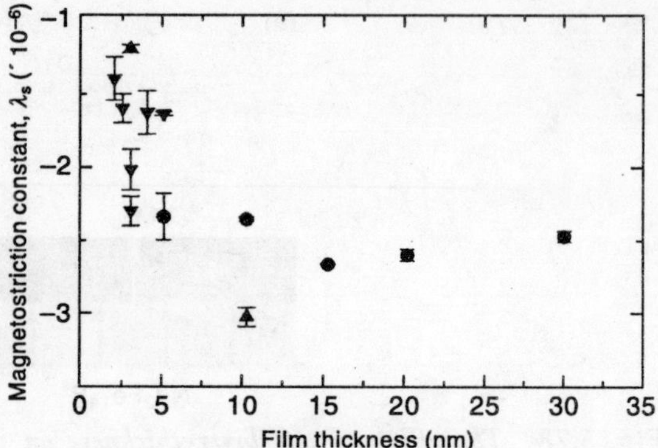

Fig. 5.80 *Plot of magnetostriction constant λ_S against film thickness for NiFe on SiO_2. The different symbols refer to different data sets.*

Geometrical permanent nanomagnet materials rely on nanoscale geometrical effects for their properties example Alnico, an alloy of Fe with Ni, Co and Al additives.

There is an energetic preference of particle elongation along the ⟨100⟩ direction during production. Thus driven by minimization of the magnetic charge density at the interface between particle and matrix. This is obtained by applying a field in a chosen direction during spinodal decomposition.

It is shape anisotropy of the particle that produces the necessary anisotropy to produce hard magnetic properties. Figure 5.81 is a TEM micrograph showing the elongated particles in the matrix. The magnitude of shape anisotropy is not great, so that the performance of the Alnico magnets is not as impressive as of the NdFeB magnets.

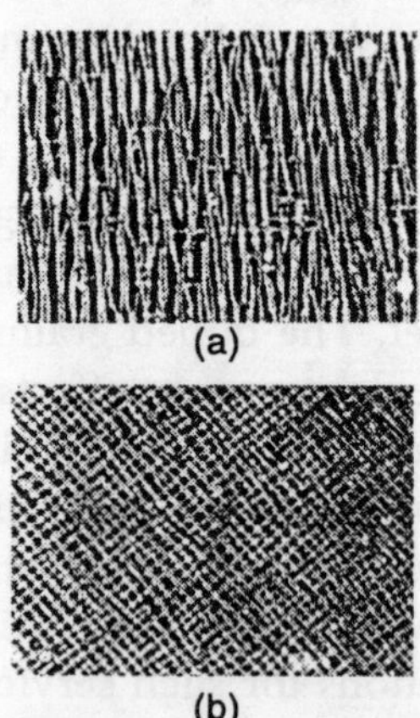

Fig. 5.81 *The nanostructure of Alnico 8 after isothermal heat treatment in a magnetic field for 9 h at 1073 K. (a) view perpendicular to the direction of the applied field, (b) view along the direction of the applied field. The microstructure has a periodicity of approximately 32 nm.*

4. Magnetic Nanoparticles

Magnetic nanoparticles show a variety of unusual magnetic behaviors when compared to the bulk materials, due to surface/interface effects, including symmetry breaking, electronic environment/charge transfer, and magnetic interactions. Ultrafine magnetic particles are made by several techniques, including vapor deposition, sputtering, and chemical synthesis. Spherical and magnetically hard Fe, Co, and Fe(Co)-B-based particles in the size of 720 nm, that possess an effective anisotropy and coercivity on to two orders of magnitude higher than in the bulk materials (Fig. 5.82), have been prepared the highest coercivities are obtained in passivated ultrafine particles that a have a core/shell morphology with a metallic core surrounded by a Fe_2O_3/Fe_3O_4 oxide shell consisting of small grains. The high coercivity of the ultrafine particles is based on single-phase, uniform ellipsoidal particles, without magnetic shells surrounding them and without any particle interactions. The large coercivities are attributed to an exchange interaction between the core and shell moments at the interface. This interaction leads to an anomalous temperature dependence of coercivity and no shifted hysteresis loops in field-cooled samples and is known as *unidirectional exchange anisotropy.* A large core/shell interface interaction must be taken into consideration in order to model and fit the data of temperature dependence of coercivity. The core/shell morphology is also present in the magnetic granules of Fe/SiO_2 films. Higher coercivities in particles embedded in an SiO_2, are there rather than B N matrix (Figs. 5.83 and 5.84).

$Fe_{30}(SiO_2)_{70}$ and $Fe_{30}(BN)_{70}$ granular solids show that the magnetic granules in the above matrices are not pure Fe particles but rather a mixture of crystalline Fe and Fe-Si in SiO_2 and amorphous Fe-B in BN thus there is the lower coercivities in the latter because of the lower anisotropy of Fe-B). But there is presence of an Fe-Si-O phase in the Fe/SiO_2 films with higher H_c.

The magnetic hysteresis behavior of small magnetic particles is that their high coercivity is due to both size and surface/interface effects, with the latter being predominant. However, it is difficult to separate the two effects in particles prepared by either vapor deposition or sputtering, because it is difficult to prepare pure (single phase) particles with either of these techniques. In the former, the particles must be passivated to be protected from oxidation, while in the latter there is often significant alloying of Fe with Si and B a technique that allows to fabricate pure particles separately but in the same chamber where sputtering takes place is by using there particle guns (one uses a resistive heater, the second uses spark erosion and the third uses sputtering) to make nanoparticles (Figs 5.85 to 5.87). This technique allows to make composite films with pure or treated uniform magnetic nanoparticles of any material in any matrix by using simultaneous particles deposition and sputtering, with a rotating substrate.

To understand intrinsic size effects, surface/interface effects, and the effects of intra- an winter-particle interactions on the magnetic properties of nanoparticles made by vapor deposition, first particles are made which later treat to modify their surfaces., Also nanoparticles of rare earth metals and intermetallic compounds with high anisotropy have been made which are expected to show interesting and unique properties.

Magnetic recording media, since the drive for higher density media requires isolated particle with size below 10 nm and coercivity greater than 3 kOe, similarly as Transition metals and their alloys have a low anisotropy and become superparamagnetic below this size.

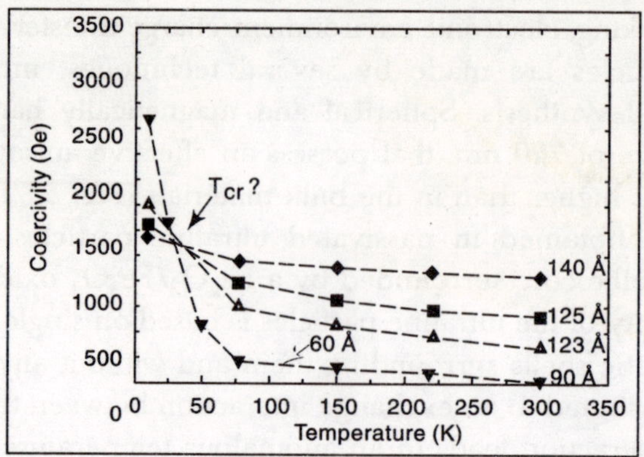

Fig. 5.82 *The temperature dependence of the coercivity in passivated Fe particles.*

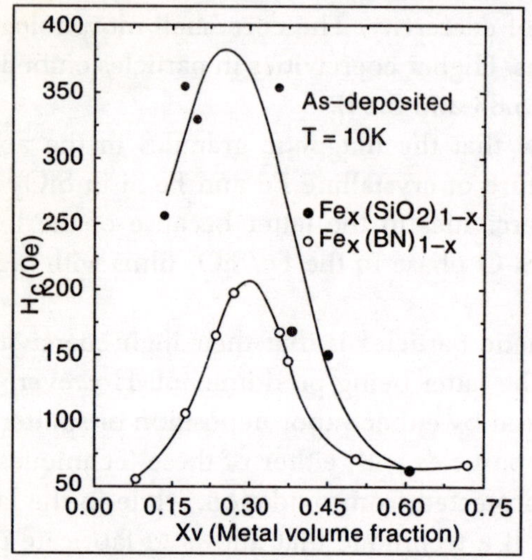

Fig. 5.83 *The dependence of the coercivity on the metal volume fraction.*

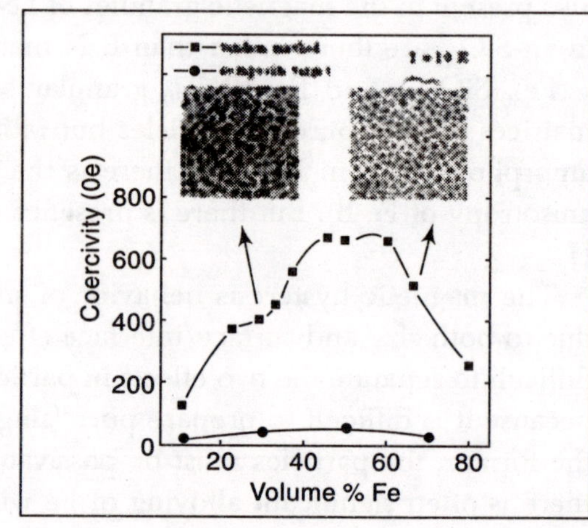

Fig. 5.84 *Coercivity versus Fe content Fe/SiO_2 granular solids.*

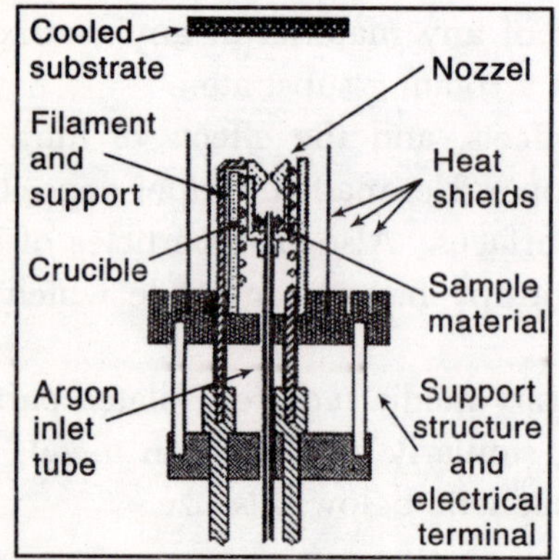

Fig. 5.85 *Particle gun based on evaporation.*

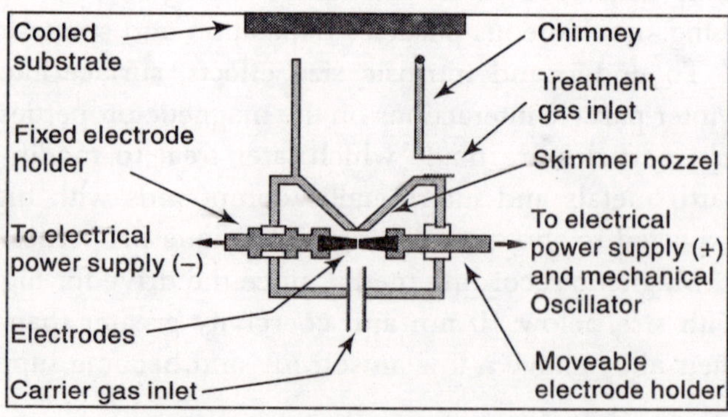

Fig. 5.86 *Particle gun based on spark.*

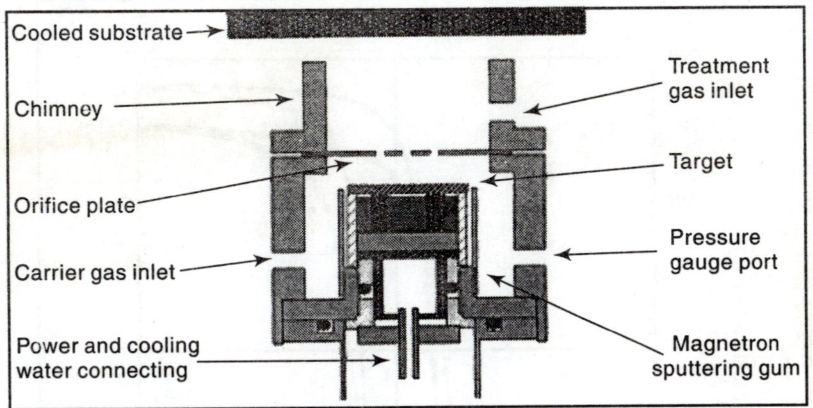

Fig. 5.87 *Particle gun based on high pressure sputtering.*

5. Nanocomposite Permanent Magnets

Nanocomposite magnets consisting of a uniform mixture of exchange coupled magnetically hard and soft phases, have useful a hard magnetic properties. High energy products and relatively high coercivities are developed in these nanocomposite magnets.

A small grain size (10-20 nm) and a uniform mixture of the two phases is a prerequisite for exchange coupling. This coupling leads to a smooth hysteresis loop (Fig. 5.88). A suitable microstructure is prpared by nonequilibrium metallurgical techniques such as melt-spinning, mechanical alloying, and sputter deposition.

Different systems including Nd_2 $(Fe\text{-}Co)_{14}B$, $Pr_2(Fe\text{-}Co)_{14}B$, and $Sm_2(Fe\text{-}Co)_{17}C_X$ are used. Melt-spun ribbons with off-stoichiometric compositions are used to ensure a microstructure consisting of fine mixture of hard (2:14:1, 2:17:C_X) and soft((-(Fe-Co)) phases.

The coercivity of these is found to decrease with the amount of soft phase, but the reduced remanence increases with values greater than 0.7. In addition, the Curie temperature of the hard phase is found to be affected by the presence of the higher T_c soft phase (Fig. 5.89).

Properties obtained so far in this type of magnet are inferior of those of existing magnets mainly because of the larger grain size of both phase and their non-uniform distribution.

6. Soft Magnetic Nanocrystalline Alloys

Soft magnetic nanocrystalline alloys have high coercivity and low remanence magnetization. Two important factors to improve the remanent magnetization are the nanocrystalline grain size and the degree of coherence across interphase boundaries. Fe-Si-B alloys containing small amounts of Nb and CU are melt spun.

Thus the Fe-Si-B-Nb-Cu amorphous phase transform to Fe-Si solid solution with grain sizes of about 10 nm during annealing at temperatures above the crystallization temperature. The presence of small amounts of Cu helps to increase the nucleation rate of the bcc phase while Nb retards the grain growth. These alloys provide low core losses (even lower than amorphous soft magnetic alloys such as Co-Fe-Si-B), exhibit saturation induction of about 1.2 T, and exhibit very good properties at high frequencies, comparable to the best Co-based amorphous alloys.

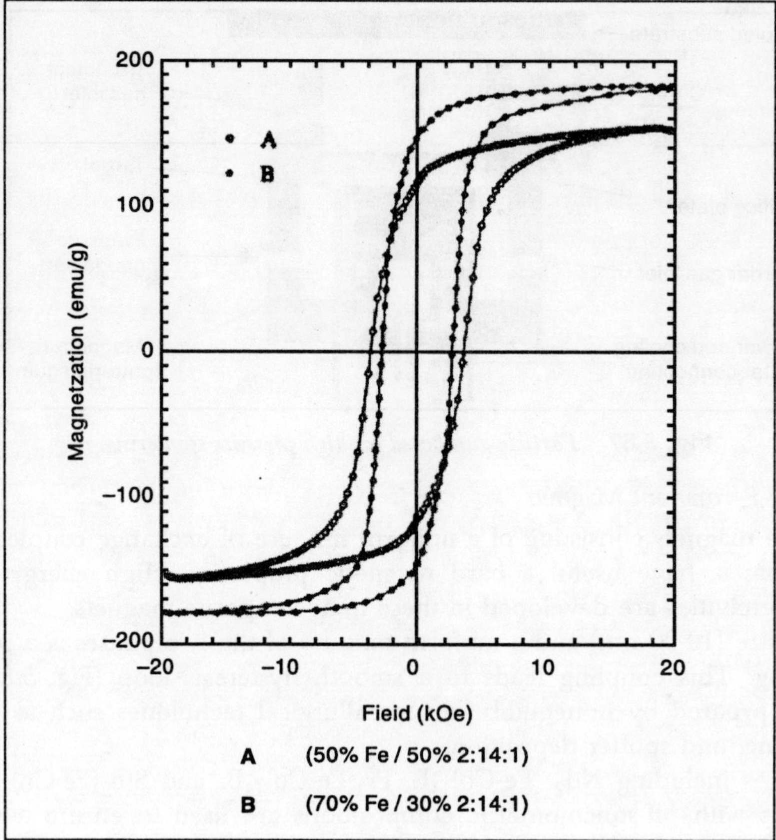

Fig. 5.88 *Hysteresis loops of Fe-rich Nd-Fe-B magnets.*

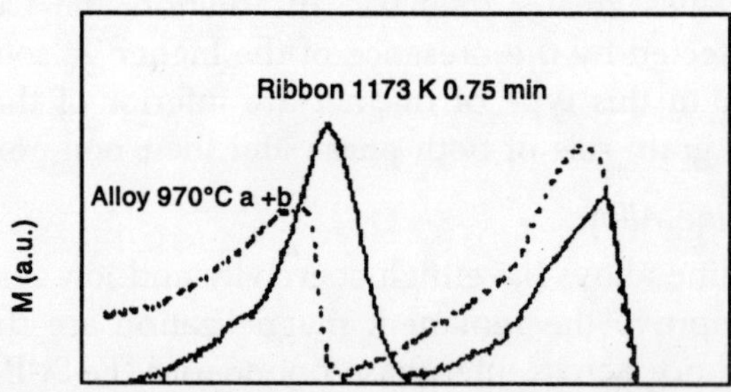

Fig. 5.89 *M vs T $Sm_2Fe_{17}Ga_2C_X$ magnets consisting of 2:17:C_X + a-Fe phases.*

Soft magnetic materials are used for data storage applications that are dependent on the microstructure and geometry of the material. Magnetic films are used in a variety of applications, including recording media and heads, magneto-optical storage and sensors. The behaviour of the magnetic domains and single domain particles, magnetic resistance, and magnetic anisotropy of the films are influenced by factors that include the grain size and orientation, the presence of the nonmagnetic phases at grain boundaries, non-magnetic interlayers and magnetostriction.

While many of soft magnetic properties of noncrystalline alloys are superior, they exhibit lower saturation induction than Fe-metalloid amorphous alloys, due to lower Fe content, to attain amorphization; and because of the addition of N and Cu (or other elements) to control the nucleation and growth kinetics.

These alloys are based on the Fe-Zr-B system. They contain larger concentrations of Fe (83-89 at.%) compared to the Fe-Si-B-Nb-Cu alloys (~74 at.% Fe) and have higher values of saturation induction ~1.6-1.7 T). They have very low energy losses at power frequencies (60 Hz), making them potentially interesting for electrical power distribution transformers Figure 5.90.

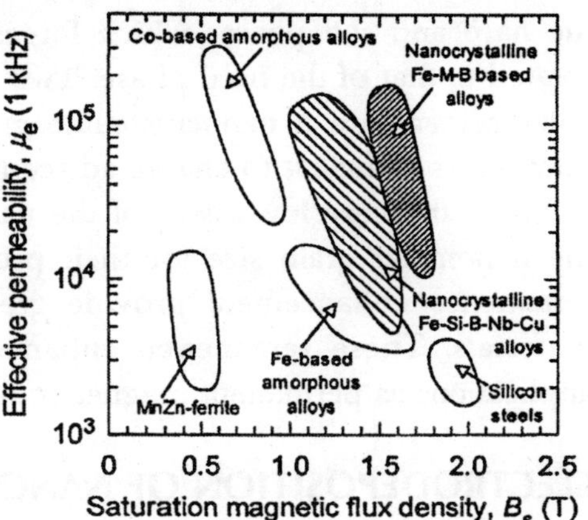

Fig. 5.90 *Effective permeability, μ_c, vs. saturation magnetic flux density, B_s, for soft ferromagnetic materials.*

Another from of nanostructured magnetic materials useful in power electronics and sensors, are obtained by crystallising precursors as amorphous alloy ribbons. The amorphous ribbons crystallise in two stages: a magnetically desirable *bcc* (Fe, X) phase appears first, followed by a boride phase. The separation between the two crystallization phases is very large (~150 K), so that crystallising heat treatments is conducted above the temperature for the first phase, while safely avoiding the onset of the other. Attractive soft magnetic properties are observed for nanocrystalline Fe-base alloys obtained by crystallization of an amorphous ribbons alloy ribbon produced by rapid solidification. The crystallization of amorphous Fe–Si–B alloys containing Nb and Cu exhibits good soft magnetic properties of 1.2 to .14 T for saturation induction (B_s) and 10×10^4 for effective permeability (μ_e) at 1 kHz. Higher Fe content amorphous alloys of type Fe–M–B (M = Zr, Hf, or Nb) are found to exhibit excellent soft magnetic properties.

The mechanisms proposed for good soft magnetic properties of nanoscale *bcc* Fe–M–B alloys are as follows.

(i) High B_s resulting from the increased Fe content and the magnetic coupling between the nanoscle *bcc* particles via the ferromagnetic amorphous phase;
(ii) reversible magnetization due to magnetic homogeneity from the nanocrystalline bcc Fe phase being smaller than the width of magnetic domain walls;
(iii) stability of the nanocrystalline structure enhanced by the enrichment of solute elements in the amorphous phase.
(iv) reduction of the saturated magnetostriction, λs, resulting from their redistribution of the solute elements between the nanocrystalline *bcc* and amorphous phases.

The reduction of Co to these alloys can enhance the values of B_s and μ_c and balancing additions of Zr and Nb can achieve nearly zero λs.

Hard magnetic materials: Magnetic induction is improved by utilizing exchange coupling in magnetically hard and soft phases. Nanoscale two-phase mixtures of hard and soft magnetic phases can exhibit values of remanent magnetization, M_r, significantly greater than the isotropic value of 0.5, *Ms*. This "remanence enhancement" is associated with exchange coupling between the hard and soft phases which forces for the magnetization vector of the soft phase to be rotated to that of the hard phase. Two important requirements for alloys to exhibit remanence enhancement are a nanocrystalline grain size and a degree of coherence across interphase boundaries sufficient to enable adjacent phase to be exchange coupled.

While the very low losses of the nanocrystalline soft magnetic materials are dependent on the nanometer grain size for their properties, the hard magnetic nanocrystalline alloys with remanence enhancement provide greater flexibility in processing, especially with powder materials. These remanence enhanced nanocrystalline hard magnetic alloys find many applications as permanent magnet components.

ELECTRODEPOSITION OF NANOSTRUCTURES

Electrodeposition is one of the techniques used to produce nanostructured magnetic multilayers as well as single thin films. Nanostructure materials such as metal/metal multilayers and multi-layer nanowires by electrodeposition are produced. The properties of such structures depend on many electrochemical parameters such as the electrolyte concentration, the electrolyte pH, the deposition potentials of metals and the current efficiency. The growth mechanisms and the structural and magnetotransport properties of ferromagnetic multilayers grown by electodeposition from a single electrolyte are given below.

Growth Mechanisms

The parameters that play particularly important roles in the growth of an electrodeposited film are the deposition potentials, the pH of the electrolyte, the current density and the electrolyte temperature. In addition, the concentration of ions of the metal being deposited in the electrolyte, the presence of additives or impurities, and whether the current is continuous or pulsed are important variables in electrodeposition. All these factors can control whether epitaxial growth takes place or not. Some parameters especially potential and concentration of metal ions may need to be varied until the optimum conditions to obtain high quality films are found.

In electrodeposition of multilayers, two techniques are used: One is the dual-bath technique, the other is the single-bath technique. In the dual-bath technique, a conducting substrate is immersed onto two different electrolytes alternately while controlling the applied voltage or the current. During the growth process, the sample is transferred rapidly from one solution to the other. Although pure layers of metals are obtained, the dual-bath technique is not usually preferred because the sample is exposed to atmospheric contamination during transfer between solutions.

Technique

Fig. 5.91 shows the diagram of an electrodeposition technique.

It consists of a potentiostat with three electrodes, a computer, a digital to analog converter (DAC), an analog to digital converter (DAC) and an electrochemcial cell. The potentiostat keeps the working electrode (cathode) potential fixed at the selected value with respect to the reference electrode (R.E.). The DAC and ADC are used to interface the computer and the potentiostat. The computer controls the process and also integrates the current to obtain the charge passed between the anode and the cathode of the cell.

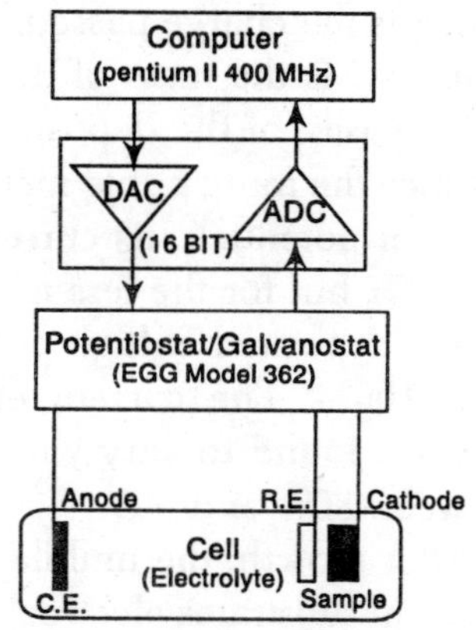

Fig. 5.91 *Diagram of experimental system used for electrodeposition.*

Polycrystalline copper plates or single crystal (100) copper disks are used as substrates because films growth in electrochemcial deposition depends on the choice of substrate material and its preparation. The polishing of the substrates is the first step in the growth of a multilayer. Therefore, one face of each substrate is mechanically polished using emery paper, and then covered with electroplating tape, except for a circular area (a few cm^2) of the polished face. The uncovered area, which is exposed to the electrolyte, is electropolished in a glass dish containing a 50% (w/o) orthophosphoric acid electrolyte at a suitable voltage vs. a counter electrode, for example, carbon rods. After electropolishing, the substrate is immersed, in turn, in 25% H_3PO_4, 10% H_3PO_4, distilled water, 10% H_2SO_4 and distilled water.

As soon as the polishing procedure is completed, the substrate is placed immediately in a electrolyte used to deposit desired metals. Ni/Cu (or CoNi/Cu) multilayers are grown from a single electrolyte containing Ni^{2+} and Cu^{2+} (or Ni^{2+}, Cu^{2+} and Co^{2+}) ions using the potentiostatic deposition method. During deposition, the electrolyte temperature is generally kept at about 303 K. The deposition is carried out at the cathode of a cell with three electrodes and monitored by the personal computer (PC). A platinum (Pt) sheet with an area of 16 cm^2 is used as counter electrode (C.E.), which is the anode terminal of the cell. The reference electrode is a saturated calomel electrode (SCE). Copper substrates are connected to the working electrode (cathode) terminal of the cell. For both Ni/Cu and CoNi/Cu multilayers, the deposition of copper layers can be made at the cathode potentials (such as –0.2 V or 0.25 V) with respect to the SCE, while Ni layers (or NiCo layers) can be deposited at the cathode potentials (such as –1.6 V, –1.7 V or –1.8 V) with respect to the SCE. At the Cu deposition voltage, only Cu which is the more noble metal is deposited while at the Ni deposition voltage both Cu and Ni or (CoNi) is deposited at rates limited by their concentrations, and hence the Ni (or CoNi) layers contain Cu as well. For this reason, the Cu layers are made of pure Cu while ferromagnetic layers are an alloy containing Cu. The Ni/Cu/Cu or CoNiCu/Cu multilayers is grown by altering the cathode potential between two values chosen for (CoNi) and Cu deposition. When the charge passed between the anode and the cathode reaches the value corresponding to the desired thickness of multilayers component being deposited, the computer changes the working electrode potential to start depositing the other component. The charge amount required for the desired layer thickness is calculated in according to $t = \left(\frac{qM_a}{A\rho Fn}\right)$, where '$t$' is the thickness of deposited film; q is the charge passed, M_a is the atomic mass; F is Faraday's constant (96487 column mole^{-1}) and 'm' is the mass of deposited metal = A.P.; A is the surface areas of the film and ρ is the bulk density of the deposited metal.

Since the more noble metal (Cu layers) is deposited at lower potentials, its current may be assumed close to 100%, but for the less noble metals (Ni and NiCO layers) deposited at high potentials it is generally less than 100%. The current efficiency for these metal layers is found to vary with the electrolyte pH value between 80% and 50%.

After growth, the multilayers can be stripped from their Cu substrates electrochemically using a chromic acid solution, and mounted on an insulator such as glass, which is required to make some measurements such as magnetoresistance.

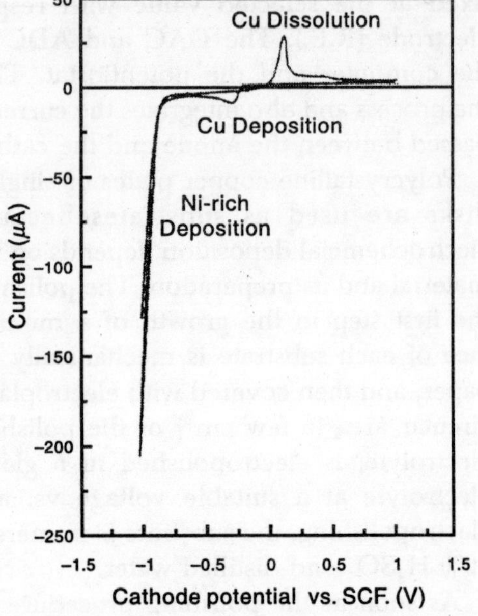

Fig. 5.92 *Cyclic voltametry of the solution used to deposite Ni/Cu multilayers.*

Characterization

The solution prepared for the deposition of Ni/Cu multilayers is characterized by the method of cyclic voltammetry. The cyclic voltammogram curve of the solution is shown in Fig. 5.92. The scan is performed on a Pt wire electrode between –1 V and +1 V with respect to SCE. In Fig. 5.92, there are two strong peaks on the cyclic voltammogram curve; one is at –0.3 V vs. SCE in the cathodic side and corresponds to Cu deposition, the other is in the anodic part, at about +0.15 V vs. SCE and is due to Cu dissolution. In the potential range between –0./4 V and –0.8 V, a current plateau with flow current occurs.

It is found that to grow "high quality multilayers" that are metallic in appearance, NiCu alloy (Ni-rich) layers must be deposited at the potentials higher than –1.5 vs. SCE.

Now, Fig. 5.93 shows the variation of current with time during the growth of a multi-layer with 100[Ni(2.0 nm)/Cu(2.0 nm)] grown on a (100) textured polycrystalline Cu substrate. Similar current transients have also been observed for multilayers grown on (100) single crystal Cu substrate. These curves show similarity to those reported for multilayers such Co/Cu NiCo/Cu and NiFe/Cu. In cathodic part, the high current pulses correspond to the deposition of Ni layers while the low current pulses to the Cu layers.

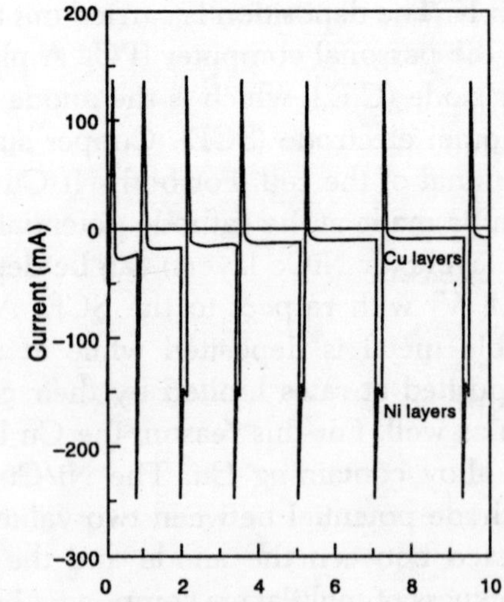

Fig. 5.93 *Current transients of the first few layers during the growth of a Ni/Cu multilayer.*

When the cathode potential is switched from the Ni deposition voltages to the Cu deposition voltage, a large anodic current is observed. This anodic peak is due to redissolution of less noble metal and/or the oxidation of absorbed or occluded hydrogen during the catholic deposition of ferromagnetic layer. For the first few Cu layers, the Cu deposition current is large because the initial Cu concentration close to the surface of the electrode is large. After the Cu layer is deposited, up on switching to the nickel deposition potential the increase in the current is due to the high concentration of Ni on the electrolyte. The Ni deposition current is almost constant for all layers, but the Cu deposition current only reaches a steady value of about 3 mA after depositing a few layers. As the Cu deposition time varies with the Cu current, it indicates the effect of the computer control that is maintaining a constant layer thickness.

The XRD pattern of a multiplayer with 50[Ni(2.0 nm)/Cu(2.0 nm)] growth on a single crystal Cu (100) is shown in Fig. 5.94. In this figure, the peak labelled as S(200) at about 50.5° belongs to the substrate, and the peak labelled as ML (200) corresponds to the mean spacing of the atomic planes within the multilayer itself and is called the main Bragg peak of the multilayer. These reflections from (200) planes of the multilayer and the Cu(100) substrate indicate that the multilayer structure is oriented in the [100] direction as its substrate. The satellite peaks up to third order, which are labelled as (+1), (−1), (+2) (−2) and so on, arising from the constructive and destructive interference of the X-rays scattered by alternate layers of the multilayer, are apparent in the XRD pattern. This shows that the multilayer has a good periodicity. The modulation wavelength (Λ) can be directly calculated from the positions of two satellite peaks having the same order using the relation

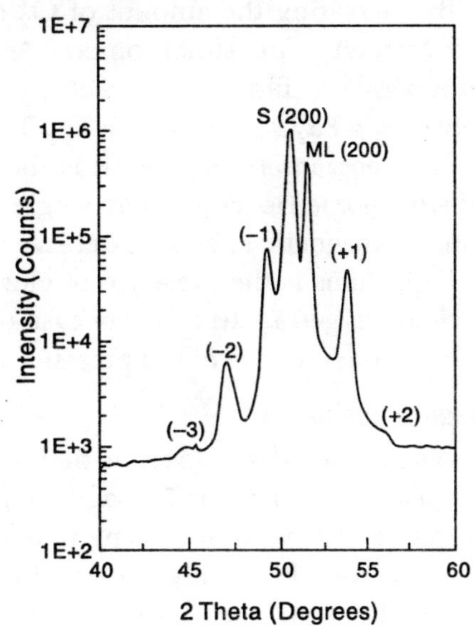

Fig. 5.94 *XRD pattern of a multilayer with 150[Ni(2.0 nm)/Cu(2.0 nm)] growth on a single crystal Cu(00) substrate).*

$$\Lambda = \frac{n\lambda}{\sin\theta_{+n} - \sin\theta_{-n}}$$

where $\lambda = 0.15406$ nm for Cu K_α radiation, and $-n$ and $+n$ refer to adjacent satellite peaks of the same order around the main Bragg peak

Applications of Nanostructured Magnetic Materials

Thin Films/Multilayers

In nanostructured materials the interaction between the electron charges and magnetic spins, have led to discoveries of new and unique phenomena that are neither observingly in traditional bulk materials nor explainable using classical theories. Examples giant magnetoresistance (GMR) in multilayers and metallic granular solids, spin valves, spin injection in ferromagnet/insulator/ ferromagnet sandwiches, etc. and in thin films as: (1) Fe-O films, bilayers/multilayers, (2) magnetostrictive materials, and (3) nanocrystalline R-Co-M films for magnetic recording.

Fe-O Films

Fe-O films are designed from small Fe(Co) particles to known if the large increase of the coercivity and temperature behaviour in small Fe (Co) particles are due to the presence of oxides. It is impossible to have good control of the stoichiometry and the thickness of the oxide shell by evaporation method of preparation. Typically, the oxide layer is about 20 Å and is comprised of a mixture of oxides.

Fabrication and characterization of thin Fe-O films is done by reactive sputtering. To deposit single phase Fe-O films with controlled stoichiometry and thickness, exchange-coupled Fe/Fe-O bilayers is prepared and the properties of Fe films and their effect in the presence of an exchange-coupled layer is measured.

By increasing the amount of O_2 during deposition, films with increased extent of oxidation are formed. The stoichiometry goes thought the following sequence: a-Fe, amorphous/ noncrystalline film, a-Fe + FeO, off-stoichiometric Fe_xO, FeO + Fe_3O_4, single-phase Fe_3O_4, $Fe3O_4$ + g-Fe_2O_3, Fe_3O_4 + a-Fe_2O_3 and single-phase a-Fe_2O_3.

The most interesting result is the observation of anomalous ferromagnetic type behavior in off-stoichiometric Fe_xO and single-phase a-Fe_2O_3 films. These two oxides are found to be antiferromagnetic in bulk. Extensive magnetic, structural, reason for the strong magnetic behavior in Fe_xO films is the presence of clusters of Fe^{3+} ions tetrahedrally coordinated in an fcc closed packed oxygen matrix. In the case of a-Fe_2O_3 films, the unusual magnetic properties are found to be a consequence of the presence of a significant amount of uncompensated surface moments.

Magnetoresistance

Magnetoresistance (MR) is defined as a change in the electrical resistance of a material when it is placed in an external magnetic field. The resistance can increase (positive MR) or decrease (negative MR) with increasing magnetic filed. There are many forms of magnetoresistance, but the most useful ones are observed in ferromagnetic thin films, usually alloys of Ni, Fe and Co. These effects are useful because magnetoresistive thin films are manufacturable in an integrated circuit progress. They are deposited on semiconductor wafers (with or without integrated circuits) and patterned into many forms of devices using standard photolithography and etching techniques.

They are extremely small, and are easily combined with many electronic functions; examples are: generic linear magnetic field sensors; special function magnetic field sensors; read heads for computer disk drives, magnetic tapes, credit cards, currency, current sensing, digital data transmission isolation, and digital data storage etc.

In order to make the magnetoresistance measurements four point contacts are applied to the corners of a square of film as shown in Fig. 5.95. This geometry is known as the van der Pauw (VDP) method. When this film is oriented R_L is defined as the potential difference $V_D - V_C$ between contact D and contact C per unit current flowing from contact *A* to contact *B* and is the "longitudinal" magnetoresistance (magnetic field parallel to the current in the film plane). Similarly, RT is defined as the ratio of the potential different VD – VA between contact *D* and contact *A* to the current flowing from contact *C* to contact *B*, and is the 'transverse' magnetoresistance.

In ferrromagnetic single layer thin films such as Ni, Fe, Co and their alloys, the resistance increases with increasing magnetic field when the field is parallel to the electric current, but decreases with increasing field when the field is perpendicular to the current. This effect is called anisotropic magnetoresistance (AMR) and is caused by the spin-orbit interaction. In order to observe this effect single layer Ni films of various thickness in the μm range were electroplated on a (100) textured polycrystalline Cu substrates from a nickel sulphate bath, which contains a small amount of Co^{2+} (~2%) in addition to Ni_2+ ions. After deposition the films are stripped from their substrates and mounted on glass, and then magnetoresistance measurements are made in the film plane by applying a magnetic field. As an example, the room temperature magnetoresistance curve of a Ni single layer films a thickness of μm is shown in Fig. 5.96. The percentage changes in the van der Pauw (VDP) resistance R_L (longitudinal M_R) and R_T (transverse MR) ($\Delta R/R$) are calculated using the relation:

$$\frac{\Delta R}{R}(H) = \left[\frac{R(H) - R_{\min}}{R_{\min}}\right] \times 100 \qquad (7.5)$$

where $R(H)$ is the resistance is any magnetic field and Rmin indicates the minimum resistance in transverse or longitudinal magnetoresistance measurements, which correspond to that in maximum magnetic field and that in coercive magnetic field respectively. As shown in this figure the longitudinal magnetoresistance increases with increasing magnetic field whereas the transverse magnetoresistance decreases, indicating that the film has the AMR effect. Both the longitudinal and transverse MR saturate in under 2000 Oersted (Oe) applied field.

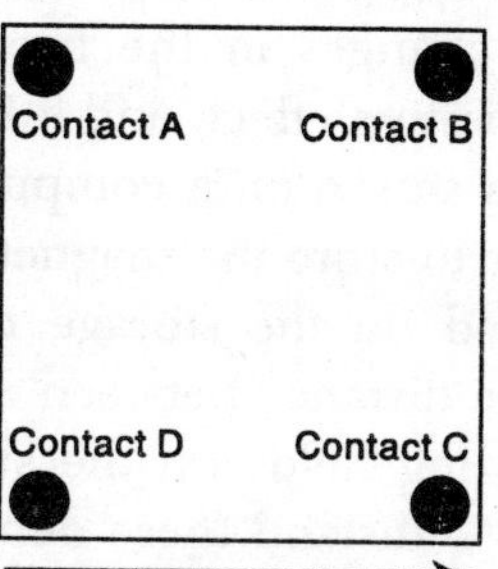

Fig. 5.95 *Van der Pauw (VDP) method used for magnetoresistance measurements of a square of film. Longitudinal magnetoresistance (magnetic field parallel to current) is measured by passing the current between contact C and contact D, and measuring the voltage between contact A and contact B, while transverse magnetoresistance (magnetic field perpendicular to current) is measured by measuring the potential difference between contact A and contact D per unit current flowing from contact C to contact B.*

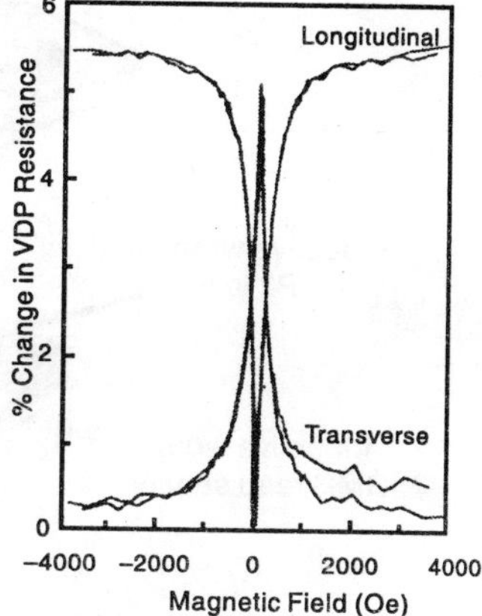

Fig. 5.96 *Magnetoresistance curves of an electrodeposited single Ni film with a thickness of 1 μm.*

(GMR) (GIANT MAGNETORESISTIVE)

Magnetic recording essentially involves detecting changes in the direction of magnetization in the storage medium. Small magnetic fields immediately above the regins of magnetization changes are associated with this divergence. These stray fields are read by inducing signals or resistance changes in the head as the storage medium is moved past them. This is called magnetoresistive effect, and it is useful for sensing magnetic fields such as those in the magnetic bits of data stored on a computer hard drive.

In order to store the magnetic transitions at high densities (e.g. 10^4 cm^{-1}), several constraints are imposed on the storage media and on the write and read heads. With a high storage density, the distance between magnetization reversals become very small. This produces strong demagnetizing fields on the stored 'bit'. For stable storage, the coercive force, must be high enough to withstand these demagnetising stray fields.

To ensure sufficient stray field for detection of the transition, a high remanence is required. A storage medium with both high coercive force and saturation magnetization is required. High permeability and high data rate (>100 MHz) place additional constraints on the write head materials, e.g., eddy currents need to be avoided. Nanostructured structures promise to solve all these constraints on media and read heads. Specially prepared layers of nanometer-thick magnetic and nonmagnetic films are found to exhibit the giant magnetoresistance (GMR) effect.

Giant magnetoresistance observed in ferromagnetic multilayers is defined as a very large change in electrical resistance in an external applied magnetic field. In such structures, the resistance decrease with increasing magnetic field when the field is both parallel and perpendicular to the current. Since this decrease is so large, the MR in ferromagnetic multilayers is widely referred to as giant magnetoresistacne (GMR). GMR is due to the spin depending scattering of electron which occurs within the ferromagnetic layers and at the interfaces between the ferromagnetic and spacer layers. In the spin valve GMR head shown in Figure 5.97, the copper

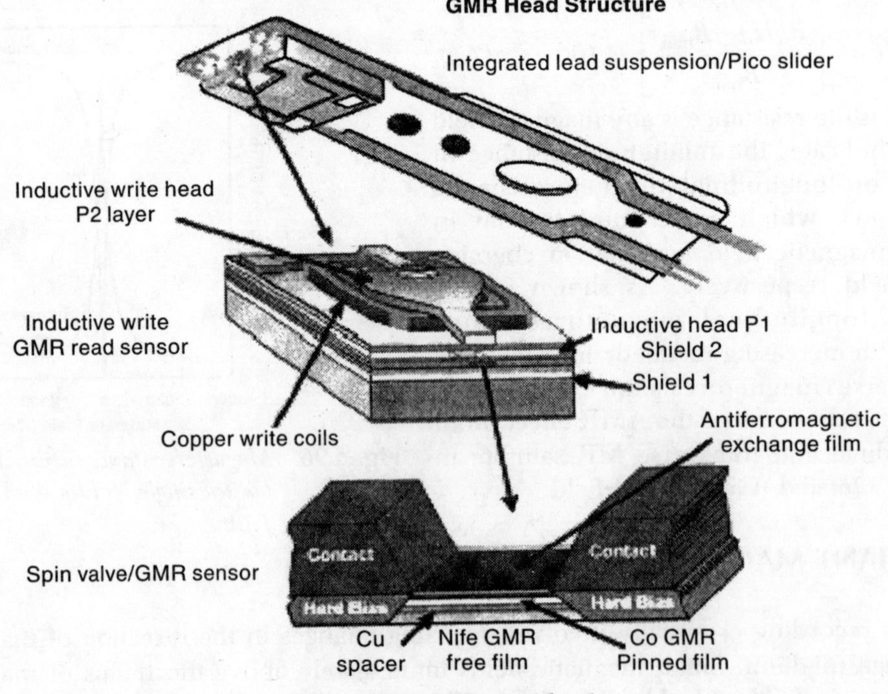

Fig. 4.97 *GMR read head.*

spacer lay is about 2 nm thick and the Co GMR pinned layer is out 2.5 nm thick. It is important to control these layers thickness with atomic precision. This novel read head has extended magnetic disk information storage from 1 to ≈ 20 Gbits.

The first multilayers exhibiting GMR were grown by MBE but now sputtered multilayers also exhibited GMR. In multilayer systems the antiferromagnetic alignment of the ferromagnetic layers in zero field becomes ferromagnetic as the field is applied and causes a decrease in resistance. Granular materials that show GMR consists of small ferromagnetic single-domain particles with randomly oriented magnetic axes in a nonmagnetic matrix. An external field rotates the magnetic axes of all magnetic particles. The rotation towards complete alignment of all magnetic axes again reduces the resistance in a similar way as for multilayers. The GMR in granular systems is isotropic. The explanation for the GMR is spin-dependent scattering of the conduction electrons at the ferromagnetic/nonmagnetic interfaces and, to a lesser extent, within the magnetic grains. The GMR scales inversely with the average particle diameter.

The electrodeposited CoNiCu/Cu multilayers have proven to exhibit significant GMR. These multilayers are grown on (100) textured polycrystalline CU substrates from a single electrolyte containing Cu^{2+}, Co^{2+} and Ni^{2+} ions. After removing their substrates, magnetotransport measurement are carried out at room temperature using the VDP geometry as described in Fig. 5.95.

Fig. 5.98 shows the magnetoresistance curves of a multilayers with 80[CoNiCu(1.5 nm)/Cu(0.8 nm)]. In this figure, the percentage change in the VDP resistance as a function of the applied field is calculated according to:

$$\Delta R/R(H) = \left[\frac{R(H)-(R_{max})}{R(H_{max})}\right] \times 100$$

where $R(H)$ is the resistance in a particular applied magnetic field and $R(H_{max})$ is the resistance in the maximum applied magnetic field which is $8kOe$. Fig. 4.96 shows that both the transverse and the longitudinal VDP resistance decrease with increasing magnetic field. In the multilayer, the magnitudes of the observed GMR for transverse and longitudinal measurements are 26% and 19% respectively, (in a magnetic field of 8kOe at room temperature).

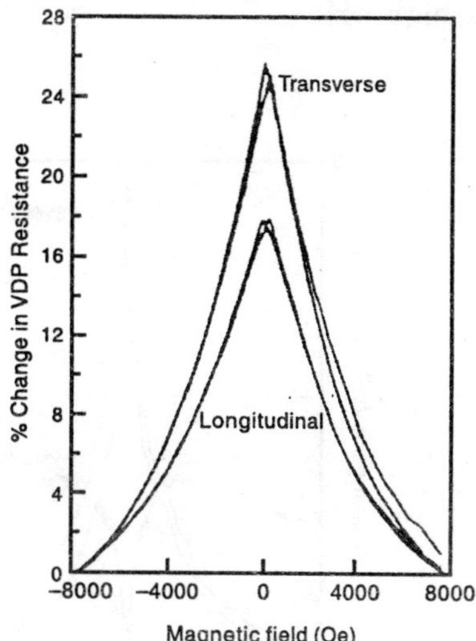

Fig. 5.98 *Magnetoresistance curves of a multilayer with 80[CoNiCu(1.5 nm)/ Cu(0.8 nm)].*

It is possible to obtain large GMR values in higher magnetic fields. The observe magnetoresistance is largely due to the multilayer structure of the film. As seen in Fig. 5.96, AMR generally causes the transverse resistance to decrease and the longitudinal resistance to increase as the applied magnetic field increase.

The GMR magnitude of electrodeposited multilayers change with parameters such as layer thicknesses and electrolyte pH. A series of CoNiCu/Cu multilayers are grown from the electrolyte with different pH values, and then their magnetotransport properties are investigated. The multilayers with the larger Cu thicknesses than 1.5 nm show GMR when they are grown both from the electrolyte with high pH (3.3) and the electrolyte with low pH (1.8). In contrast, for the smaller Cu thicknesses (0.8 nm), the multilayers grown from the electrolyte with the low pH (1.8) exhibit GMR while the multilayers from the electrolyte with high pH (3.3) have predominant AMR although their XRD measurements showed the formation of a periodic structure.

Fig. 5.99 and 5.100 show that the magnetoresistance curves for the multilayers with 107[CoNi(2.0 nm)/Cu(0.8 nm)] grown at low pH (1.8) and high pH (3.3) respectively. The percentage changes in the magnetoresistance are calculated for the sample grown at high pH and for the sample grown at low pH. The multilayers grown from the electrolyte with the low pH, both the longitudinal and transverse magnetoresistance are negative (see Fig. 5.100), but for the films grown at high pH there is a region where anisotropic magnetoresistance is dominant (see Fig. 5.99). Comparing fig. 5.99 to the MR curve of the single Ni film in Fig. 5.96, the maximum percentage changes in the resistance of single Ni film are almost the same both for the transverse and longitudinal MRs, while in the multilayer the maximum percentage change in the transverse MR is larger than the longitudinal MR. This is because, as seen in Fig. 5.99, in the low magnetic field region the longitudinal MR increases with increasing magnetic field whereas in the high magnetic field it decreases as the magnetic field increases. This indicates that the multilayer structure has a tendency to exhibit GMR even though grown at high pH.

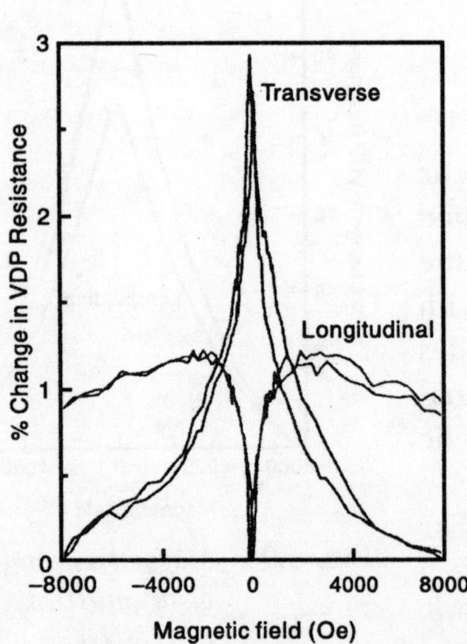

Fig. 5.99 *Magnetoresistacne curves for a multilayers with 106[CoNiCu(2.0 nm)/Cu(0.8 nm)] grown from their solution with a high pH value (3.3).*

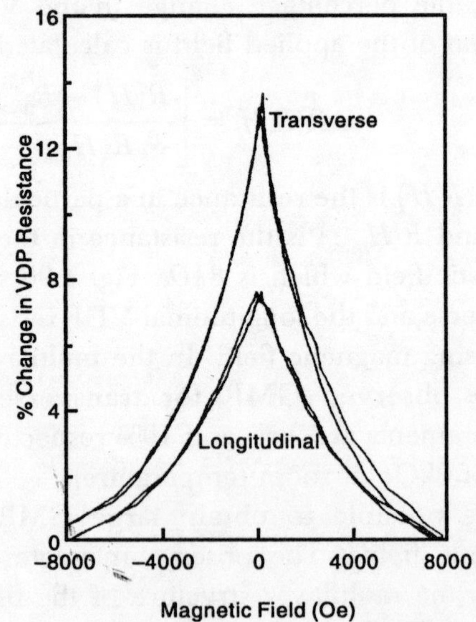

Fig. 5.100 *Magnetoresistance curves for a multilayer with 107[CoNiCu(2.0 nm)/Cu(0.8 nm)] grown from the solution with a low pH value (1.8).*

Isolators

Digital signal isolators are a relatively new application for magnetoresistive materials.

The purpose of signal isolators is to allow information to pass from one electrical circuit to another without having a direct conduction path between the two. This helps reduce noise in data transmission, and protects sensitive circuity from potentially damaging voltages.

Isolators are used to protect the integrity of both digital and analog data. Here, focus is on the digital applications because that is what NVE has on the market.

As a way of describing basic isolator functions while also explaining why magnetorestive materials are so well suited for the application, the operation of optical and magnetic based isolators is given below.

Optical. The dominant form of low cost integrated digital signal isolator is optically based. Current flowing into the input side of the isolation barrier goes through a Light Emitting Diode (LED). The light form this LED is directed across the isolation barrier to a semiconductor photodetector. The photodetector registers a signal once sufficient photons have been collected from the beam from the LED.

Magnetoresistive. The newest and fastest form of integrated digital signal isolator is based on integrated GMR devices. Current flowing into the input side of the isolation barriers goes through an on-chip planar coil that generates a magnetic field. The coil is situated so that the GMR spin valve devices are on the other side of the isolation barrier ut in the field generated by the coil. The GMR is configured in a bridge so that the voltage out switches high or low depending upon which direction the coil current is flowing. This contrast is shown below in Figure 5.101.

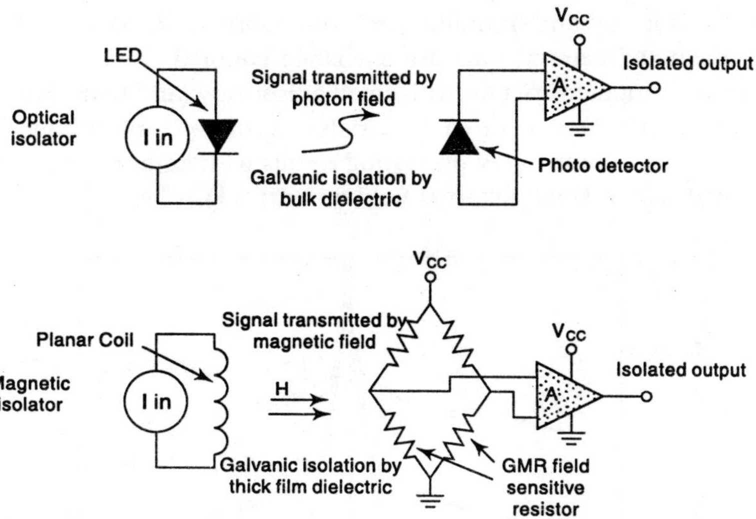

Fig. 5.101 *Contrasting GMR isolator with optical isolator concepts.*

Delays. The GMR isolator is faster than the optical isolator. Photons travelling from the LED to the photodetector are as fast as the magnetic field propagation from the coil to the GMR detector. The difference, is in the generation and detection of the signal rather than its propagation. The LED requires some time, typically 10's of ns, to generate enough photons to drive the photodetector beyond the switching threshold of the detection electronics. This speed can be increased in proportion to the amount of current passing through the LED, but is limited by practical power requirements. The current in the GMR isolator, however, generates a magnetic field instantaneously. The speed limit for the GMR, then, is how fast the magnetic sensing film's magnetization can rotate in the applied field. It is found that for reasonable amounts of current, the switching takes place in 1 ns or so.

Since the fundamental limitation for the GMR isolator switching time is related to the dynamics of the soft film's magnetization, soft films with higher "speed limits" are development. This is where the nanocrystalline ferromagnetic material becomes relevant.

Magnetostrictive Materials

Magnetostrictive Materials have been of great scientific and technological importance for many years. The magnetostrictive phenomenon refers to the deformation of a material in a magnetic field. For microsystems capable of performing functions that are not fulfilled by existing integrated electronic circuit magnetostrictive films they allow such functions to be readily obtained. The could constitute the driving elements of micro-actuators (micro-robots pumps or motors) and magnetic control of elastic properties electronic devices like a resonator with magnetically adjustable frequency and stress-controlled inductance. In these application, the efficiency depends on change in length ($l = Dl/l$) as a function of the applied field. Values Dl/l on the order of 200×10^{-6} with a field smaller than 200 Oe are desired. This is not possible with existing materials, as they require saturation fields of tens of kOe.

Nanostructured materials are potential candidates that can lead to the desired properties through atomic engineering. Such properties can be obtained in amorphous or nanocrystalline alloys with reduced anisotropy and in multilayers with alternate stacks of high magnetostrictive materials and soft magnetic materials that are exchange coupled.

It is found that giant mangnetostriction with small anisotropy (and therefore a small saturation field) can be obtained in nanostructured materials. Result on amorphous $Tb_{33}Fe_{67}/Fe_{80}B_{20}$ multilayers (Fig. 5.102) have shown very promising results with high transverse magnetostriction of 600×10^{-6} obtained with a small external field around 2 kOe.

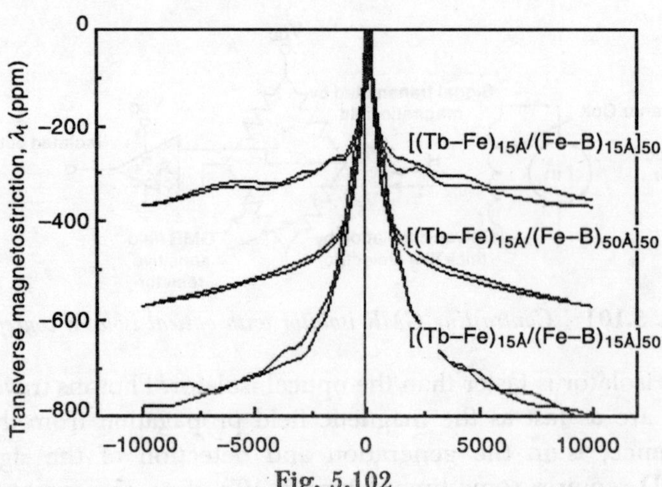

Fig. 5.102

NANOWIRES

A nanowire is a wire of dimensions of the order of a nanometer (0–9 meters). Alternatively, nanowires can be defined as structures that have a lateral size constrained of tens of nanometers or less and an unconstrained longitudinal size. At these scales, quantum mechanical effects are important–hence such wires are also known as "quantum wires". Many different types of nanowires exist, including metallic (e.g., Ni, Pt, Au), semiconducting (e.g., InP, Si, GaN, etc.), and insulating (e.g., SiO_2, TiO_2). Molecular nanowires are composed of repeating molecular units either organic e.g. DNA or inorganic (e.g. $Mo_6S_{9-x}I_x$).

Typical nanowires exhibit aspect ratios (the ratio between length to width) of 1000 or more. As such they are often referred to as 1-Dimensional materials. Nanowires have many interesting properties that are not seen in bulk or 3-D materials. This is because electrons in nanowires are quantum confined laterally and thus occupy energy levels that are different from the traditional continuum of energy levels or bands found in bulk materials. Peculiar features of this quantum confinement exhibited by certain nanowires such as carbon nanotubes manifest themselves in discrete values of the electrical conductance. Such discrete values arise from a quantum mechanical restraint on the number of electrons that can travel through the wire at the nanometer scale. These discrete values are often referred to as the quantum of conductance and are integer values of $\frac{2e^2}{h} \approx 12.9\ k\Omega^{-1}$. They are inverse of the well-known resistance h/e^2, which is roughly equal to 25812.8 ohms. Since 1990, a fixed conventional value R_{K-90} is accepted.

Example of nanowires include inorganic molecular nanowire ($mo_6S_{9-x}I_x$, $Li_2Mo_6Se_6$), which have a diameter of 0.9 nm, and can be hundreds of micrometers long. Other important examples are based on semiconductors such as InP, Si, GaN, etc., dielectrics (e.g. SiO_2, TiO_2), or metals (e.g. Ni, Pt).

There are many applications where nanowires may become important in electronic, opto-electronic and nanoelectrochemical devices, as additives in advanced composites, for metallic interconnects in nanoscale devices, as field-emitters and as leads for biomolecular nanosensors.

Conductivity of Nanowires

The conductivity of a nanowire is expected to be much less than that of the corresponding bulk material. This is due to a variety of reasons. First, there is scattering from the wire boundaries, when the wire width is below the free electron mean free path of the bulk material. In copper, for example, the mean free path is 40 cm. Nanowires less than 40 nm wide will shorten the mean free path to the wire width.

Nanowire also show other peculiar electrical properties due to their size. Unlike carbon nanotubes, whose motion of electrons can fall under the regime of ballistic transport (meaning the electrons can travel freely from one electrode to the other), nanowire conductivity is strongly influenced by edge effects. The edge effects come from atoms that lay at the nanowire surface and are not fully bonded to neighboring atoms like the atoms within the bulk of the nanowire. The unbounded atoms are often a source of defects within the nanowire, and may cause the nanowire to conduct electricity more poorly than the bulk material. As a nanowire shrinks in size, the surface atoms becomes more numerous compared to the atoms within the nanowire, and edge effects become more important

Furthermore the conductivity can undergo a quantization is energy: i.e. the energy of the electrons going through a nanowire can assume only discrete values, multiple of the constant $G = 2e^2/h$ (where e is the charge of the electron and h is Planck's constant).

The conductivity is hence described as the sum of the transport by separate channels of different quantized energy levels. The thinner the wire is the smaller the number of channels available to the transport of electrons.

The conductivity of a nanowire can be studied suspending it between two electrodes. This has been proven by measuring the conductivity of a nanowire while pulling it: while it shrinks, its conductivity decreases in a stepwise fashion and the plateaus correspond to multiples of G.

The quantized conductivity is more pronounced in semiconductors like Si or GaAs than in metals, due to lower electron density and lower effective mass. Quantized conductance are observed in 25 nm wide silicon fins which result in increased threshold voltage.

Semiconducting Nanowires

Semiconductor nanowires are one-dimensional structures, with unique electrical and optical properties, that are used as building blocks in nanoscale devices. Their low dimensionality means that they exhibit quantum confinement effects. For example, narrowing the wire's diameter increases its band gap, compared to the bulk material.

There are various ways of growing semiconductor nanowires, including laser ablation, chemical vapour deposition (CVD) and template-assisted growth. While laser ablation and template-assisted approaches provide bulk quantities of semiconductor nanowires, they do not provide much control over the composition, size or crystallographic direction of the nanowire.

Another way is to grow core-shell and core-multishell nanowire heterostructures using a CVD method that provides increased control over the structure's composition. Using their technique, the nanowires are grown by gradually building up thin, uniform shells around a nanometre-sized cluster of gold atoms.

Semiconductor nanowire are increasingly used in electronic devices including field-effect transistors, sensors, detectors and light-emitting diodes. Nanowires that are only 50 nm in diameter containing a germanium core surrounded by a silicon shell have been synthesized.

They also made "triple decker" wires of silicon silicon oxide and germanium. These so-called "core-shell" and "multi-shell" structures allow a semiconducting wire to be encased in an insulating shell.

Production of Nanowires

Nanowires are not observed spontaneously in nature and must be produced in a laboratory. Nanowires can be either suspended, deposited or synthesized from the elements.

A suspended nanowire is a wire in vacuum chamber held at the extremities. A deposited nanowire is a wire deposited on a surface of different nature: e.g. it could even be a single strip of metallic atoms over a non-conducting surface.

A suspended nanowire can be produced by chemical etching of a bigger wire, bombarding a bigger wire and with some highly energetic particles (atoms or molecules).

Another way to produce a suspended nanowire is to indent the tip of an STM in the surface, of a metal near the melting point, and retract it. This can form a wire very much like the tips of a fork does with the cheese of a pizza.

A common technique for creating a nanowire is the Vapor-Liquid-Solid (VLS) synthesis method. This technique uses a source material either laser ablated particles or a feed gas (such as silane). The source is then exposed to a catalyst. Foe nanowires, the best catalysts are liquid metal (such as gold) nanoclusters, which can either by purchased in colloidal form and deposited on a substrate or self-assembled from a thin film by dewetting. This process can often produce crystalline nanowires in the case of semiconductor materials.

The source enters these nanoclusters and begins to saturate it. Once supersaturation is reaches, the source solidifies and grows outward from the nanocluster. The final products's length can be adjusted by simply turning off the source. Compound nanowires with super-lattices of alternating materials can be created by switching sources while still in the ground phase.

STRUCTURE OF NANOWIRES

The nanowires can show peculiar shapes. Sometimes they can show nonocrsytalline order, assuming e.g. a pentagonal symmetry or a helicoidal (spiral) shape. Electrons zigzag along pentagonal tubes and spiral along helicoidal tubes.

The lack of crystalline order is due to the fact that a nanowire is periodic only in one dimension (along its axis). Hence it can assume any order in the other directions (in place) if this is energetically favorable.

E.g., in some cases nanowires can shown a fivefold symmetry, usually not observe in nature, but for clusters of few atoms. The fivefold symmetry is equivalent to the icosahedral symmetry of (small) atomic clusters: the icosahedron is often an energetically favorable shape for cluster of few atoms, but icosahedral ordering is not observed in crystals since it si not possible to stack together icosahedra (repeating infinite copies of them in each direction) and tile the whole space (fill it without holes).

1. Fabrication of Nanowires at Structures

The goal of this project is the design of artificial materials that consist of ultrafine wires or linear arrays of dots, ten to hundred times finer than those produced with commercial micro-structure fabrication techniques. In fact, we have gone all the way down to atom chains which may be viewed as the ultimate nanowires. These patterns are formed by self-assembly, where atoms arrange themselves naturally at stepped silicon surfaces.

An important aspect in fabricating nanowires is the ability to prepare wires of an any material on any substrate with any thickness. In particular, using silicon wafers as substrate is highly-desirable. To achieve this goal we suggest the following "universal" process. First, a silicon substrate with a regular arrays of steps is prepared (A). Then, stripes (B) or dots (C) of a passivating material are attached to the step edges. This apart is analogous to creating a photoresist mask in traditional lithography. As mask material we use calcium fluoride, which is lattice-matched to silicon and chemically inert. Eventually, the desired material is deposited on the remaining silicon, for example by substrate-selective chemical vapor deposition (CVD) or electroplating. Alternatively, calcium fluoride could become useful as an etch mask for producing trenches in the silicon that can be filled with new materials to achieve a planer structure.

The figure below shows the preparation of calcium fluoride masks in schematic form (top), together with actual data (bottom).

To start along this pathway, we determine the conditions for obtaining highly-regular step on structures on silicon. The images below Fig. 5.103 demonstrate the range of step arrays that can be formed on silicon surfaces by self-assembly. Typically, the step spacing is comparable to the size of a virus. These images are taken with a scanning tunneling microscope (STM). They show the derivative of the tip height. That gives the impression of a surface illuminated from the left, with the steps casting dark shadows to the right.

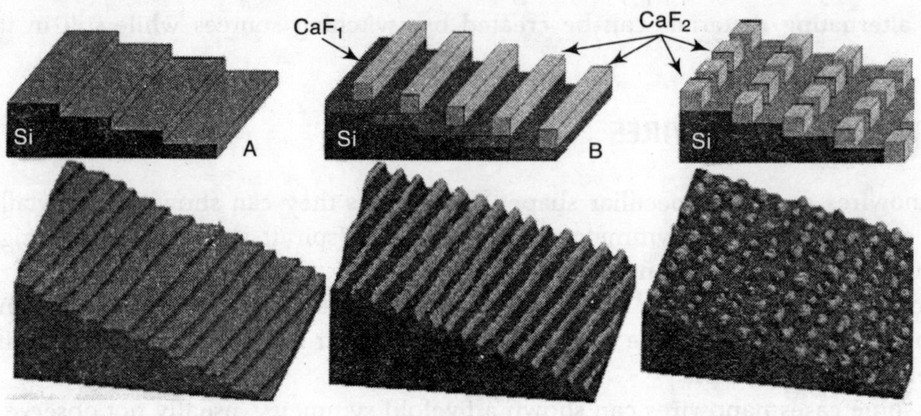

Fig. 5.103

Particularly perfect step arrays could be achieved on the **Si(111)** 7×7 surfaces. The 7×7 structures causes steps running along the [011]-direction to become extremely straight, because each kink requires generating 14 new rows of silicon atoms (7 row, two layers deep). The step edges are atomically straight with a kink spacing as low as a single kink in 20,000 atoms, as seen in the image below. These are taken with a scanning tunneling microscope (STM). The x-derivative of the topography is displayed, which makes steps appear as dark lines. The image on the right zooms in one the terrace between two steps (heavy dark lines). The atomic pattern of the 7×7 structure is resolved, which has fine grooves built in that run parallel to the step edges. These grooves are 2.3 nm apart and determine the location of possible step edges. They may be viewed as the LEGO blocks of the aspiring nanoengineer.

Fig. 5.104

Our "universal process" can be carried further by producing a calcium fluoride mask, as shown in the image below. A stepped silicon surface is coated with a layer of CF_1 and CaF_2 stripes are formed on top of that layer. These stripes are continuous and do not touch each other, because adjacent stripes cannot bond to each other. The stripe width of 7 nm achieved here is well below the resolution of 180 nm achieved in lithography for chip fabrication Fig. 5.104.

The third step of the "universal process" involves selective deposition or etching between the masked areas. The picture below shows that molecules can be deposited selectively in the $CaF_2$1 grooves between CaF_2 stripes.

Using organometallic molecules, such as ferrcene, it is possible to fabricate iron wires that are only 3 nanometers wide.

2. Atoms Chains, the Ultimate Nanowire

It is possible to go all the way to the atomic limit and create atomic wires by depositioning a fraction of a monolayer of metal atoms onto stepped silicon. An example is the Si(557)-Au surface shown below Fig. 5.105. It contains a step every five silicon atom rows and a row of goals atoms in the middle of the terrace. The STM image below shows two rows of fine white dots, which are magnified in the inset. They correspond to silicon atoms with dangling bonds.

Being able to fabricate atomic wires opens up the possibility of studying one-dimensional electrons moving along the atom chains. The properties of electrons become more and more exotic when progressing from the three-dimensional world into lower dimensions. In a two-dimensional electron gas one observes surprising phenomena already, such as fractional charge and statistics with the fractional quantum Hall effect. Predictions for one-dimensional electrons are even farther out. The concept of a single electron fails because electrons constantly penetrate each other when moving along the same one-dimensional line. They lose their identity and separate into two quasiparticles, a spin on that carries spin without charge, and a holon that carries the positive charge of a hole without its spin. That is the theoretical prediction, and the hunt is on whether such strange behavior can be observed experimentally.

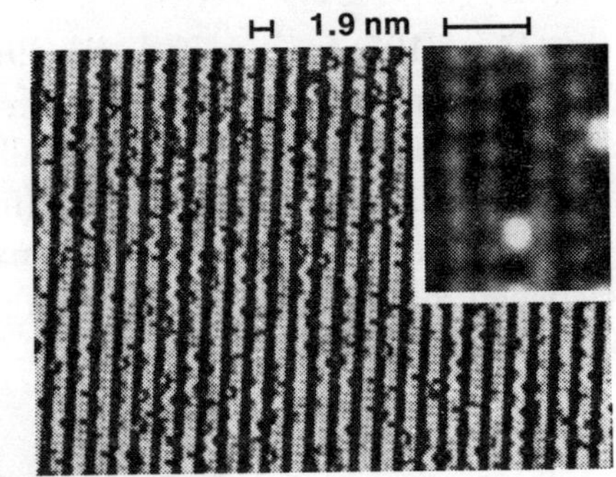

Fig. 5.105

The complete set of quantum numbers of electrons in a solid is revealed by **angle-resolved photoemission** with tunable *synchrotron radiation.* The two principal quantum numbers energy and momentum are combined in a band structure plot, as the picture below. The dark bands represent regions with high photoemission intensity. With multidetection techniques for energy and angle it is now possible to see the band structure to see the band structure directly on a television screen.

Some of the atomic wires are found to be metallic (above left). The metallic behavior is deduced from the observation that the bands extend all the way up to the Fermi level E_F for Si(557)-Au, as in a metal. (See Fig. 5.106)

By way of contrast, the flat SI(111)-Au surface in the panel on the right shows a band that does not reach EF. Even though the metal atoms are strongly coupled to the substrate, metallic electrons do not interact with the silicon substrate because their energy lies in the band gap of silicon.

Electrons at the Fermi level are particularly interesting because they assemble themselves into exotic phases in low dimensions. These electrons are best characterized by the Fermi Surface, which describes their momentum distribution. Such Fermi surfaces are shown below

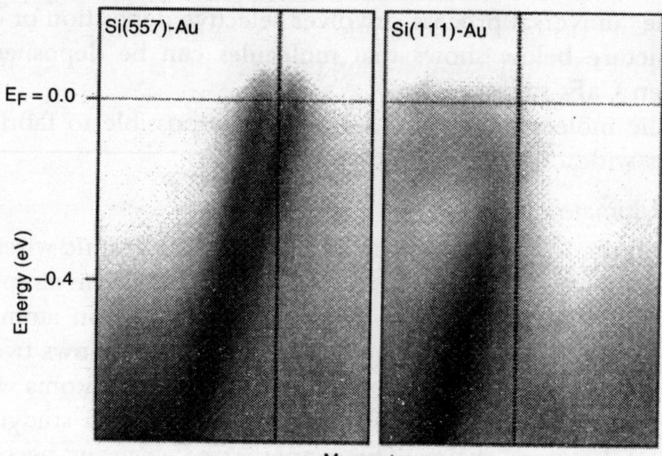

Fig. 5.106

(see Fig. 5.107) for two-and one-dimensional electrons at silicon surfaces with extra metal atoms (typically a monolayer for 2D structure and 1/5 of a monolayer for a 1D structure). Two-dimensional fermi surfaces typically consist of circles (top), which may form intricate patterns by replication at a superlattice (middle). One-dimensional Fermi surfaces are straight lines, which become modulated by small wiggles if a weak two-dimensional coupling is present between the atom chains (bottom).

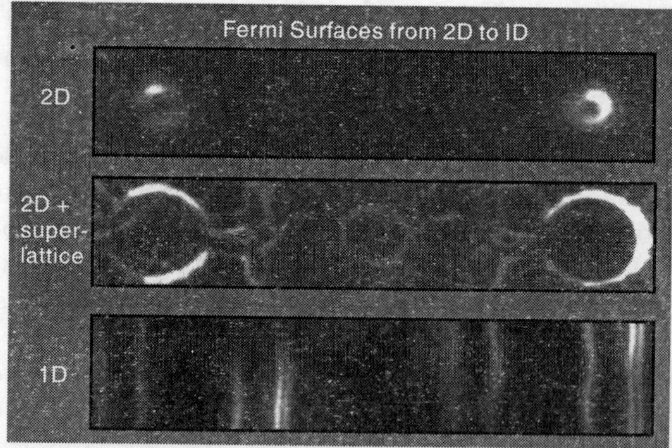

Fig. 5.107

Currently, we are exploring silicon surfaces with a variety of step spacings. That makes it possible to vary the dimensionality between 2D and 1D. For example, the coupling ratio parallel/perpendicular to the chains can be varied from 10:1 to >70:1. In addition, structures with fractional filling can be produced, such as for the one-dimensions Fermi surface shown above. Fractional filling is achieved by embedding the chains into a 2D superlattice where silicon atoms act as dopants, as in modulation doping of semiconductors. That raises interesting questions about an analog to the fractional quantum Hall effect in a one-dimensional electron gas. Chains can be formed not only with gold atoms, but also with a large variety of other metal atoms.; These include magnetic atoms, such as Gd and other rare earths, which open up possibilities of fabricating spin chains. In fact, even non-magnetic gold chains form spin-split Fermi surface, as in the example above. That is due to spin-orbit interaction in the reduced symmetry of a surface. Another chain structure spontaneously decomposes into metallic and semiconducting segments, which can be viewed as the one-dimensional analog of stripes, which have been a hot topic in the context of high temperature superconductivity. All these atomic chain structures form an ideal playground to search for interesting new phases of matter in one dimension.

USE OF NANOWIRES

Nanowires still belong to the experimental world of laboratories. However, some early experiments have shown how they can be used to build the next generation of computing devices.

To create active electronic elements, the first key step was to chemically dope a semiconductor nanowire. This has already been done to individual nanowires to create *p*-type and *n*-type semiconductors.

The next step was to find a way to create a *p-n* junction, one of the simplest electronic devices. This was achieved in two ways. The first way was to physically cross a *p*-type wire over an *n*-type wire. The second method involved chemically doping a single wire with different dopants along the length. This method created a *p-n* junction with only one wire.

After *p-n* junctions were built with nanowires, the next logical step was to build logic gates. By connecting several *p-n* junctions together, researchers have been able to create the basis of all logic circuits: the AND, OR, and NOT gates have all been built from semiconductor nanowire crossings.

It's possible that semiconductor nanowire crossings will be important to the future of digital computing. Though there are other uses for nanowires beyond these, the only ones that actually take advantage of physics in the nanometer regime are electronic.

Nanowires are being studied for use as photon ballistic waveguides as interconnects in quantum dot/quantum effect well photon logic arrays. Photons travel inside the tube, electrons travel on the outside shell.

When two nanowires acting as photon waveguides cross each other the juncture acts as a quantum dot.

Because of their high aspect ratio, nanowires are also uniquely suited to dielectrophoretic manipulation.

Striped Nanowires

Two-dimensional microchips are made by depositing layers of different types of semiconductors with different electrical properties on a water of silicon. It is the interfaces between these different layers that control the flow of electrons and anable transistors and other electronic components to function in a one-dimensional nanoscale platform. Nanowires composed of two different semiconductors–silicon and silicon germanium alloy– in discrete alternating segments have been developed Nanowires made out of a single tape of semiconductor, have created interfaces by semiconductors. The striped nanowires much more ideal interfaces and are much better suited.

Working inside a 2.5 centimeter (one inch) tube furnace, thy grow their wires using a hybrid "pulsed laser-ablation/chemical vapor-deposition" process.

A silicon wafer coated with a thin layer of gold is introduced into the furnace. When the furnace is sufficiently heated, the gold film forms a liquid alloy with silicon and spontaneously breaks up into nanometer-sized droplets. Vapors of two semiconductors will then condense around the gold droplets as deposits, which become nanowire segments. Chemicals are used as the source of the silicon vapor and a laser is used to vaporize the germanium. When the laser is off, only silicon is deposited on the gold particles; when the laser is on, both silicon and germanium are deposited.

"By periodically turning the laser on and off–and this can be readily programmed–we can form a silicon and silicon germanium superlattice on every individual nanowire in a block-by-block fashion. The entire growth process resembles the living polymerization synthesis of a block copolymer."

The use of two independent vaporization techniques, enables to achieve "remarkable periodicity" for their blocks of silicon and silicon germanium. In just one hour millions of striped nanowires can be made.

These striped or "superlatticed" nanowires function as transistors, LEDs (light-emitting diodes) and other optoelctronic devices, biochemical sensors, heat-pumping thermoelectric devices, or all of the above, along the same length of wire.

The potential of these tiny wires is enormous. Striped nanowires can be assembled into complex circuits and faster computer chips. They also serve as nanoscale light-emitting sources that could be integrated into light-speed optocommunication systems. Striped nanowires could even serve as powerful sensors of chemical and biological agents. Contact with extremely low concentrations of a chemical or biological agent is enough to cause a change in the electronic state of a striped nanowire, which can be used to identify the agent.

Nanospring

A nanowire wrapped into a helix.

Wires tens of nanometers in diameter are not actually wrapped to make springs, they are grown that way through a process known as vapor-liquid-solid (VLS) growth mode. VLS growth occurs when a catalyst droplet resisting on a surface absorbs wire-building material from a surrounding vapor. Once the concentration of the building material reaches super-saturation in the droplet, a portion of the material is secreted out the droplet base, and a wire gradually forms. Under some circumstances, the material deposition is asymmetrical and the wire develops into a helical nanospring.

Nanowire Formation a small catalyst droplet, roughly the same diameter as a growing nanoqire, remains centered on top of the wire, and the resulting growth is linear. If the droplet exceeds the wire diameter, its balance atop the structure is precarious and a small perturbation can bump the droplet to one side, abruptly jolting the growth pattern from straight to helical eg. the growth of first boron carbide nanospring is produced this away. (see Figure 5.108)

Nanosprings may someday make highly sensitive magnetic field detectors, perhaps finding applications in hard drive read heads. Alternatively, nanosprings could serve as positioners, or even as tin conventional springs, for nanomachines of the future.

A scanning electron microscope captured this image of the transition from linear to helical growth in a boron carbide structure built up through a process known as vapor-liquid-solid growth mode. A small perturbation disturbed the droplet responsible for depositing material as it passed the point labeled B, transforming the nanowire in to a nanospring.

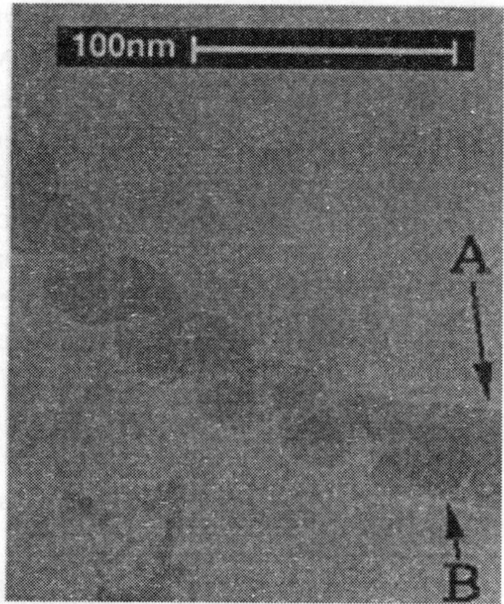

Fig. 5.108

Helical Silica Nanosprings

To make the springs in a dual flow-tube furnace is piace a silicon wafer substrate in the inner tube, and iron and Si/SiO_2 powders inside the outer flow tube. Synthesis of the springs is done in two hours at 1433 K in the presence of methane.

In this way amorphous helical SiO_2 nanosprings up to 8 μm long with diameters of 80-140 nm is made. The nanosprings, which are like coiled telephone cords, a joined at either end to straight amorphous SiO_2 nanowires Fig. 5.109. The ribbon making up the nanosprings was rectangular, rather than cylindrical, in cross-section.

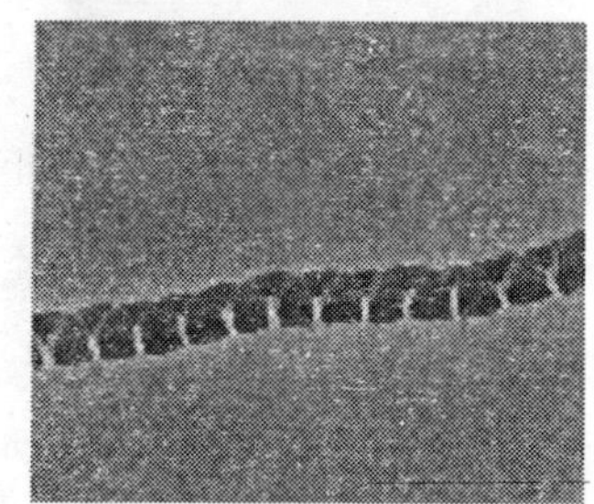

Fig. 5.109

On testing the spring-like qualities of the helical wires by focusing the electron beam of a transmission electron microscope (TEM) onto various points along their length, causes local heating and expansion of the spring, with corresponding contraction of the spring away from the heating point on applying pressure to the nanosprings using the tip an atomic force microscope (AFM). Nanosprings bent and slightly stretch.

"Thus, helical nanosprings possess the same characteristics as micro/macro springs, such as the storage of mechanical energy and flexibility". "This is an important step for the use of nanosprings in nanodevices."

The spring have applications in composites, and nanomechanical and nanoelectromagnetic devices.

Nanoshells

"The nanoshells act as an amazingly versatile optical component on the nanometer scale. Nanoshells are particles with an insulating core coated by a thin shell of gold i.e." Nanoshells are nanoscale metal spheres, which can absorb or scatter light at virtually any wavelength in the visible or infrared ranges depending upon the dimensions of the particle's core and shell (see Fig. 5.110). These are a new type of optically tunable nanoparticle composed of a dielectric (for example, silica) core coated with an ultra-thin metallic (for example, gold) layer. Gold Nanoshells possess physical properties similar to gold colloid, in particular a strong optical absorption due to the collective electronic response of the metal to light. The optical absorption of gold colloid yields a brilliant red color, which has been of considerable utility in consumer-related medical products such as home pregnancy tests. In contrast, the optical response of gold Nanoshells depends dramatically on the relative sizes of the nanoparticle core and the thickness of the gold shell. By varying the relative core and shell thickness, the color of gold Nanoshells can be varied across a broad range of the optical spectrum that spans the visible and the near-IR spectral regions. Gold Nanoshells can be made either to absorb or scatter light preferentially by varying the size of the particle relative to the wavelength of the light at their optical resonance (See Figure 5.110 below) they can absorb or scatter light in the infrared. This range, particularly around one micron wavelength.

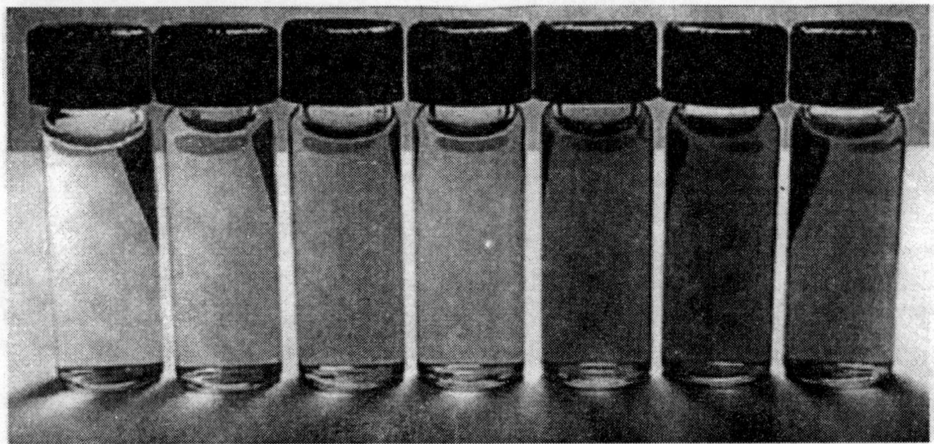

Fig. 5.110 *Nanoshells designed to absorb various wavelengths of light (the six vials of the right), including infrared (vial at far right) compared to gold colloid (far left).*

Figure 5.111 below depicts a Mice scattering plot of the Nanoshell plasmon resonance wavelength shift as a function of Nanoshell composition for the case of gold/silica Nanoshell with a 100 nm core. In this figure, the core and shell of the nanoparticles are depicted to relative scale directly beneath their corresponding optical resonances.

The ability of "tune" Nanoshells to a desired wavelength is critical in vivo therapeutic applications. Human blood and tissue minimally absorb certain near-infrared wavelengths of light, enabling to use an external laser to deliver light to Nanoshells either in a tumor (for thermal destruction of imaging), a wound (for wound closure or tissue repair) or whole blood (to diagnose disease. The relative absorption of certain tissues and water (a principal component of tissue and blood) are illustrated in Figure 5.112 below.

Metal nanoshells may be embedded in solid-state materials in solid-state materials and films, such as plastics or glasses, or attached directly to surfaces for special-purpose coatings. New products could include energy efficient paints and windows, coatings for cars, airplanes or

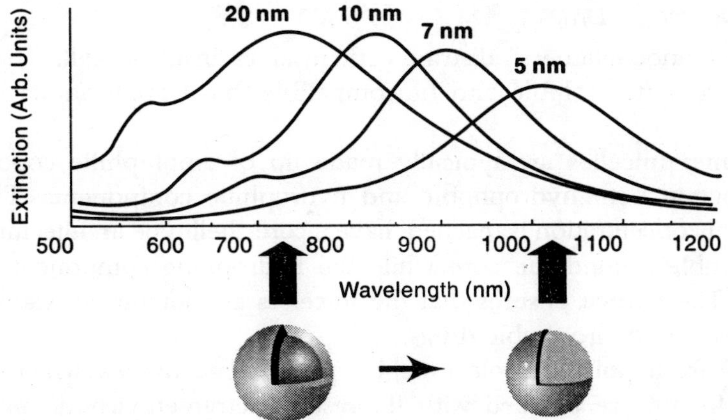

Fig. 5.111 *Gold Nanoshell plasmon resonances for a 120 nm core with the shell sizes shown.*

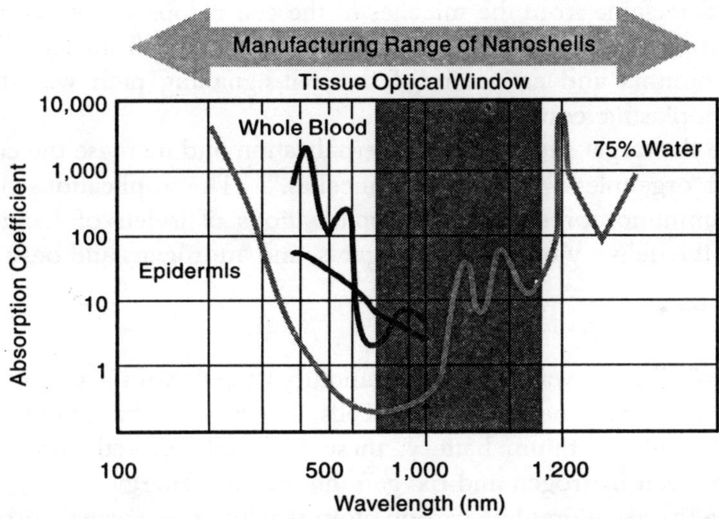

Fig. 5.112 *Illustration of relative absorption of tissue components at various wavelengths, including the "tissue optical window:"*

buildings, solar collection materials, or even fabrics. Metal nanoshells may also prove quite valuable in chemical and biosensors and in optical switches. They have been used to enhance signals in Raman spectroscopy, a chemical monitoring technique extensively use in industry.

Their ability to absorb is so great that if you put just one thousandth of an ounce of nanoshells into auart of water, the water would be completely opaque. The nanoshells currently range in size from 50 to 1,000 nanometers in diameter (one nanometer is one-billionth of a meter).

Nanocontainers

"Micellar nanocontainers" or "Micelles", these are nanoscale polymeric containers used to selectively deliver hydrophobic drugs specific sites within individual cells. example

Nanocontainers Deliver on Drugs

How miceller "nanocontainers" distribute themselves inside a cell. The block copolymer micelles, which are water soluble and biocompatible, have applications in local, controlled drug delivery.

Blocks copolymer micelles are typically made up of amphiphilic copolymers, which are polymers that contain both hydrophobic and hydrophilic components. The micelles, which form once a critical concentration is reached, have a core-shell-type architecture: the hydrophobic components assemble around the core while the hydrophilic components form a protective shell, or corona. The corona ensures that the miceless are soluble in water, while the hollow core can accommodate hydrophobic drugs.

Miceless made from poly (caprolactone)-b-type(ethylene oxide), which are spherical with diameters of 20-45 nm are labelled with fluorescent tetramethylrhodamine-5-carbonyl azide (TMRCA) molecules attached to the hydrophobic PCL end of the polymer.

On examining the distribution of the labelled micelles inside rat pheocheomocytoma cells. Fluorescence was detectable from the micelles in the cell cytoplasm–in cytoplasmic organelles such as mitochondria and the Golgi apparatus–but not in the cell nucleus. This is relevant for the delivery of inhibitors and activators of the cell-signalling path was that are commonly activated in the cytoplasmic compartment.

"We can change the target site, mode of internalization and increase the concentration of the drug in subcellular organnles when we use micelles,". "The applications of micelles are: the local delivery of immunosupressants in transplantations of iselets of Langerhans (pancreatic tissue] in diabetes, the delivery of anticancer agents, and androgens and oestrogens in hormonal disorders."

Nanohorns·

One of the SWNT (single walled carbon nanotube) types, with a irregular horn-like shape, which component of a new generation of fuel cells. The fuel cell has about 10-times the energy capacity compared with a lithium battery, these fuel cells, directly transform the chemical reaction energy between hydrogen and oxygen into electric energy.

Nanohorns have the same graphite carbon atom structures as normal carbon nanotubes (See Fig. 5.113). The main characteristic of the carbon nanohorns is that when many of the nanohorns group together an aggregate (a secondary particle) of about 100 nanometers is created. The advantage being, that when used as an electrode for a fuel cell, not only is the surface area

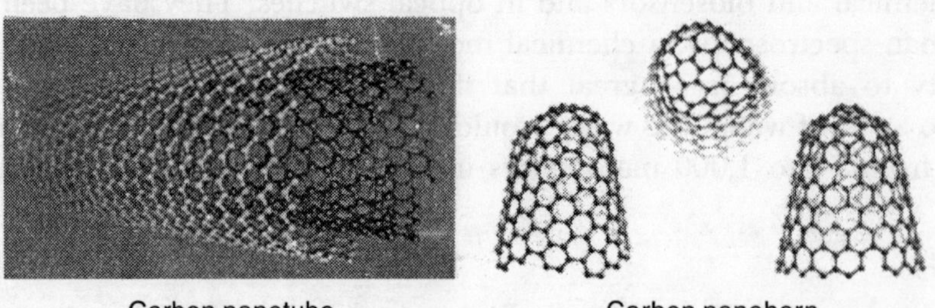

Carbon nanotube Carbon nanohorn

Fig. 5.113

extremely large, but also, it is easy for the gas and liquid to permeate to the inside. In addition, compared with normal nanotubes, because the nanohorns are easily prepared with high purity it is expected to become a low-cost raw material.

ELECTROMAGNETIC FIELDS IN ARRAYS AG NANOPARTICLES

Linear arrays of very small AG nanoparticles (diameter ~10 nm, spacing 0–4 nm) were fabricated in sodalime glass using an ion irradiation technique. The mode of confinement and plasmon coupling in nanostructures with smaller length scales, composed of linear chain arrays of Ag nanoparticles with diameters in the 01 nm range and interparticle spacing as small as several nanometers. Linear AG nanoparticle chain arrays with small length scales are formed in silica glass by ion irradiation. Other methods for generating very small ordered metal structures include pulsed-laser irradiation and biologically templated assembly.

Optical extinction spectroscopy data that show evidence for strong plasmon coupling in ion-beam-synthesized Ag nanoparticle chain arrays is taken & compare with full-field 3D electromagnetic simulations for arrays with various chain lengths and particle spacings. The simulations corroborate the experimental data and reveal large local-field enhancements in arrays of strongly coupled nanoparticles.

Nanoparticle Array Fabrication

Linear nanoparticle arrays in glass are formed by use of a high-energy ion irradiation technique as follows. First, ionic Ag is introduced into sodalime silicate glass by immersion in a melt of $AgNO_3$, 10% by mass, in $NaNO_3$ at 623 K for 10 min. Silver displaces the constituent sodium via an ion exchange interaction, resulting in a Ag content of ~6 at. % near the surface. Next the sample is irradiated with 1-MeV Xe ions to a fluence of 1×10^6 cm^{-2} at normal incidence to induce the nucleation and growth of Ag nanoparticles. The typical particle diameter is in the range of 5–5 nm, and the particles preferentially form in a ~80-nm-thick near-surface region of the silica glasses.[26] Finally, the sample is irradiated with 30-MEV Si ions at an angle of 60° with respect to the surface normal while cryogenically cooled to 77 K.

Figure 5.114 shows a plan-view transmission electron microscopy (TEM) image on a sample irradiated with 2×10^{14} Si/cm^2. A polydisperse Ag particle size distribution si found, with a typical diameter of 091 nm and upper bound of about 20 nm diameter. The majority of Ag particles appear to have been incorporated into quasilinear chain arrays, aligned along the ion beam direction. This observation is confirmed by spatial fast Fourier transform of the image, inset in Fig. 5.114(a). The redistribution of AG is ascribed to the effect of the thermal spike caused by silicon ion' electronic energy loss. Figure 5.114(b) shows a magnified view of the particle arrays. While in these plan-view images it is not possible to separately identify individual arrays, as they overlap in the image, we estimate a typical array length of up of ten particles, with particles either touching or very closely spaced.

Optical Absorption Spectroscopy

The inset to Fig. 5.115 shows optical extinction spectra taken under normal incidence of a sample before ("control") and after irradiation with 2×10^{14} cm^{-2} 30 MeV Si. Data is derived

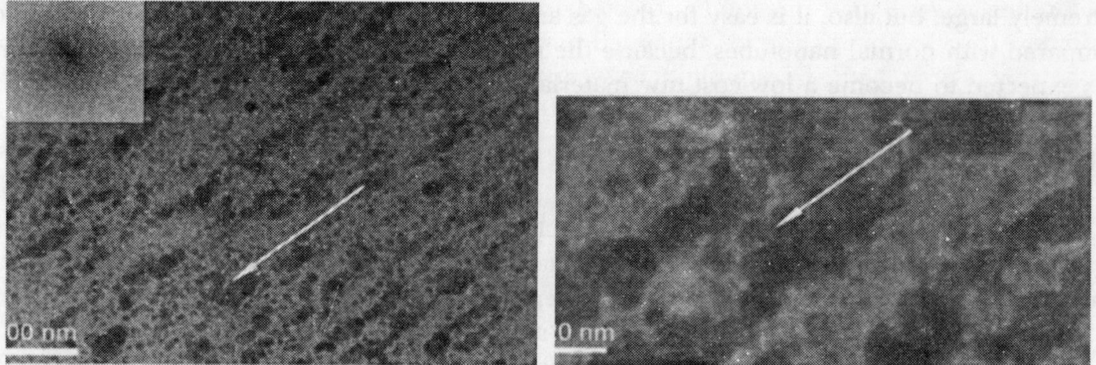

Fig. 5.114 *(a), (b) Plan-view Tem images of AG nanoparticles in sodalime glass after 30-MeV Si-ion irradiation at two different magnifications. Nanoparticle array are observed along the ion beam direction (indicated by arrows). The inset in (a) shows a spatial Fourier transform image of the micrograph confirming this anisotropy. The typical particle diameter in the micrographs is 01 nm, albeit with significant size polydisperisity. Touching or closely spaced particles are observed, in arrays with a length of up to ~10 particles.*

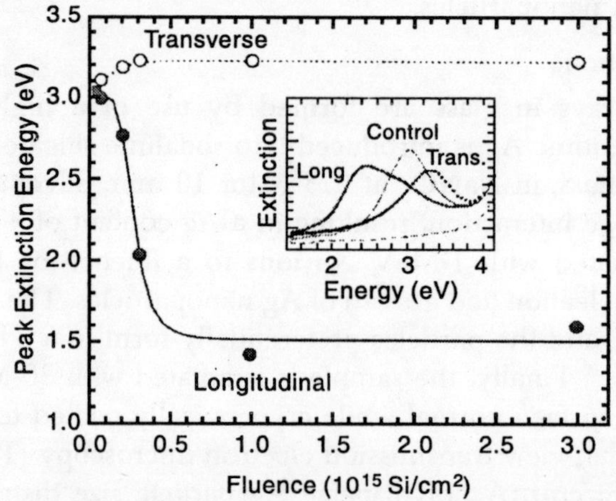

Fig. 5.115 *Measured optical extinction resonance peak energy for silver nanoparticle arrays in a glass as a function of 30 MeV Si fluence. The polarization of the incident light is transverse (open circles) or longitudinal (solid) to the projection of the ion beam direction into the normal plane. Inset: typical extinction spectra for both polarizations in a sample before irradiation ("control") and after irradiation with* 2×10^{14} 30 MeV Si cm^{-2}.

from optical transmission spectra measured using a spectroscopic ellipsometer with the incident beam perpendicular to the sample surface. The wavelength is scanned from 300 nm to 100 nm in 5-nm steps. Before irradiation, an extinction peak is observed at an energy of 3.0 eV (free-space wavelength 410 nm), corresponding to the surface plasmon dipole excitation of isolated small Ag nanoparticles in a sodalime glass matrix (refractive index $n = 1.60$). After irradiation, two distinct spectra are observed for incident light polarized either parallel to perpendicular to the ion beam incidence direction projected into the sample surface plane. The splitting of plasmon extinction bands can be explained by "collective particle plasmon" resonances which result from electromagnetic coupling between neighboring particles in linear arrays. When incident light is polarized transverse to the array axis, repulsion between like surface charges on neighboring particles increases the energy required to drive a resonant oscillation and therefore results in a spectral buleshift. Conversely, attraction between nearby unlike surface charges under longitudinally polarized incident light will result in an extinction redshift.

The main panel of Fig. 5.115 shows the peak energy of the transverse (open circles) and longitudinal (solid) mode extinction spectra as a function of Si ion fluence up to 3×10^{15} cm^{-2}. The transverse branch shows a modest blueshift that saturates at 3.2 eV (390 nm) while the longitudinal absorption peak can be tuned over a wide range, from 3.0 eV (401 nm) to .5 eV (839 nm), into the near infrared.

The. longitudinal resonance redshift of over .5 eV is much greater than that previously recorded in chains of noble-metal particles with relatively large diameter and spacing, for which a redshift of 010 meV is observed. The 1.5-eV shift is attributed to the very strong particle coupling in the present arrays. The ability to tune the resonance frequency into the near infrared is clearly valuable, as it enables applications in the important telecommunications band around μm. Also, the strong interparticle coupling implies a large enhancement of electromagnetic fields[31] in the interparticle gaps, as will be discussed further on. In the context of nanoparticle waveguides, the extraordinarily large splitting of the plasmon bands indicates a large bandwidth and high group velocity for transport. However, very strongly coupled nanoparticles are suitable for waveguiding only over short distances, as significant spatial overlap of the mode with the metal particle leads to severe damping.

Finite Integration Simulation Procedure

The influence of geometrical parameters on the collective plasmon resonance of linear chains of Ag nanosphere is studied by three-dimensional full-field electromagnetic simulations which employ finite-element integration techniques to solve. Maxwell's equation. The particle size constant is kept at 0 nm diameter and the particles are spaced evenly along a line with perfect axial symmetry. Particle spacing is varied from 0 nm (touching particles) to 4 nm. The optical constants of Ag are modeled

$$\varepsilon(\omega) = 5.45 - 0.73\frac{\omega_p^2}{\omega^2 - i\omega\gamma}, \tag{1}$$

with $\omega p = .72 \times 10^{16}$ rad s^{-1} and $\gamma = 8.35 \times 10^{13}$ s^{-1}, which provides a good fit to tabulated experimental data[36] throughout the visible and infrared. The index of the surrounding glass matrix is set to $n = 1.60$. The simulation volume is a rectangular solid which extends at least 100 nm beyond the nanoparticle surfaces in all directions. The mesh in linearly graded with a 10:1 ultimate ratio; i.e., if the grid cells in the immediate vicinity of the particle are ~0.25 nm on each side, those on the outer boundary of the simulation volume are ~2.5 nm on each side. Technical limitations constrain the mesh size under these conditions to about 2×10^6 total grid cells. In all simulation the incident light is polarized longitudinally relative to the array. We focus on the longitudinal plasmon resonance because the optical resonance has a strong functional dependence on geometrical parameters and because the extraordinary tunability of the longitudinal resonance may be of interest of a variety of applications.

A two-step process is used to find the collective particle plasmon resonant mode and its frequency for each array. First, the simulation is illuminated by propagating plane wave with an off-resonance frequency that allows the particle chain to absorb energy. Second, the incident field is switched off and the electric field amplitude is observed in the time domain as any particle excited by the incident plane wave resonantly decay or "ringdown." The time response of the electric field in interparticle gaps and other points of lower symmetry are also observed to provide additional information about the mode structure. A fast Fourier transform (FFT) of these data gives the spectral response that enables the resonant frequencies to be identified. Since the absorption and ringdown are resonant phenomena, the frequency at which the peak FFT response occurs is directly comparable to the frequency of maximum extinction in an optical spectrum. Once the spectrum outlined in this way, on-resonance excitation is used to excite individual modes to examine the corresponding spatial distribution of the field intensity. Such distributions are used to discriminate between spectral features which correspond to the collective dipole excitation and other physical resonances or, in some cases, unphysical artifacts of the simulation or frequency domain transform which can be eliminated once identified. In some cases small "hot spots" were observed in the intensity maps, which occurred at slight faceted corners of the rendered nanospheres. By varying the grid size and geometry these are identified as artifacts of the simulation. When the grid cell linear dimensions are reduced to 0.25 nm in the immediate vicinity of the metal nanoparticles, these hot-spot artifacts have a negligible effect on the overall field distribution. This requisite fine mesh density has the indirect effect of constraining the maximum simulation size to linear arrays of about 21 nanoparticles.

Figure 5.116(a) shows peak-normalized Fourier transform amplitude spectra of particle ringdown of particle ringdown for four-particle chains of 10-nm-diam Af particles, with interparticle spacings of 1 nm, 2 nm, and 4 nm, and for an isolated particle. It is found that the peak frequency value is robust against small changes to the mesh cell density, sphere smoothness, or other minor variations in simulation procedure. In contrast, the apparent linewidth and spectral shape of the resonance were found to be somewhat dependent on arbitrary factors such as the total ringdown time and the absorption cross section of the structure at the off-resonance excitation frequency. The Fourier transform spectra often also contain several peaks with normalized amplitudes less than one. Investigation of spatial energy density profiles using selective-frequency excitation revealed that low energy secondary peaks [e.g., around 2 eV in Fig. 5.116(a)] correspond to artifact modes particular to the 3D polygonal representation of the nanoparticle. The weak features observed in the spectra of Fig. 5.116 at energies slightly greater than that of the primary peak are due to real multipolar nanoparticle resonances. Selective-frequency excitation simulation of these features proves impossible due to the low scattering strength of these resonances and spectral proximity to the primary peak. The primary peak, however, it found to correspond to the dipole excitation in all cases. From the above analysis it is concludes that the peak energy of the Fourier transform spectra is a good metric for the dipole resonance energy and can thus be compared with experimental peak extinction energies.

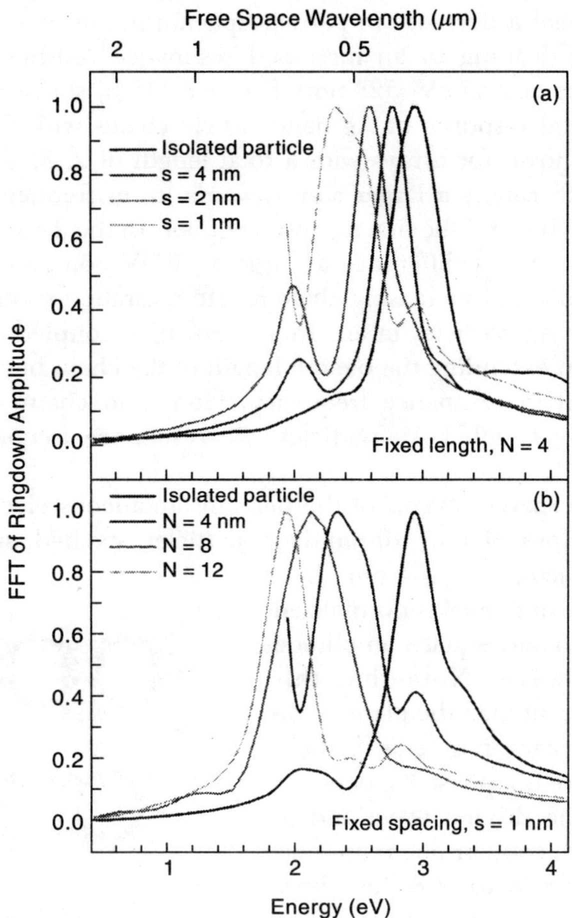

Fig. 5.116 *Simulated longitudinal extinction spectra for linear arrays of 10-nm-diam Ag particles in glass. (a) Array of four particles spaced by 1, 2, or 4 nm. (b) Array of 4, 8, or 21 particles with interparticle spacing fixed at 1 nm. A reference spectrum for a single AG particle is also shown in each panel. Each spectrum is normalized to the height of the strongest extinction peak, which we identify as the collective dipole resonance.*

The resonance peak for isolated particles in Fig. 5.116(a) occurs at 2.93 eV (free-space wavelength 424 nm), in good agreement with the experimental data (Fig. 5.115), indicating that the initially prepared Xe-irradiated sample (i.e., without ordered nanoparticle arrays) consists of uncoupled nanoparticles. This is consistent with earlier work showing that under these Xe irradiation conditions about 33% of the incorporated Ag is a agglomerated in nanoparticles with a mean diameter of 5–01 nm, while the remaining 67% remains embedded in the glass network as Ag ions. The corresponding interparticle distance is about three particle diameters, too large for significant interparticle coupling.[26,37]

Figure 5.116(a) shows that a decrease in particle spacing in the four-particle array increases the interparticle coupling, leading to an increased resonance redshift. A 1 nm spacing the simulated resonance occurs at 2.35 eV (528 nm). Figure 5.116(b) shows the effect of total chain length on simulated spectral response of Ag nanoparticle chains with the interparticle spacing fixed at 1 nm. Data are shown for arrays with a total length of 2, 8, 4, and 1 particles(s). So, increasing the chain length causes a larger shift towards lower frequencies, implying that the particle interaction extends beyond the first nearest neighbor. In the 2-particle chain the observed frequency is 1.92 eV (647 nm), a difference as large as .0 eV compared to the single-particle plasmon resonance. The effect of increasing chain length saturation chains of not more than 20 particles, as calculated for Ag particles in air. In, less strongly coupled chains, with 2 nm or 4 nm spacing, it is found that extending the overall length of the chain beyond four particles does not lead to a further shift in resonance frequency. Hence, in chains of widely spaced and therefore relatively weakly coupled Au particles, the resonance frequency was not a strong function of array length.

Figure 5.117 shows the spatial images of the peak instantaneous electric field intensity (i.e., E^2) at steady state for arrays of four 10-nm-diam particles, excited on resonance. Data are shown for interparticle spacings of (a) 4 nm, (b) 2 nm, or (c) nm. The background level is normalized to the maximum instantaneous square amplitude of the incident plane wave. Note that this normalization level is different than the plane-wave background at the same instant in time as the field profile snapshot, since there is a phase lag between the amplitude peaks in the driving wave and in the resonant response. Each contour line represents an intensity difference by a factor 1.8 (four lines correspond to one order of magnitude). The maximum local field is observed in the dielectric gap between the two metal particles at the midpoint of the array. The factors by which the field intensity at the array midpoint exceeds the background level is ~80 in the weakly coupled array, Fig. 5.117(a), and ~5000 in the more strongly coupled array (c). For comparison, the maximum field intensity enhancement near an isolated nanoparticle drives resonantly is typically 30. The giant 5000-fold intensity enhancement is consistent with previous reports of 10^{16}–10^{8}-fold enhancement of the effective Raman scattering cross section near metallic nanostructures, given the fact that Raman scattering is quadratic in field intensity.

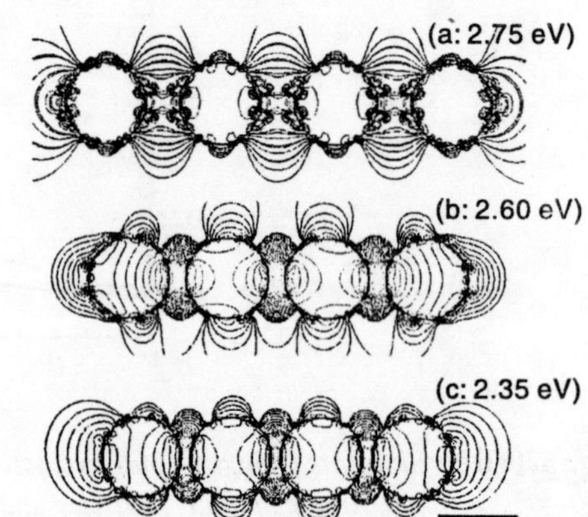

Fig. 5.117 *Two-dimensional spatial images of the electric field intensity in a plane through the particle centers of four Ag nanospheres with interparticle spacing of (a) 4 nm, (b) 2 nm, and (c) 1 nm at resonant excitation. The background is normalized to maximum instantaneous intensity of the incident plane wave. Each contour line represents an intensity difference by a factor of .8 (four lines represent one order of magnitude).*

Figure 5.118 shows the field intensity along a lone normal to the particle axis through the interparticle gap at the midpoint of a four-particle array. The field confinement is most pronounced for the array with the smallest spacings, 1 nm and 2 nm, where the lateral distance (from the array axis) at which the field diminishes by 01 dB is 3 nm; it vanishes by 30 dB in less than 6 nm. Given that the resonant excitation wavelength for this mode in bulk glass is 330 nm, this clearly demonstrate the giant field enhancement and localization in these closely spaced nanoparticle arrays. Figure 5 also shows that for 4 nm spacing the field is less concentrated, with a 10 dB decay distance of 5.5 nm and peak field intensity nearly 100-fold lower then for the 1-nm-spaced array. This demonstrates that true nanoscale engineering is required to advantage of these high-field-concentration of effects. Indeed, the ion-irradiation-induced nanoparticle arrays in Fig. 1 are an example of that.

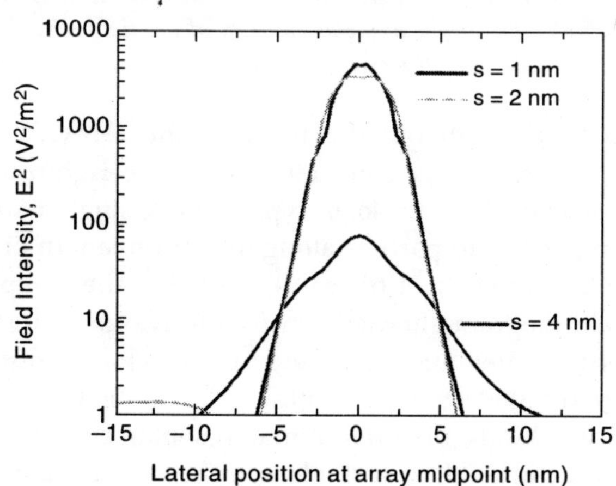

Fig. 5.118 *Electric field intensity on a line through the dielectric gap at the midpoint of an array of four Ag nanospheres with interparticle spacing of (a) 4 nm, (b) 2 nm, and (c) 1 nm at resonant excitation. Maximum instantaneous intensity of the incident plane wave is* $V^2\ m^{-2}$.

In TEM images such as Fig. 5.114 it is observed has in some arrays there is a strong possibility of the interparticle separation which has been reduced to the point that nanoparticles just touch their neighbors. Assume, these arrays as having "0 nm interparticle spacing." In simulations the touching spheres are defined to overlap each other by about 0.25 nm (one grid cell depth) and therefore share a circular boundary surface with a radius of about 1 nm. In this case two distinct longitudinal modes are found in the spectrum. For an array length of 12 particles there occurred at 0.35 eV (free-space wave-length 3500 nm) and 1.65 eV (750 nm). Selective excite and study of the spatial distribution of the electric field for the two cases. It is found that the additional low-energy peak is not an edge or corner artifact but has real physical significance. This id demonstrated in Fig. 5.119, which shows the longitudinal component (E_x) of the electric field in a system which consists of a linear array of 12 touching Ag spheres excited at two resonance frequencies. Areas colored red have positive field amplitude, while areas colored blue has negative field amplitude. In a snapshot of the chain driven by a longitudinally polarized plane wave at 0.35 eV [panel (a)], regions of positive E_x are observed at

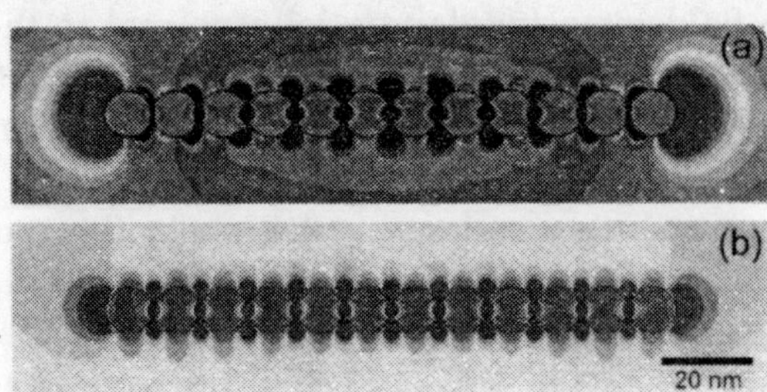

Fig. 5.119 *(Color Distribution of the longitudinal component of the electric field (E_x) in the vicinity of an array of 12 Ag particles with 10 nm diameter, illustrating two distinct modes. In panel (a), an antennalike mode resembling that of a single elongated wire is excited resonantly at 0.35 eV; in panel (b), a coupled-particle-like mode resembling that of a chain of independent particles is excited resonantly at 1.65 eV. The slight axial asymmetry of the field distribution is caused by superposition of the resonant mode with the exciting plane wave.*

either end, with negative E_x through out the body of the array. This electric field pattern indicates that positive surface charge is concentrated on the rightmost particle and negative charge on the leftmost particle. The mode is typical of a single-wire antenna and requires surface charge to flow from particle to particle along the entire length of the array. Alternatively, when the same structure is driven at 1.65 eV [panel (b)] the coupled-dipole resonance is selectively excited. The field diagram alternates from positive in each dielectric gap to negative inside each particle. This indicates an alternating surface-charge distribution in which each individual particle is polarized but electrically neutral. Thus, in the touching-particle configuration, the system can support two kinds of longitudinal resonance: the particles can still act as individual coupled dipoles or, instead, as a single continuous-wire antenna.

Although particle like and wirelike modes of toughing-particle chains have strongly shifted peak resonance frequencies, they do not in general exhibit an extremely high degree of local-field amplification. The lack of a dielectric gap mean that the interparticle interaction is only weakly capacitive in nature, and therefore only a small magnitude of opposing surface charge builds up on neighboring particles. However, a contrary influence also comes into play. The sharp "crevice" formed between two intersecting spheres contributes shape-induced enhancement to the local fields. For the specific geometric configuration and material properties input to our simulation, it is found that the former effect dominates.

Figure 5.120 shows a compilation of longitudinal resonance energies for linear arrays of 10-nm-diam Ag particles, plotted as a function of chain length. The data series represent interparticle spacings of 4, 2, 1, or 0 nm. In the case of zero nanometer spacing, the energies of "particlelike' and "wirelike" modes are plotted separately. The particle like resonances are relatively weak, leading to some uncertainty in their position, as indicated by error bars on that series. The figure shows that for 12-particle array length the resonance of the most strongly coupled array (0 or 1 nm spacing) shifts by about 1 eV a peak resonance energy around 2 eV. More weakly coupled chains saturate at peak energies above ~2.5 eV.

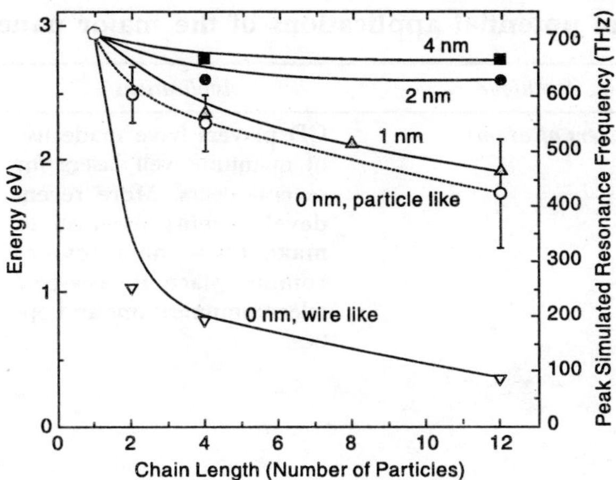

Fig. 5.120 *Resonance frequency for linear arrays of various overall length, ranging from 1 to 21 Ag particles each 10 nm in diameter. Series represent interparticle spacings from 0 to 4 nm. In the case of zero nanometer spacing, the frequencies of both coupled-nanoparticle-like and wire-like modes are plotted.*

Comparing these data with experimental particle distributions and resonance energy measurements in Figs. 5.115 and 5.116; the nanoparticle distributions are inhomogeneous, touching and nearly touching particles are observed in the TEM, which according to our simulations lead to large plasmon shifts. The decrease in plasmon energy with increasing fluence (Fig. 5.115) can be attributed to either (or both) a gradual growth in particle array length or a decrease in interparticle spacing. The experimentally observed peak resonance for high-fluence irradiation in Fig. 5.115 is 1.5 eV. The resulting resonance energy is thus lower than the lowest coupled-particle mode energy calculated in Fig. 5.120 (1.8 eV). This may indicate that long chains (N>12 particles) form at large fluence. Alternatively, wirelike modes in small chains could be dominating as the interparticle spacing approaches zero. Further investigation of the relative interaction cross section for the wire and particle modes would be necessary to support this hypothesis.

For applications of field enhancements at the important telecommunication wavelength of 1.5 μm (0.8 eV), very strongly coupled arrays of nanoparticles are requited. Comparing our data with those our results from arrays of larger particles it appears that it is the combination of particle center-to-center spacing and diameter, rather than interparticle spacing along, that is a key parameter determining the coupling strength. Finally, we note possible applications of the wirelike mode in the THz-frequency domain, in particular for very long array lengths.

The great utility of wavelength tunability and local-field enhancement for applications such as nonlinear optics and sensing of small volumes, nanosized ordered or quassiordered ensembles of very closely spaced metal particles serve as an ideal platform for active device regions in integrated plasmonic networks. Innovative nanoscale engineering and fabrication (such as the ion irradiation technique used here) are required to synthesize particle arrays with these interesting properties.

Summary of potential applications of the major nanomaterials

Materials	*Technique*	*Applications*	*Informatics*
Quantum well structures	***Clusters of atoms***	CD players have made use of quantum well lasers for several years. More recent developments promise to make these nanodevices commonplace in low-cost telecommunications and optics.	Telecommunications/optics industry. Potentially very important applications in laser development for the data communications sector. The aim is to use fibre optic communications in building and computers. The problems are cost and high temperature operating conditions. Quantum well/dot structures can potentially solve this problem
Quantum dots		Telecommunications, optics.	
Polymers		Computing, energy conversion.	Display technologies–this sector is driven by the electronics consumer market.
Nanocapsules	***Grains that are less than 100 nm in size***	Many applications envisaged e.g. nanoparticulate dry lubtricant for engineering (Saxi, 2000).	
Catalytic nanoparticles		Wide range of applications, including materials, fuel and food production, health and agriculture (Hay and Shaw, 2000).	
Carbon nanotubes	***Fibres that are less than 100 nm in diameter***	Many applications are envisagested: space and aircraft manufacture, automobiles and construction, Multilayered carbon nanotubes are already available in practical commercial quantities. Bucktubes some way off large-scale commercial production (Saxi, 2000).	
Carbon nanotubes in nanoelectronics. These hold promise as basic components, for nanoelectronics—they can act as semiconductor, semiconductors and insulators			Memory and storage; commercoal prototype nanotube-based (nonvolatile); RAM; display technologies; E-paper.

(*Contd.*)

(*Contd.*)

Self-assembling monolayers (SAMs)	***Films that are less than 100 nm in thickness***	A wide range of applications, based on properties ranging from being chemically ranging from being chemically active to being wear resistance (Saxi, 2000).	
Nanoparticulate coatings		Sensors, reaction beds, liquid crystal manufacturing, molecular wires, lubrication and protective layers, anti-corrosion coatings, tougher and harder cutting tools (Holiser, 2002).	
Nanocomposites	***Nanostructured materials***	A number of applications, particularly where purity and electrical conductivity characteristics are important, such as in microelectronics. Commercial exploitation of these materials is currently small, the most ubiquitous of these being carbon black, which finds widespread industrial application, particularly in vehicle tyres (Hay and Shaw, 200).	
		Military, lifestyle.	
Photonic crystal technologies	**Textiles**		Optical communication sector, i.e. fiber optics.
			Photonic integrated cir-cuits can be nearly a million times denser than electronic ones. Their tighter confinement and novel dispersion properties also open up opportunities for very low power devices
Spintronics—the utilisation of electron spin for significantly enhanced or fundamentally new device functionality			Ultra-high capacity disk drives and computer memories.
Molecular nano-electronics (including DNA computing)			Circuits based on single molecule and single electron transistors will appear, initially in special applications.

Applications for materials and devices resulting from self-assembly and self-organisation.

Name	*Technique*	*Application*
Newmaterials		
Sol-gel technology	Inorganic and organic component combination	The design of different types of materials; functional coatings.
Intercalation of polymers	Intercalation of polymers with other materials (DNA, drugs).	Toxicity testing, drug delivery and drug performance analysis.
Nano-emulsions	Nanoparticle size and composition selected.	Production of required viscosity and absorption characteristics.
Biommimetics	Design of systems, materials and their functionality to mimic nature.	High strength, structural applications, such bas artificial bones and teeth.
New devices		
Field-sensing devices	Combination of molecular films with optical waveguides and resonators.	Biosensing and optical switching.
Material-sensing devices	Surface of liquid crystals or thin membranes and other organic compounds can be used to detect molecules via structural or conductive changes.	Gas chemical sensing.

Summary of applications for energy processing.

Materials/technique	*Applications*
Power generation (PV technology)	
Polymer materials (organis).	Solar cells (pers, comm., Jenny Nelzon, Imperial College London, 2 Dec 2002). Current developments aim to balance moderate efficiency with low cost. Another big advantage is that these layers can easily be incorporated into appliances. Current problems stem from the material's instability.
Combination of organic and inorganic molecules.	Dye-sensitised solar cells made from a thin hybrid layer (pers. Comm., Jenny Nelson, Imperial College London, 2 Dec 2002). These cells are potentially very cheap because fabrication is from cheap, low purity mate rials by simple and low cost procedures (Saxi, 2000).
Quantum wells (inorganic).	Quantum-well solar cells (pers. Comm., Jenny Nelson, Imperial College London, 2 Dec., 2002). Current research is taking place in high-efficiency applications because the infrared part of the solar may be absorbed.
Nanorods	These structures can be tuned to respond to different wavelengths of light forming cheap and efficient solar cells (Holister, 2002).
Fuel conversion/storage	
Improved fuel catalysts through nanostructuring.	Fuel conversion (Saxi, 2000).
Nanotubes	Fuel storage. Eg. A methane-based fuel cell for powering mobile phones and laptops is currently being developed. (Holister, 2002).
Nanoparticles.	Vastly increased (e.g. ×10) charge and discharge battery rate (Holister, 2002).

Summary of application for nanoscale pharmaceuticals and medicine.

Materials	*Property*	*Applications*
Diagnostics		
Nanosized markers i.e. the attachment of nanoparticles to molecules of interest	Minute quantities of a substance can be detected, down to individual molecules.	Eg. Detection of cancer cells to allow early treatment.
'Lab-on-a-chip' technologies	Miniaturisaiton and speeding up of the analytical process.	The creation of miniature, portable diagnostic laboratories for uses in the food, pharmaceutical and chemical industries; in disease prevention and control; and in environmental monitoring.
Quantum dots	Quantum dots can be tracked very precisely when molecules are 'bar coded' by their unique light spectrum.	Diagnosis
Drug delivery		
Nanoparticle in the range of 50-100 nm	Larger particles cannot enter tumour pores while nanoparticles can easily move into a tumour.	Cancer treatment.
Nanosizing in the range of 100-200 nm	Low solubility.	More effective treatment with existing drugs.
Polymers	These molecules can be engineered to a high degree of accuracy.	Nanobiological drug carrying devices.
Ligands on a nanoparticle surface	These molecules can be engineered to a high degree of accuracy.	The ligand target receptors can recognise damaged tissue, attach to it and release a therapeutic drug.
Nanocapsules	Evading body's immune system whilst directing a therapeutic agent to the desired site.	A Buckyball-based AIDS treatment is just about to enter clinical trials (Ho, 2002a).
Increased particle adhesion (Holister, 2002).	Degree of localised drug retention increased.	Slow drug release
Nanoporous materials	Evading body's immune system whilst directing a therapeutic agent to the desired site.	When coupled to sensors, drug-delivering implants could be developed.
'Pharmacy-on-a-chip'	Monitor conditions and act as an artificial means of regulating an maintaining the body's own hormonal balance.	E.g. Diabetes treatment.
Sorting biomolecules	Nanopapers and membranes are capable of sorting, for example, left- and right-handed versions of molecules.	Gene analysis and sequencing.

(*Contd.*)

(*Contd.*)

Tissues regeneration, growth and repair		
Nanoengineered prosthetics	Increased miniaturisation; increased prosthetic strength and weight reduction; immproved biocompatibility.	Retinal, auditory, spinal and cranial implants.
Cellular manipulation	Manipulation and coercion of cellular systems.	Persuasion of lost nerve tissue to grow; growth of body parts.

Nanolayers

Film Formation

The manipulation of single-molecule-thick films at the water air interface and their transfer onto solid substrates are among the earliest efforts setting the foundation for nanotechnology. The investigation of Langmuir-Blodgett (LB) films dates back to the work of Langmuir and Blodgett at General Electric. A renewed interest in LB films began with the work on energy transfer between dye molecules incorporated in fatty acid LB films and has continued unabated ever since.

The first step in the preparation of an LB film is the spreading of a monolayer of molecular, polmeric or particulate material (non-amphiphilic) by depositing drops of a dilute solution of the material in an "spreading solvent" onto the water surface. The concentration of solvent is in millimolar or less, and the solvent should be one that spreads across the surface rapidly and evaporate without remaining at the surface or dissolving into the subphase. Common spreading solvents include chloroform, benzene, and hexane, sometimes with a small fraction of ethanol or methanol added to alter the polarity. For vertical LB transfer, the monolayer must be able to get compressed to a stable surface pressure (Π) during transfer to the substrate.

A schematic of LB trough is shown in Figure 6.1. The main components are the Teflon trough, barrier(s) to smoothly compress the monolayer, a surface pressure transducer, and the dipper for raising and lowering a substrate smoothly. The simplest LB troughs have one barrier, with a dipping well at one end of the trough, while others have two opposing barriers for compression of the monolayer and the dipping well in the middle of the trough. Most LB troughs allow for temperature control of the subphase. Larger surface areas of water are need for deposition of LB films of many layers. Typical LB trough areas are in the 200–1500-cm^2 range. The trough does not need to be more than 2–3 mm deep. The dipping must be deep enough and wide enough to accommodate the desired substrate. Also available are "alternate-layer" LB troughs, with two separate subphase compartments allowing for transfer of layers of two different materials in controlled sequences.

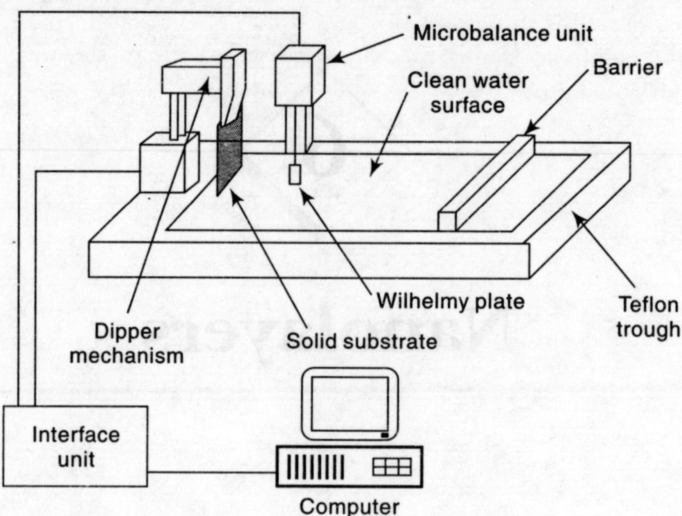

Fig. 6.1 *Major components of a Langmuir-Blodgett trough.*

The majority of LB throughs use (for measurement of surface pressure (P)), the cut pieces of filter paper, which are completely wetted by water. But in Langmuir float method, plate used is a piece of high-purity platinum metal, that be cleaned in a flame and gives a reproducible contact angle with water of 60°.

Langmuir-Blodgett films have been deposited on many different substrates example: different types of glass such as quartz for UV-visible spectroscopy; CaF_2 plates for transmission infrared spectroscopy; silicon, germanium, and ZnSe plates for internal reflection infrared spectroscopy. For electrochemical application, *LB* films have been transferred to evaporated metal films, highly oriented pyrolytic graphite (HOPG), glassy carbon, glass coated with a transparent layer of indium-tin oxide (ITO glass), semiconductor, and metal surfaces. For examination by scanning probe microscopy, freshly cleaved mica and HOPG are used because of their flatness at the molecular level.

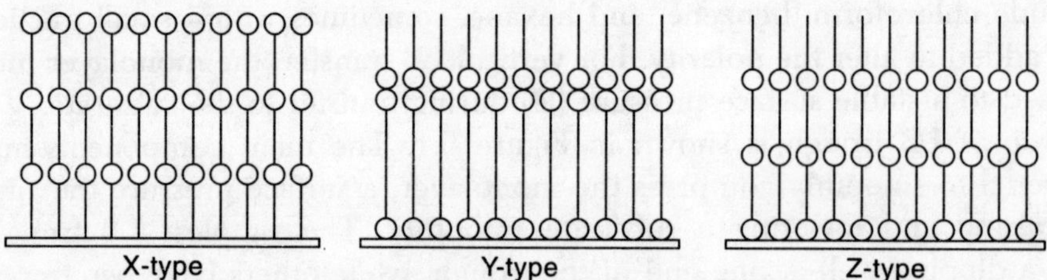

Fig. 6.2 *Common mode of Langmuir Blodgett film transfer: X-type, Y-type, ands Z-type.*

The deposition of the LB film is made in a vertical or horizontal mode. In the vertical dipping mode, the substrate is immersed (lowered) and emersed (raised) through the interface. The transfer of a single layer onto a hydrophilic substrate is done by immersing the substrate prior to spreading of the monolayer. The film is then compressed to the desired target surface pressure, and the substrate is drawn up through the interface while the trough feedback system, adjusts the barrier position to maintain a constant surface pressure. If multilayer deposition is desired, then immersion followed by emersion is repeated for the desired number of cycles. A pause is needed before reimmersion of the substrate to allow for evaporation of the entrained water. If a layer is transferred on every upward and downward pass through the interface, then the transfer is known a Y-type. Or a layer is transferred only the downward passes (X-type) or only on the upward passes (Z-type). The LB films resulting from these three typed of transfer are depicted Figure 6.2. The transfer for each pass is described by the *transfer ratio*, which is defined as the decrease in area of the monolayer divided by the area of the substrate. A transfer ratio near unity indicates uniform transfer. Transfer ratios below unity indicate partial transfer, and those much greater than unity indicate monolayer collapse, structural relaxation, or dissolution during the transfer process. Dipping speeds of ~0.1 to several millimeters per second are used. In the horizontal mode of transfer, the substrate is held parallel to the water surface and lowered until it touches the compressed monolayer; it is then gently lifted up. Horizontal transfer has been applied to films that cannot maintain a constant surface pressure or are unsuitable for vertical dipping.

Formation of Metallic Film Nanoparticle

Silver-particle films are formed by electrocrystallization at the water-air surface. Monolayers of a dialkylphosphate surfactant are spread in a dish containing a solution of silver nitrate. A silver electrode is immersed in the subphase, and a platinum electrode is floated on the water surface. A potential of 1.8–1.9 V is applied. The water surface shows a film of silver particles starting at the Pt cathode and growing concentrically, covering the water surface at rate of 12 cm^2 $hour^{-1}$. Silver-particle films are formed under arachidate monolayers and those of another dialkylphosphate, but not under cationic surfactants. No silver-particle film is observed on bare water. Scanning tunnel microscopy on horizontally transferred films show interconnected silver spheriods with 20-nm short and 30-nm long axes and 10-20 nm in height. A monolayer of surfactant-capped silver particles, prepared by photoreduction of silver nitrate in reverse, micelles, absorbed onto highly oriented phrolytic graphite (HOPG) is slowly removed from the solution.

For the electrocrystallization of silver beneath monolayers of 13 different surfactants. The monolayers are spread in a trough containing an saturated calomel electrode (SCE), Ag wire counterelectrode, and a silver (or Pt/Ir) working electrode brought into contact with the surface. The spread monolayer is compressed to the desired area. The potential below −150 mV is applied. It results in the formation of shiny silver-particle films, that are observed under arachidic and monolayers. Also observed are two long-chain dialkylphosphates, and three phospholipids. Transmission electron microscopy (TEM) observation of horizontally transferred films show interconnected 40–90-nm-size Ag particles, (seen Figure 6.3). The resistivity in the plane of the film is 5–20 Ω cm^{-1}. The mass of silver reduced during a given period of film growth is determined coulometrically from the total charge passed. As the silver-particle films grow radially outward from the working electrodes, it is possible to estimate a film thickness of 60 nm using the mass of the reduced silver, the film radius, and the density of silver.

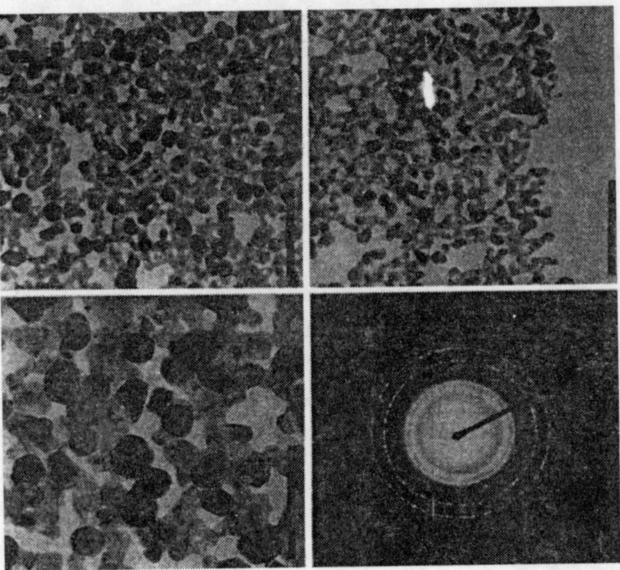

Fig. 6.3 *TEM images of silvber particulate film formed under dihexadecylphosphate monolayers (0.452 molecule^{-1}, 51 mN m^{-1}) by electrocrystallization on a 0.01 M silvber nitrate subphase at –0.200 V vs. SCE. The top left shows a representative image from the middle of the film; the top right shows a representative image from the edge of the film. The bottom left shows an image at high magnification. The bottom right shows the diffraction pattern.*

Silver-particle films are also formed beneath monolayers using a reducing agent. Monolayers of four different surfactants are examined as templates for silver-particle film formation and the most successful is *N,N*-bis(2-aminoethyl)-2hexadecyl-1,3-propanediamide. The monolayers are spread on silver nitrate subphases in a circular trough covered by a jar with a dish of formaldehyde placed next to the trough. The formaldehyde vapor causes slow growth of silver-particle films. *In-situ* optical reflectivity is used to monitor the film growth and film thickness is found to be in the range 8–50 nm. The films are noi transferred by vertical dipping; however; horizontal transfer is possible. The optical micrographs of the films growing beneath monolayers of dihexadecyl phosphate for 8 hours show separate fractal treelike aggregates. After a long period, the growth patterns show a dendritic appearance. The absorbance spectra of the films transferred on to a quartz after 12 hours show a peak at 270 nm and a growing plasmon absorbance at 440 nm. Silver films grown beneath monolayers of the propanediamide surfactant looked different. After 12 hours of exposure, well-separated circular islands were seen, that subsequently enlarged into irregularly shaped islands. The silver ion reduction is faster under these monolayers. The absorbance spectra after 12 hours showed maxima at 270 nm and 380 nm,. associated with nonmetallic Ag_4^{2+} clusters and with 5-nm-diameter silver particles. After one work, "interband" transition, arising from contact of silver particles, are observed. The lateral resistivity of the films is dependent upon the thickness and connectivity of the silver clusters and islands.

Spreading of Metallic Nanoparticles

(i) Spread monolayers of silver nanoparticles capped by actadecanethiol are generated in sodium bis(2-ethylhexyl) sulfosuccine (AOT) reverse micelles in isooctane with added aqueous silver nitrate by sodium borohydride reduction. Capping by addition of octadecanethiol precipitated the particles, which are collected and redispersed in chloroform. The mean particle diameters varies from 3.0 to 6.8 nm.

The spread monolayers of the capped particles are rigid for vertical deposition, so horizontal deposition is used. The areas per particle in the close-packed monolayers of the three capped particle preparations are 20(w = 2.5), 58(w = 5), and 130(w = 10) nm^2 $particle^{-1}$, which shows the dependence on micelle core size. Brewster-angle microscopy of spread uncapped particles show smaller aggregates than for spread monolayers of the capped particles. They show large islands merged to a uniform film or compression. The attraction between the hydrophobic capping molecules favor extensive aggregation at the water surface.

(ii) An extraction method is employed for the preparation of stabilized silver nanoparticles. A silver-particle sol is added to oleic acid in hexane emulsified. On standing, the silver particles are transferred into the organic layer and the oleic acid is absorbed onto them. Dilution of the oleic acid-stabilized silver particles (d = 10.3 ± 0.4 nm) in hexane is used to prepare the spreading solution.

(iii) The extraction method is applied to the preparation of surfactant-stabilized nanoparticles of platinum palladium, and platnium/palladium alloys using dis-tearyldimethylammonium chloride. Platinum particles coated by the surfactant are prepared by mixing the surfactant in chloroform with an aqueous solution of H_2PtCl_6. The metal salt is extracted into the reverse micelles and reduced by formaldehyde. Surfactant-coated palladium particles are prepared using $PdCl_2$ reduced by sodium borohydride. Monolayers are spread from chloroform solutions of the surfactant-stabilized metal particles. The isotherms show plateulike features related to collapse of the surfactant regions of the films followed by a rise to higher pressure until collapse of the particle monolayers. Vertical transfer is successful for compressed films if a slow dipping rate is used. Brewster-angle microscopy showed islands, and TEM show separate regions of surfactant-coated particles and regions of surfactant

Platinum nanoparticles formed by the reduction of H_2PtCl_6 by sodium borohydride aere capped by polyvinylpyrrolidone. The hydrosol is centrifuged, dried, and re-dispersed in chloroform prior to spreading at the water air interface. The average size of the polymerstabilized Pt particles is determined using TEM to be 29 nm. The TEM micrographs of the transferred films clearly showed the Pt clusters.

(iv) The production of fatty acid-capped silver nanoparticle is done by heating. Heating of the silver salts of fatty acids (tatradecanoic, stearic, and oleic) under a nitrogen atmosphere at 523 k results in the formation of 5–20-nm-diameter silver particles. Monolayers of the capped particles are spread form the toulene and transferred onto TEM grids. An ordered two-dimensional array of particles is observed. The oleic acid-capped particle arrays has some void regions not present for the other two fatty acids.

(v) The spreading of Au_{55} clusters capped by triphenylphosphine has been described. The cluster $Au_{55}(PPh_3)_{12}Cl_6$ has a metallic core of 1.4 nm and a 0.7-nm thick shell. Monolayers of the clusters were spread from freshly prepared solutions in methylene chloride; the II-A isotherm showed a compact area of 3.7 nm^2 per cluster. The films were compressed above 30 mN m^{-1} without evidence of collapse. And BAM observation showed large islands, hundreds of microns in size, merging into a uniform film under compression. The monolayers were transferred onto mica, silicon and gold. Atomic force microscopy observation of the boundary of the film edge and bare mica showed a 1.9-nm step. Scanning tunneling microscopy observation of the film on Au showed some region of locally ordered clusters. The clusters $Au_{55}(PPh_3)_{12}Cl_6$ was also spread mixed with two different polymers and the lateral arrangements of the clusters compared using Tem. For the case of a polymer ionizable carboxylic acid groups and incorporation of cysteamine in the subphase, a homogeneous distribution of particles was observed.

(vi) Two-dimensional arrays of metallic nanoparticles separated by electron tunneling barriers could potentially exhibit single-electron conductivity effects. Monolayers of do-decanethiol-capped 3-nm gold clusters are compressed and examined with optical microscopy. The large extinctions coefficient of the particles make direct observation possible. The surface pressure begin to rise near 18 nm^2 $particle^{-1}$, and jagged islands are seen that do not fully merge until a surface pressure above 5 mN m^{-1} is reached. Langmuir-Blodgett monolayers are transferred onto mica at 12 mN m^{-1}. Observation by AFM confirmed the 3.5–3.7-nm height expected for a monolayer, although bilayers covered about 10% of the surface. Monolayers transferred onto Si/SiO_2 substrates with interdigitated gold electrodes were treated with a bifunctional capping compound, 2,5″-bis(acetythio)-5,2′, 5′, 2″-terthienyl. The room-temperature in-place conductivity is 5×10^{-6} S cm^{-1} before exposure to the bifunctional capping compound, and increased to 5×10^{-3} S cm^{-1} after exposure. The bifunctional capping compound displaced dodecanethiol to some extent and lowered to tunneling barrier between neighbouring particles. And STM measurements of current from the tip through a single particle for cluster monolayers on gold/mica showed features in the *I-V* curve, suggesting single-electron tunneling. Clusters of $Au_{55}(Ph_2PC_6H_4SO_3H)Cl_6$ self-assembled onto LB monolayers of eicosamine showed single-electron tunneling in *I-V* curves measured at 90 K using STM.

THIN FILMS

Preparation of Thin Layers

1. Layer Deposition

The deposition of various thin layers of different materials, (thickness below 100 nm) and the ultrathin layers (about 1–10 nm), is of importance in nanotechnology.

The conditions on the substrate surface determines the quality of the deposition process. The chemical composition of the surface is influence by the binding between the substrate and the layer, thereby determining the adhesion of the layer and the chemical resistance of the interface.

It establishes the starting conditions for the layer preparation, and the properties of the interface zone determine the functional properties of the layer. The preprocessing and cleaning of the surface and the control of the chemical and physical surface properties is an essential prerequisite

A general feature of the technique is the transport of precursors for the layer preparation through the gas phase to the substrate surface, where they are involved in layer growth. In the case of two-dimensional deposition (all techniques without direct-writing deposition), the trajectories of the particles are statistically distributed, as are the points of contacts of the particles. The morphology of the resulting layer depends significantly on the mobility of the layer-building particles on the surface. If the particles have no mobility, layers with a rather unordered arrangement of the individual particles develop. These layers are amorphous. The fixation of the particles and therefore the tendency towards an amorphous layer correlates with the reciprocal of the ratio of the melting temperature of the deposited material and the absolute temperature, and with the decrease in bonds between deposited atoms. In the case of a low surface mobility, a solid-state structure exhibits only a low order in small dimensions based on particle diameter and the interaction between the nearest neighbors.

An increased mobility of the particles on the surface leads to condensation and crystallization centers as seeds for the layer development. Only in certain cases does the substrate induce the creation of adjacent seeds into monocrystalline layer elements. Every seed usually leads to an individual grain, which is either amorphous, an elementary crystalline, or consists of several other crystallites. The resulting layers are layers are always heterogeneous, including a variety of morphological domains. Depending on the local arrangement of atoms or molecules, one can differentiate between an amorphous, partially crystalline and polycrystalline arrangement. Grain boundaries demonstrate changes in orientation of crystal planes and anisotropy axis, and lead to increased free volumes. In the case of heteroatom layer composition, chemical differences can also be observed.

Size and density of the grains depends on particle mobility. The lower the mobility, the denser and smaller the structure of the grains and wider leads the effects of the starting seed arrangement into the grains. For low mobility, this effect leads to column-like structures in thin layers, with column diameter correlated with mobility. An increased surface mobility of the particles results in denser structures. Individual particles come to a halt after the contact with several neighbours. For an overall low mobility (i.e., the removal from given positions exhibits a significant activation barrier) the layers become denser, but the thermodynamically most stable positions are usually not reached. On the in the vicinity of atoms is a rather ordered arrangement achieved, on the atomic scale it is not ordered. So the state is amorphous or glassy. Grain boundaries and domains are found inside such amorphous layers.

Nanotechnology is aimed at layers with very small domains and amorphous materials without grains, such as glassy layers or single-crystalline structures.

The fabrication of single-crystalline layer requires epitactical layer growth. Such growth is important for the production of stacks of layer for compound semiconductor. Layer arrangement of compound semiconductors are applied in optoelectronics. Epitaxial depositions start from a planar, single-crystalline surface. Seed induction is minimized, the layer growth occurs on the faces of the atomic layers, thus being created plane by plane (Fig. 6.4). This process requires two preconditions: the lattice distance of both substrate and deposited materials should be comparable; and there should be certain particle mobility to allow for minimum binding on lattice planes but maximum growth at lattice steps.

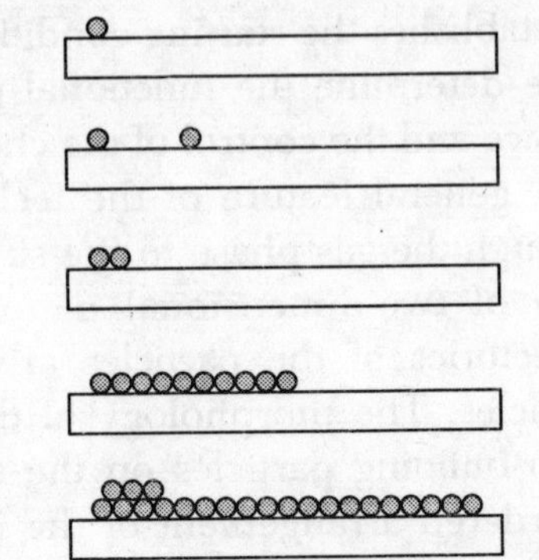

Fig. 6.4 *Creation of single-crystalline layer resulting from the deposition of particles with very high surface mobility.*

2. Chemical Vapor Deposition

Chemical vapour deposition (CVD) is based on the addition of the starting substances for layer deposition in the gaseous phase. The layer composition is determined by the parameters of the activating plasma and, most notably, the types and the preparation of the gases. The conditions ensure that the condensing products are created on the substrate but not in the gas volume. CVD processes normally exhibit higher pressure compared with sputtering. Particles arriving on the substrate have a broader angular distribution of orientation, and their energy is low. Because reactive components are involved and the layer creation is base on the chemical conversion at the substrate surface, in the presence of functional groups the particles from bonds with the substrate, despite the low kinetic energy of the arriving particles.

Incomplete conversion of the gases into layer material leads to gaseous or at least volatile by products. Thus silane (SiH_4) and ammonia (NH_3) react to give silicon nitride (Si_3N_4 as a layer) and hydrogen.

The preparation of monocrystalline layers on monocrystalline substrates is a key process for many applications, such as in solid-state electronics and optoelectronics. Such epitaxial layers exhibit a high quality and provide optimal conditions for the control of charge transfer in miniaturized devices. The preparation of multi-component epitaxial layer arrangements prepared from mono atomic layers of different materials are denoted as atomic layer epitaxy (ALE). Thus, the next layer develops only on the previous layer and every layer has as self-limited growth based on the chemical properties of the substrate material. The first layer of substance A is deposited on the substrate. Because this material does not react with itself, the growth stops after the formation of a monolayer. Subsequently, a second material *B* is deposited in the substrate, which forms a chemical compound with A but not with itself. The third step again includes deposition of *A*, and so on. The process can be controlled so that the so that not only is the composition and degree of ordering of individual layers determined, but even the exact number of individual atomic layers, example, Fig. 6.5 shows the preparation of an epitaxial zinc sulfide layer through ALE from alternating reactions with $ZnCl_2$, H_2S. In principle, layers containing just one element are also fabricated by ALE, when the valences of the deposited atoms required for the subsequent construction are blocked by protecting substituents. The deposition and activation steps occurs in an alternating pattern.

ALE is also suitable when a high precision of layer thickness and/or layer composition is require, without the need for epitaxial layers. This wider application field, with numerous examples of ALE utilizing a wide range of materials, is detailed in Table 6.1.

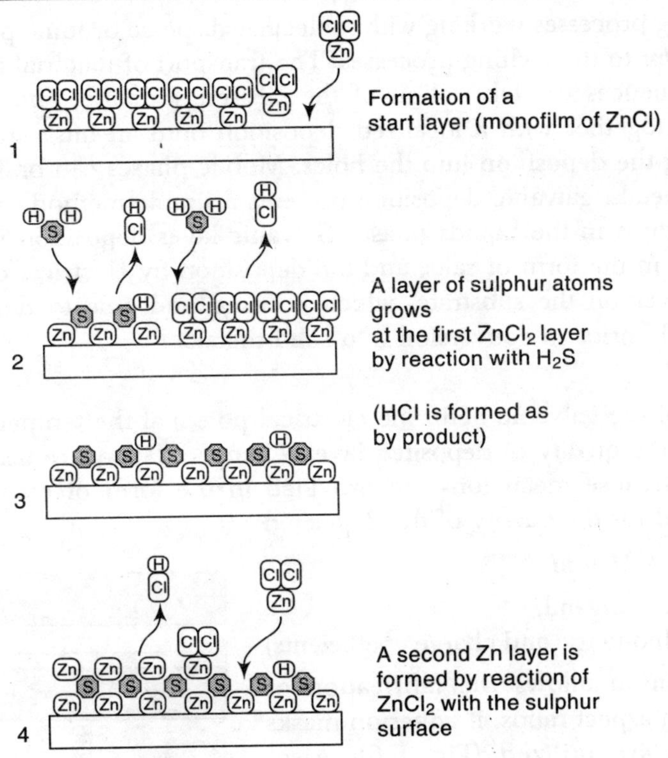

Fig. 6.5 *Generation of zinc sulfate layer by a reactive synthesis based on alternating of hydrogen sulfide and zinc chloride using an ALR prove.*

Table 6.1 *Example of compound materials that are created by ALE processes (including polycrystalline and amorphous materials)*

Metals	Ge, Cu, Mo
Semiconductor	Si
Carbide	SiC
Nitrides (metallic)	AlN, GaN, InN, SiNx
Oxides, dielectrics	TlN, TaN, Ta_3H_5, NbN, MoN
Oxides (transparent conductors and semiconductors)	In_2O_3, ITO $In(O_xF_y)$, ZnO, Ga_2O_3, NiO, CoO_x
Oxides (supraconductor)	$YBa_2Cu_3O_{7-x}$
Sulfides	ZnS, ZnSe, ZnTe, CaS, SrS, CaS, SrS, BaS, CdS, CdTe, MnTe, HgTe, La_2S_3, PbS, In2S_3, $CuGaS_2$
Fluorides	CaF_2, S2F_2, ZnF_2
III/V semiconductors	GaAs, AlAs, AlP, InP, GaP, InAs

3. Galvanic Deposition

Galvanic deposition can be applied for the creation of continuous thin layers as well as for the direct preparation of structures. It is an important additive structuring technique.

Additive shaping processes working with molecular disperse or ionic phases are the material-constructive pendant to the etching processes. The transport of material from the mobile phase to the surface influences the fabrication of the desired structure, and is be controlled by a directed transport together with a localized deposition onto an unsaturated area, or by mask structures directing the deposition into the holes. Mobile phases can be fluids or gases.

In the case of metals, galvanic deposition presents a simple method for the additive buildup of nanolocal structures in the liquids phase. Galvanic layer deposition includes the supply of the layer materials in the form of salts, and the deposition by electrical current. A prerequisite is a conductive layer on the substrate, which is wired and able to deposit metal ions from solution cathodically prior to conversion into solid metal:

$$M^+ + e^- \rightarrow M_s$$

The composition of the galvanic bath, the electrical potential the temperature and the current density determine the quality of deposited layer. Compact layers are usually the result of low deposition rates. Because metal ions are provided in the form of complex compounds, the choice L is essential for the quality of the deposited

$$[ML_n]^m + ke^- \rightarrow M + nL^{[(m-kn]}$$

(M = metal, L = Ligand,

k, n, m = stoichiometric and charge coefficients)

Galvanic deposition allows the fabrication of structures with high aspect ratios, if adhesion masks with steep flanks are utilized (Fig. 6.6). Also, geometrically defined nanostructures can be fabricated for amorphous or microcrystalline layers, and even sub-micrometer and nanometer structures with high aspect ratios are possible.

Nanogalvanic deposition uses mask of non-conductors, and deposits metal from the solution in the windows onto the conductive substrate. Mask materials are polymer resists, preferably PMMA or similar compounds, which are nanostructured by electron, ion or X-ray lithography. Galvanic deposition transfers the mask window as a plastic negative onto the metal structure. Therefore, the flank geometry of the mask window determine both the lateral dimensions and (in connection with the thickness of the deposited material) the three-dimensional shape of the metal structure. The mask can usually be removed later by using an isotropic solvent, leading to for example, structures with large aspect ratios and structures with undercut edges (Fig. 6.7). One example is the fabrication of metal

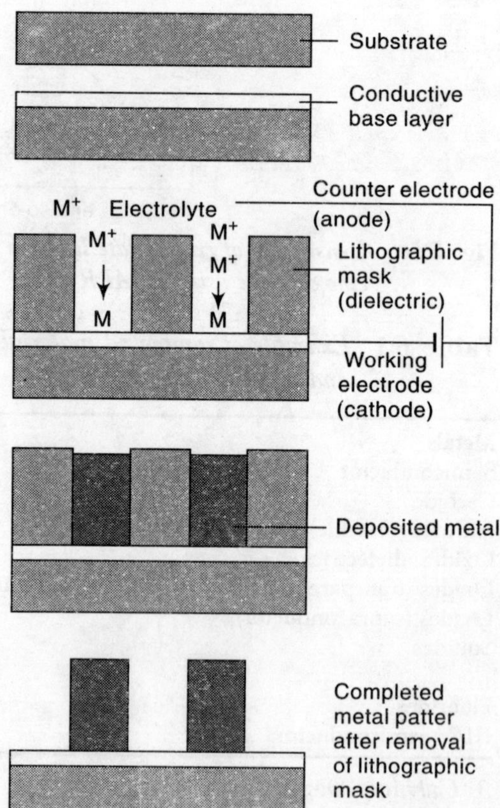

Fig. 6.6 *Additive preparation of microstructure layers by galvanic deposition in the windows of a lithographic mask.*

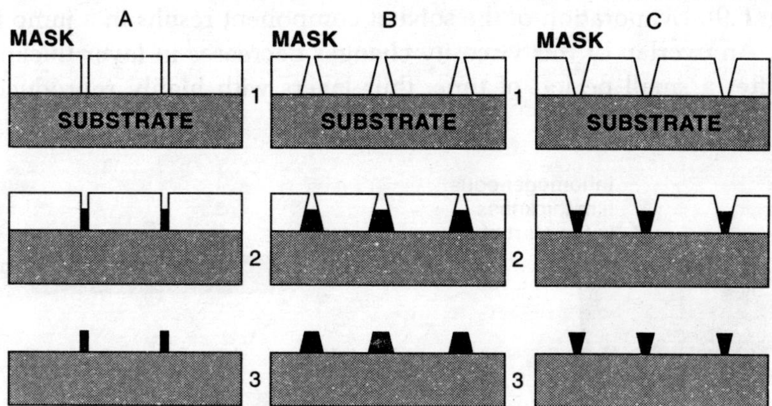

Fig. 6.7 *Additive preparation of nanostructures with different flask angles by galvanic deposition in resist structures with various flask profiles.*

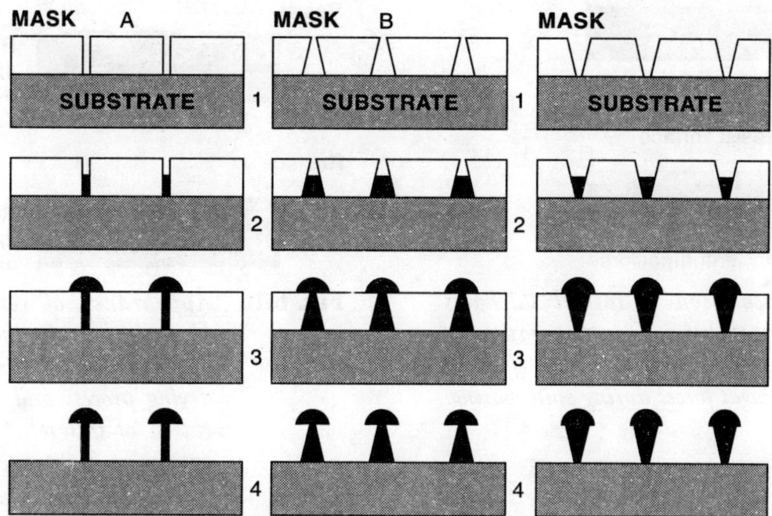

Fig. 6.8 *Preparation of mushroom-like nanostructures with various profiled by galvanic deposition.*

columns of 75 nm diameter and 700 nm height (aspect ratio 9.3) by cathodic deposition of nickel in an electronic beam structured array of holes in a resist layer. Complex flank profiles can be prepared by partial galvanic overgrowth of the mask layer surface or the use of multiple layer resist technology (Fig. 6.8).

4. Deposition by Spinning (Spin Coating)

For spin-coating, a solution of the layer-building material is placed onto a plane and centrally rotating substrate. Fast rotation of the substrate results in a centrifugal floating of the material, leading to a decrease in resist thickness. Differences (even significant ones) in substrate height are leveled (Fig. 6.9). Evaporation of the solvent component results in a jump in the viscosity of the resist layer. An overlap of this viscosity changes decreases in layer thickness and the resist floating stops after a small period of time, thus layers with highly reproducible thickness are prepared.

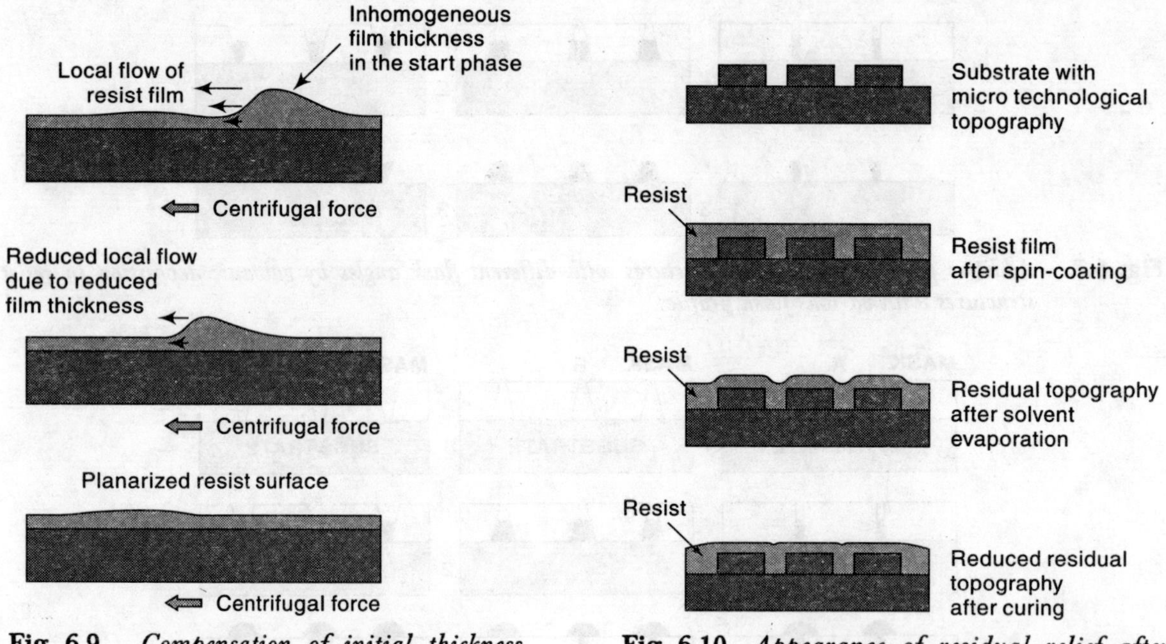

Fig. 6.9 *Compensation of initial thickness inhomogeneities and preparation of homogeneous layers by the effect of centrifugal forces during spin coating.*

Fig. 6.10 *Appearance of residual relief after planarization of substrate topography due to shrinkage of resist layers during the drying process and subsequent minimization of this relief by a local floating process at elevated temperature (curing).*

Spin-coating delivers layers of high thickness homogeneity and extends down below a thickness of 100 nm. Spin-coating planarizes structured surface better than the other deposition technique. There is only a residual relief due to the evaporation of the solvent after solidification of the layer. This shrinkage is approximately proportional to the local resist thickness. Because heating decreases the viscosity of the resist, subsequent tempering steps can reduce the observed relief in the case of densely located structures (Fig. 6.10). But, topographic failures are higher than the final resist layer, because they lead to long-ranging inhomogenities in the thickness. For low topographic relief, spin-coating is an interesting deposition technique.

A precondition for an application of spin-coating is the feeding of the layer-building material as a spinable resist. It requires viscous floating of the resist material. The layer material has to be soluble in a solvent which exhibits an adequate evaporation rate during the spinning process, so that a stabilization of the chosen layer thickness is achieved. It requires the application of a mixture of solvents with different evaporation rates. The solvent with the lower evaporation rate facilitates the layer stabilization, whereas that with the higher rate leads to a homogeneous, failure-free layer. The remaining solvent is usually in an subsequent drying step.

The requirements of solubility and solvent parameters limit the choice of spinable layer materials significantly. Organic polymers, and also bipolymers, are easily deposited by this method. Inorganic layers are accessible when the atoms of the planned layer are chemically integrated into the resist and the organic components of the layer are later removed, e.g., by thermal treatment or plasma processing. One example of this approach is the so called spin-on glass layer. Normally oxide layer are involved, because the polymers are removed in an atmosphere of oxygen.

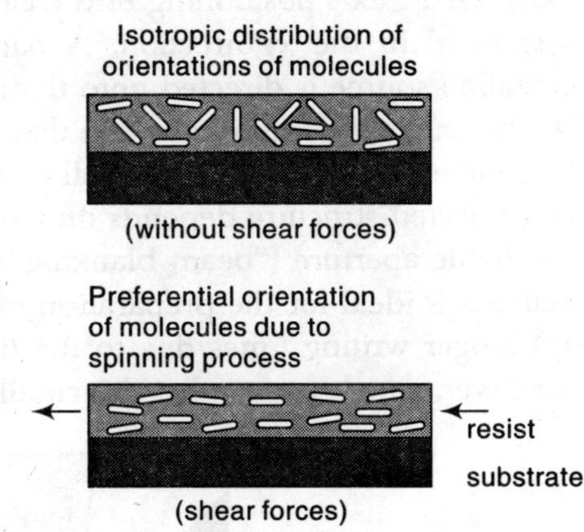

Fig. 6.11 *Alignment of rod- or chain-like molecules within resist layers by shear forces during spin coating.*

An important point for nanotechnological applications is the stretching and orientation of chain-like molecules such as polymers in spin-coating process (Fig. 6.11). Although the layers are solidified as glassy or rubber-like, they are not fully isotropic.

5. Shadow-mask Deposition Techniques

In nanostructures, a local deposition of material from the gas phase results instead of a homogeneous layer. One simple approach to this problem is the application of an aperture at a defined distance form the substrate surface. Because the structures are created at the moment of material deposition (and without an etching step) this process represents an additive technique.

Local deposition of directed particles allows the fabrication of structures with very small dimensions. This technique applies a shadow mask which contains the shape of the final structure as a hole, which is positioned in close proximity to the substrate. A material beam with low divergence is required. Because the structure is generated below the mask holes, the technique represents a negative process (Fig. 6.12).

The structure generated on the substrate can be significantly smaller than the mask opening, as long as the distance between mask and substrate is optimal and the material beam s highly focused. Structures of down to 15 nm diameter are prepared using micro beads with a diameter of 1.6 μm as distance mounts and mask openings of about 50 nm. Slit-shaped masks yielded lines of similar dimensions.

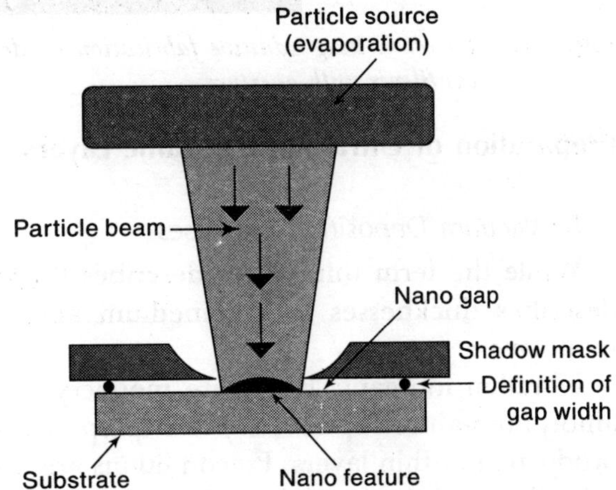

Fig. 6.12 *Additive fabrication of nanostructure by directed depositions through a shadow mask with a small distance to the substrate.*

A special shadow-mask process utilizes a small slit created in the cantilever of a scanning force microscope (Fig. 6.13, for scanning probe technique). Using the scanning tip and the laser-based *z*-axis positioning of the cantilever, the cantilever is aligned over the surface, which is scanned in the *xy*-direction. A particle beam of low divergence from an evaporation or ionization source is directed onto the cantilever and through the slit onto the substrate surface. The beam diameter has to be less than the lateral cantilever dimensions to shield he beam from the cantilever, except for the small part which is shining through the mask opening. The size of the projected structure depends on the aperture of the opening in the cantilever. An additional switchable aperture ("beam blanking") allows any pattern to be created on the substrate. This methods is ideal for the preparation of some small structure, but is problematic for large areas and longer writing times due to the time requirements and the deposition of material on the cantilever, which means that the cantilever properties are changed.

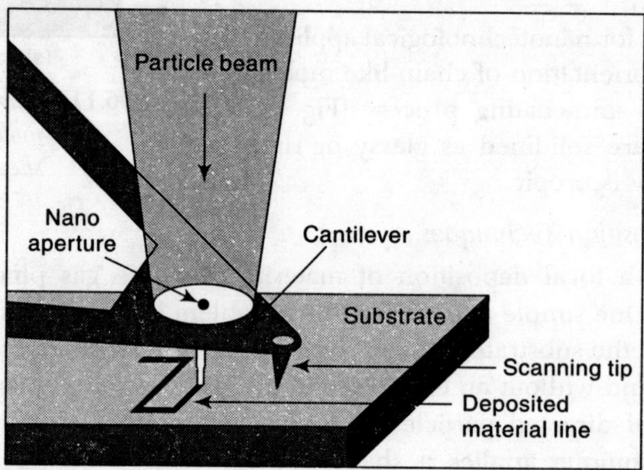

Fig. 6.13 ***Direct-writing additive fabrication by deposition based on shadow masks using a scanning probe cantilever with aperture.***

Preparation of Ultrahigh Inorganic Layers

1. By Vacuum Deposition Processes

While the term thin layers describes thicknesses of about 1 μm, the term ultrathin layers describes thicknesses in the medium and lower nanometer range. It refers to molecular monolayers.

Ultrathin inorganic layers are monocrystalline, amorphous-isotropic or polycrystalline and, amorphous-anisotropic. They are prepared by standard vacuum process known from the production of thin layers. Preconditions are homogeneous, smooth and clean substrate surfaces as well as high homogeneity in the deposition process and in film formation.

Epitaxial deposition techniques allow for the preparation of a highly defined single atomic layer on a monocrystalline substrate. Ultrathin layers consisting of several atomic layers normally exhibit atomic steps. These layers consisting of a few atomic layers, are prepared by sputter processes. This method yields no ordered monolayers, but the precision of the average layer thickness is in the size range and the fabrication of fairly homogeneous, closed layers with thicknesses down to the sub-nanometer range is possible.

Ultrathin inorganic layers are used for magnetic multilayer systems. The magnetic decoupling of a ferromagnetic layer by ultrathin layers of non-ferromagnetic metals yields materials with a high magnetic field coefficient with respect to electrical conductivity, with an application as magnetoresistive sensors (GMR sensors).

Another preparation technique for ultrathin inorganic layers is the CVD process. The precursors for a CVD process are provided in the gas phase. Extra thin layers are achieved by layer deposition induced by external energy sources (PE-CVD), such as HF or microwave excitation. The rather inactive ducts are transformed into highly reactive particles, e.g., by ionization or homolytic bond cleavage with the formation of reactive radicals.

A change of gas composition in between periods of energy irradiation allows the realization of a stack of different layers. Owing to the influence of the type of electronic excitation and the potential distribution near the substrate on the layer deposition, the control the energy input yields defined layer properties perpendicular to the surface. The layer elements with, different electrical isolation or mechanical stress will have thicknesses down to lower nanometer range.

2. Deposition of Ultrahigh Films for the Liquid Phase

The choice of the type of deposition for ultrathin inorganic materials form the liquid phase depends strongly on the material properties. Metal layers are made by electrochemical methods, such as cathodic, or by electroless deposition. A cathodic deposition requires a conductive substrate. An even layer formation is required to realize homogeneous layers. Surface-controlled processes are preferred, due to the occurrence of autoenhancement of inhomogenieties in transport-controlled deposition processes. A high quality of the substrate is important for the creation of a homogeneous layer deposited by cathodic ablation.

But for electroless deposited metal layers, the fabrication of layers in the liquid phase represents an electrochemical process. Heer, the electrons required for the reduction of the metal ions are provided by a reducing agent and not an external current source. So only a local current flow is observed (Fig. 6.14).

To achieve the deposition of material on the substrate, spontaneous reactions between metal ions and reducing agent are avoided. The composition of the metallization solution and the deposition protocol have to ensure the metal seed formation occurs only on the surface or even only on selected elements, such as nanoparticle or individual molecules. These seeds grow to large clusters or closed thin layers.

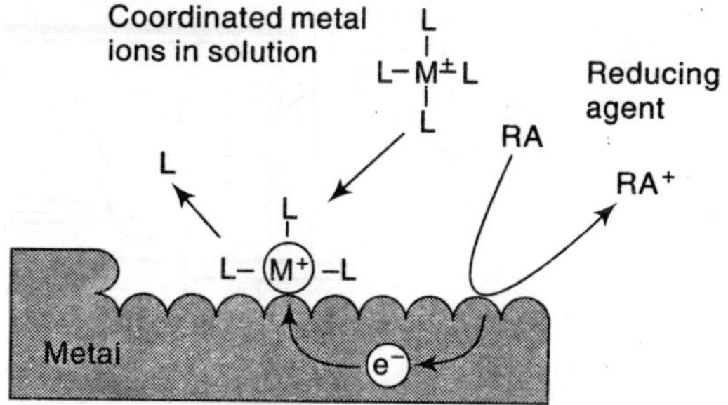

Fig. 6.14 **Principle of electrodeless deposition of metal- or semiconductor layer (M, metal; L, ligand; RA, reducing agent.**

Production of Nanolayers

1. Physical Vapor Deposition (PVD)

Physical vapor deposition (PVD) from the gas phase is subdivided into four groups, namely (i) evaporation, (ii) sputtering, (iii) ion plating, and (iv) laser ablation. The first three methods occur at low pressures. See Figure 6.15 below.

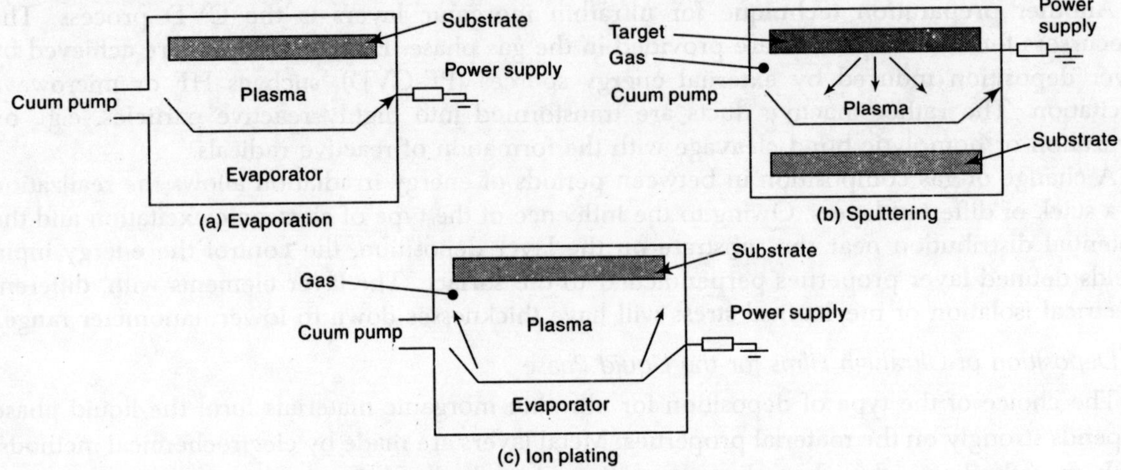

Fig. 6.15 *Three fundamental PVD methods.*

(i) *EVAPORATION.* This procedures is carried out in a bell jar as depicted in Fig. 6.16. A crucible is heated up by a resistance or an electron gun until a sufficient.

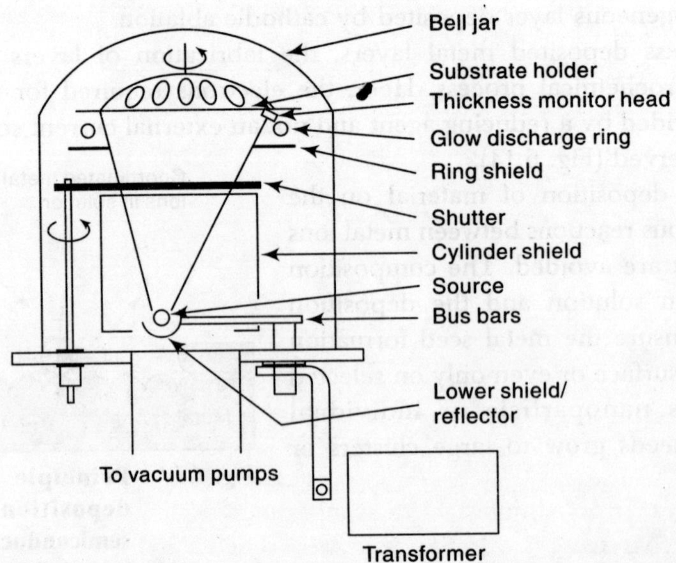

Fig. 6.16 *Vacuum system for the vaporization from resistance-heated sources. When replacing transformer and heater with an electron gun, vaporization by means of an electron beam occurs.*

Vapor pressure develops. As a result, material is deposited on the substrate. The resistance is wrapped around the cruicible, or a metal wire is heated up by a current and vaporized. The electron gun (*e-gun*) produces an electron beam of, e.g., 10 keV. The beam is directed at the material intended for the deposition on the substrate. The gun's advantage is its unlimited supply of evaporating material and applicability of non-conductive or high-melting materials. It shortcomings lie in the production of radiation defects, for instance in the underlying oxide coating. See Figure 6.17.

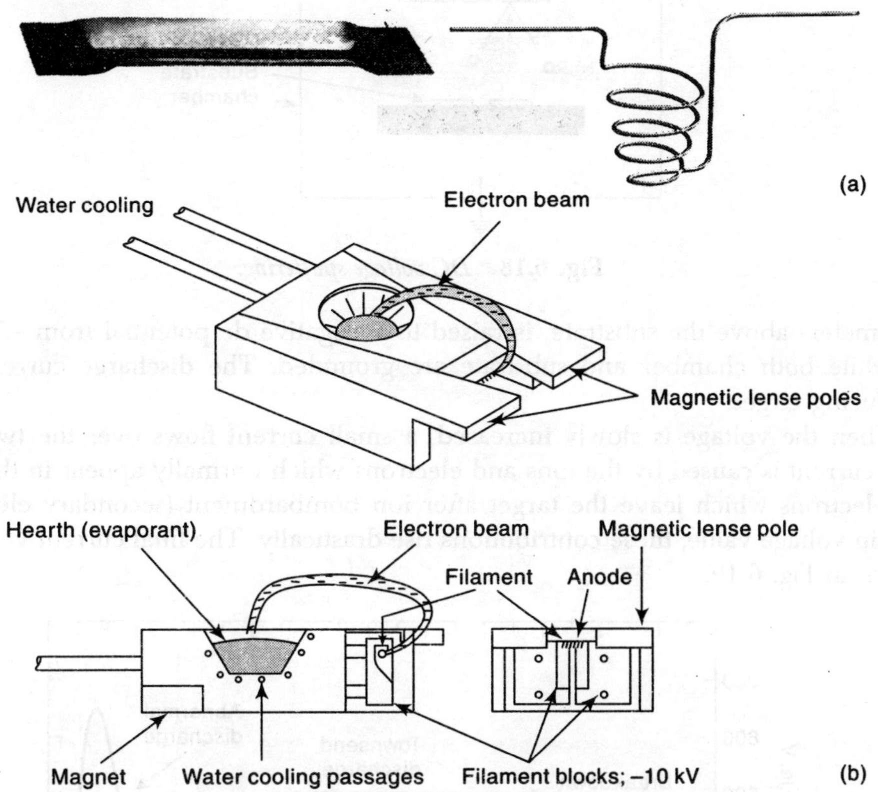

Fig. 6.17 ***Evaporation by means of (a) resistance-heating with tungsten boat and winding electron gun.***

(ii) *SPUTTERING.* When an atom or a molecule, in its ionized form, hits a solid state (***target***) and knocks out surface atoms; the erosion is accompanied by a second process, namely the deposition of the knocked out atoms on a second solid state (***substrate***). The latter process is relevant when forming thin layers and is known as sputtering. Following processes are used for sputtering:

(a) *Glow discharge.* In its simplest form, sputtering is achieved by a glŏw discharge with dc voltage. A cross section of arrangement is schematically represented in Fig. 6.18 After mounting the samples on a holder, the chamber is rinsed repeatedly with Ar. Eventually, a constant gas pressure of some 100 mPa is built up. The target, being attached a few

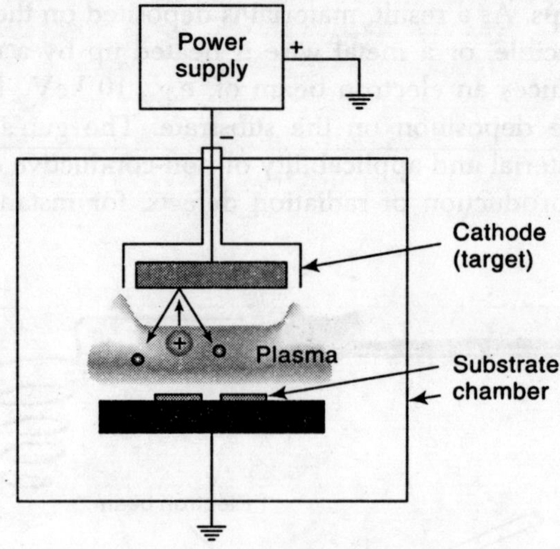

Fig. 6.18 *DC voltage sputtering.*

centimeters above the substrate, is raised to a negative dc potential from –500 to –5,000 V, while both chamber and substrate are grounded. The discharge current requires a conducing target.

When the voltage is slowly increased, a small current flows over the two electrodes. This current is caused by the ions and electrons which normally appear in the gas and by the electrons which leave the target after ion bombardment (secondary electrons). At a certain voltage value, these contributions rise drastically. The final current-voltage curve is shown in Fig. 6.19.

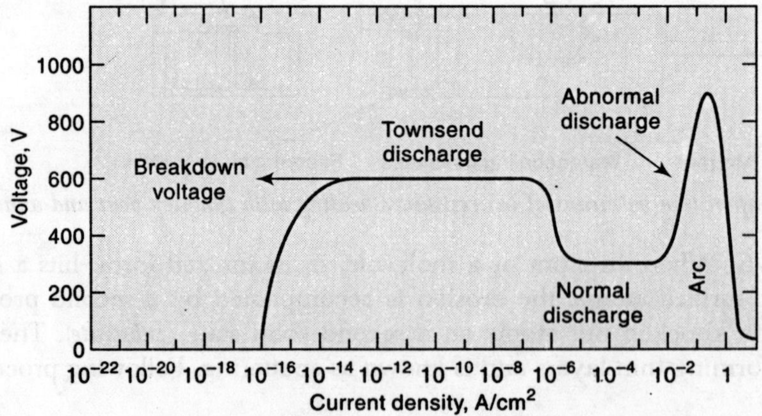

Fig. 6.19 *Applied voltage vs. discharge current.*

The first plateau at 600 V of the discharge current is referred to as Townsend discharge. Latter the plasma passes through the "normal" and "abnormal" ranges. The latter is the operating state of sputtering. A self-contained gas discharge requires the production of sufficient secondary electrons by the impact of the ions on the target surface and conversely, the production of sufficient ions in the plasma by the secondary electrons.

(b) *High frequency discharge.* When replacing the dc voltage source from Fig. 6.18 with a high frequency generator (radio frequency, RF, generator), the target and the substrate erode alternately depending on the respective polarity. But even with these low frequencies, a serious shortcoming becomes apparent: due to the substantially small target surface (compared to backplate electrode consisting of the bell, the cable shield, etc.) a proportionally large ion current flows if this backplate electrode is negatively polarized. This would mean that the substrates are covered with the material of the bell, which is not intended.

For, this shortcoming, a capacitor is connected in series between the high frequency generator and the target, and/or the conducting target is replaced with an insulating one fig. 6.20. During the positive voltage phase of the RF signal, the electrons from the discharge space are attracted to the target. They impact on the target and charge it; current flow to the RF generator is prevented by the capacitor. During the negative half-wave of the RF signal, the electrons cannot leave the target due to the work function of target material. Thus, the electron charge on the target remains constant.

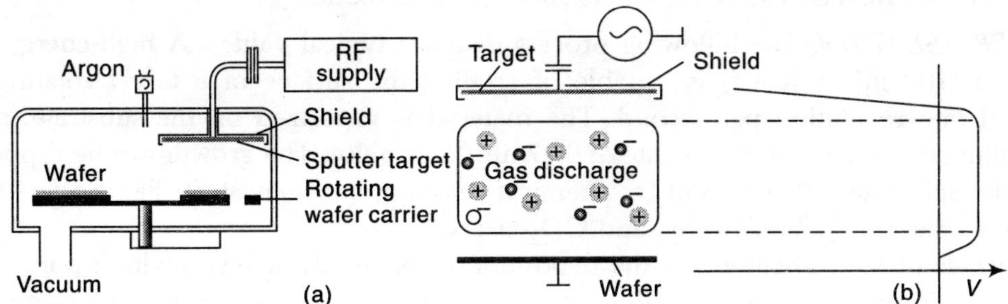

Fig. 6.20 *(a) RF sputter system and (b) distribution of the potential in and RF plasma.*

Due to their mass, the positively charged ions are not capable of following the RF signals with frequencies above 50 kHZ. Therefore, the ions are only subjected to the average electrical field which is caused by the electron charge accumulated on the target. Depending on the RF power at the target, the captures charge leads to a bias 1,000 V or more and causes on ion energy within the range of 1 keV.

When using a capacitively coupled target, the limitations of the glow discharge can be overcome, i.e., a conductive target is no longer required. Therefore, the number of layers which can be deposited by sputtering is greatly increased.

(iii) *ION PLATING.* This process is classed between resistance evaporations and glow discharge fig. 6.21. A negative voltage is applied to the substrate, while the anode is connected with the source of the metal vaporization. The chamber is subsequently filled with Ar with a pressure of a few Pa, and the plasma is ignited. After cleaning the wafer by sputtering, the e-gun is switched on and the material is vaporized. The growing of the layer on the substrate is improved by the plasma in some properties such as adhesion and homogeneity compared to as sole PVD.

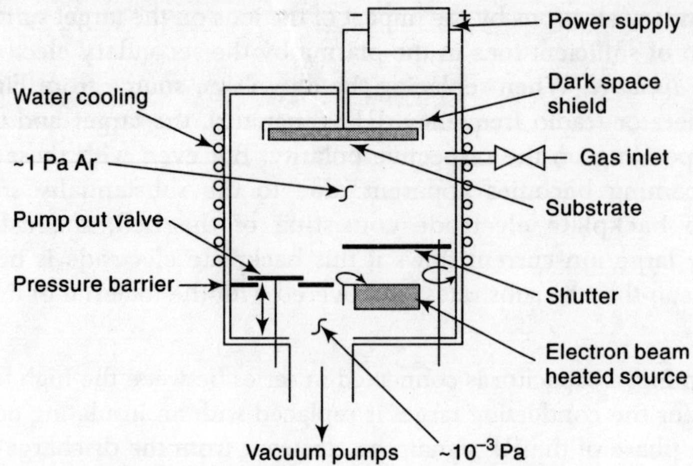

Fig. 6.21 *Ion plating system slightly modified.*

The advantages of ion plating are higher energies of the vaporized atoms and therefore better adhesion of the produced films. The disadvantage is heating of the substrate and plasma interactions with radiation-sensitive layers such as MOS oxides.

(iv) *LASER ABLATION.* The following process data are typical values. A high-energy focused laser beam (100 mJ, 1 Jcm^{-2}) is capable of eroding the surface of a target rotating with a velocity of one revolution per second. The material is vaporized on the substrate, and as a result, a film is produced on it at a rate of 0.07 nm/laser pulse. The growth can be supported by heating the substrate (1023 k) and by chemical reactions (oxygen at 50 Pa). So far, the used lasers are excimer, Nd: YAG, ruby, and CO_2 lasers.

Advantages of laser ablation are the deposition of materials of high-melting points, a good control over impurities, the possibility of the vaporization in oxidizing environments, and stoichiometric vaporization. A shortcoming is the formation of droplets on the vaporized layer. See fig. 6.22.

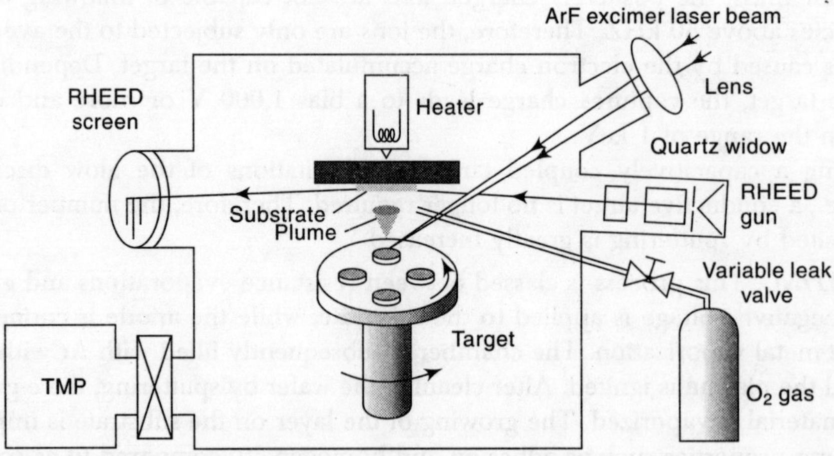

Fig. 6.22 *Laser ablation system under O_2 pressure.*

2. Chemical Vapor Deposition (CVD)

CVD, is a layer structure without the continuation of the underlying lattice. The reverse case is called vapor phase epitaxy. The CVD process is performed in an evacuated chamber. The wafer is put on a carrier and heated to a temperature between 623 k and 1073 k. Four possible versions of the chamber are presented in Fig. 6.23.

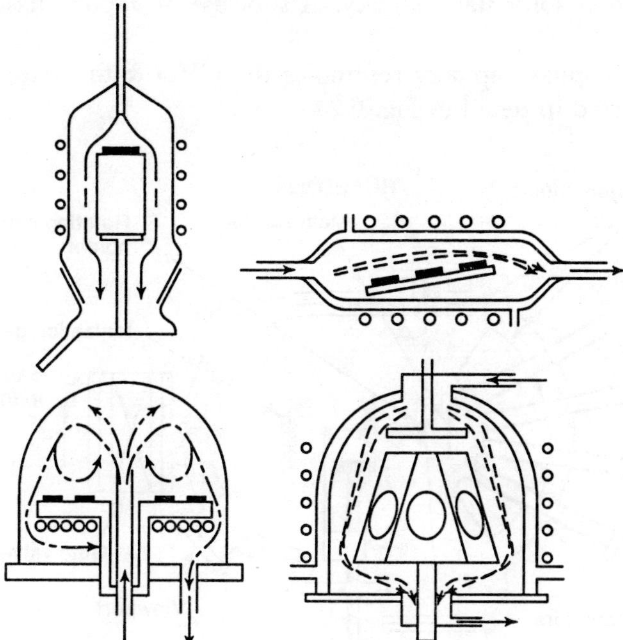

Fig. 6.23 *Four versions f CVD chamber.*

One or several species of gases are let in so that a gas pressure if formed between very low and normal pressure. The gas flow hits the wafer at a normal or a glancing incidence. A dissociation (in the case of a single gas species) or a reaction between two species takes place. In both cases, a newly formed molecule adheres to the wafer surface and participates in the formation of a new layer. Consider silane (SiH_4), it disintegrate into elementary silicon, which partly adheres to the surface, and to hydrog n, which is removed by the pumps. In the second case SiH_4, reacts with N_2O to form SiO_2. The process is accompanied by other types of gases which act as impurities in the deposited layer. Examples are phosphine (PH_3) or diborane (B_2H_6) which also disintegrate and deliver effective phosphorous or boron doping of the deposited silicon.

CVD deposition is also supported by an RF plasma. The major difference in CVD is the addition of Ar for the addition of the plasma and of H_2. The degree of the SiH_4 content in H_2 determines whether amorphous of microcrystalline silicon is deposited. In the first step, both types are deposited. However, a high concentration of H_2 etches the amorphous portions, and only in microcrystalline component remains. The etching process is favored if higher frequencies (e.g., 110 MHz).

3. Epitaxy

Epitaxy is a layer deposited ion a (crystalline) substrate in such a way that the layer is monocrystalline. The layer is often referred to as film. In many cases, the film takes 99.9% of the entire state; example (Czochraski crystal). If films and substrate are form the same material, it is homopitaxy (e.g., silicon-on-silicon), otherwise heteroepitaxy (e.g., silicon-on-sapphire). An other distinction is made by the phase from which the film is made: vapor phase epitaxy, liquid phase epitaxy (LPE), and solid state epitaxy. A subclass of vapor phase epitaxy is molecular beam epitaxy (MBE).

The setup of a vapor phase epitaxy resembles the CVD & the setup of a molecular beam epitaxy (MBE) is depicted in detail in Fig. 6.24.

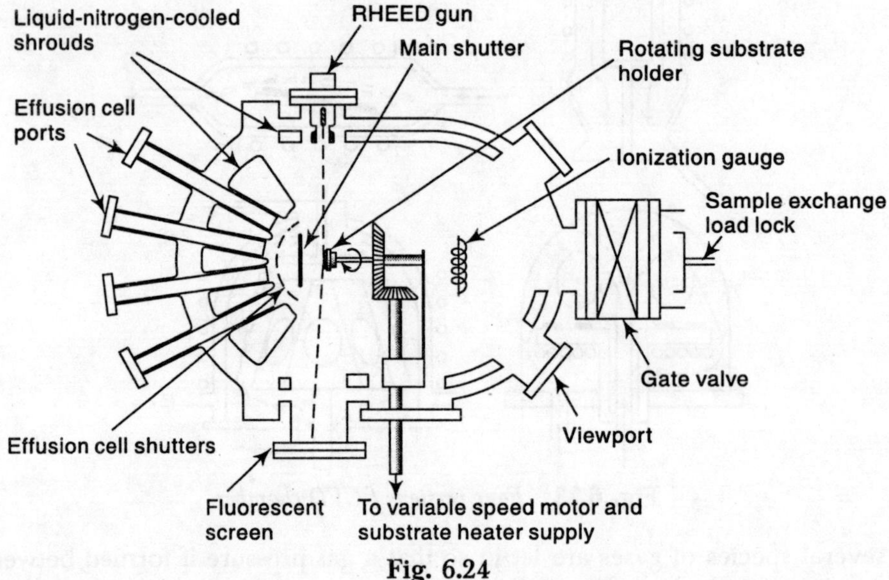

Fig. 6.24

The constituents of the deposited film are contained in mini furnaces as elements, the so-called Knudsen cells, which are discussed below. During heating some vapor pressure develops and an atom beam is emitted, which is bundled by successive apertures. The beam hits the wafer surface to which the atoms remain partially adhered. There, they can react with atoms of a second or third beam, which if also directed toward the wafer surface. A favorable reaction and finally the film deposition depend on the selection of the parameters, i.e., wafer temperature, the ratios of the beam densities, the purity of the surface, etc. As shown in same figure, the effusion cell can be replaced by an evaporation with the electron gun. The chamber contains many devices for inspection of the growing layers, for example low energy electron diffraction (LEED), secondary ion mass spectroscopy (SIMS), and Auger and Raman spectroscopy. The quality of the vacuum is controlled by a residual gas analyzer. The effusion (Knudsen) cell is seen in detail in Fig. 6.25.

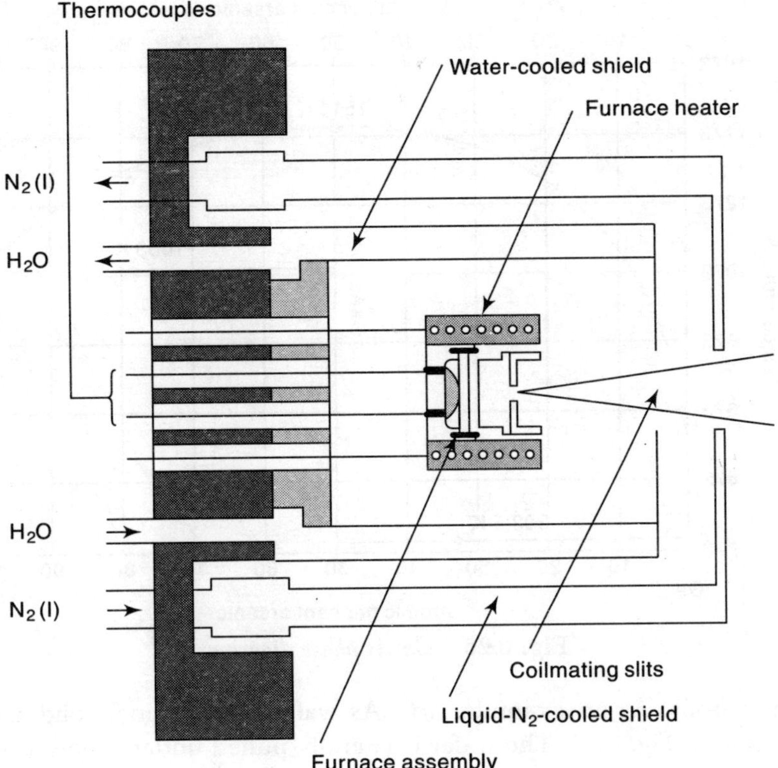

Fig. 6.25 *Effusion cell (schematic).*

The material to be deposited is contained in the innermost cell which is heated up. Its temperature is controlled by a set of thermocouples and resistance heaters. The high temperature leads to molecular desorption from all warned up surfaces, toe the emission of impurities into the substrate, and in the worst case, to the breakdown of the vacuum. Therefore, a screen cooled with liquid air is installed around the internal cell. Conversely, in order to avoid high thermal flows between furnace and screen, a water-cooled shield is inserted between them.

For the *operation of the LPE,* knowledge of the phase diagram is required. Consider the phase diagram of Ga and As as an example (Fig. 6.26). This phase diagram represent the case of congruent phase transition. The mixture of Ga and As in a ratio of 50:50 takes over the role of a constituent, i.e., the phase diagram disintegrates into a Ga-As and second GaAs-As phase diagram. In other words: GaAs forms a stable compound which is not subjected to any chemical change during the change of temperature.

Now consider, a melt of Ga_xAs_{1-x} and a solid state of GaAs, both in close contact. In order to avoid sublimation of the As, begin with a melt enriched with Ga so that left side of the phase diagrams above the liquids is operated. Pure GaAs freezes out on the solid GaAs by decreasing the temperature and the melt becomes richer in Ga until finally only pure Ga remains. In the reverse case (operation on the right of GaAs), solid GaAs is again deposited but the melt is rich in As. Doping atoms can be added to the melt before the deposition, for instance, when

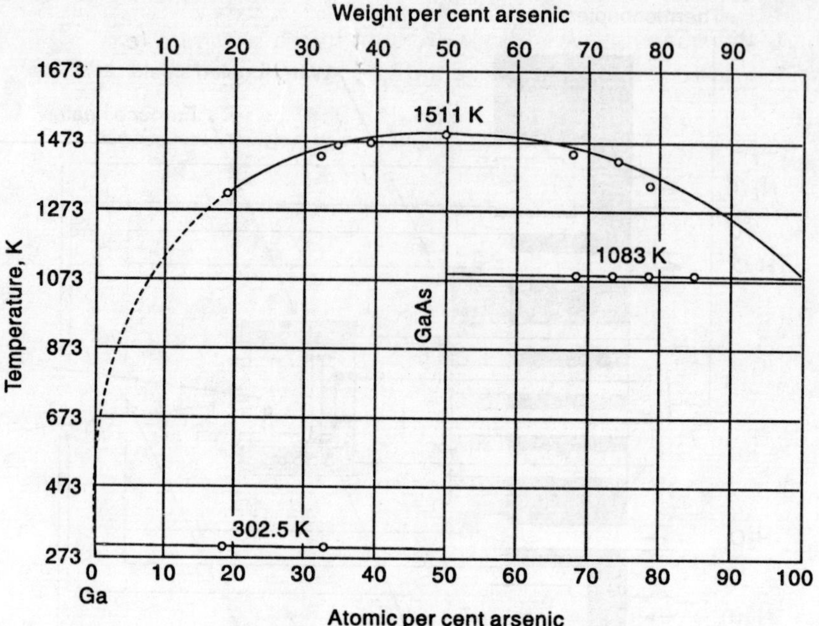

Fig. 6.26 *Ga-As phase diagram.*

fabricating *p-n* junctions. As an example, a GaAs wafer (the original solid state in the above example) is depicted in Fig. 6.27. The wafer is laterally pulled under different melts, so that the above mentioned deposition process can be repeated with different material layers (Ga, GaAlAs) and different doping types. This process is used for fabricating an injection laser.

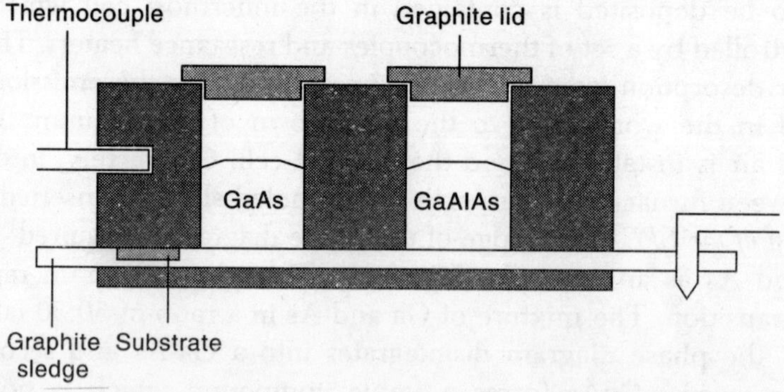

Fig. 6.27 *Sledge arrangement for the fabrication of injection lasers using LPE.*

Solid state epitaxy is a phenomenon which appears, for example as a recrystallization of amorphous silicon. Ion implantation destroys the crystalline structure of silicon upto complete amorphization. If the water is annealed, the structure can be reorganized beginning from the unimpaired bulk of the wafer. This phenomenon is of less interest for nanotechnology.

It should be mentioned that there are some versions of the classical epitaxial procedures as *in-situ* implantation or laser-enhanced deposition.

4. Ion Implantation

It is a doping technique with which ions are shot into a substrate (e.g., a silicon wafer) using an accelerator. The basic principle is presented in Fig. 6.28.

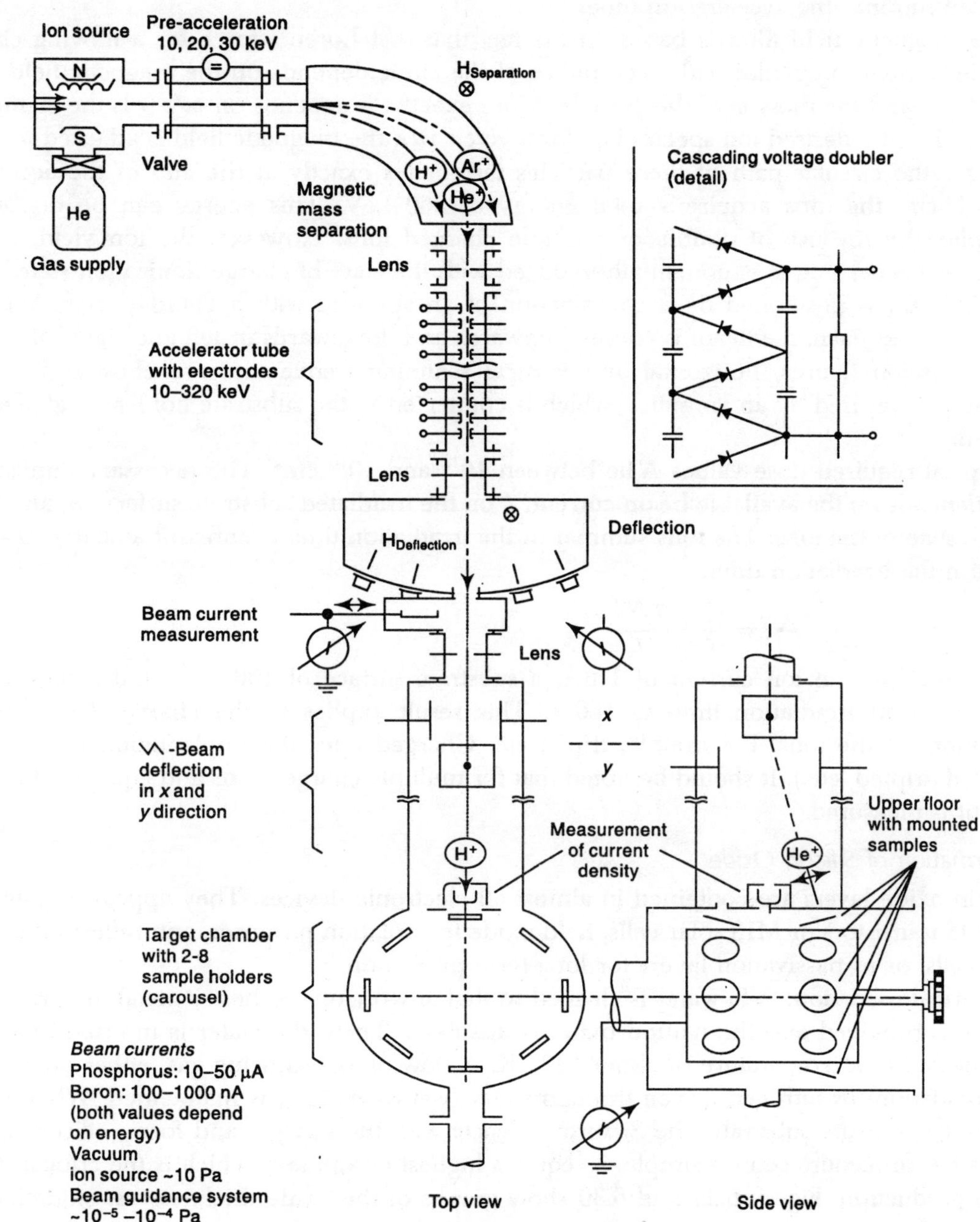

Fig. 6.28 *Ion implantation equipment.*

The desired ion species is let in as a gaseous compound through a needle. The compound is dissociated and ionized with an electron beam. The arising ions are pulled out of the source area and brought to an energy of 30 keV by a preliminary acceleration. Then, the ions are passed through a magnetic field filter which is adjusted in such a way that only the desired ion type can run into the acceleration tube.

The magnetic field filter is based on the fact that that Lorentz force for a moving charged particle compels a circular path. The radius of the circle depends on the magnetic field B, the velocity, v, and the mass m of the particle. More exactly, it depends on m/e (e is the elementary charge). For the desired ion species i.e., for a given m/e, the magnetic field is adjusted in such a way that the circular path of these particles terminates exactly at the end of the accelerator tube. There, the ions acquire a total energy of 360 keV. This energy can be doubled or multiplied by the use of double or multiple charged ions. However, the ion yield, i.e., the available ion current, is exponentially reduced with the state of charge (ionization state).

This beam is positioned by a combination of an aperture with a Faraday cage. With two capacitor disks each, the beam is scanned upwards and downwards or left and right. In order to avoid Lissajou figures, horizontal and vertical scanning frequencies are chosen. The beam current is measured by an ammeter, which is connected to the substrate holder isolated against ground.

Typical required dose values N lie between 10^{12} and 10^{16} cm^2. The necessary implantation time depends on the available beam current, I, on the irradiated substrate surface, A, and on the charge state of the ions. The ions summer in the irradiation time t represent a charge $Q = qNA$. Thus, in the irradiation time:

$$t = \frac{Q}{I} = \frac{q\,NA}{I}$$

As an example, an ion current of 1 μA, a substrate surface of 100 cm^2, and a dose of 10^{13} cm deliver an irradiation time to 160 s. This result applies to the charge state 1 (simple ionization) of the ions. For double, triple, etc. Charged ions, the implantation time must be doubled (tripled, etc.). It should be noted that for multiple charged ions and equivalent multiple current is measured.

5. Formation of Silicon Oxide

Thin oxide layers are contained in almost all electronic devices. They appear as gate oxide in MOS transistors or MIS solar cells, field oxide for isolation purposes, anti-reflection layers in solar cells, or as passivation layers for long-term protection.

Thermal oxidation: A Si wafer is cleaned so that any organic or heavy metal impurity on the surface is removed and the natural oxide is dissolved. Then, the wafer is inserted in a quartz tube heated to a temperature of about 1373 K. A flow of an oxidizing gas, either pure oxygen (dry oxidation) or nitrogen driven through water (wet oxidation), is maintained. When oxygen penetrates into the substrate, the Si surface reacts with the oxygen and forms silicon dioxide. While has procedure sounds simple, it requires highest cleanliness, which is the critical step for MOS production. Figure 6.29 and 6.30 show graphs of the oxide thickness vs. oxidation time for several temperatures.

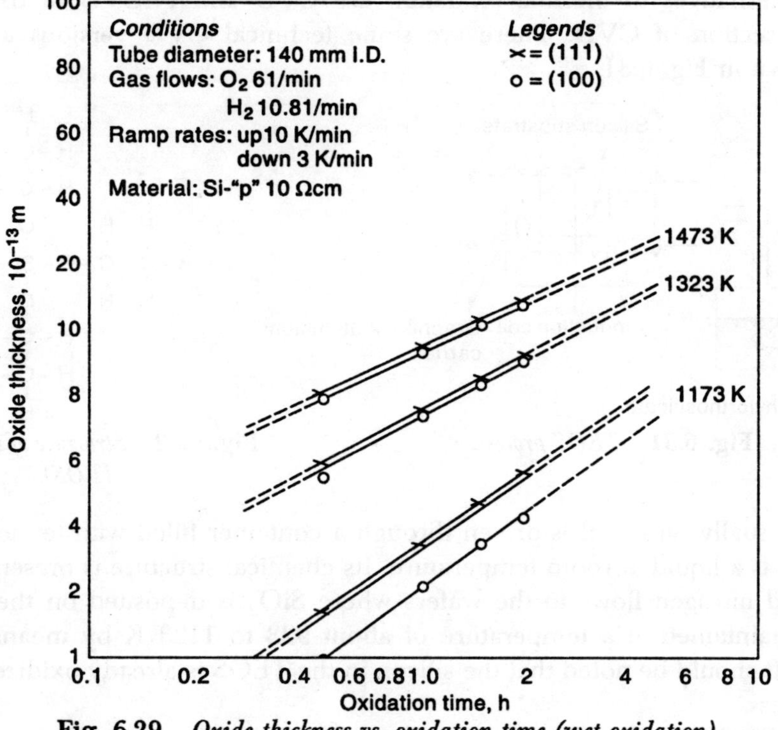

Fig. 6.29 *Oxide thickness vs. oxidation time (wet oxidation).*

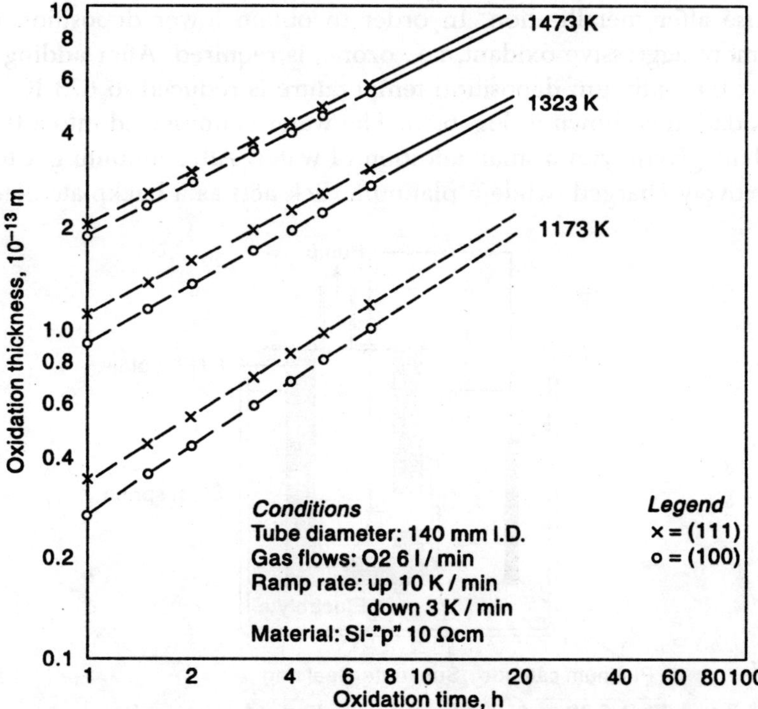

Fig. 6.30 *Oxide thickness vs. Oxidation time (dry oxidation).*

Technical alternatives to thermal oxidation are CVD and PECVD of oxides. These are treated in the section of CVD. There are some technical CVD versions a such as TEOS deposition shown in Fig. 6.31.

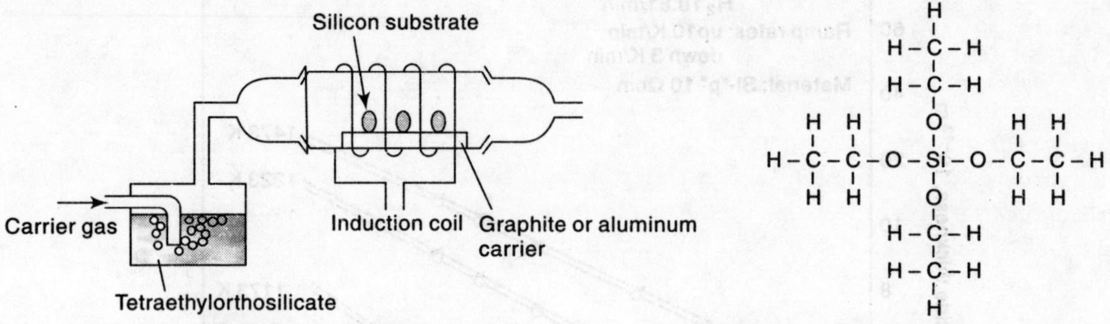

Fig. 6.31 *TEOS process.*

Fig. 6.32 *Structure of tetrathylorthosilicate (TEOS).*

A feed gas (usually nitrogen) is driven through a container filled with tetraethylorthosilicate (TEOS). TEOS is a liquid at room temperature. Its chemical structure is presented in Fig. 6.32.

The enriched nitrogen flows to the wafers where SiO_2 is deposited on their surfaces. The deposition is maintained at a temperature of about 923 to 1123 K by means of an external induction coil. It should be noted that the silicon in the TEOS is already oxidized, in contrast to the silane process.

In this state, TEOS finds only limited application because the deposition temperature (> 923 K) prevents its use after metallization. In order to obtain lower deposition temperature, the application of a more aggressive oxidant, i.e., ozone, is required. After adding some few molar per cent of ozone, the optimum deposition temperature is reduced to 673 K.

The anodic oxidation is shown in Fig. 6.33. The wafer is immersed into a 0.04 M solution of $KMNO_3$ in ethylene glycol with a small addition of water. After mounting it to a holder with a vacuum, it is positively charged, while a platinum dick acts as a backplate electrode.

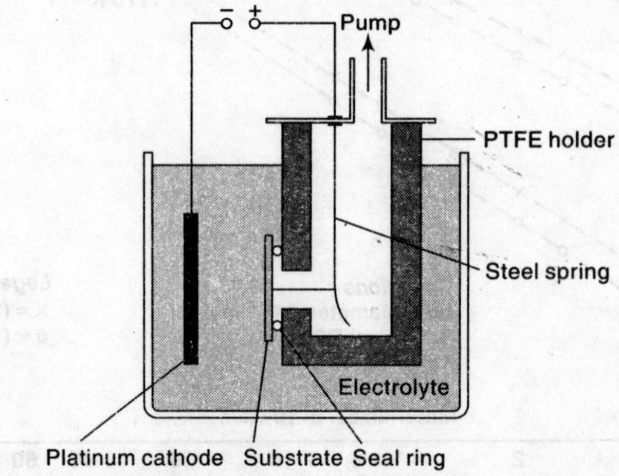

Fig. 6.33 *Assembly of the anodic oxidation of silicon.*

The current causes a reaction on the surface of the silicon:

$$Si + 2H_2O + 4H^+ \rightarrow SiO_2 + 4H^+$$

$$CH_2OH - CH_2OH \rightarrow 2HCHO + 2H^+ + 2e^-$$

$$H_2O \rightarrow \frac{1}{2}O_2 + 2H^+ + 2e^-$$

Technically, anodic oxidation is of little importance. The quality of the oxide is too low, and the process is time-consuming.

Rarely, sol gels are used to manufacture oxide layers. During this process, a suspension of oxide particles in an organic solvent is distributed over the water. A centrifuge facilitates homogenous distribution of the liquid on the wafer. Then, the wafer is baked and the solvent evaporates. The required temperatures range from 773 to 1073 K. The oxide can have the quality of a gate oxide.

Application of Nanolayers

Among the three nanostructures; nanodefects, nanolayers and nanoparticles, the nanolayers have found the most widespread applications. They are mostly used for electronic or protective purposes:

- MOS gate oxide
- Field oxide
- SOI oxide
- Amorphous layers for heterojunction solar cells, TFTs, optical sensors
- MOS channels
- Counter-doped Si layers for p-n junctions, transistors
- Recrystallized layers in dielectrics for device production
- Oxide as implantation and diffusion masks
- Oxide for the photolithography
- Silicides or metals for connections
- Epitaxial layers for transistor, laser quantum detectors
- ITO for anti-reflection and charge collection in solar cells
- Back side surface field (BSF) layers in solar cells
- Metal layers for glasses, lenses, beam splitters, interferometers
- Anti-corrosion and passive layers

This enumeration is by no means complete, and each item can be subdivided into numerous applications. For instance, counter-doped layers produced by ion implantation are used for MOS source and drain, CMOS wells, isolation in integrated circuits, buried channel CCDs, layers for the suppression of blooming in the CCD, emitter for solar cells, etc. The number of applications in corrosion and metallization is vast as well. Epitaxial layers enable the production of high frequency, opto-electronmic, and quantum layer devices.

Following are some examples which are briefly described for future research.

(i) The scaling and reduction of all geometrical sizes of electronic devices has also lead to application of thinner gate oxide. It has to be considered, though, that a substantial aspect of gate oxide based MOS technique is based on it high isolation ability. However, when thickness falls short of 4 nm, tunneling currents removing the blocking capability come into action. The only remedy so far consists of accepting these tunneling currents and simply refreshing the signals. So far it is an unsettled question whether more stable dielectrics can be developed.

(ii) The operability of image-delivering CCD devices on the basis of isolation oxides. In this case the signal (the accumulated charge) can flow over the oxides as tunneling current. Thus, apparently the necessary refreshment times are increased but the signal is nonlinearly falsified.

(iii) When dealing with thinner oxides, the above field eventually increases. Effects such as oxide charge trapping and associated bias point shifts can occur then. Current blocking mechanisms such as Coulomb blockbade, telegraph noise become apparent among other things.

(iv) Contradictory demands are made concerning the desired size of the dielectric constants of insulators. On the one hand, capacitance values become increasingly small by reducing dimension. Thus, the desire for a high dielectric constant exists. On the other hand, electrical connections are packed more closely together so that the danger of cross talk increases. The capacitive cross-talk is almost proportional to the dielectric constant. Therefore, insulators with small dielectric constant should be used.

(v) In the production of solar cells, antireflection coatings are used as outer layers. A good optical transparency and a small ohmic resistance are required. At present a transparency of 92% and a resistivity of 2×10^{-4} Ωcm are usual.

CHARACTERIZATION OF NANOLAYERS

1. Thickness, Surface Roughness

Following are some methods to determine nanometer thin layer thickness.

(a) Admittance Bridges

A metal oxide semiconductor (MOS) diode is produced by the oxidation of a Si wafer with, for instance, an oxide of 10 nm thickness and metal evaporation on the oxide (this structure being the middle part of the so-called MOS transistor). From the specifications of the evaporation (or the lithography), area *A* of the metal point is known. The measurement of the oxide thickness is performed wit the help of an admittance (or capacitance) bridge (Fig. 6.34). Additionally, the substrate is operated in accumulation by applying a bias voltage (attraction of majority carrier to the silicon boundary surface). The modulating frequency of the bridge should be 200 Hz ion order to suppress the resistance of the silicon bulk.

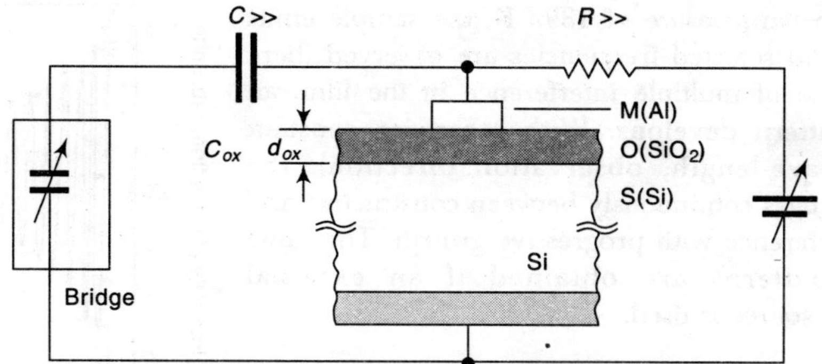

Fig. 6.34 *Measurement of the oxide layer with an admittance bridge.*

The measured capacitance is caused by the oxide alone. Therefore, it is called $C_{ox}^{\cdot}$. Now the equation for the parallel-plate capacitor can be applied to give the oxide thickness:

$$d_{ox} = \varepsilon_{ox} \frac{A}{C_{ox}}$$

where A is the area of the metal point, and $\varepsilon_{ox} = 3.4 \times 10^{-13}$ F cm^{-1}.

(b) Michelson Interferometer Measurements

Assume that the film can be etched locally in such a way that (i) sharp edge develops, and (ii) the exact thickness of the film to be measure is removed at the etched position. If SiO_2 is etched by HF with high selectivity against Si the etching stops when the Si is reached a thin layer of slightly-doped silicon on a carrier of heavily doped silicon. The structure, is made reflective by metal evaporation and inserted into a Michelson interferometer as shown in Fig. 6.35.

The parallel sets of bright and dark interference fringes appear to the viewer. They are shifted against each other by ΔN. The height of the step and thus the thickness of the film d is given by:

$$d = \Delta N \frac{\lambda}{2}.$$

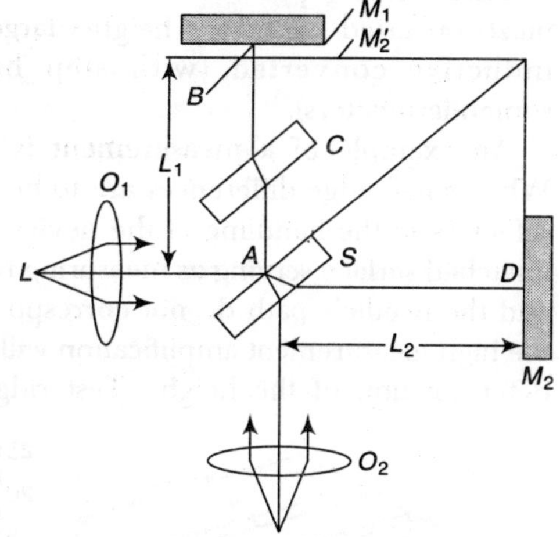

Fig. 6.35 *Michleson interferometer. L is the light source, M_1 and M_2 are mirrors, C is a disk for the compensation of the double way A–D–A by the beam splitter S. M_2' is the virtual image of M_2. O_1 and O_2 are the objective lenses.*

(c) Method for Polycrystalline Silicon

An example of measurement of the layer thickness during the deposition of polycrystalline silicon on silicon nitride is given in Fig. 6.36.

By flowing $SiCl_4/H_2$ os SiH_2/H_2 gas over a Si_3N_2 substrate, polycrystalline silicon is deposited on the substrate by the dissociation of the silane.

Due to the high temperature of 1393 K, the sample emits infrared light, and selected frequencies are observed (here 2.0 μm). Because of multiple interference in the film, an interference pattern develops. With otherwise constant parameters (wave-length, observation direction), the interference changes continuously between constructive and destructive interference with progressive growth. The same interference patterns are obtained if an external monochromatic source is used.

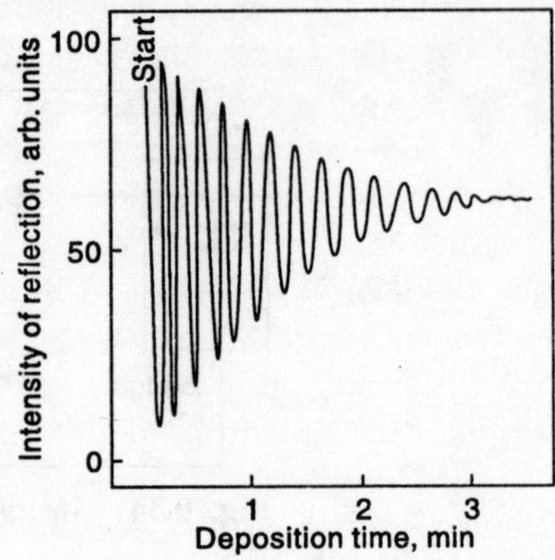

Fig. 6.36 *Interference patterns of the emitted infrared radiation by 0.55 μm polycrystalline silicon deposition on Si_3N_4 [70, slightly changed]. Reactant: SiH_2Cl_2–H_2. Deposition temperature: 1120 °C. Detector: PbS 2.0 μm filter. Evbidence: infrared radiation.*

(d) Profilometer

A thin film, whose thickness is to be measured, is grown on a substrate. The sample is partly covered with wax or photoresist so that a sharp edge appears between the covered and untreated surface. The untreated surface is etched in acid, which removes the film but does not attack the substrate (selective etching). Consequently, the edge deepens in the sample. Now, the cover (wax) is removed. The edge's depth is measured with a needle, which is moved across the edge and which is sensitive to changes in the surface. (Fig. 4.37).

Than vertical position of the needle is checked with a piezoconverted (with step heights larger than 2 nm) or an inductive converted (with step heights larger than somemicrometers).

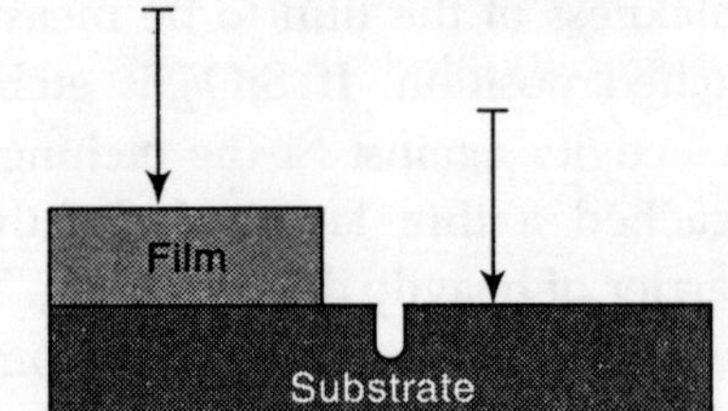

Fig. 6.37 *Thickness measurement with a needle.*

An example of a measurement is shown in Fig. 6.38. When small edge differences are to be measured, the major difficulty in the handling of the device is the leveling of the unetched surface serving as measuring reference. If this plane and the needle's path do not corresponds, the noise during the high measurement amplification will prevent a reasonable determination of the height. Test ridges are produced by

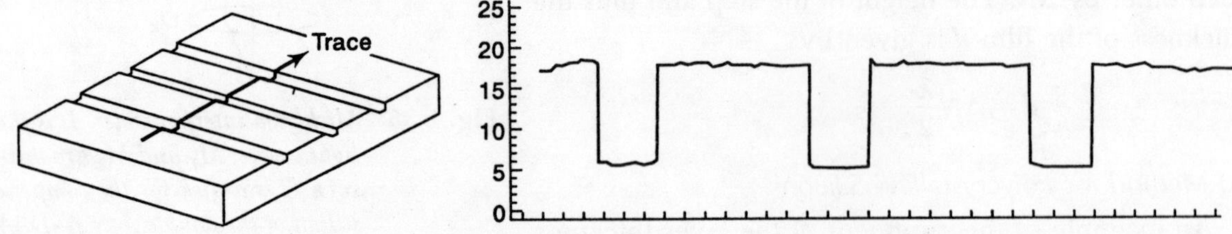

Fig. 6.38 *Measuring an edge with an inductive needle.*

removing the mask. Vertical magnification, 1,000,000 fold, one small division represents 2 nm. Horizontal magnification 200 folds, one small division represents 0.0256 mm. Thickness of the deposited layer (mean value) approximately 26 nm (25.9 mm on the diagram).

(e) Scanning Tunneling Microscopy (STM), and Atomic Force Microscopy

STM. The surface of a conducting material is covered with an electron could, whose density reduces with increasing distance form the surface. If a metal tip is within a close distance to this surface, a current flows between the tip and the surface. The current flows begins from a distance of about 1 nm, and it decreases by a factor of 10 for every reduction of 0.1 nm in the distance. This phenomenon can be used for an *x-y* presentation of the roughness. When moving the tip laterally, a constant current is maintained by following the distance of the tip. The necessary adjustment is a measure of the roughness. The fitting is done with piezoelectric actors. These piezoelements can displace the tip with a minimum increment of 10^{-7} mm/V.

AFM. Basically, the system consists of a cantilever with a tip, a deviation sensor, piezoactor, and a feedback control (Fig. 6.39). If the tip is within a small distance to the surface, an atomic force develops between the tip and the surface that the centilever is bent upwards. A regulator is laser light, which is reflected by the cantilever and which is sensitive to its position.

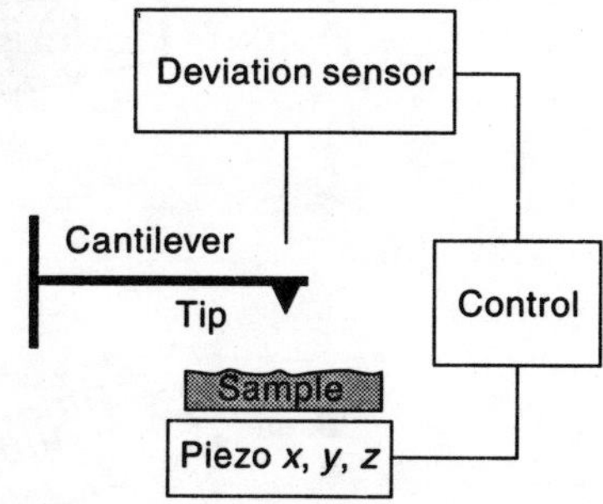

Fig. 6.39 *Atomic force microscope.*

2. Crystallinity

The crystalline state of a nanolayer is investigated with a *X-ray diffraction* (XRD), i.e. *x-ray topography*. The fundamental idea consists of aligning the crystal which to be examined in such as way that reflection is measured under a certain angle. A perfect crystal should maintain the diffraction intensity if the beam (or rather the crystal) is shifted. Every imperfection of the crystal violates the diffraction equations. Usually, the structure is created in such a way that a first reference crystal is carefully adjusted so that a sharp monochromatic beam is produced. This one is in turn directed toward the crystal, which should be measured, shifted perpendicularly to the beam. An example is given in Fig. 6.40.

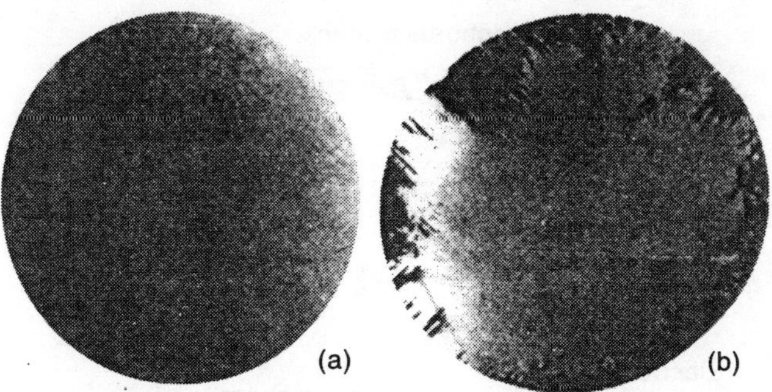

Fig. 6.40 *(a) Wafer before heat treatment and (b) after formation of dislocations by oxidation at 1473 K. examined with x-ray topography.*

Additionally, the system can be improved by "rocking" the crystal perpendicularly to the plane of incidence (Fig. 6.41). As an advantage of this rocking setup, extremely small deviations in the lattice constant can be picked up. If, heteroepitaxial films are deposited, two (a double line) instead of only one signals are sometimes found. This means that the film still different from the substrate.

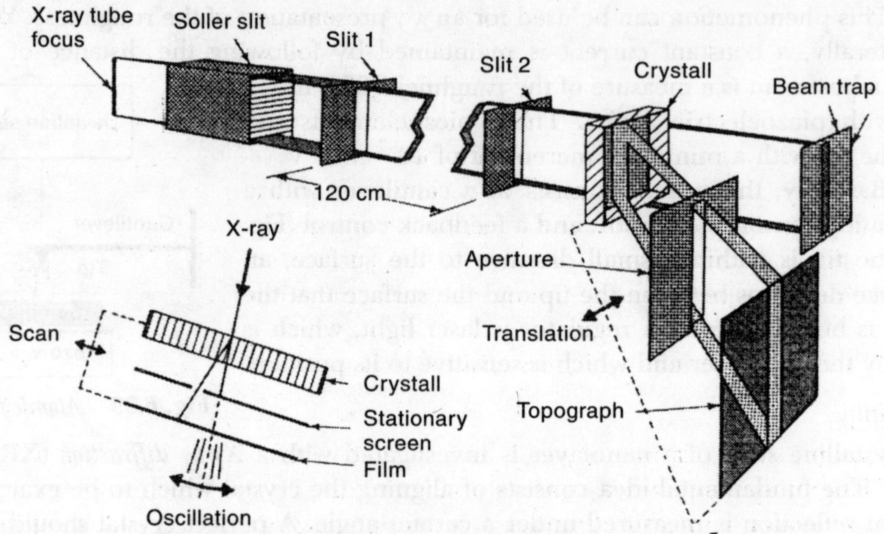

Fig. 6.41 *X-ray topography with a rocking setup.*

Electron diffraction is performed in vacuum with a monoenergy beam and stationary samples.

In a first version known as low energy electron diffraction (LEED), voltages between 10 V and 1 kV are used. The diffracted electrons are observed in reflection with a fluorescent screen (Fig. 6.42). In order to avoid wrong signals due to surface impurities, the vacuum must be maintained under 10^{-9} Pa. LEED is used in order to measure the so-called surface reconstruction, i.e, the termination of a surface by a new type of lattice ("superlattice"). A typical LEED pattern is shown in Fig. 6.43.

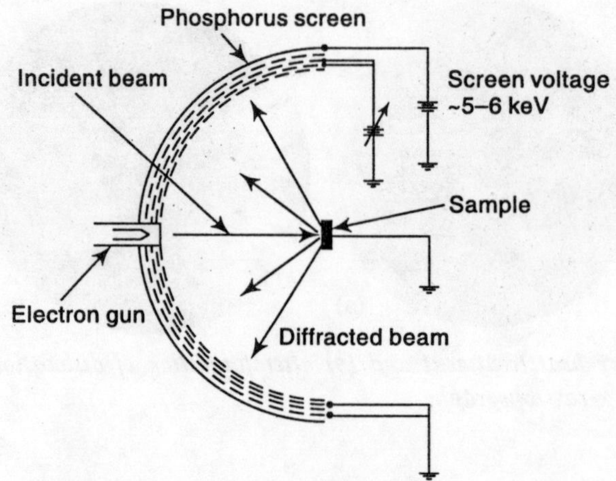

Fig. 6.42 *Electron diffraction at low energy.*

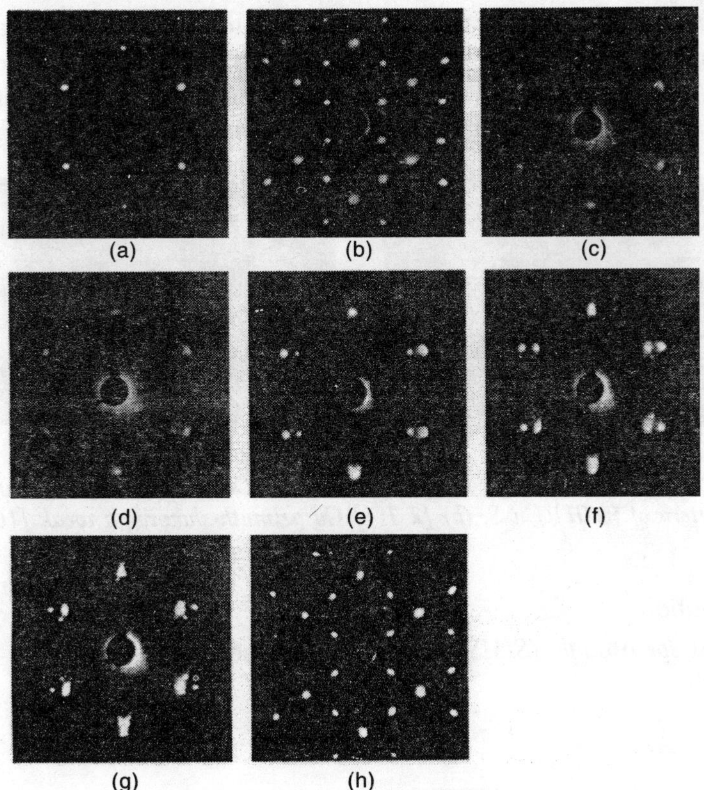

Fig. 6.43 *LEED pattern showing the effect of oxygen on the epitaxy of copper on a tungsten (110) surface. (a) Clean tungsten, (b) half a mono layer on the tungsten, (c) two mono layers of copper–note the appearance of poorly oriented diffraction spots in the outer copper layer, (d) ten layers of copper, (e) heating to 573 K for 5 min (resulting in some improvement of the orientation and in the reappearance of the tungsten beams), (f) heating to 823 K for 15 min, (g) heating to 1123 K for 1 min (tungsten oxide as evidenced by the diagonal rows of beams about the tungsten beam position, (h) heating to 1323 K for 1 min and returning to half a monolayer of oxygen on the tungsten.*

Another well-known version is the reflection high-energy electron diffraction (RHEED) as depicted in Fig. 6.44.

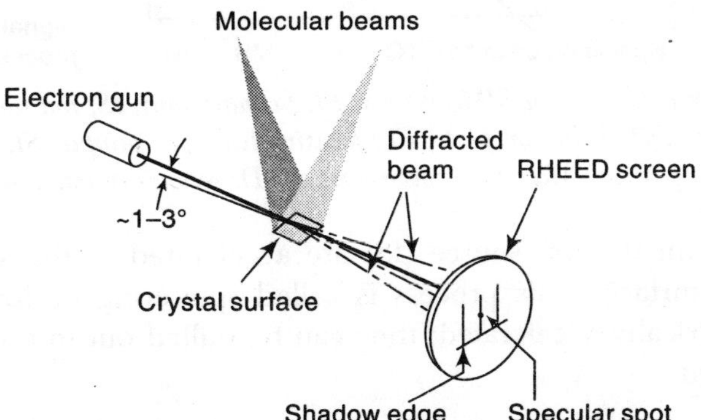

Fig. 6.44 *Reflection high-energy electron diffraction.*

Applied energies range from 10 to 100 keV. In order to avoid the penetration of these high-energy electrons into the lower-lying substrates, the operation is done under narrow glancing angles. (RHEED is often used for in-situ monitoring of the growth of an epitaxial film (See the section on epitaxy). An example of an RHEED measurement is shown in Fig. 6.45

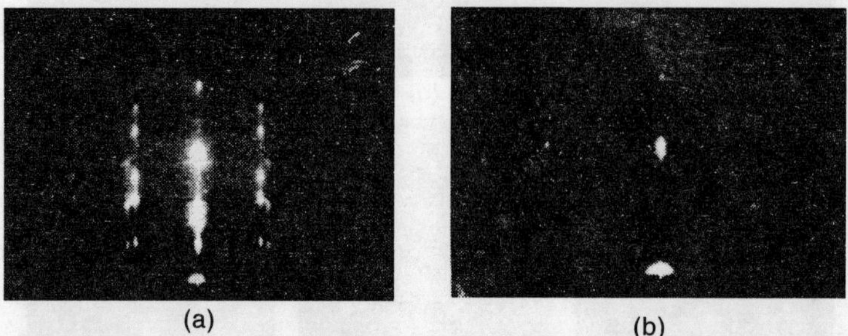

Fig. 6.45 *RHEED pattern; (00.1) Cu_2S overgrowth on (111) Cu. (a) [1 1 0] Cu azimuth showing a weak [12.0] pattern of (0.01) Cu_2S, (b) [2 1 1] Cu azimuth showing a weak [10.0] pattern of (00.1) Cu_2S.*

3. *Chemical Composition*

Secondary ion mass spectroscopy (SIMS) is usually done to detect impurities and their profiled solid state (Fig. 4.46).

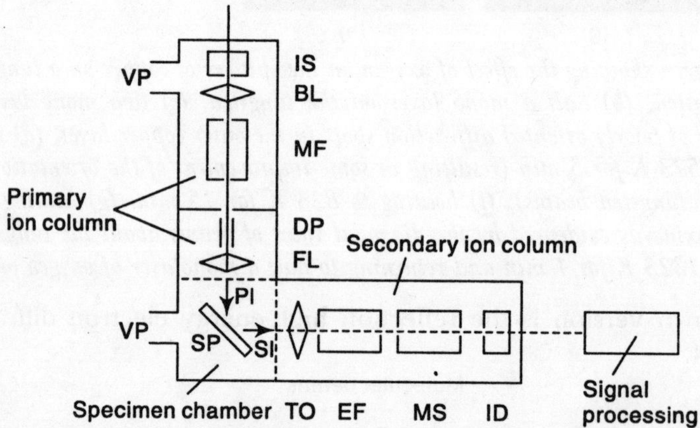

Fig. 6.46 *Schematic representation of a SIMS system: PI: primary ions, IS; ion source, BL: beam forming lens, MF: mass filter, DP: deflection plate, FL: focusing lens, SP: sample, SI: secondary ions, TO: transfer optics, EF: energy filter, MS: mass spectrometer, 1D: secondary ion detector, VP: vacuum pumps.*

Primary ions (PI) from the ion source (IS) are accelerated to the sample (SP) in order to knocks atoms off the surface. This process is called sputtering or ion milling. Some of the sputtered ions are electrically accelerated; they can be pulled out to a mass spectrometer (MS) where they are analyzed.

Essentially, SIMS can be operated in two ways: (i) The composition of the wafer (more precisely, that of the surface zone) is determined by the analysis of the secondary ions. This is done with the mass spectrometer (Fig. 6.47). (ii) If the spectrometer is set to depths at a fixed mass, the measured intensity vs. sputtering time is a measure of the impurity density vs. depth. Calibrated samples are used for the transformation of the ion signal (current from a signal processor) into an ion density. Similarly, the removal rate must be known for the transformation of the sputtering time into a depth.

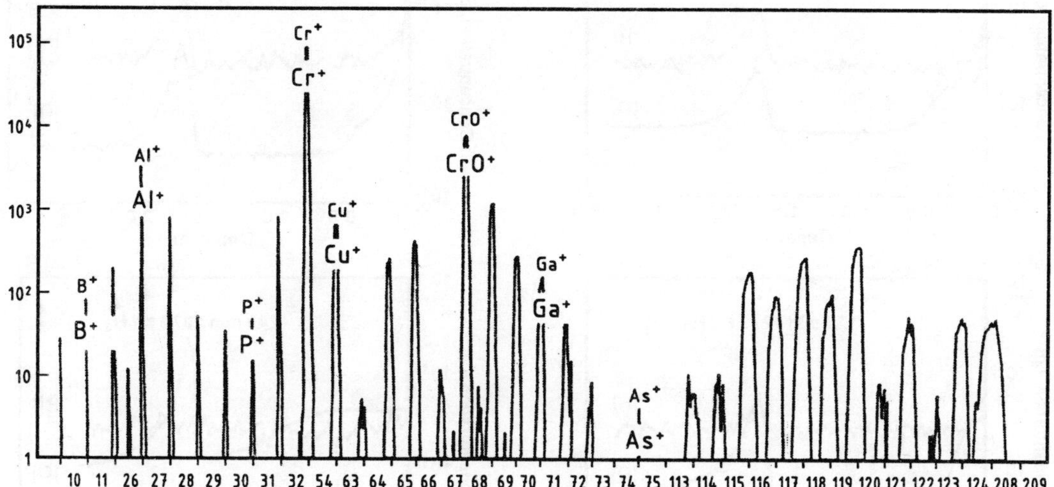

Fig. 6.47 *Mass spectrum of a Cr layer of 50 nm thickness on Cu substrate.*

The great advantage of the method is the possibility to measure all elements (even those in interstitial position) and the high detection sensitivity of between 10^{15} and 10^{18} cm^{-3}, depending on the element. To obtain a constant atomization rate, it is necessary to use a stable ion gun. The atomization can be achieved either by a finely focused beams, which can be scanned on the surface to be analyzed, or by a beam with a constant current density distribution throughout the radius. In the first case, one has additionally the possibility to record an "ion image" by synchronizing the deflecting plates with an oscilloscope.

The ionization rate strongly depends on the gad coverage on the sample. In particular, oxygen can increase the ionization rate up to a factor of 100. Therefore, sputtering is done with oxygen in some systems. If not, there is always a starting effect always, i.e., an apparently foreign atom concentration on the surface due to the unavoidable oxygen allocation.

Noble gases are used as primary ions. The selection of the primary energy is a comprosime. On the one hand, it must be large enough for a sufficient sputtering yield. On the other hand, a too high energy simply causes ion implantation and a small sputtering yield. Typical SIMS energies range from 5 to 10 eV. The problems of the procedure are the adjustment of sputtering rates, of plane surface, and of signals uniparied by noise. Two examples of the possibilities of SIMS are shown below.

In the first example (Fig. 6.48), an oxygen-rich Cz wafer is implanted with hydrogen and annealed in hydrogen for 15 to 240 min. The depth profiles of water and oxygen, recorded simultaneously, reveals hydrogen to work as a getter center for oxygen. After 120 min the original hydrogen is almost evenly distributed, while the oxygen maintains its sharp profile.

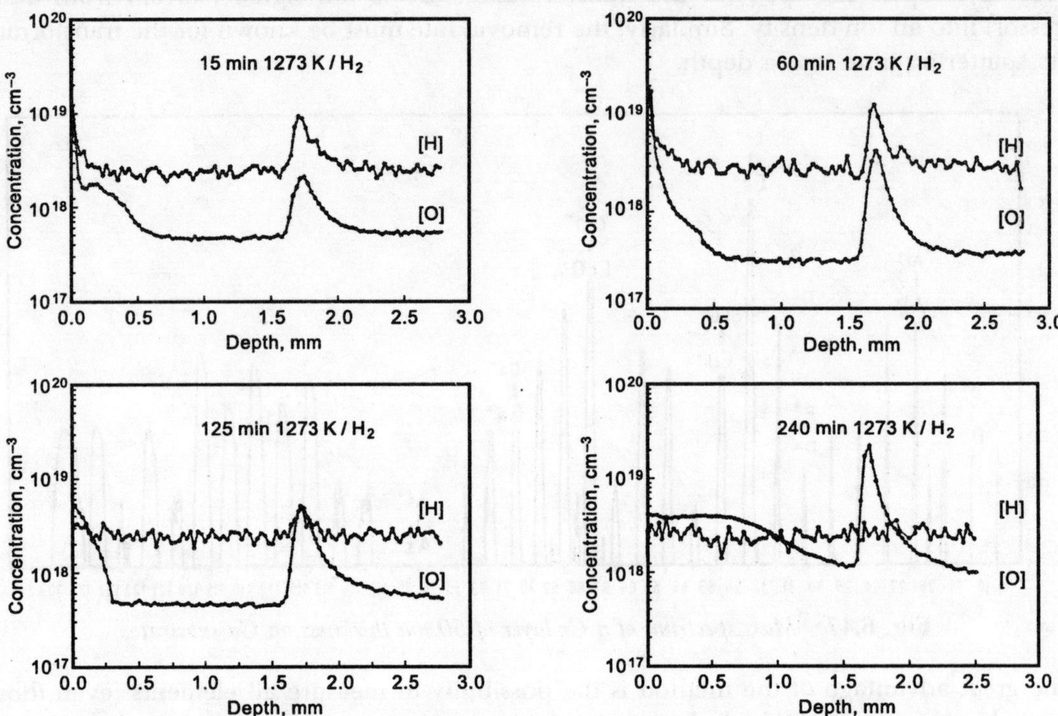

Fig. 6.48 *Hydrogen and oxygen profiles after hydrogen implantation in Cz silicon.*

The second example (Fig. 6.49) proves the feasibility of nanolayers by ion implantation. Conventional boron implantation with an energy of 500 eV and plasma-assisted diborane implantation with an energy of 350 eV are done. Activation occurs with rapid thermal annealing in order to prevent the widening of the original Guassian curve or outdiffusion.

A similar profiling procedure is based on the Auger electron emission. Again, sputtering is applied in order to drive through the depth of the layer. However, the signal is now gained by a process that is schematically shown in Fig. 6.50.

An incoming x-ray quantum of a high-energy electron removes an electron from the inner shell (step 1). The deficiency is compensated by an electron from an outer shell (step 2), as shown in the center section of the figure. In a third step, a

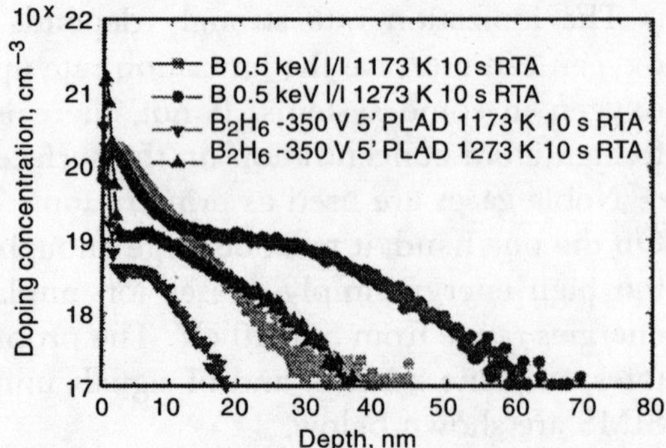

Fig. 6.49 *SIMS depth profiles of boron.*

part of the free energy is transferred to another electron of the outermost shell the gains the remainder as kinetic energy (alternatively, an x-ray quantum can be emitted, right part of Fig. 6.50). The energy in step 2 is the difference of discrete energies, the ionization energy in step 3 is a discrete energy. Therefore, the remaining kinetic energy is likewise a discrete energy. All energies are specific for the respective atoms in whose shell the process happen. Thus, the kinetic energy identities the material under investigation like a fingerprint (the same applies to the emitted x-ray quanta). A typical Auger electron spectrum is depicted in Fig. 6.51.

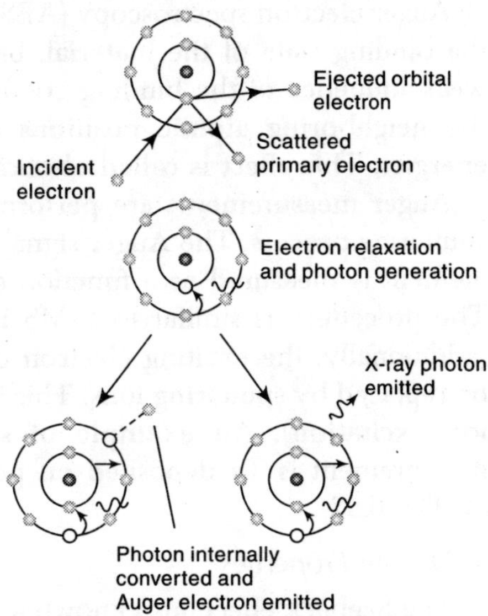

Fig. 6.50 *Auger emission process.*

Since the Auger lines are put on a rather broad background, they are detected more easily by differentiating the energy distribution $N(E)$. Therefore, the ordinate of the usual Auger spectrum is converted to the function dN/dE. Electronic differentiation can easily

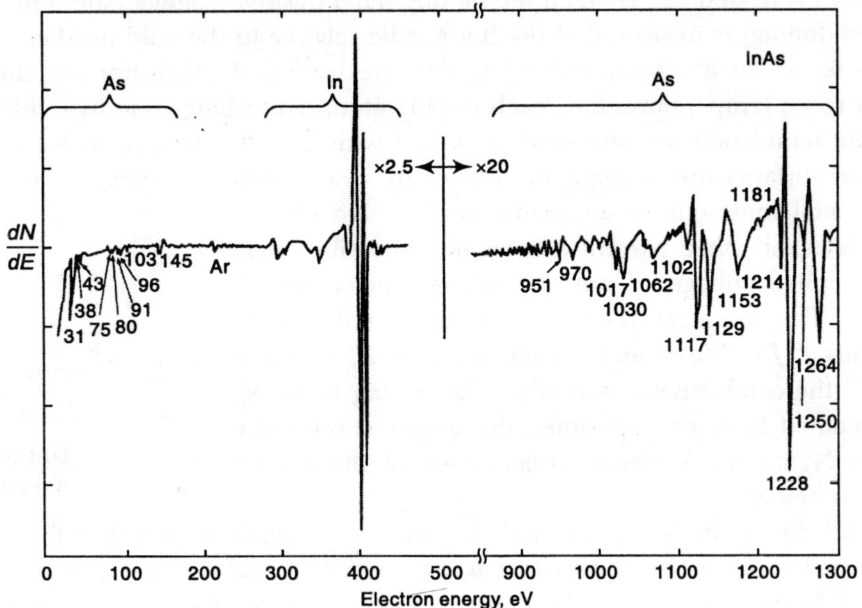

Fig. 6.51 *Auger spectrum of InAs.*

be done with velocity analyzers superimposing a small alternating voltage to the energy-selecting dc voltage and with synchronously recording the output of the electron multiplier. The line height of the Auger signal is usually proportional to the surface concentration of the element causing the Auger electrons.

Auger electron spectroscopy (AES) also verifies the binding state of the material, because of the weak influence of the binding configuration and the neighboring atomic positions on the shell energies. This effect is called *chemical shift*.

Auger measurements are performed during a sputtering process. The Auger signal of a selected element is measured as a function of the depth. The procedure is similar to SIMS measurement (additionally, the exciting electron or x-rays can be replaced by sputtering ions. This is called AES ion excitation). An example of such a deep measurement is Ta deposited on polycrystalline Si (Fig. 6.52).

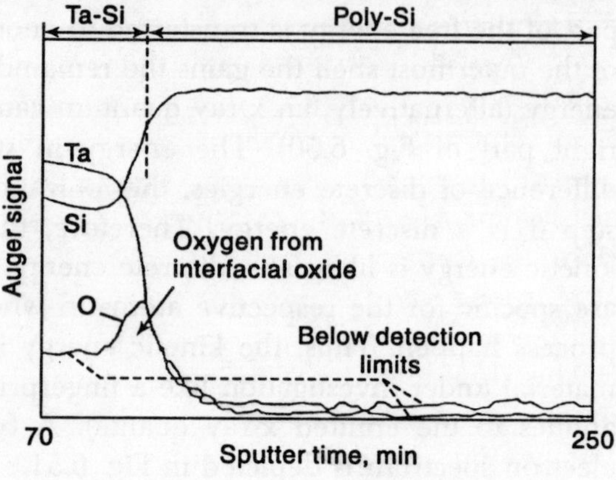

Fig. 6.52 *Ta, Si, and O profiles of a $TaSi_2$ film on Si by means of AES.*

4. Doping Properties

The Seebeck effect, also known as thermoelectric effect, can be used to determine the doping type. Two metal contacts, for example needle tips, are mounted on the semiconductor. One of the two needles is heated up. With an n-type doping, a positive voltage (and a negative voltage with a p-type doping) is measured at the hot needle relative to the cold needle.

Here, the terms *carrier density* and *doping level* are used without distinction since silicon, in saturation at room temperature, (i.e., each doping atom) contributes one free charge carrier.

Ideally, the semiconductor represents a cuboid with the edge lengths *a*, *b* and *c* (Fig. 6.53). Two opposite surfaces are brought in contact by evaporation or coating with a conductive paste. The metal must ensure an ohmic contact. No electric rectifying effects or other current-voltage non-linearities may occur. The applied voltage, *V*, and the subsequent current, *I*, are measured. The two measured values are converted into the current density J, $J = I/(a\ c)$, and the electric field E, $E = V/b$. From $J = \sigma E$, the conductivity, σ, results. The doping level, N_D or N_A, is obtained from an experimentally acquired reference table (N_D or N_A *vs.* σ). A direct measurement of the doping level is shown below.

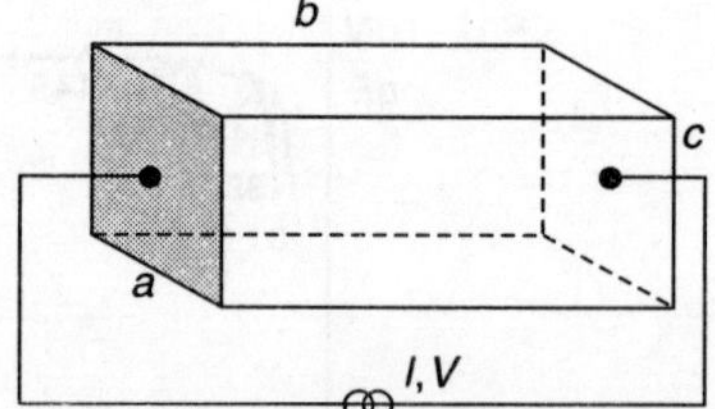

Fig. 6.53 **Determination of the conductivity.**

A semiconductor available as a round disk or a rectangular chip with a thickness of some 100 μm is used. Four-probe measurement are used (Fig. 6.54). Four parallel needles at a distance of 0.635 mm from each other are mounted on the semiconductor. A current, *I*, is fed through the outer needles resulting in a voltage drop in the semiconductor. The voltage between the inner needles, *V*, is measured with a high-ohmic voltmeter. The quotient *V*/*I* is a direct measure of the specific resistance of the semiconductor.ρ. The correction factor between *V*/*I* and ρ is determined by the potential theory measurement. The current and voltage distributions and thus the correction factor depend closely on the distance of the measuring needles and the radius of the disk. For a disk with an infinite radius, the correction factor is

4.53, so that the surface resistance $R_{\square} = 4.53\,V/I$, and the resistivity $\rho = R_{\square}d$ (with a disk thickness d).

In some cases (e.g., when measuring an imhomogeneous doping process), the advantages of the four-probe measurement, and the two-probe measurement is applied. Here, the feed current and the applied voltage are measured in only one pari of points. A correction factor between V/I and ρ is also needed.

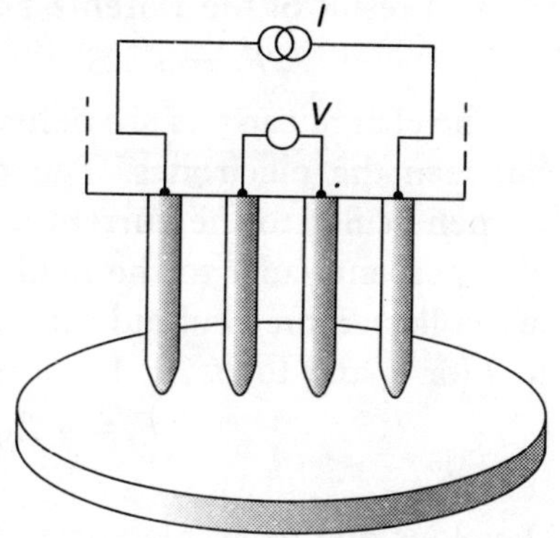

Fig. 6.54 *Four-probe measurement (not to scale). If the current feed is also put on the inner needles, a two-probe measurement is performed.*

The doping values are determined from the specific conductivity. In a simplified physical consideration, each doping atom contributes exactly one electron to the increase of the electron density, n, which leads to an increase of the conductivity because. (q is the electron charge, μ the mobility).

$$\rho = q\,\mu\,n \tag{4.35}$$

The question of which conductivity is obtained with a certain doping (or which measured conductivity corresponds to which doping). The conductivities of several sample of different doping are measured, and the curves of ρ *vs.* N_D and ρ *vs.* N_A are plotted, (for historical reasons, the specific densities for electron and hole doping. These curves are called *Irvin curves* as shown in fig. 6.55.

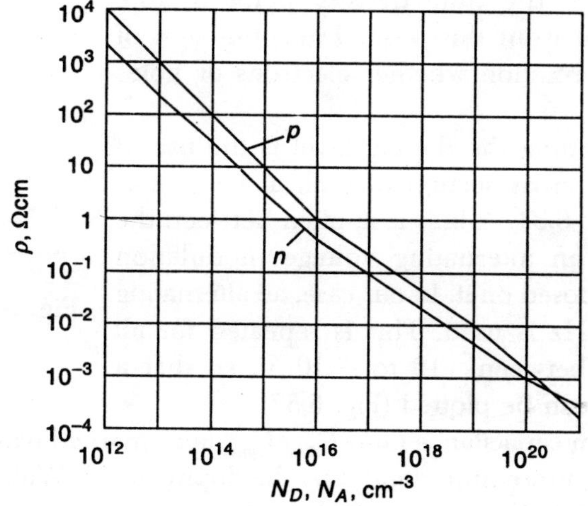

Fig. 6.55 *Specific resistance as a function of the doping.*

It should be clear that a current-voltage measurement delivers only the product of the charge carrier n (or p) and the mobility of the electrons (or holes) when considering Eq. $p = q\,\mu n$. To separate the product, a second equation (second measurement) is necessary: the Hall experiment. Assume the parallelepiped sample shown in Fig. 6.53. However, a magnetic induction B is applied to the sample in a direction perpendicular to the current flow (Fig. 4.56).

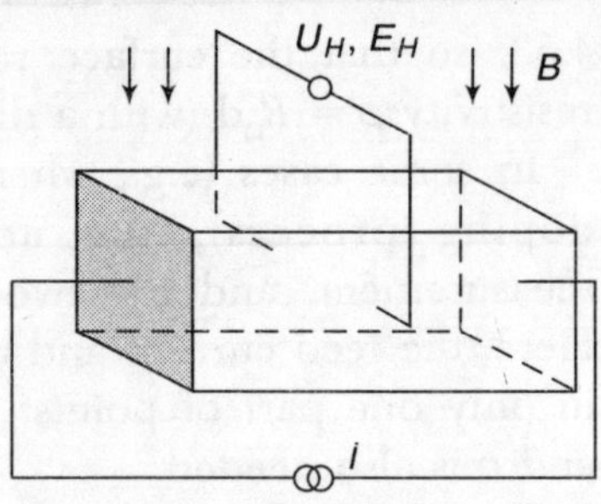

Fig. 6.56 *Experimental setup of the Hall effect (Schematic).*

As a result of the Lorentz Force

$$F_L = q\, v\, B,$$

The charge carriers are deflected from their straight-line course between the electrodes. The direction of the Lorentz force is perpendicular to the current and original path (given by v) and also perpendicular to the field direction (B). The charge carriers are collected on a cuboid surface, whose normal is perpendicular to I (or v) and to B. At the same time, a backward force.

$$F_e = q\, E_H,$$

develops due to an electric field caused by the accumulation of the electrons on the cuboid's surface (this field must not be mistaken with the field causing the current I). In the stationary case, the forces are equal so that $q\, v\, B = qE_H$, or after multiplying by the electron density n:

$$(q\, n\, v)\, B = q\, n\, E_H$$

The product $q\, n\, v$ represent the current density. The term R_H determined from the measurement is known as the Hall constant:

$$R_H = \frac{E_H}{j\, B} = \frac{1}{q\, n}.$$

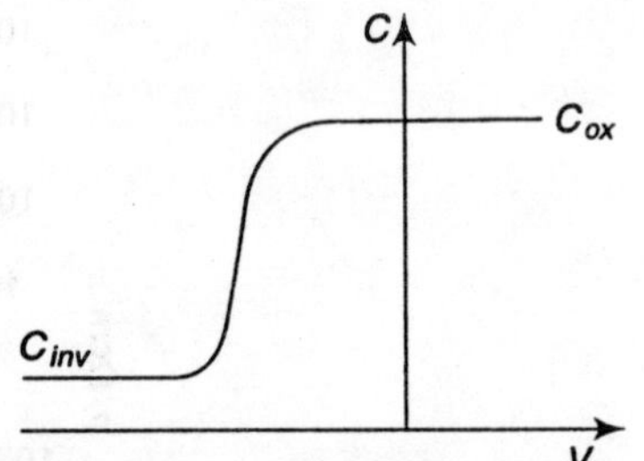

Fig. 6.57 *High frequency C-V curve in equilibrium (n-type semiconductor).*

The electron density n (the hole density p for a hole semiconductor) is obtained from this term. From the sign of the Hall constant, the information whether electrons or holes are present is obtained.

Another way of determining the doping level is the use of an MOS diode. A typical cross section of such a device has already been shown in Fig. 6.34. A bias is applied between the two metal contacts, and an alternating voltage modulation (usually 25 mV) is superimposed on it. In our case, an alternating voltage frequency of 1 MHz is used. This is repeated for all bias values, for instance, between +10 to – .10 V, so that a capacitance-voltage curve can be plotted (Fig. 6.57).

The minimum/maximum capacitance ratio C_{min}/ C_{max} (also marked with C_{inv}/ C_{ox}) is determined by the oxide thickness and, more importantly; by the doping level. With known oxide thickness, this relation can be calculated and presented in the form of a nomograph with the oxide thickness as an additional parameter (Fig. 6.58).

Doping Profile

The doping value changes with the depth in the semiconductor, (after diffusion, implantation or oxidation). Important measuring methods are:

(i) *Beveling and two-probe measurement.* In the first step, a probe tip is mounted on the surface of a semiconductor; the lower surface of the semiconductor is grounded (Fig. 6.59). The theory shows that the applied potential of the needle drops about to zero in the semiconductor within a length of approximately three spherical radii of the needle tip. Thus, the resistance between needle and back side is determined only by the resistance within three spherical radii of the needle tip. Thus, the resistance between needle and back side is determined only by the resistance within three spherical radii or the doping in this volume. But often there is a high impedance layer, a *p-n* junction, or an oxide within the semiconductor or on its backside. In this case, the hemispherical potential distribution is violated. Thus, a second needle is usually mounted on the semiconductor surface as counter electrode in a second step. Here, almost half of the applied potential drops symmetrically within three radii (please note the relation to the four-point measurement: the two-points can be compared to the outer needles of the four-point measurement. Thus, the two additional points of the four-point measurement brought in between detect only a small fraction of the potential applied on the outer points.

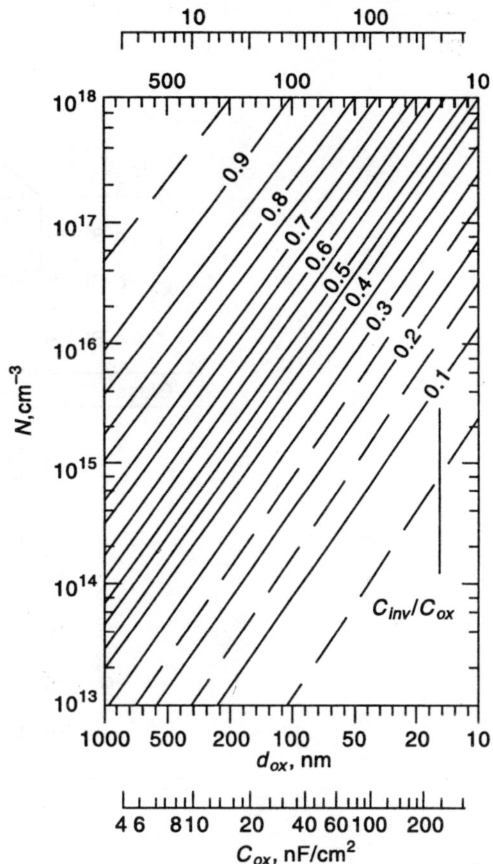

Fig. 6.58 *Namograph for the determination of doping from the ratio C_{min}/C_{max}.*

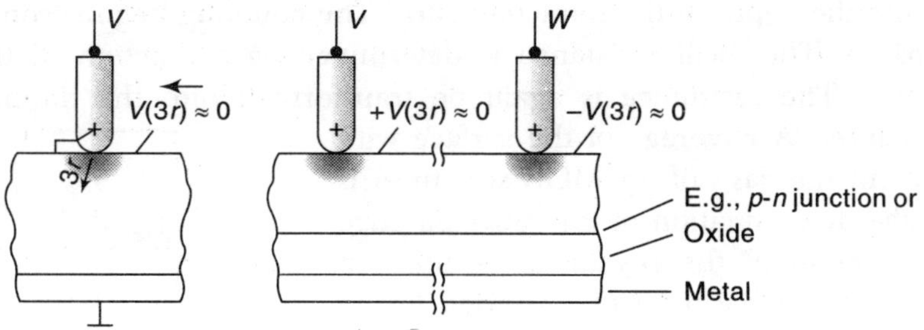

Fig. 6.59 *One and two probe measurements.*

When performing a profile measurement, that means: the sample, which has been implanted, diffused, etc., is cut into chips and beveled in a third step (Fig. 6.60). The beveling angle depends on the depth profile; typical values are between 0.5° and 10°. The two points are place on the beveled surface parallel to the beveled edge, and the local resistance is measured. Afterwards, the two needles are moved over the beveled surface and the measurement is repeated. The measurement path along the beveled surface is

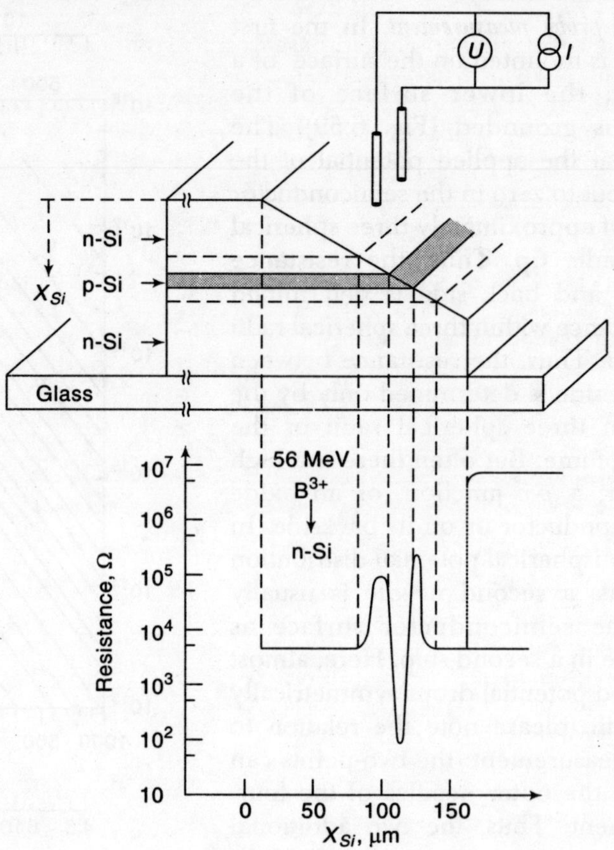

Fig. 6.60 ***Bevel of an n-type sample implanted with 56 Me V boron and the related resistivity profile. The boron has been activated in nitrogen atmosphere for 30 minutes at 1073 K.***

converted into the depth in the semiconductor. The counting begins from the edge of the beveled surface. The local resistance is determined as a function of the depth of the semiconductor. The resistance is again be transformed into the doping density using calibration curves. A coverage of the surface with an oxide (as in the case of an MOS structure) is helpful for the determination of the beveled edge since a clear jump in the resistance occurs. An example of such a measurement (a buried boron layer in n-type silicon) is presented in Fig. 6.60.

(ii) *MOS profile measurement.* The MOS capacitance is the result of a series circuit for the oxide capacitance C_{ox} and the space charge capacitance of the semiconductor C_{sc} (Fig. 6.61).

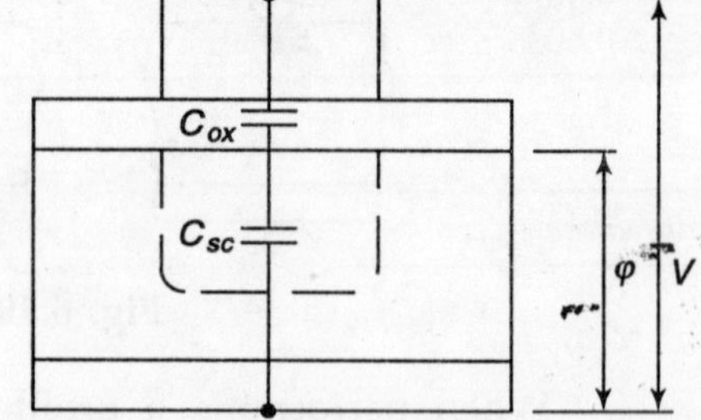

Fig. 6.61 ***High frequency MOS equivalent circuit (simplified) and voltage definitions.***

5. *Optical Properties*

Interference colors of thin layers. A first estimate of the thickness of a thin transparent layer deposited on an opaque substrate is obtained based on its color. The colour charts for SiO_2 and Si_3N_4 are given in Table 6.3 and 6.4 respectively.

Table 6.3 *Colors of SiO_2*

Film thickness, μm	*Color*	*Film thickness, μm*	*Color*
0.05	Tan	0.63	Violet-red
0.07	Brown	0.68	"Bluish"
0.10	Dark violet to red violet	0.72	Blue-green to green
0.12	Royal blue	0.77	"Yellowish"
0.15	Light blue to metallic blue	0.80	Orange
0.17	Metallic to very light yellow-green	0.82	Salmon
0.20	Light gold or yellow-slightly metallic	0.85	Dull light-red-violet
0.22	Gold with slight yellow-orange	0.86	Violet
0.25	Orange to melon colored	0.87	Blue-violet
0.27	Red-violet	0.89	Blue
0.30	Blue to violet-blue	0.92	Blue
0.31	Blue	0.95	Dull yellow-green
0.32	Blue to blue-green	0.97	Yellow to "yellowish"
0.34	Light green	0.99	Orange
0.35	Green to yellow-green	1.00	Carnation pink
0.36	Yellow-green	1.02	Violet-red
0.37	Green-yellow	1.05	Red-violet
0.39	Yellow	1.06	Violet
0.41	Light orange	1.07	Blue-violet
0.42	Carnation pink	1.10	Green
0.44	Violet-red	1.11	Yellow-green
0.46	Red-violet	1.12	Green
0.47	Violet	1.8	Violet
0.48	Blue-violet	1.19	Red-violet
0.49	Blue	1.21	Violet-red
0.50	Blue-green	1.24	Carnation pink to Salmon
0.52	Green(broad)	1.25	Orange
0.54	Yellow-green	1.28	"Yellowish"
0.56	Green-Yellow	1.32	Sky blue to green-blue
0.57	Yellow to "Yellowish" (not)	1.40	Orange
0.58	Light-orange or yellow to pink	1.45	Violet
0.60	Carnation pink	1.46	Blue-violet

Table 6.4 *Colors of Si_3N_4*

Film thickness, μm	*Color*	*Film thickness, μm*	*Color*
0.01	Very light brown	0.095	Light blue
0.077	Average brown	0.105	Very light blue
0.025	Brown	0.115	Light blue-brownish
0.034	Brown-pink	0.125	Light brown-yellow
0.035	Pink-lilac	0.135	Very light yellow
0.043	Intensive lilac	0.145	Light yellow
0.0525	Intensive dark-blue	0.155	Light to middle yellow
0.06	Dark-blue	0.165	Average yellow
0.069	Average blue	0.175	Intensive yellow.

Polymers

Polymer materials are characterized by their bulk properties including stiffness, elongation, gas permeability, impact and modulus. Because the bulk polymers are homogeneous, characterization of substrate is not important. Advances in polymer science require advanced methods for polymer micro and nanostructure characterization.

Polymer nano-composite materials are created in clusters, layers, rods; and networks of nanometer sized materials are blended with a bulk polymer. These materials, such as Nylon-6/clay nano-composites, have desirable properties such as a high modulus and low gas permeability. Ideally, nano-composite polymer materials show a dramatic increase in a physical property with a small amount of nano-particle additive.

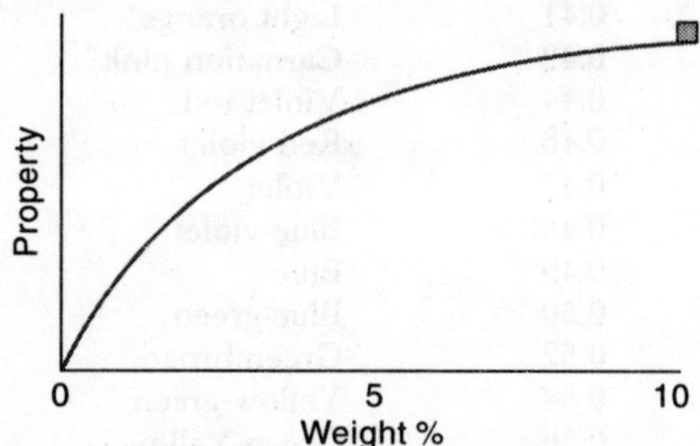

Fig. 7.1 *The properties of nano-composite materials can change dramatically with the weight percentage of a materials blended or mixed with the bulk polymer.*

A unique aspect of the new class of nano-composite polymer materials is the high interfacial volume of the materials. The chemical and physical properties of the interfaces give rise to the unique and desirable properties of nano-composite polymer materials.

Polymer materials are being developed for a host of new applications in nanotechnology, example, low *K* materials derived from polymers are replacing SiO_2 in semiconductor devices.

Polymer coatings are used in packaging applications in the food industry. Such applications require characterization of surface structure and surface properties, using optical and electron microscopes. Besides giving a 2-D projection of the sample topography, the optical microscope gives contrast associated with polarization. However, the optical microscope is limited to resolution of greater than 0.5 μm. Electron microscopes such as SEM and TEM give the high-resolution images but require sample coating to make the polymers electrically conductive. The Atomic Force Microscope is well suited for the characterization of nano-composite polymer materials when compared to traditional microscopes such as the electron and optical microscope.

IMAGING

Two imaging modes, continuous and vibrating, are used for imaging polymers. If the polymer sample is "hard", then "contact" mode is used. However, if it is "soft", then "close contact" mode is used. The forces placed on the polymer sample surface with "close contact" mode are about 10-100 times less than in "contact" mode.

Direct visualization of 2-D and 3-D images of polymer surfaces is helpful understanding the effects in changes in process conditions. Because the AFM does not require conductive samples, there is little or no sample preparation required. Below are three polymer samples prepared from the conduction of polypeptides. In figure 7.2a, 7.2b, and 7.2c, the density of peptides is 1%, 10%, and 100%.

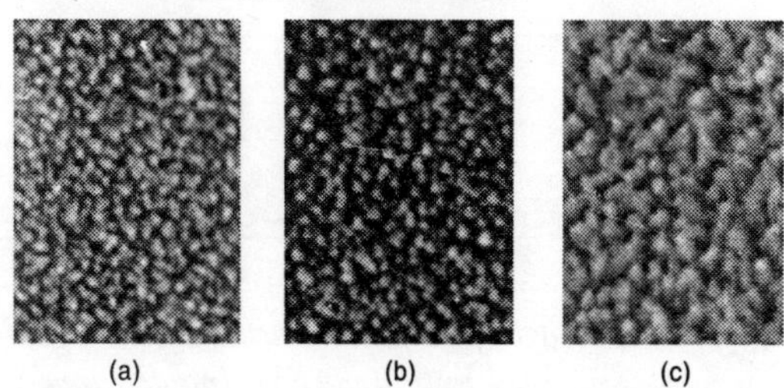

Fig. 7.2 *Atomic force microscope are ideal for visualizing the surface texture of polymer material. In comparison to a scanning electron mciroscope, nano coating is required for an AFM. Images a, b, and c are of a sot polymer material and were measured with close contact mode. Field of view: 4.85 µm × 4.85 µm.*

Many methods are available for measuring the size of particles suspended in a liquid. Detailed visualization and metrology of nano-particles is possible with traditional electron microscopy. However, SEM and TEM require sample preparation such as coating and microtoning. AFM allows the direct measurement of particle metrology. Sample preparation requires finding a substrate to which the particle naturally adhere.

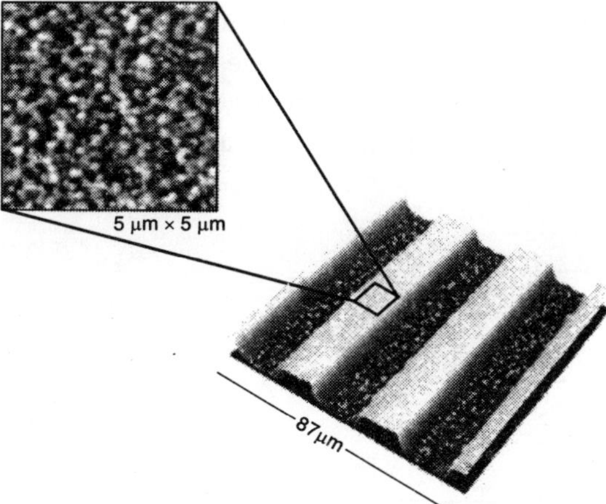

Fig. 7.3 *At bottom is an illustration of 3-Dimensional image of a patterned polymer material deposited on silicon substrate. At the top is a 2-Dimensional image of the polymer structure at the top of one of the lines.*

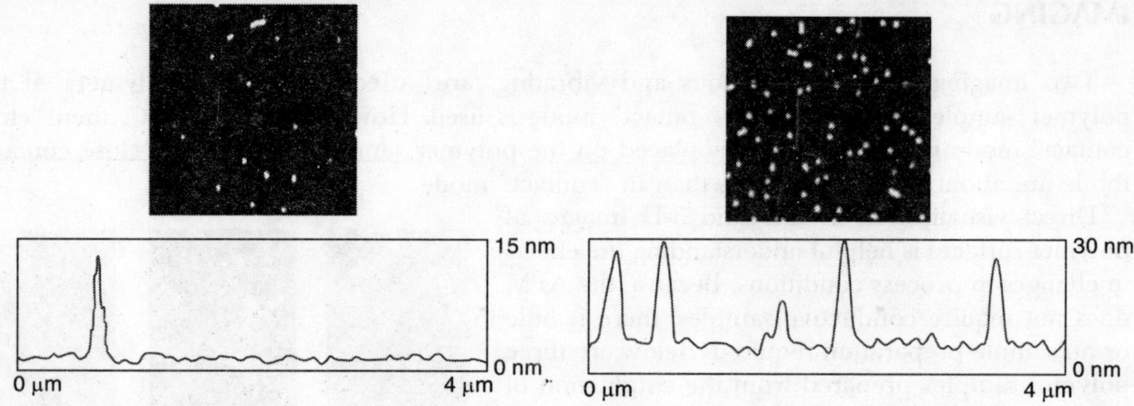

Fig. 7.4 *Image of 14 nm diameter colloidal gold sphere imaged with an AFM. Image range is 4 μm × 4 μm.*

Fig. 7.5 *Image of 28 NM colloidal gold spheres imaged with an AFM. Image ranges is 4 μm × 5 μm.*

Above fig. 7.4 adn 7.5 are images of two sizes of latex spheres measured with the AFM. The size of the particle can be directly measured with a line profile.

MATERIAL SENSING

The AFM makes it possible to measure not only the topography of the sample but also changes in the mechanical properties of the surface. The two most common types of material sensing modes are Lateral Force Microscopy (LFM) and Phase Contrast Mode (PCM).

The torsion of the cantilever is monitored with the AFM scanning head in lateral force microscopy. As the cantilever is scanned across the surface fig. 7.6(a) to (c), the amount of torsion changes depending on the types of materials at the surface. It is possible to acquire both an LCM image and a topographic image simultaneously with an AFM.

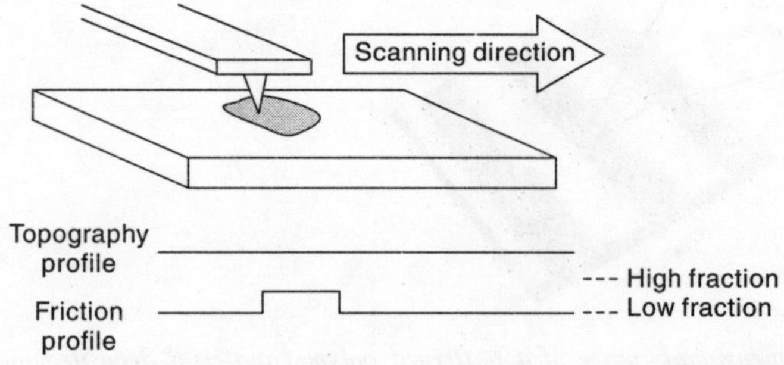

Fig. 7.6(a) *As the probe is scanned across a surface, any change in chemical composition can be detected by a change in the lateral torsion motion of the cantilever.*

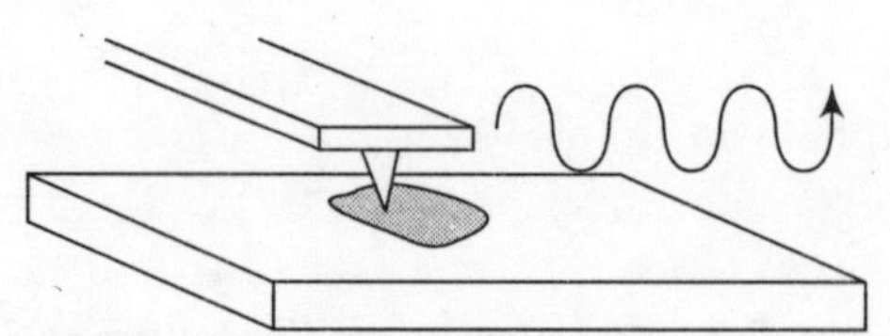

Fig. 7.6(b) *The probe is vibrate as it is scanned across the surface.*

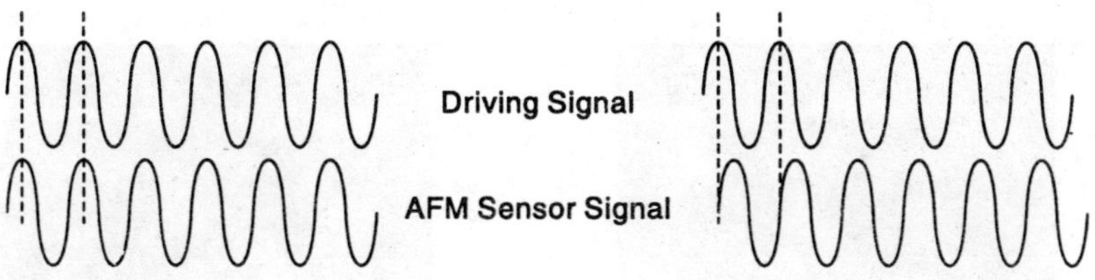

Fig. 7.6(c) *Phase contrast mode measures the changes in the phase of a vibrating cantilever as the cantilever is scanned across a surface Areas of varying "hardness" result in changes in the phase of the vibrating cantilever.*

Although this technique is not quantitative, it provides a contrast mechanism for differentiating chemical materials at a sample surface. LFM is best applied to samples fig. 7.7 with small topographic features because changes in topography can also cause changes in the cantilever torsion.

Fig. 7.7 *Topography (left) and Lateral Force image (right) of a structured polymer material. Both images were measured simultaneously. Only the topography is seen in the top image while the changes in chemical composition are readily observed in the LFM image.*

It is possible to measure changes in the mechanical properties at a polymer surface with PCM fig. 7.8. Close contact mode operates by maintaining the amplitude of a vibrating cantilever constant while scanning over a polymer surface. At the same time, monitoring changes in the phase of the cantilever gives an image that is related to the localized "hardness" of the surface. A harder section of a sample will give more (or less) phase shift of the vibrating cantilever as it scans across a sample.

POLYMER NANOFIBERS

The development of high performance polymer fibers has been achieved in fiber spinning techniques such as wet spinning, dry spinning, melt spinning, and gel spinning. One of the areas gaining significant attention is the electrospinning of nanoscale polymer fibers.

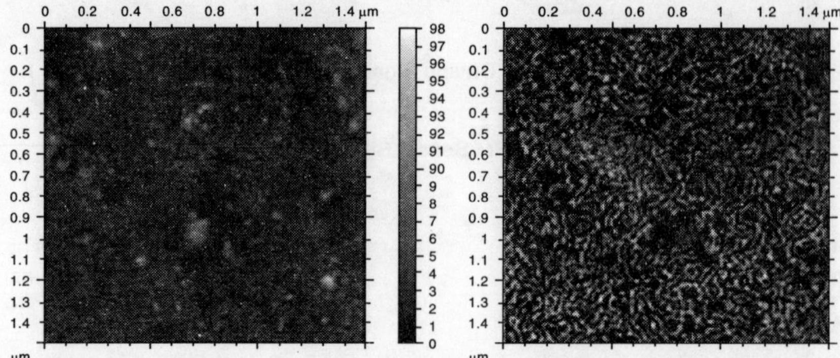

Fig. 7.8 *Topography, right, and Phase Contrasts, left, images of a co-blank polymer. In the phase contrast image the contrast is derived form changes in the mechanical properties of the surface. Features as small as 50 nm are readily resolved with phase contrast mode.*

Electrospinning technique is an inexpensive and easy way to produce nanofibers from many types of polymers, having the advantages of low basis weight, small fibers diameter, small pore size, and high surface area.

A distinct advantage of electrospinning is that the nanofibers are produced at great length because electrospinning is a continuous process limited only by the amount of solution. This compares favorably to other 1D nanostructures produced by other chemical or physical methods. These long fibers are then assembled into other three-dimensional, nonwoven mats as a result of the bending instability of the electrospinning jet.

A schematic diagram to interpret electrospinning of polymer nanofibers is shown in Figure 7.9. There are basically three components required to implements the process; a high voltage supplier, a capillary tube with a pipette or needle of small diameter, and a metal collecting screen. In the electrospinning process, high voltage is used to create an electrically charged jet of polymer solution, or melt out of the pipette. Before reaching the collecting screen, the solution jet evaporates or solidifies, and is collected as an interconnected web of small fibers. Meanwhile, one electrode is placed into the spinning solution/melt and the other

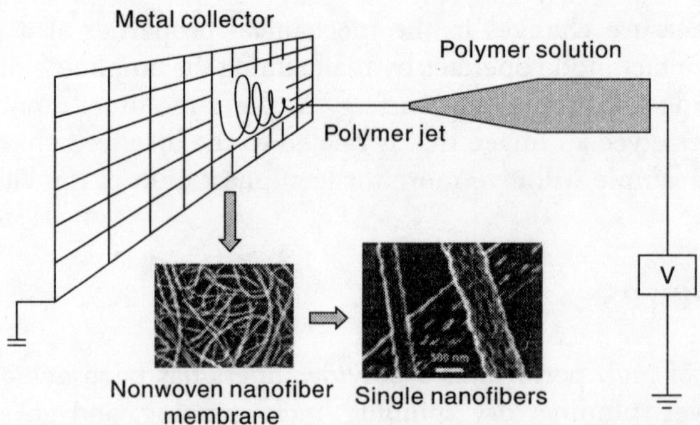

Fig. 7.9 *Schematic diagram showing polymer nanofibers by electrospinning.*

is attached to the collector. In most cases, the collector is simply grounded, indicated in Figure 7.9. The electric field is subjected to the end of the capillary tube that contains the solution fluid held by its surface tension. This induces a charge on the surface of the liquid. Mutual charge repulsion and the contraction of the surface charges to the counter electrode cause a force directly opposite to the surface tension. As the intensity of the electric field is increased, the hemispherical surface of the fluid of the capillary tube elongates to form a conical shape known as the Taylor cone. Further increasing the electric field, a critical value is attained where the repulsive electrostatic force overcomes the surface tension, and the charged jet of the fluid is ejected from the tip of the Taylor cone. The discharged polymer solution jet undergoes an instability and elongation process that allows the jet to become very long and thin. In this process, the solvent evaporates, leaving behind a charge nanosized polymer fiber.

Electrospinning Parameters

The following are the main parameters for electrospinning.

1. Voltage
2. Feedrate
3. Temperature
4. Effect of collector
5. Diameter of pipette orifice
6. Distance between top & target
7. Effect of Environment

1. Voltage

The voltage applied to electrospinning jet is one of the parameters, as electrospinning involves the movement of charged ions and jet initiation occurs when the electrostatic force in the solution overcomes the forces associated with surface tension. A voltage of 6 kVB to 20 kV is necessary for electrospinning. At a higher voltage, the acceleration of the jet also increases as the ions within gain more energy and the solution volume drawn away from the tip of the source increases.

When the draw rate of the electrospinning jet becomes faster than the feed rate, the origin of the electrospinning jets may recedes into the needle. At higher voltage, bead formation to occurs and the beads change form spindle-like to spherical-like at increasing voltage.

The voltage applied has an influence in the diameter of the fibers obtained. When the voltage is increased, the fibers obtained tend to be finer. This is due to increased electrostatic stresses on the jet, which exerts a larger drawing force on the solution of a similar volume. When a solution with lower viscosity is used, a high voltage favors the formation of several jets. However, at a lower voltage, the fiber diameter decreases. This is due to the slower flight speed of the jet; allowing time for the charged jet to split and elongate. A voltage close to the critical value for electrospinning is yields finer diameter fibers. Thus, voltage has influence on the crystallinity of the fiber. The crystallinity of the fiber initially increases with increasing the voltage of the electrostatic field, however, further increase in the voltage subsequently reduces the crystallinity.

2. Feedrate

At a higher feedrate, the diameter of the fiber increases with a corresponding increase in the size of the beads. This due to greater volume of solution available for the electrospinning jet to carry away. However, there is a limit to the increase in the diameter size at a higher feedrate. Junctions are also formed at a higher feedrate; the electrospun fiber rakes a longer time to dry at the same flight time. Thus, when the fibers are collected, the residual solvents dissolve part of the fibers where they meet, causing them to fuse together. At a lower feedrate, uniform fibers are obtained as the solvent has more time to evaporate.

3. Temperature

For the polyurethane solution, electrospinning at a high temperature of 343 K results in several jets from the drop at the tip of the needle, and the fibers produced have a highly uniform diameter. The deposition rate of fibers at high solution temperature is significantly higher than electrospinning solution at room temperature therefore introducing high temperature may not be applicable for all electrospinning solutions. When protein and biological substances such as the growth factor and enzymes are blended with polymer solution, a higher temperature causes the enzymes to denature, thus losing their desired property not because there an increase in fiber production, but the diameter of the electrospun fibers seems to decrease with increased solution temperature. The decrease in fiber diameter is the result of the expansion of the polymer molecules, which results in less chain entanglement. This causes a reduction in the solution viscosity, and the Coulombic stretching force causes a decrease in the fiber diameter.

4. Effect of Collector

In the electrospinning, a conducting target is commonly used as the collector because an opposing charge can be applied to it or it can be electrically grounded. This is necessary as charge accumulate on the surface of non-conducting materials and fewer are deposited on its surface. The fibers deposited on non-conducting materials are more loosely packed then fibers deposited on conducting materials. A conducting material would dissipate the residual charges on the fiber surface upon contact so that subsequent fibers can be packed closely or, residual charges on the fiber surface deposited on a non-conducting material will remain and repel subsequent fibers, thus giving the overall mesh a lower packing density.

The accumulation of the charges on a non-conducting collector assist in the formation of 3D fiber structures. When enough charges are accumulated on the surface of the collector, the new electrospun fibers are prevented from depositing directly due to electrostatic repulsion. Thus, a honeycomb structure is formed when the fibers are allowed to dry before being deposited on the collector.

5. Diameter of Pipette Orifice

In electrospinning, the source is either a syringe needle or a pipette. In both cases, the diameter of the tip has an effect on electrospinning. Jet initiation occurs when the charged ions in the electrostatic field are storing enough to overcome the surface tension of the polymer solution. When the diameter of the tip decreases, the radius of the droplet decreases, thus increasing its surface tension. At a constant electrostatic field, the initial acceleration of the jet decreases due to the increased surface tension. The resultant decrease in the velocity of the jet means a longer flight time for the jet before reaching the target. The increase in flight time allows more time for the charges solution jet to split and elongate, resulting in smaller and more uniform fiber diameter.

6. Distance Between Tip and Target

The distance between the source tip and the target has an influence on the flight time of the jet. This, determines the amount of residual solvent on fiber when it is collected on the target. However, exists there effect of the diameter of the fiber, due to the distance between tip and target. When the distance between the source tip and target is below a critical distance, there is much residual solvent left on the fiber during deposition. The excess solvent causes the fibers to merge to form junctions, resulting in interlayer and interlayer bonding, as shown in Figure 7.10. Depending on the application, such junction may provide additional strength to the scaffold as the fibers are interlinked. Thus, there are seems different effects on the fiber diameter with increasing tip to target distance.

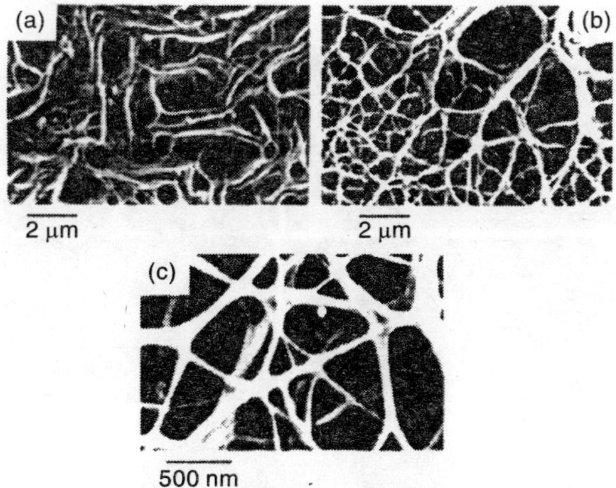

Fig. 7.10 *PEO deposited at (a) 0.5 cm deposition distance, (b) 0.75 cm deposition distance, (c) 1.0 deposition distance.*

There is no significant change in the diameter for Polystyrene (PS), dissolve in tetrahydrofuran (THF) when the distance is varied, although beads began to form when the distance is reduced to only 5 cm. An increase in average diameters for poly(vinyl alcohol) (PVA) dissolved in water when the distance between the tip and the target was increased. It is due to reduction in the effective voltage when the distance is increased. In other parameters, an increase in the flight time of the jet causes a reduction in the fiber diameter due to greater opportunity for the splitting of the electrospinning jet.

7. Effect of Humidity on Electrospinning

There are changes in fiber morphology due to the humidity in the atmosphere when the electrospinning is carried out. This happens when polymer is dissolved in volatile solvents. Polysulfone (PSO) dissolved in tetrahydrofuran (THF) shows changes in its surfaces morphology when electrospun in different humidity levels. At humidity of less than 50%, the fiber surfaces are smooth. However, as the humidity increases, the number of circular pores on the fiber surfaces and their diameters increase, until the pores became so numerous that they coalescence to form larger, non-uniform shaped structures, as shown in Figure 7.11. Above a certain humidity, the number of pores seems to saturate and there is only a small increase in diameter of the pores. The humidity did not have an effect on the shape or diameter of the electrospun fiber. Although the depth of the pore also increases with increasing humidity, it also saturates above a certain humidity, as determined by atomic force microscopy (AFM). There are several reasons offered for the formation of pores on the surface of the fibers. As the electrospinning jet travels through the air, water from the air condenses on the surface of the fiber, creating the pores as both water and solvent evaporates. However, unlike pores formed by the breath figures, pores on electrospun fibers are less uniform. This is due to different conditions between pore formation on static polymer film surface and the dynamic movement of the electrospinning jet. The charges on the surface of the electrospinning jet also influence the pore formation on the fibers. Another factor that causes the pore formation is be due to the phase separation.

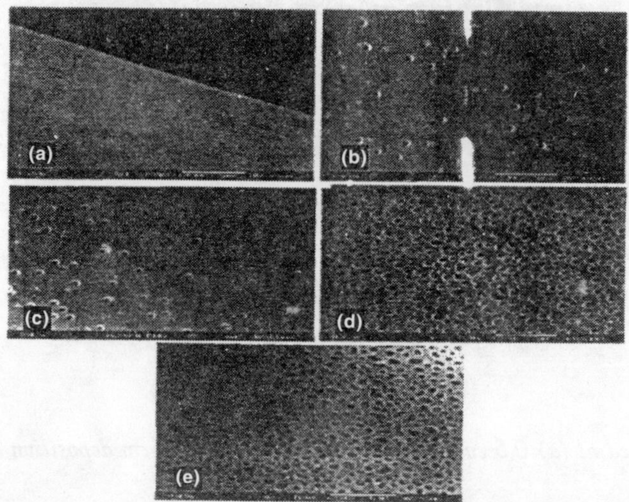

Fig. 7.11 *FESEM micrograph of 180 000 g/mol polysulfone/tetrahydrofuran fibers electrospun under varying humidity: (a) <25% (b) 31–38%, (c) 40–45%, (d) 50–29%, (e) 60–72%.*

Generally, humidity has an influence on the solution droplet at the tip of the source. The extent of influence depends on the solvent used because this affects the evaporation rate of the solvent or it may interact with a certain solvent. In dry air of less that 5% humidity, electrospinning of acrylic dissolved in dimethyl formamide (DMF) is carried out in a few minute because it dries rapidly at high humidity of over 60% Fibers start to tangle above the founded target as they are not able to dry properly electrospinning in different atmospheres. In helium, electrospinning is not possible because the gas starts to break down under a high electrostatic filed. When electrospinning is carried out in Feron gas, the fiber diameter is twice that spun in air with all other conditions kept consistent. Secondary jets were seen erupted from the primary electrospinning jet. This was said to be due to higher breakdown of voltage of Feron gas compared to air, which allows the fiber to retain its electrical charge for a longer duration.

Types of Nanofibers

Different type of nanofibers are:

1. Solid Nanofibers
   (a) Uniform and beaded nanofibers
   (b) Flattened or ribbon nanofibers
   (c) branched nanofibers
2. Hollow Nanofibers
3. Porous Nanofibers
4. Composite Nanofibers

1. Solid Nanofibers

(a) Uniform and beaded nanofibers

The electrospun polymer nanofibers are influenced by a number of parameters like solution properties, processing conditions, and ambient conditions, as shown in Figure 7.12. The parameters under solution properties, are closely related to each other. For instance, the change of polymer concentration, molecular weight and solvent composition affects solution viscosity. Solvent composition and salt addition occasionally change electrical conductivity of the polymer solution. Hence, precaution is necessary to consider whole solution properties. While making electrospun polymer nanofibers, beaded structured fibers formation, as shown in Figure 7.13 is avoided.

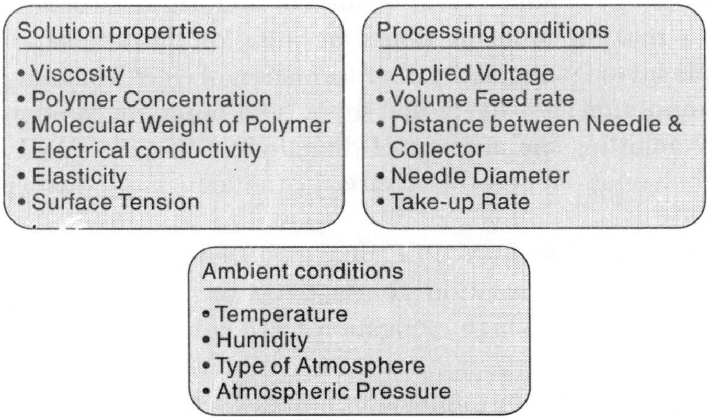

Solution properties
- Viscosity
- Polymer Concentration
- Molecular Weight of Polymer
- Electrical conductivity
- Elasticity
- Surface Tension

Processing conditions
- Applied Voltage
- Volume Feed rate
- Distance between Needle & Collector
- Needle Diameter
- Take-up Rate

Ambient conditions
- Temperature
- Humidity
- Type of Atmosphere
- Atmospheric Pressure

Fig. 7.12 *Three main categories of electrospinning parameters which affect uniform and beaded fibers.*

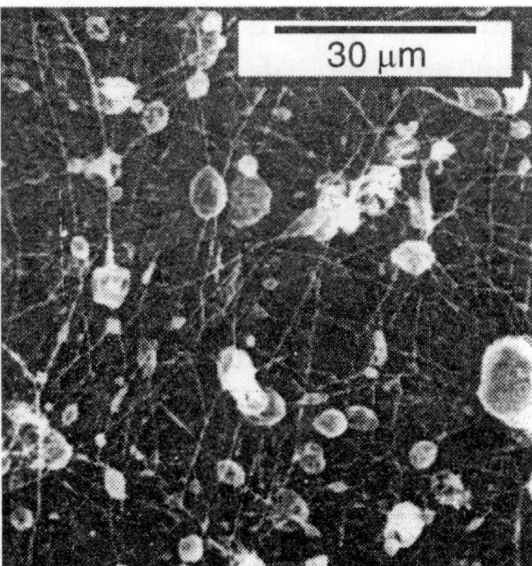

Fig. 7.13 *Beaded poly(e-caprolactone) (PCL) nanofibers formed from PCL 10 wt% solution.*

Polymer-type, less viscous solution with low polymer concentration leads to the formation of beaded fibers, while viscous solution with high polymer concentration leads to the formation of uniform fibers. When increasing the polymer concentration, the spherical morphology of beads changes to spindle-like and finally to uniform fibers (see Fig. 7.14). Now, optimum polymer concentration depends on the molecular weight of the polymer and polymer chain structure. As seen in Figure 7.15, poly(vinyl alcohol) (PVA) nanofibers with three different molecular weights are electrospun from PVA/water solution. Beaded fibers are formed at low molecular weight (Mw = 9,000 ~ 10,000), while uniform fibers are formed at higher molecular weight (Mw = 13,000 ~ 23,000) even with the same polymer concentration.

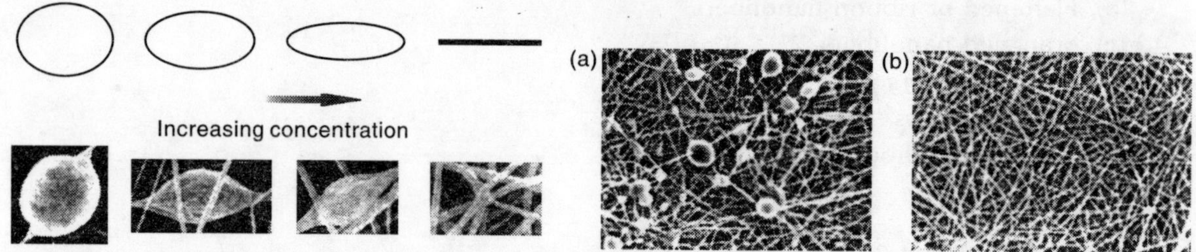

Fig. 7.14 *Morphology change form beaded fibers to uniform fibers by increasing polymer concentration.*

Fig. 7.15 *Morphology change of poly(vinylalcohol) nanofibers with (a) low molecular weight (Mw = 9,000 ~ 10,000) and (b) high molecular weight (Mw = 13,000 ~ 23,000) at a 25 wt% polymer concentration. 30 kV applied voltage, and 120 mm tip-to-collector distance.*

Solution viscosity due to solvent composition is also key for uniform and beaded nanofibers, eg: poly(ethylene oxide) nanofibers, fiber formation is investigated with varying weight ratio of water and ethanol solvents. Increasing ethanol concentration led to higher solution viscosity and lower surface tension, and fiber morphology changed from beaded to uniform in structure. The addition of ethanol contributed to making uniform fibers because of quick ethanol evaporation in the air. In the other example investigating of the fiber formation of electrospinning of acrylic in DMF by varying solvent composition. It is likely that there is an optimum solution viscosity to make uniform nanofibers by adjusting the amount of dimethyl-formamide (DMF) solvent. DMF possesses high electrolytic behavior. Hence, the electrical conductivity of polymer solution is another important parameter.

Also uniform fibers without beads are elctrospun from PEO and collagen-PEO solutions with a small amount of NACl addition. Enhanced solution conductivity by NaCl addition results in a high charge density of the polymer jet and high elongation force contributed to the disappearance of the beaded structure.

For processing condition uniform and beaded fiber formation is by volume feed rate, applied voltage and inner needle diameter. Collagen-PEO nanofibers electrospun with a higher feed rate resulted in uniform fiber formation. It is likely that smaller inner needle diameter is preferable in making uniform nanofibers form poly(L-lactide-co-ε-caprolactone) P(LLA-CL) 5 wt% solution. For poly (D,L-lactic acid) (PDLA) nanofibers beaded nanofibers are formed.

It is found that the diameter of beads and the distance between beads decreased with an in crease in the applied voltage. As the polymer concentration of PDLA solution is very high (30 wt%), the electrical force generated by 30 kV applied voltage is not high enough to elongate the polymer jet for form uniform fibers.

(b) Flattened or Ribbon-like Fibers

Flattened or Ribbon-like Fibers have been reported from several polymers under certain spinning conditions. Figure 7.16 shows the SEM photos of flattened poly(vinyl alcohol) (PVA) nanofibers and ribbon-like elastic-mimetic peptide polymer fibers. The mechanism to make flattened or ribbon-like fibers is related to solvent evaporation during the electrospinning process. In the case of poly(vinyl alcohol) (PVA) nanofiber, flattened nanofibers are created with higher molecular weight and higher polymer concentration. It is solution viscosity, due that wet fibers that reach the fiber collector are flattened by the impact. The mechanism of forming the ribbon-like fibers is considered, as shown in Figure 7.17.

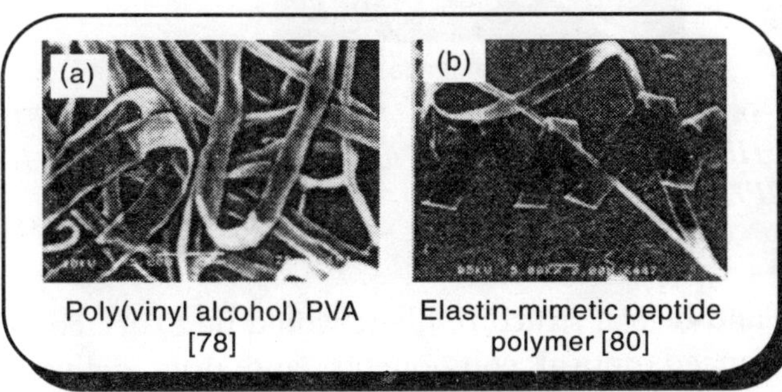

Fig. 7.16 *(a) Flattened and (b) ribbon-like electrospun nanofibers.*

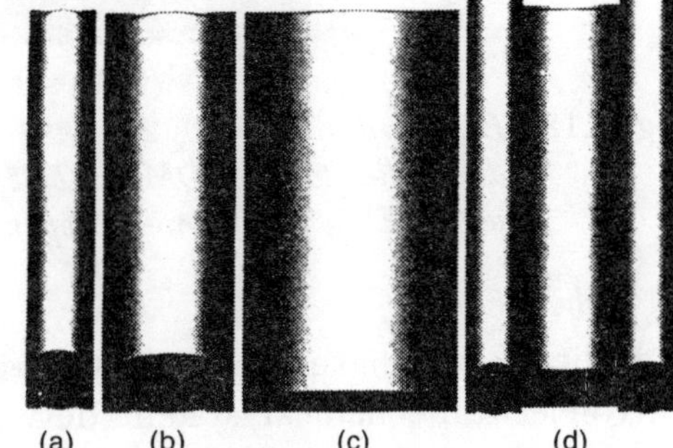

Fig. 7.17 *Mechanism of forming ribbon-like fibers.*

Thin polymer skin is formed on the liquid jet (Fig. 7.17(a)), and the solvent inside escaped. Atmospheric pressure tends to collapse the tube formed by the skin as the solvent evaporates. The circular cross section becomes an elliptical shape (Fig. 7.17(b)) followed by a ribbon-like shape (Fig. 7.17(c)). Sometimes small tubes formed at each edge of the ribbon and a web made from the skin is collected the two tubes (Fig. 7.17(d)).

(c) Branched Fibers

Branched Fibers (Fig. 7.18) can be formed by the ejection of smaller jets from the surface of the primary jets, which is smaller to ejection of an initial jet from the surface of a charged droplet. Split fibers are also formed by the separation of a primary jet into two smaller jets. The elongation of the jet and evaporation of the solvent change the shape and the charge per unit area carried by the jet. The balance between the surface tension and electrical forces shifts so that the shape of a jet becomes unstable. Such an unstable jet can reduce its local charge per unit surface area by ejecting a smaller jet from the surface of the primary jet or by splitting apart into two smaller jets.

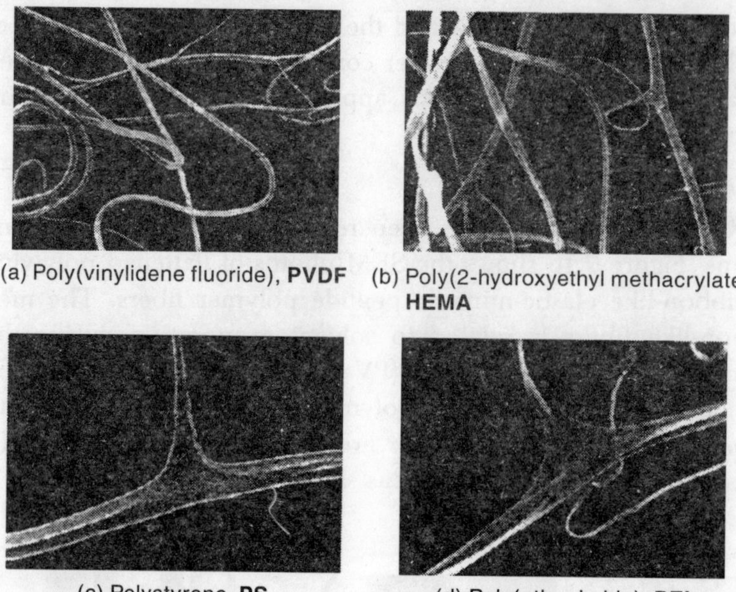

Fig. 7.18 *Branched fibers (a) 20 wt% PVDF in a 50:50 mixture of dimethylformamide (DMF)/ dimethylacetamide (DMA), (b) 16 wt% HEMA in ethanol, (c) 30 wt% PS in DMF and (d) 10 wt% PEI in hexafluoro-2-propanol (HFIP).*

(d) Helical Fibers

Helical Fibers have been recently reported and helical structure fibers would be expected as micro-electromechanical system devices, advanced optical components, and drug delivery systems. Figure 7.19 shows helical fibers made of a mixture of poly(ethylene oxide) (PEO) and poly(aniline sulfonic acid) (PASA). Loop diameter of helical structures is varied with PEO concentration. There is curiosity as to why such a structure is created during the electrospinning process. It is considered that the electrochemical property of conductive PASA polymer plays an important role, as helical fiber structure is not achieved from PVA solution. The possible mechanism is that there is equilibrium between the electrostatic repulsive force and the viscoelastic restoring force in an electrospun polymer jet. When a fiber is collected at a conductive collector, the charges from PASA regions are transferred to the fiber surface, causing a force imbalance, and the fiber undergoes a structural rearrangement to keep force equilibrium.

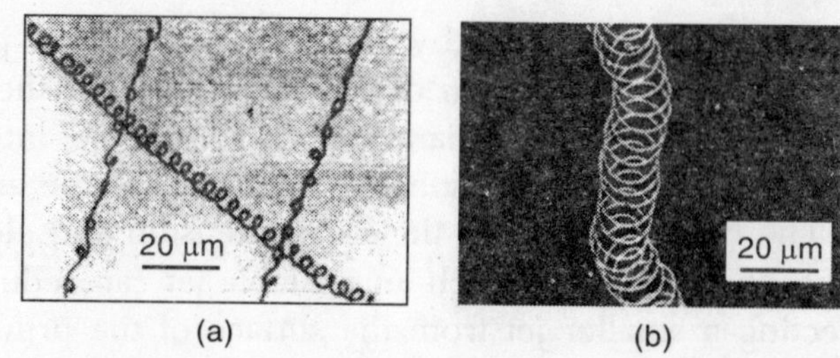

Fig. 7.19 *Helical fibers electrospun from (as) 6wt% PEO solution and (b) 8.5 wt% PEO solution with 1.5 wt% PASA in each solution.*

2. Hollow Nanofibers

Nanoscale tubes made from various materials including carbon, ceramics, metals, and polymers are important in industrial applications such as separation, gas storage, and energy conversion. The electrospinning process, the chemical vapor deposition (CVD) method and the direct coaxial spinning method are available to create hollow nanofibers. The principle of the CVD method is described in Figure 7.20. Template polymer nanofibers are first electrospun, and then the template nanofibers are coated by sheath material using CVD. Hollow nanofibers are formed when the template is removed through annealing. For this purpose, the template nanofiber must be stable during the coating and be degradable or extractable with out destructing the coating layer eg. Fig. 7.21.

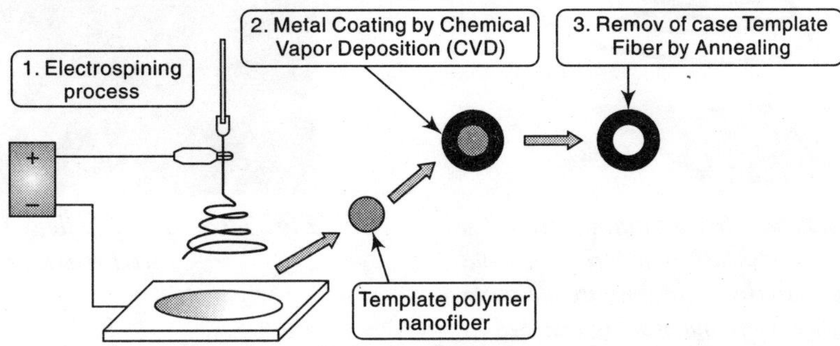

Fig. 7.20 *Hollow fibers made by coating of template polymer nanofibers using chemical vapor deposition method.*

Poly (L-lactic acid) PLLA) nanofibers are used as template and poly (p-xylylene) (PPX) as coating material by CVD. The PPX/PLLA (sheath/core) fibers formed are subsequently annealed to remove core PLLA template, at 453 K under vacuum, which leads to PPX hollow nanofibers. The outer and inner diameters of PPX hollow nanofibers are 55 nm and 7 nm, respectively. On the other hand, the direct coaxial spinning method supplies a one-step fabrication process of hollow nanofibers, utilizing mineral oil as a core material (see Fig. 7.22)

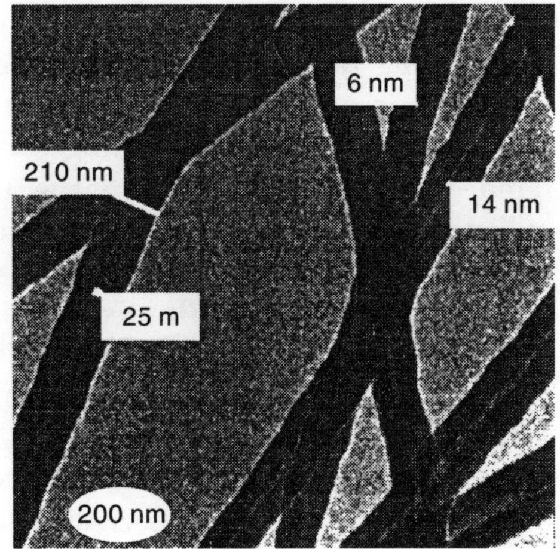

Fig. 7.21 *Poly(p-xylene) (PPX) hollow fibers fabricated by chemical vapor deposition (CVD) method.*

Poly (vinyl pyrrolidone) (PVP) solution, which contained the precursor of Titania [$Ti(O\,iPr)_4$] as sheath material is prepared and electrospinning is simultaneously conducted with oil eject. Hollow PVP/ Ti(OiPr)4 nanofibers are formed after the core oil is extracted with octane.

TiO_2 ceramic hollow nanofibers are fabricated by direct coaxial electrospinning and as shown in Figure 7.23.

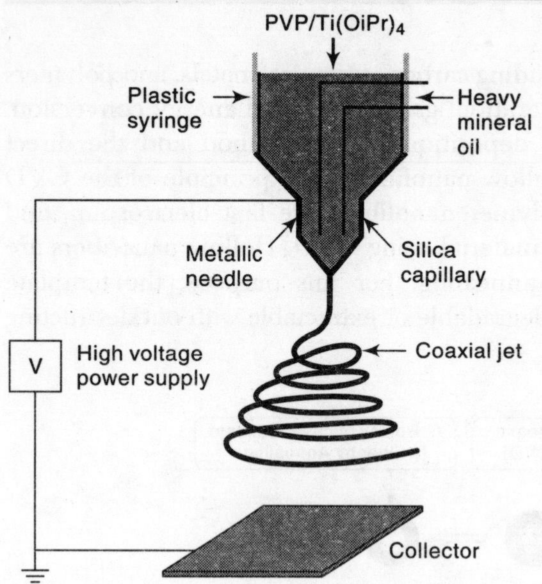

Fig. 7.22 *Direct coaxial electrospinning method to create a core/sheath structure. Core and sheath materials are immiscible with each other. Core material is removed by chemical solvent or annealing.*

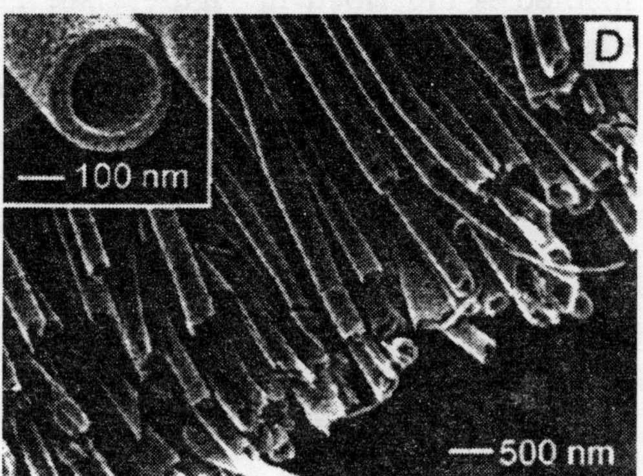

Fig. 7.23 *TiO_2 hollow fibers fabricated by direct coaxial electrospinning.*

3. Porous Nanofibers

Unique nanometer-scale porous structures have been created on the surface on the polymer fiber during the electrospinning process, as shown in Figure 7.24. The size of the surface structures, in a range from a few tens nanometer to 1 μm, are dependent on type of polymer, solvent used, and electrospinning conditions. Electrospinning of polymer solutions involves a rapid evaporation of the solvent as the jet travels toward the fiber collecting target. During millisecond jet flying curiosity is placed on the mechanisms to create a porous surfaces on nanofibers. The mechanism to form a porous surface on polymer casting film is applicable to the phenomenon on electrospun nanofibers. When a polymer solution is cast on a support, evaporation takes place under convective conditions. During solvent evaporation, the solution becomes thermodynamically unstable and phase separation occurs, resulting in a polymer rich and a polymer lean phase. The concentrated phase solidifies shortly after phase separation and forms the matrix, while the polymer lean phase forms the pores. In this regard, the solvent vapor pressure has a critical

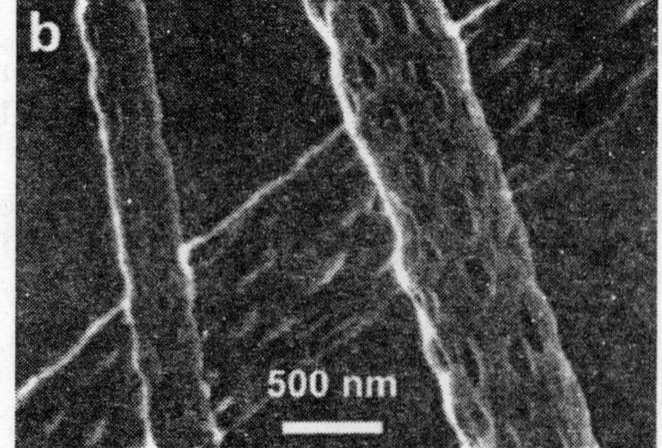

Fig. 7.24 *PLLA porous nanofibers electrospun from a solution of PLLA in dichloromethane.*

influence on the process of pore formation. On poly(L-lactic acid) (PLLA) porous nanofibers the replacement of dichloromethane by chloroform with lower vapor pressure significantly reduced the pore formation. Another mechanism of pore formation on the fiber surface is evaporative cooling due to rapid solvent evaporation. Hence, the surface of the electrospinning jet is significantly cooled during journey toward the fiber collecting target. As a result, moisture surrounding the electrospinning jet is also cooled down and forms water droplets. The droplets remain as individual entities acting as hard spheres due to convection currents on the surface of the jets. As the jet dries leaving dried fibers, the water droplets leave an imprint on the surface of the fibers in the form of pores. That humidity is also a concern for pore formation on electrospun nanofibers higher humidity caused larger pore size on the nanofiber surface.

4. Composite Nanofibers

Polymer nanofibers that contain inorganic materials such as carbon nanotubes, clay, ceramics and metals have been fabricated. From polymer/CNT composite nanofibers, excellent mechanical properties and electrical conductivity of CNTs are properties of concern on polymer nanofibers.

It is found that efficient reinforcing nanofibers rely on even dispersion and well controlled alignment of CNTs along fiber direction. To align single-wall CNTs or their bundles along the fiber direction due to the combination of dielectrophoretic forces caused by dielectric or conductivity mismatch between CNTs and the polymer solution and high shear forces induced by the spinning. The dispersed CNTs posses 1.2–1.6 nm in diameter and 3 μm long in a polyimide (CP2) solution for electrospinning. The resulting ultra fine fibers had diameters ranging from 50 nm to 2 μm. Dispersing single-wall CNTs in poly(acrylonitrle) (PAN) solution that was electrospun into ultrafine fibers; compression force is applied to a single composite nanofiber for load-displacement response. It is seen that PAN/CNT composite nanofibers indicate stiffer behavior as compared to PAN nanofibers.

Also, electrical conductivity of polymer nanofibers is enhanced by single-wall CNTs. CNTs are incorporated into a piezoelectric fluoropolymer, poly(vinylidene fluoride) (PVDF), and electrospun composite nanofibers are obtained. Electrical conductivity of PDVF/CNT composite nanofibers improve with 0.1 wt% amount of CNTs in the range from 1×10^{-12} S/cm^{-1} to 7×10^{-6} S cm^{-1}.

Physical Properties

1. Size

With a decrease in particle size, the relative number of atoms at the surface and interfaces increase.

The large ratio of surface atoms to the total number is attributed to other properties of nanomaterials. Many fibers are in the micron size level. However, when the diameter of the fibers is reduced to the nanometer scale, the most apparent change is the huge increase in the surface area to volume ratio, as shown in Figure 7.25.

Consider a fiber diameter of 100 nm 2% of the molecules in the fiber are at the surface. Thus reactive groups of a polymer molecule chain will be more accessible to chemical reactions.

2. Electrical Properties

When the size of an electrically conducting materials is decreased to nanoscale, its electrical properties change from that of the bulk. The electrical conductivity of a free-standing thin film consisting of uniaxially aligned carbon nanofibers and SnO_2 nanofibers (slightly doped with Sb) has an anisotropic ratio greater than those of thin films containing aligned carbon nanotubes. For nanofibers made of conducting polyaniline doped with camphorusulfonic acid (PAn.HCSA) blended with insulating polyethylene oxide, fiber diameters less than 15 nm are electrically insulating. Thin fibers have a lower conductivity than thick ones with the conductivity of 20 nm to 70 nm fiber is 10^{-3} and 10^{-2} S cm^{-1}, respectively, which are much less than bulk material with 1 S cm^{-3}. The insulating property of nanofibers with diameters less than 15 nm is due to the full de-doping of PAn.HCSA fibers by atmospheric gases or by phase separation of PAn.HCSA and PEO in the nanofiber, leading to discrete nanoscale conducting polymer grains separated by insulating PEO regions.

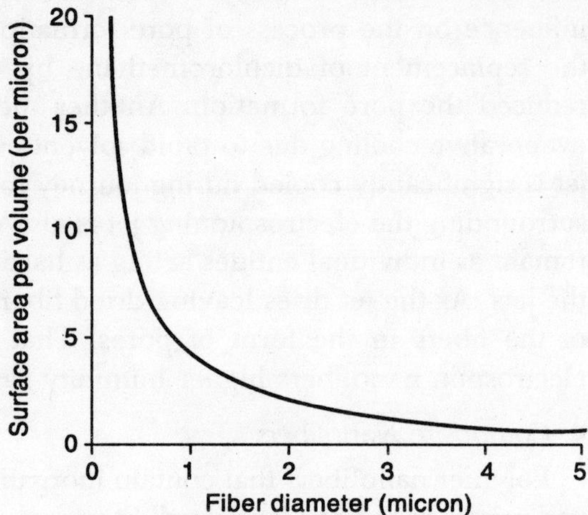

Fig. 7.25 *Comparing surface to volume ratio with decreasing fiber diameter.*

3. Mechanical Properties

Using nanoparticles of polyethylene (PE) as an example of mechanical properties that are different from bulk, the yield point is always smaller than the modulus in bulk PE. However, the molecular dynamics study of nanoparticles show a compressive modulus that is an order of magnitudes smaller than the bulk values, and the yield point is much larger than the modulus. The initial deformation of the PE particles caused by the compression is not reversible. The irreversible phase is the result of the force required to deform the spherical particle to an abiate top. Thus, the modulus of compression in this region is more a measure of the force to deform the spherical particle. Further deformation is reversible up to the point of rupture. Which is more closely related to the bulk modulus because the stress is due to the cohesive energy and not the microstructure. This leads to significantly higher yield point that the modulus. In micron fibers, there are defects within the fiber in the forms of voids between the molecule chains (see Fig. 7.26).

In nanofibers, because the diameter of the fiber is so small, there may be considerably fewer defects within the fiber in the form of voids. As a result, nanofibers may exhibit better mechanical properties than micron fibers when processed using the same method.

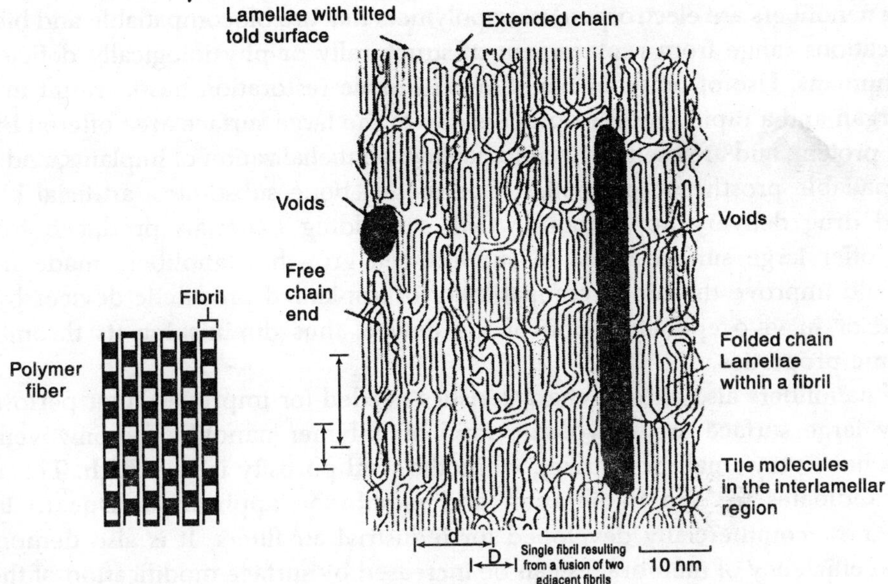

Fig. 7.26 *(left) Polymer fiber consists of smaller fibrils. (right) Defects in the fibril, in the form of voids, may affect the mechanical properties of the polymer fiber.*

4. Optical Properties

Due to the fine diameter of electrospun fibers, there is a high degree of light scattering produced by mat made of electrospun fibers, resulting in a white appearance. A solution containing, carbon black has a black appearance. The resultant mat, after electrospinning the black polymer solution, is white due to the scattering of light. When heat is applied, the mat changes color form white black as it loses its fibrilar morphology. Uniaxially aligned nanofibers can function as an optical polarize. Incident light with polarization parallel to the alignment of nanofibers is made extinct by three times as much when compared with light polarized perpendicular to the long axes of the aligned nanofibers.

Another interesting aspect of the optical properties of nanofibers is their use in composite applications. In general composite materials, the composite becomes opaque due to light scattering of the particle embedded with the composite. However, by using fibers with a diameter significantly smaller than the wavelength of visible light, is possible to obtain composite films that are fully transparent.

APPLICATIONS FOR POLYMER NANOFIBERS

Nanofibers are fibers with diameters one or two orders of magnitude smaller than conventional fibers. These fibers have a uniquely large surface area-to-mass ratio, in the range of 10 m^2 g^{-1} to 1000 m^2 g^{-1} (when the diameter is approximately 500 nm). This makes them suitable for a broad range of applications in life science, filtration, and sensor application.

Polymer nanofibers are electrospun using polymers that are biocompatiable and biodegradable; their applications range from replacement of structurally or physiologically deficient tissues to organs in humans. Use of polymer nanofibers in tissue restoration has to result in an efficient, compact organ and a rapid recovery process due to the large surface area offered by nanofibers made from protein; and are used for wound healing, epithelialization of implants, and construction of biocompatiable prostheses, cosmetics, face masks, bone substitutes, artificial blood vessels, valves, and drug delivery applications. Also, scaffolding materials produced from polymer nanofibers offer large surface areas to support cell growth. Nanofibers made from silk-like proteins could improve the blood compatibility of implanted prosthetic devices by promotion in *vitro* and/or in vivo epithelilization of the device, thus diminishing its thrombogenic and immunogenic properties.

Polymer nanofibers also exhibit characteristics needed for improved filter performance. Due to the very large surface area-to-volume ratio of polymer nanofibers, nonwoven mats from these fibers have very small pores sizes, while the total porosity is very high. This makes them excellent candidates for use in filtration and membrane application. Due to the filtration devices are now commercially developed for industrial air filters, It is also demonstrated that the filtration efficiency of membranes can be increased by surface modification of the nanofiber.

Polymer Nanofibers for Biomedical Applications

The fabrication of a scaffold by electrospinning biomaterials such PLA, poly(glycolic acid) (PGA), poly(ethylene-co-vinyl acetate) (PEVA), and Type I collagen, is achieved. Particles have been encapsulated into electrospun polymeric nanofibers, by adding insoluble particles to the polymer solution. These are particles of were gold, pollen, spores, alginate, and silver sulfadiazine for burns. Biomaterials for wound healing, or soluble drugs, antibacterial agent are also added. A skin mask is made by directly electrospinning fibers onto the skin surface to protect and heal wounds. Electrospun fiber mats are also explored as drug delivery vehicles. Mats are made from PLA, PEVA, or from a 50:50 blend of the two, using tetracycline hydrochloride as a model drug. Poly(lactide) (PLA) is used in various biomedical applications due to its biodegradability, biocompatibility, good mechanical properties, and its ability to be dissolved in common solvents for processing. Silicone rubber, nylons, polyesters, polyurethanes, acrylics, and other polymers are have been used in biomedical applications. Some of the uses of biomaterials include dental implants, bone replacements and cements, heart valves, cosmetic surgeries, and vascular grafts

Electrospun nanofibers structures also meet the essential design criteria of an ideal tissue with the ability to support and guide cell growth. Cells seeded on this structure tend to maintain phenotypic shape and guides growth according to nanofiber orientation. An electrospun structure for tissue engineering applications is developed. It is composed of poly(D,L-lactide-co-glycolide) fibers ranging from 500 to 800 nm in diameter. The structural features are similar to the extracellular matrix of natural tissue, which is characterized by a wide range of pore diameter distribution, high porosity, and effective mechanical properties. That electrospinning is used to produce tissue-engineering scaffolds composed of collagen nanofibers (a matrix composed of 100 nm fiber) is also achieved. The structural properties of electrospun collagen are formed to vary with the tissue of origin, the isotope, and the concentration of the collagen solution used to spin the fibers. Blended components with collagen have are for electrospinning. Type I collagen-PEO fibers and nonwoven fiber networks are produced by electrospinning a weak acid solution of purified collagen at ambient temperature and pressure. The process provides a convenient, non-toxic, non-denaturing approach for the generation of collagen-containing nanofibers (diameter range of 100–150 nm) and nonwoven fabrics that have potential application in would healing, tissue engineering, and as hemostatic agents. Also produced are elastin-mimetic fibers from peptide polymer to mimic protein fibers that are found in the arterial wall. Electrospinning is used to create biocomparticle thin films with useful coating design and surface structure that can be deposited on implantable devices to facilitate the integration of these devices with the body. Silk-like polymer with fibronectin functionality (extracellular matrix proteins) are electrospun to make biocompatible films used on prosthetic devices implanted in the central nervous system.

Electrospun membranes composed of elastomeric fibers are used in several protective clothing applications, eg. to develop garments that reduce a soldier's risk for chemical exposure. The idea is to interlace several different types of polymers and fibers to make protective ultra-thin layers that would enhance chemical reactivity and environmental resistance.

Polymer Nanofibers for Tissue Engineering

Blood Vessel

Smooth muscle cells (SMC) are a functional element of a number of cardiovascular, gastrotestinal, and urinary tissues, and typically reside in a mechanically dynamic environment. Engineering blood vessels under pulsatile conditions resulted in increased mechanical strength, increased production of collagen, and expression of the contractile SMC phenotype. The resultant blood vessels are found to be accessible for longer periods after implantation, than flood vessels engineered without mechanical stimulation. The application of cyclic mechanical strain during the smooth muscle tissue engineering process shows increased elastin and collagen production and tissue organization. This has resulted in an increase in the mechanical properties of the engineered smooth muscle. Vascular grafts are intended to replace damaged or occluded blood vessels. Poly(glycolide-co-ε-caprolactone) scaffolds posses elastic mechanical properties that promote SMC adhesion and subsequent tissue formation. Typical materials used for vascular grafts include: poly(ethylene terephthalate) (PET) and poly(tetrafluoroethylene) (PTFE). These materials have a problem with blood clotting in areas of low blood flow, however, materials like poly(carbonate-urea) ureathane has improved it. The preparation and evaluation of acrylic copolymers are replacements for PET and PTFE in vascular grafts applications.

Natural and healthy blood vessels in the body possess a lining of endothelial cells that actively resist thrombosis and plasma protein deposition, by releasing prostacyclin, heparin sulfate, and other anticoagulants. It is found that endothelial cells when electrostatically seeded onto PTFE grafts, the cells remain in place for at least one week. Another method used for the promotion of endothelial cell attachment and spreading on vascular grafts, is to attach molecules to the graft, that bind to receptors on endothelial cells; example: the RGD peptide sequence, found in many adhesion proteins in the extracellular (EC) matrix. It is found that RGD is the sequence responsible for endothelial cell binding in the adhesive domains of many of these proteins. Also, when the RGD peptide sequences are grafted to polymers, the adhesion and spreading of endothelial cells is greatly enhanced. It is found that the methacrylate-based materials resist cells adhesion.

That synthetic polymer scaffold (poly(llactid-cos ε-caprolactone) is composed of aligned and random nanofibers. This scaffold shows favorable behavior of SMC and EC cell interaction in the scaffold (Figs. 7.27 and 7.28). The electrospinning methods are used to fabricate the scaffold is suitable for mass production, demonstrating a huge potential for scaffold application in blood vessel engineering. Surface-induced thrombosis is one of the major drawbacks. It also hampers the applications polyurethane (PU) and chitosan, in blood-contacting. To provide a bioactive and biological graft interface, *in vitro* endothelization on grafted surfaces (such as polyurethane) has shown promising results in improving blood compatibility. Various methods are developed to support the seeding and growth of endothelial cells on PU or other biomaterial surfaces; eg: a surface modification by plasma treatment and photochemically grafted GRGD peptide on a modified PU or chitosan surface. The above mentioned surface modifications are categorized as chemical or biological modifications enhancing endothelialiaztion on biomaterial surfaces. On the other hand, the modification of physical properties of surfaces such as enhancing hydrophilic properties, changing porosity of materials, and increasing roughness may also enhance endothelialization of biomaterials. Hence, the construction of a blood vessel scaffold is created (to functional tissue engineered blood vessel) as a substitute for the replacement of diseased organs.

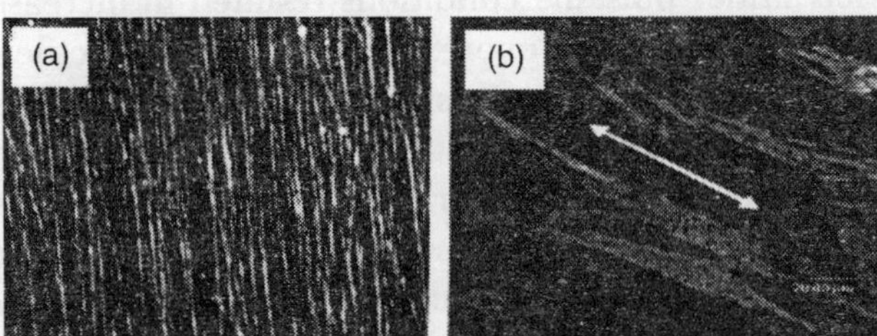

Fig. 7.27 *Aligned P(LLA-CL) nanofibers scaffold (200×) and (b) smooth muscle cells on aligned nanofibers (LSCM α-acting protein).*

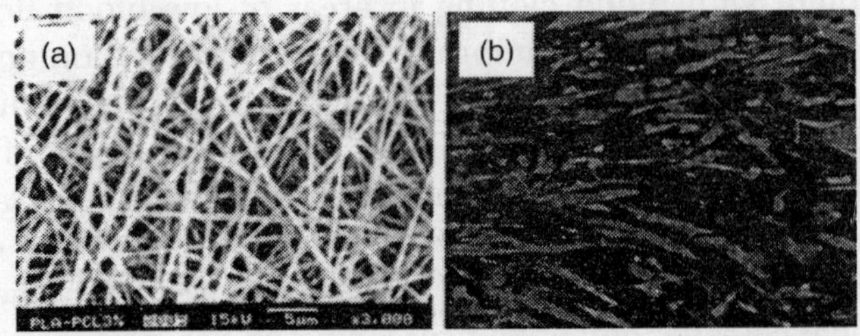

Fig. 7.28 *(a) Random (P(LLA-CL) nanofibrous scaffold (SEM) 3,000×), (b) SMC attached on random nanofibers (LSCM: CFMA dye).*

Cartilage

Polymer scaffolds are primarily used for the delivery and retention of chondrogenic cells in cartilage tissue engineering. Many naturally derived and synthetic polymers are used as scaffolds for the regeneration of articular cartilage. Natural polymers contain ligands that are recognized by cell surface receptors and have the advantage of known biocompatibility and fewer regulatory constraints. Drawbacks of these materials are: lack of large quantities for clinical applications, difficulty of processing into scaffolds, concerns about antigenicity and disease transfer of all grafts, and variable degradation rate from host to host. Collagen is a natural polymer for articular cartilage regeneration.

PGA fibrous, nonwoven scaffolds, with porosity higher than 90% are used for tissue engineering of articular cartilage. Better integration of the repair with the host tissue, as well as increased remolding of the subchondral bone, are observed. Compared to fibrin, collagen, poly(L-lactic acid), and poly(d,L-*lactic-co-*glycolic acid), PGA to provides a better scaffolds for *in vitro* cartilage regeneration. This is also demonstrated by cell densities found in natural tissues. Poly(ethylene oxide) (PEO) gels are utilized as templates for chondrocyte transplantation. Histological and biochemical characteristics demonstrate necartilage formation 12 weeks after subcutaneous injection in nude mice. PEO and chondrocytes are also injected subcutaneously in athymic mice, and photopolymerized transdermally. Formation of neocartilage is demonstrated 7 weeks following implantation. Recently developed hydrogels based on oligo (poly ethylene glycol) fumarate (OPF) is also used in cartilage tissue engineering.

Bone

Bone grafts have been used to fill defects in bone caused by clinical disease states or trauma such as fractures with bone loss, bone infection, or bone tumors. Autografts have the distinct advantage of histocompatibility, without risk of transfer of disease, and are still the best materials for bone repair. In contrast, use of allogenic bone may transmit diseases and cause an immune response that can lead to a high failure rate. It does have advantages over autografts, however, including better availability, which avoids the need for a second surgical procedure to provide bone graft and the possibility of reconstruction large segmental defects in bone.

Engineering living tissue for reconstructive surgery requires an appropriate cell source, optimal culture conditions, and a biodegradable scaffolds as the basic elements. While specialized cells remain an important source, stem cells have emerged as a promising new alternative. Recent advances in stem cell biology have shown that mesenchymal stem cells (MSC) can differentiate into cells of mesenchymal tissue, such as bone, cartilage, muscle, tendon, ligament, and fat. They are expected to play an important role in the repair of skeletal defects. Osteoblasts derived from MSC of neonatal rats were cultured on poly(D,L-lactide-co-glycolide) foams; mineralization, as well as three-dimensional bone formation, was observed.

Biodegradable polymer often have been combined with biceramics to produce materials for bone repair. Biodegradable polymers such as PGA, PLA, and PLGA have found application as reasonable sutures, bone fixation screws, plates, and drug carriers. They are used as bone graft substitutes, with positive results. Poly(ε-caprolactone) (PCL) was used mainly for long-term implants for drug release and to support mineralized tissue formation, and is now used for the treatment of bone defects. An improvement in the mechanical properties of PCL has been achieved by copolymerization with PLA, enabling its use for orthopedic applications such as the repair of bone defects. There has been widespread use of bioceramics, such as hydroxyapatite (HA) and tricalciumphosphate (TCP) for bone regeneration applications. Their biocompatibility is thought to be due to their structural similarity to the mineral phase of bone. Biodegradable ceramics are used in dental and orthopedic surgery as fillers for bone defects and as coatings on metallic implants to improve implants integration in host bone. Biodegradable polymers have been used as binders for HA and TCP to overcome the problems of brittleness and difficulty in shaping hard ceramic materials to fit bone defects. At the same time, bioceramic reinforcement of polymer has been shown to improve mechanical properties and osteoconductivity. Biodegradable polymer/bioceramic composites are promising materials for bone graft replacement and this has been intensity investigated in the last decade.

Carbon nanotubes and nanofibers have several properties that suggest these materials may be of value in the development of novel devices for bone reconstruction. Carbon fibers with nanometer dimensions simulate dimensions of collagen fibrils (0.1–8 μm) in bone. Smaller diameter carbon nanofibers, without the pyrolytic layer (i.e., PR-24 PS fibers), are suitable for use in orthopedic/dental implant material designs due to the increase in osteoblasts, and decrease in osteoblast competitive cell line (fibroblasts, chondrocytes, smooth muscle cells) adhesion. More osteoblast adhesion and less competitive cell adhesion leads to faster integration of the bone to the implant surface. In addition, increased osteoblast alkaline phosphatese activity and increased calcium deposition on nanometer carbon fibers, is there. Thus, nanometer carbon fibers enhance osteoblast activity for possible increased integration of bone on the orthopedic material surface.

Skin

Dressing for human wounds have been aimed at protection, removal of exudates, inhibition of exogenous microorganism invasion, and improved appearance. However, in ancient times, wound care and dressing strategies primarily relied on empiricism. It has been reported that work on occlusive wound dressing actually began with observation that a blister healed faster if left unbroken. Covering a wound with a polyethylene film, decreased the time to epithelialize the wound. Also, protection is accomplished by covering the wound with wound dressing. Wound areas that are kept just damp heal faster but accumulation exudates under the dressing cause infection. Wound dressing electrospun nanofibrous membrane meet the requirements such as higher gas permeation and protection of wound from infection and dehydration. The goal of wound dressing is the production of an ideal structure, which gives higher porosity and good barrier. To reach this goal, wound-dressing materials must be selected carefully and the structure must be controlled to confirm that it has good barrier properties and oxygen permeability.

Composite films comprising non-cross-linked collagen mats stabilized by PCL are found to support higher numbers of human ostroblasts in cell culture in comparison with PCL films. The preparation and characterization of collagen: PCL biocomposite and nanofiber membranes for support of human dermal fibroblast (see Fig. 7.29) and keratinocytes in tissue engineered skin in regenerative medicine. The PCL nanofibers support the fibroblast cell culture compared to PCL films. PCL is characterized by a resorption time in excess of one year but known to be susceptible to enzymatic degradation. A variety of collagenase enzymes are secreted by macrophages, epidermal cells, and fibroblasts during wound healing, including matrix metalloproteinases, gelatinase A, and stromelysin-1, which are implicated in the process of cell migration and repair.

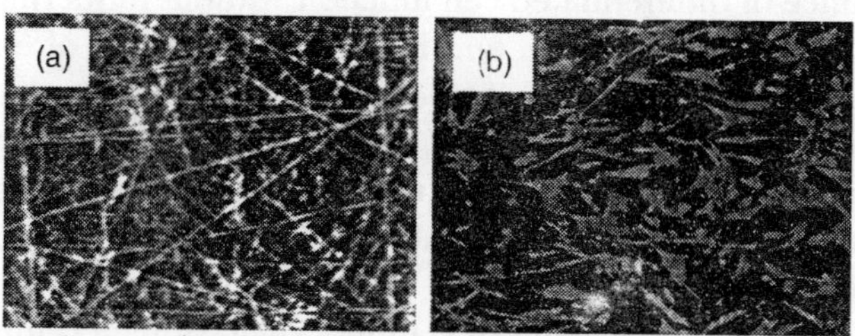

Fig. 7.29 *(a) PCL random nanofibers (SEM 1000X), (b) Human dermal fibroblast on PCL nanofibers (LSCM; CMFDA dye).*

Polyurethane (PU) and PCL nanofibers (see Fig. 7.29) are frequently used in wound dressings because of their good barrier properties and oxygen permeability. Semipermeable dressing, many of which are PU, are healing dressings. The permeability to water is also important so that fluid form the wound does not build up between the wound and the dressing and that wound desiccation does not occur. Occlusive dressing prepared by polyurethane for healing of wounds have a draw backle. Significant fluid accumulation occurs under them, particularly after a few days of use. This requires aspiration of the wound to prevent leakage and infection.

Electrospun nanofibers membrane show good and immediate adherence to wet wound surfaces. The membrane attain uniform adherence to the wound surface without any fluid accumulation. The dermis of wound covered with Tegaderm™ was inflammatory, whereas the rate of epithelialization increased and the dermis was well organised in electrospun nanofibrous membrane group. The result indicates that the electrospun nanofibrous membrane provide a good means for wound healing. This dressing showed controlled evaporative water loss, excellent oxygen permeability, and promoted fluid drainage ability due to the nanofibers with porosity and inherent property of PU. Histological examination confirmed that epithelialization rate was increased, and the exudates in the dermis were well controlled by covering the wound with the electrospun membrane. Thus, nanofibrous PU and PPDO/PLLA-b-PEG membrane prepared by electrospinnig can be used as would dressings.

Nerve

Nerve tissue repair is a previous treatment concept in human health care as it directly impacts on the quality of life. It is well known that the adult central nerve system (CNS) can not be regenerated on its own after trauma disease. Therefore, restoring its function is a challenge for neurobiologists and neurologists. Traditionally, tissue transplantation or peripheral nerve grafting are used top repair damaged or diseased regions at the CNS, out often encountered donor shortage and immunological problems associated with infectious disease. Cells are living in a complex mixture of pores, ridges, and fibers of ECM at nanometer scales.

Hence, a nanostructured porous scaffold with interconnective pores and large surface area is needed as an alternate to natural ECM for better cell ingrowths in a three-dimensional (3-D) fashion. Several methods to fabricate polymeric nanofibers or submicron fibers by using phase separation electrospinning and self-assembly are there. Polymers are used as scaffold to promote cell adhesion, maintenance of differentiated cell function without hindering proliferation, template to organize and direct the growth of cells, and help in the function of extracellular matrix. The potential of nanofibrous porous scaffold is prepared by phase separation using poly(1-lactic acid) with a similar structure of natural ECM in nerve tissue engineering. Cells from nerve expants are able to construct and restructure the walls of porous collagen-GAG matrices, *in vitro*, thus, suggest that contractile cells are capable of restructuring the extracellular component of the nerve wound environment *in vivo.*

The non-degradable conduits are made of stainless steel, poly(acrylonitrile-co-vinylchloride), polycarbonate, and silicone. However conduits fabricated with these durable materials remain *in situ* as a foreign body and elicit an inflammatory response or cause compression injury, thus limiting nerve regeneration. Degradable materials provide an alternative and are used for the repair of nerve defects These temporary conduits are composed of materials such as polyurethane, polyorthoester, glycolide trimethylene carbonate, poly(lactic-co-glycolic acid), poly(glycolic acid), poly(DL-lactic acid) and poly(lactic acid-co-ε-caprolactone). Also, seeding a reasonable implant (trimethyencarbonate-co-ε-caprolactone) with Schwann cells significantly increases growth, ensheathment, and peripheral nerve myelination as been demonstrated.

Polymer Nanofibers for Filter Applications

Filtration is necessary in many engineering fields. Fibrous material used for filter media provide advantages of high filtration efficiency and low air resistance. Filtration efficiency, which is closely associated with the fiber fineness, in one of the most important concerns for the filter performance. In the industry, coalescing filter media are studied to produce clean compressed air. These media are required to capture oil droplets as small as 0.3 micron. It is realized that electrospinning is rising to the challenge of providing solutions for the removal of unfriendly particles in such submicron range. Because the channels and structural elements of a filter must match the scale of the particles or droplets that are to be captured in the filter, one direct way of developing high efficient and effective filter media is by using nanometer-sized fibers in the filter structure. In general, due to the very high surface area-to-volume ratio and resulting high surface cohesion, tiny particles of the order of less than 0.5 μm can be easily trapped in the electrospun nanofibrous structured filters and, hence, the filtration efficiency can be improved.

A method for making a dust filter bag, which constitutes a plurality of layers including a carrier material layer and a nanofiber nonwoven tissue layer is developed. Nanofibers for applications in pulse-clear cartridges for dust collection and in cabin air filtration of mining vehicles are also used. Polymer nanofibers are electrostatically charged to modify the ability of electrostatic attraction of particles without increase in pressure drop to further improve filtration efficiency. Here, the electrospinning process is integrated for, the spinning and charging of polymer into nanofibers in one step.

In addition to fulfilling the more traditional purpose in filtration, the nanofiber membranes fabricated from some specific polymers or coated with some selective agents are used (e.g., molecular filters). For instance, such filters can be applied to the detection and filtration of chemical an biological weapon agents.

Polymer Nanofibers for Sensing

Nanosensors are sensing devices in which recognition elemenc is made up of material that is nanosize. When the size of the material shrinks to nanoscale, there are evident advantages such as an increase in surface area and a marked difference in the physical properties. Both these advantages assume a major role in increasing the sensitivity and resolution levels of a sensor. Thus when material is scaled down to the nanosize, the advantages mentioned also inflate thus creating a potential substance for sensor fabrication.

Membranes are one of the essential components of a biosensor. They are used for (1) preventing fouling; (2) eliminating interference; and (3) controlling the operating regime of the sensor. In some configuration they form the sensing element itself, as in the case of a nanofiber-based sensor. When a small molecule is the analyse, macro-molecules such as a proteins can be prevented to enter the active sensing zone by using a membrane that has nanosize pore size. The transport of charged molecules can be controlled by using ion exchange type membranes. A combination of different discriminative membranes can also be used for blocking the passage of different interferents.

Polymer Electrolyte Fuel Cell (PFEC)

A fuel cell, which uses a polymer ion exchange film for an electrolyte, and compared with other types of fuel cells, has a low operation temperature of 373 K for less. The solid type also very lightweight and has a high energy-conversion efficiency of 50%. For this reason, it is a fuel cell suitable for small mobile applications or automobiles. Fluoride polymer films are used because of the high mobility of hydrogen ions and high heat resistance.

The developed tiny full cell, classified as a polymer electrolyte fuel cell (PEFC), utilizes the carbon nanohorns as electrodes for catalyst support. It is observed that very fine platinum catalyst particles are dispersed on the surfaces of the carbon nanohorns. The size of the platinum particle is less than half of the supported on the ordinary activated carbon (acetylene black) by the same method. The size of the catalyst particle is one of the most important factors that determine the performance of the fuel cell, and it is considered that, the finer the size the better performance.

Although the reason a catalyst particle becomes fine is still not clear in the case of the carbon nanohorn, because of the unique shape of the aggregate it is thought that contact and grains growth of catalyst particles will be prevented. It is also expected that by further altering the form of the carbon nanohorn the dispersed state and the battery characteristic of the catalyst particle will improve. In addition, because a carbon nanohorn is produced by the laser ablation method, if a platinum catalyst is also simultaneously evaporated it is observed that a platinum particle will naturally adhere to the surface of a carbon nanohorn. If this method is used, the complicated catalyst supporting process through the conventional wet process can be omitted resulting in a large cost reduction.

Until now, since the discovery of the carbon nanotube, although it has been acknowledged as having a high possibility of being applied to semiconductors, flat-panel displays, lightweight and high-strength raw materials an fuel cells etc.,

POLYMER MICRO/NANO-STRUCTURE MANUFACTURING

Micro/nano-structure inspired from biological gecko foot-hairs are fabricated by three different manufacturing techniques.

Micro/nano-structures are becoming crucial for biomimetic dry adhesives, smart surfaces, type of nanostructures. Mostly low aspect ratio structures are achieved. Three novel techniques are developed which are given below. Atomic force microscope (AFM) nano-probes are used to emboss the probe tip shape to a soft wax surface. By this technique, single and arrays of silicone rubber nano-pyramids are manufactured by a step-and-repeat process. This method has the flexibility of fabricating different orientation and nonsymmetric nano-structures over almost any surface topography while it enables prototyping of only certain tape of low aspect ratio nano-structures, and it is relatively slow due to its serial control process. Therefore, two parallel, methods are introduced.

The first parallel technique is based on molding self-organised nano-pore membrances with liquid polymers, curing the polymer, and etching away the membrance, or peeling off the cured polymer from the membrane. As the next technique, micro/nano-structures are grown using directed self-assembly. In this technique, a thin liquid polymer film is coated on a flat conductive substrate, and a closely spaced another metal plate is used to apply a DC electric field on the polymer film. Due to the instabilities on the film, nano-pillars start to grow until they touch to the upper metal plate.

Micro/Nano-manufacturing Methods

Three manufacturing methods are used for manufacturing micro/nano-structures with following specifications.

(1) High aspect ratio cylindrical micrometer (1:10-30) and nanometer (1:20-25) scale structure fabrication with diameters of 3-10 μm and 50-500 nm respectively, (2) Maximize micro/nano-hair density (number of hairs in a given area, e.g. 1 cm^2) of higher adhesion. (3) Maximize nano-hair stiffness to prevent matting, (4) Orient the micro-and nano-hairs with 15-60° to enable directional attachment and detachment and (5) Material properties of synthetic hairs: Young's modulus of 0.110 GPa, hydrophobic, and high tensile strength.

At first, a master template with high aspect ratio micro/nano-holes is modeled with polymers. These templates are manufactured by nano-embossing or using self-organized nano-pores.

Method I: Molding Nano-Embossed Templates

At first, an atomic force microscope (AFM) and scanning tunneling microscope (STM) nano-prob tip, high aspect ratio mico/nano-pillars fabricated using optical or electron-beam lithography type of methods, or an array of ultra sharp glass pipettes are embossed mechanically on a soft wax surface to copy their negatives on to the surface. Then, this template is modeled with a liquid polymer such as silicone rubber, and the polymer is cured and peeled off from the surface. Previously, nano-tip arrays are used as templates for important patterning of a polystyrene surface with 0.87 μm diameter and 3 μm depth holes. Also, PMMA is indented by an AFM probe for data-storage applications or by a micro-fabricated tip array and molded with a metal layer for patterning down to 40 nm metallic contacts.

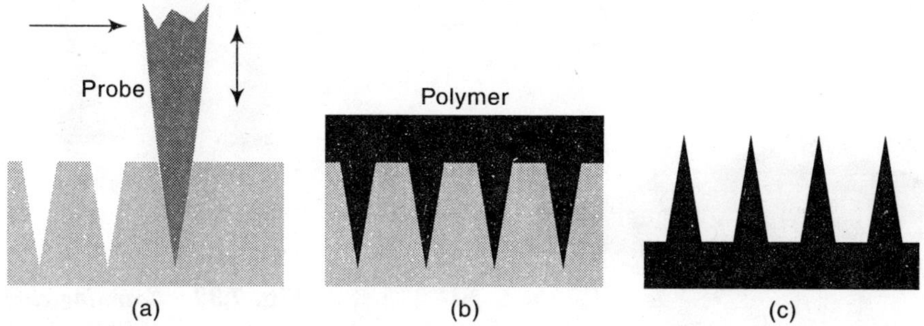

Fig. 7.30 *Synthetic hair fabrication by Method I: (a) Indenting a flat wax surface using a micro/nano-fabricated probe or structure, (b) molding it with a polymer, and (c) separating the polymer form the wax by peeling.*

By this precisely controlled technique, a silicon AFM probe (with around 20-30 nm tip size) and tungsten SATM probe (with around 30-40 nm tip size) are used to indent a flat wax surface. A single probe indenting can give the flexibility of even fabricating oriented hair. By indenting the wax surface as can be seen in Figure 7.30a, the template given in Figure 7.31a was obtained. The profile of the indented wax surface is shown in Figure 7.31b. By molding the wax template with silicone rubber, curing the rubber and separating it from the template by peeling, synthetic nano-bumps of Figure 7.31c are obtained. Moreover a wax surface indented using the STM probe by a step-and-repeat process gave the high aspect ratio micro-bumps shown in Figure 7.32.

This nano-robotic method has the flexibility of fabricating different orientation and non-symmetric nano-structures over almost any surface topography and area. However, it enables prototyping of only certain type of low aspect ratio nano-structures, and it is relatively slow due to its serial control process. Therefore, parallel and higher aspect ratio manufacturing methods are given below.

Method II: Molding Self-organised Nano-pore Membranes

For the second method, a membrane with self-organized high aspect ratio polycarbonate micro/nanopores is used as the master template and molded with a liquid polymer as illustrated

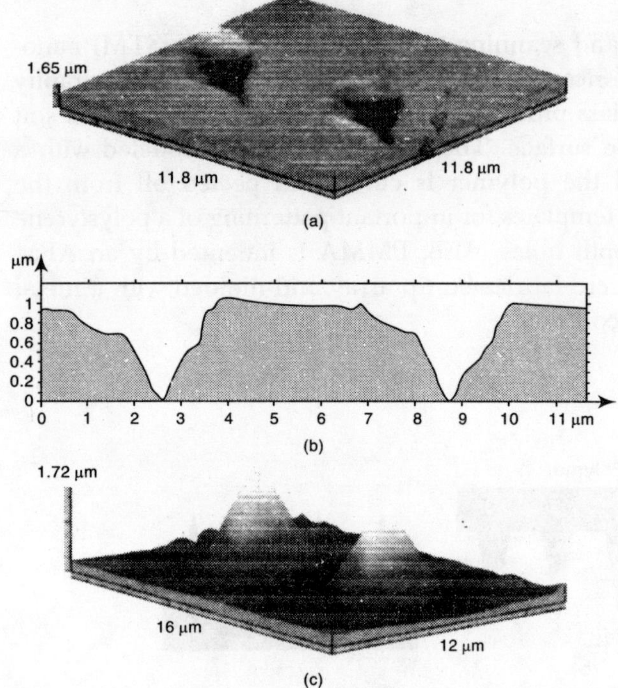

Fig. 7.31 *3-D AFM tapping mode image of (a) the AFM probe based indented flat wax surface, (b) profile of the indented wax surface, and (c) molded and peeled off silicon rubber nano-bumps.*

Fig. 7.32 *Scanning electron microscope (SEM) micro-graph of micro-bumps (around 10 μm × 20 μm size) molded from the STM probe based indented wax surface.*

in Figure 7.33. This membrane can have pore size of 0.02-20 μm, thickness of 5-10 μm, and pore density of 10^5-10^8 pores cm^{-2}. Moreover these membranes have a random orientation of the nano-pores (±15°) created by a nuclear track etch. The SEM micrograph of polycarbonate membrane is displayed in Figure 7.34. 5 and 0.6 μm diameter polycarbonate membranes are molded with polydimethy siloxane (PDMS) under vacuum and the PDMS hairs are peeled off from the membrane after curing at 338 K for 4 hrs. Resulting hairs with 5 and 0.6 μm diameters and 10 and h μm lengths, respectively, are shown in Figure 7.35.

The real gecko foot-hairs are hierarchical integration of micron and nanometer diameter high aspect ratio structures for adapting to micron and nanometer scale surface roughness. For enabling the similar hierarchical geometry, two micron and nanometer size pore membranes are bonded, and molded with the liquid polymer as shown in Fig. 7.36. This method enables high volume, low cost and large area manufacturing of synthetic gecko foot-hairs.

Here, the molded macro/nano-hairs are randomly spaced and oriented on the surface due to the random distribution of the self-organized pores, and regularly spaced and oriented hairs are hard to get.

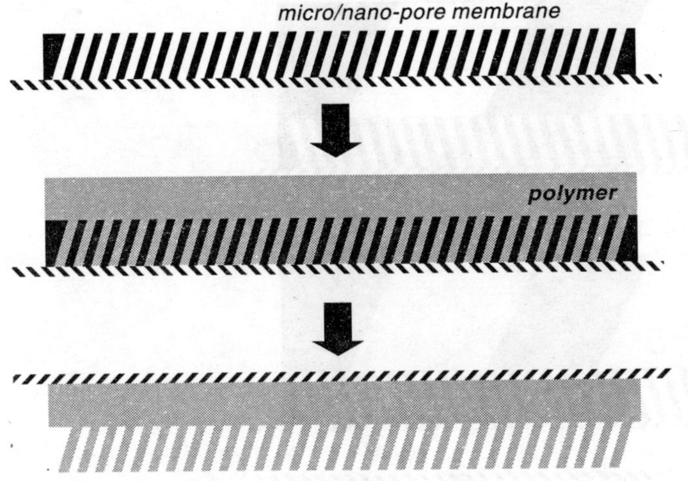

Fig. 7.33 *Molding a micro/nano-pore membrane template: bonding the membrane to a substrate, molding the liquid polymer under vacuum, and peeling off the cured polymer of etching away the membrane.*

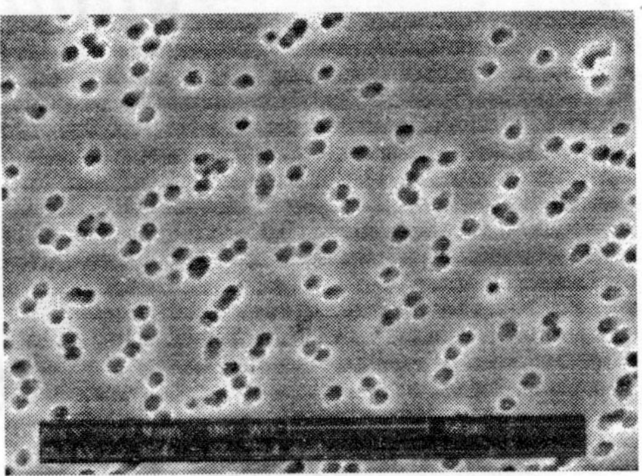

Fig. 7.34 *SEM micrograph of a polycarbonate micro/nano-pore membrane top-view image (5 μm pore size here).*

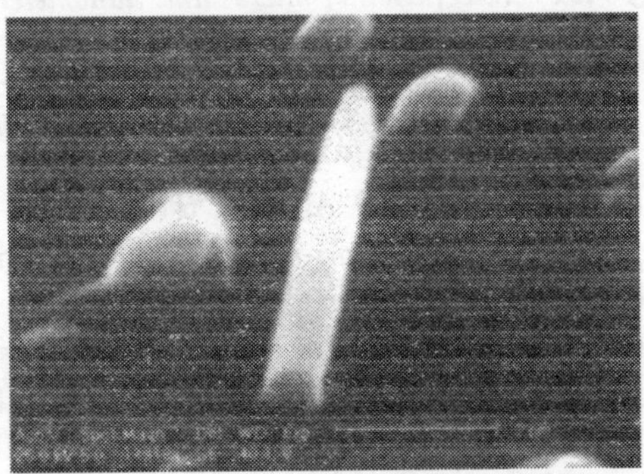

Fig. 7.35 *Side view SEM micrographs of PDMS micro/nano-pillars with 5 μm (upper image, 1:2 aspect ratio) and 0.6 μm (lower image, 1:9 aspect ratio) diameters, respectively fabricated by the Method II.*

Method III: Directed Self-assembly based Micro/Nano-hair Growth

As a possible solution to the regularly spaced and oriented micro/nano-hairs, a directed self-assembly manufacturing technique is proposed. Here, a thin liquid polymer film is coated on a flat conductive substrate, and a closely spaced another metal plate is used to apply a DC electric field on the polymer film. Due to the instabilities on the film, micro/nano-pillars are started to grow until touching to the upper metal plate. By precise timing control using optical microscope feedback, micro/nano-pillars are stopped on the top plate.

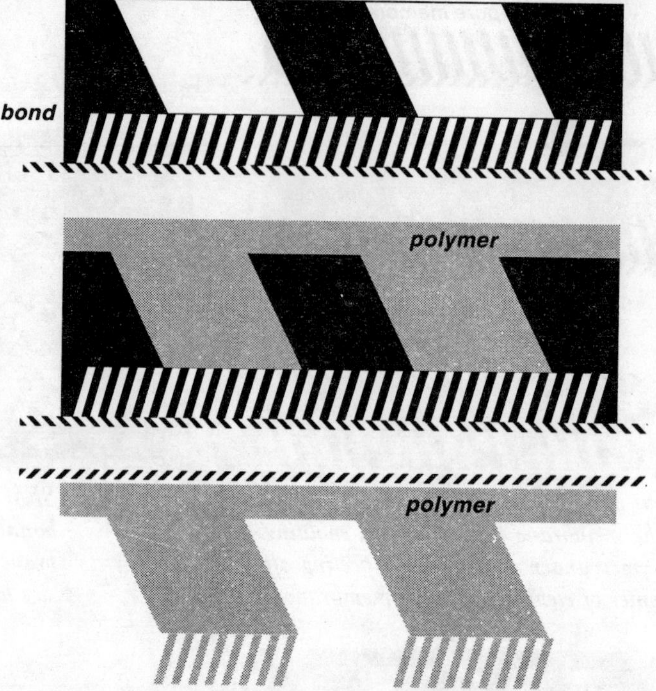

Fig. 7.36 *Integration of micro- and nanometer size structures for manufacturing hierarchical high aspect ratio gecko foot-hairs; bonding the micron and nanometer size pore membranes (uppermost), molding with the liquid polymer through the micro-pore membrane site (middle), and curing the polymer and etching away the both membranes (lowermost).*

As an extension of this technique, the self-organized micro/nano-pillars are stretched in z-direction and sheared along x-direction after soft-baking the polymer in this paper. These tensile and shear forces are applied by moving the upper electrode plate precisely for controlling the expected aspect ratio and orientation of nano-pillars. The growth and stretch/shear schemes are illustrated in Figure 7.37. The desired structures are held by hard-baking the final structures, and peeling the polymer from the upper electrode. Pulling length is limited by the volume of the nano-pillar, surface energy of the polymer, and the pulling speed.

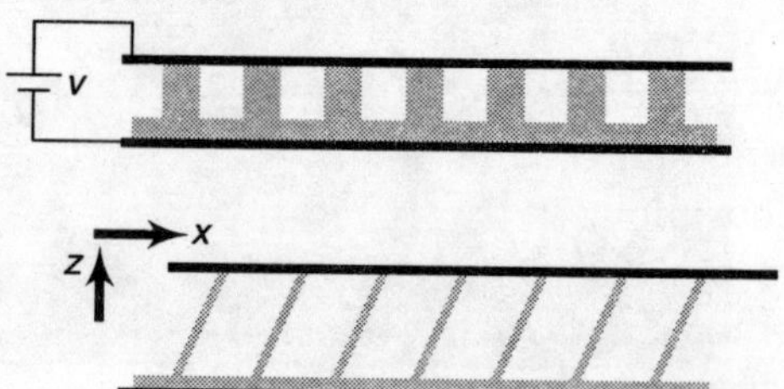

Fig. 7.37 *Directed self-assembly based growing of high aspect ratio micro/nano-hair scheme: (a) Growing low aspect ratio micro/nano-pillars by applying a DC electric field to a thin-film polymer, and (b) stretching and shearing the pillar structure stuck to the upper gold electrode using a precision stage.*

POLYMERS NANOTUBES

Nanotubes are among the most promising structures in nanotechnology. Carbon nanotubes are already finding applications in composite materials requiring high strength-to-weight ratio, and has many more applications.

Nanotubes also can naturally occur in biology. Transport of genetic material from phage viruses to bacteria occur via protein nanotubes, or phospholipid nanotubes between live cells. The use of nanotubes for transport of biological molecules has potentially significant applications in biotechnology. Gold nanotubes in polymer membranes are used to separate small biological molecules on the basis of molecular size, and DNA fragments according to base recognition. The movement of DNA in nanofabricated structures shows that new transport mechanisms occur when spatial confinement in one or more dimensions is less than the radius of gyration. Similar effects also have been observed with DNA fragments passing through multiwalled carbon nanotubes. All of these nanotubular structures have an interior aqueous environment suitable for biological molecules, with the exception of the length of an individual channel which is only on the order of 1 μm.

Extremely long,, water-filled nanotubes can be formed from self-assembling amphiphilic molecules, example, phospholipids can spontaneously form nanotubes under appropriate conditions. Vesicle membranes composed of self-assembled amphiphilic molecules can exhibit both elastic and viscoelastic properties, which, under the application of a localized force, can result in the formation of a nanotube.

The formation of phospholipid nanotubes by using a micropipetted to pull on phosphollpid vesicle (liposome) membranes has been demonstrated and elaborate networks of liposomes interconnected by phospholipid nanotubes have been created, which have demonstrated transport through these networks, optical tweezers are used to create phospholipid nanotubes by pulling directly on membranes of liposomes or by pulling on a microsphere attached to a lipid membrane or by simultaneous hydration and suction driven flow is used to create extremely long phospholipid nanotubes in microfluidic devices.

Phospholipid nanotubes, are transient and tend to collapse unless steps are taken to make them more stable. Attempts to cross-link self-assembled systems composed of lipids (or other low-molecular-weight amphiphiles) have met with limited success. Conversely, supramolecular structures from the self-assembly of higher-molecular-weight amphiphiles, such as block copolymers, are more amenable to stabilization by cross-linking. Polymersomes (polymer vesicles) have been chemically cross-linked to form structures that are stable indefinitely. Worm like micelles of diblock copolymers with diameters as small as 10 nm and lengths exceeding 1μm are cross-linked without any disruption of the cylindrical morphology, spontaneous formation and subsequent cross linking of water-filled nanotubes from the self-assembly of amphiphilic tribloc copolymers has also been achieved. Polymersomes are composed of amphiphilic diblock copolymers, and the nanotubes formed have an aqueous core connected to the aqueous interior of the polymersome. The ability to form networks of polymer nanotubes and polymersomes by optical manipulation is given below:

The aqueous core of the polymer nanotubes together with their robust character makes them interesting candidates for nanofluidics and other applications in biotechnology, water-filled nanotubes are formed directly by pulling or by using optical tweezers (or a micropipette) on bilayer membranes of polymersomes (composed of amphiphilic diblock copolymers). The stretching of nanotubes from polymersomes by using optical tweezers to pull on a microsphere is achieved by attaching as a handle to the diblock copolymer membrane. Thus, generating sufficient forces to form a nantubes by pulling directly on the polymer membrane.

An alternative way is to modify the membrane such that it is easier to pull, which is achieved by the inclusion of surfactants such as detergents or charged lipids. Charged lipids increases the fluidity of lipid membranes such that optical tweezers can easily stretch nanotubes from lipid membranes. Detergent-like surfactants are used to alter the membrane properties of polymersomes, to increase water permeability and susceptibility. Incorporating a detergent-like, triblock copolymer in the formation of the polymersomes, the membrances are made sufficiently fluid to form nanoutbes by pulling directly on the membranes with optical tweezers. Then by subsequently cross-linking the polymers stable and robust nanotubes are which can be removed.

Fig. 7.38 is a sequence of images that shows a polymer nanotube being pulled from the membrane of a giant polymersome using optical tweezers. These pulled nanotubes are up to ≈1 cm in length they retract back to the parent polymersome when the optical trap is switched off.

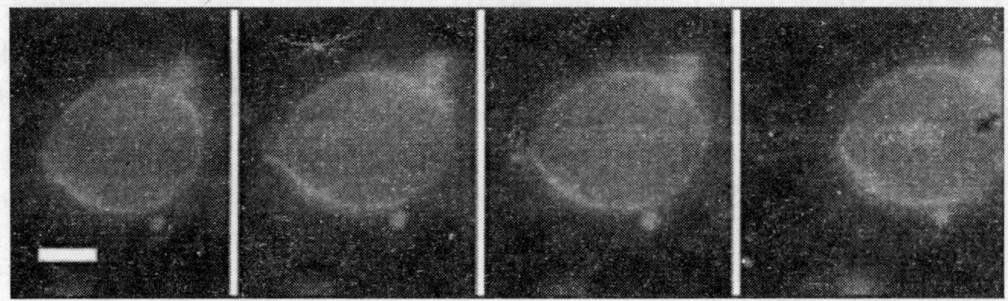

Fig. 7.38 *Video images of the fluorescence from the membrane howing the creation of a nanotubes from a polymersome by directly pulling on the polymer membrane with optical tweezers. The nanotubes are stretched from the membrane at of ≈20μm/s.*

The images of Fig. 7.38 are taken by using video fluorescence microscopy in the polymer membrane. The apparent diameter of the nanotube is ≈250 nm. The retraction of the nanotube when the end of the nanotube is released, is indicative of the system (i.e., transient nature of pulled structures from self-assembled membranes) which wants to relax to the lower energy configuration. This behavior is readily observed for nanotubes pulled from lipid bilayer membranes. Although such nanotubes can be kept form retracting by attaching them to surfaces, the resulting structure is quite fragile and typically stable for only a few hours.

Although there have been attempts to stabilize these pulled lipid nanotubes by imbedding, in the lipid bilayer, macromolecules that contain cross-linkable monomers, this strategy is not successful for small (<200nm) diameter lipid nanotubes. Conversely, polymers can be readily cross-linked to form stable structures.

The pulled polymer nanotubes are stabilized by tree radical polymerization of the hydrophobic butadiene tails by introducing the cross-linking reagents (using a flow-cell device). The nanotubes become cross-linked and stable within 1 min of adding the reagents. Fig. 7.39 is a fluorescence

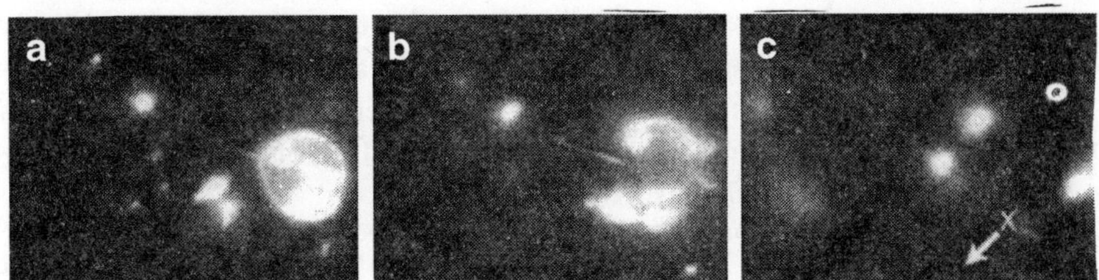

Fig. 7.39 *Video fluorescence microscopy images of a pulled nanotube after cross-linking showing attachment to the parent polymersome (a), severing from the parent polymersome by the application of a short, high-energy UV laser pulse (b), and deflection (in the direction indicated by the arrow) of the rigid, cross-linked polymer nanotube by optical tweezers (c). The optical tweezers is not on in a and b, whereas in c the location of the optical tweezers is indicated by X.*

image of a cross linked polymer nanotubes attached to the parent polymersome. The nanotube is curved because of the fluid flow as the cross linking reagents are introduced. (Straight, cross linked nanotubes are produced by reducing the flow velocity.) A pulsed UV laser (optical) is used to cut the nanotube free from the similarly cross linked, parent polymersome. Fig. 7.39*b* is a fluorescence image of the polymer nanotube after application of UV pulse. For the rigidity of the polymer nanotube; before cross-linking the nanotubes retracts to the parent polymersome (at a rate of $\approx 10 \mu m/s$) but after release of the end of the nanotube from the optical trap, the cross-linked nanotube does not change shape even though its connection to the polymersome is severed. Further evidence of the rigidity of the nanotube is seen in Fig. 7.39c, in which pulling on the end of the severed nanotube with optical tweezers (in the direction indicated by the arrow) displaces the nanotube without changing its shape. The cross-linked, polymer nanotubes are extremely robust, maintaining their shape even after several weeks of the storage at room temperature. In contrast to lipid nanotubes pulled from liposomes, cross-linked polymer nanotubes can also be moved from the chamber in which they are formed to another container with a different buffer.

The rigidity of the nanotube suggests that the entire nanotube is cross-linked; however, the use of the pluronic F-127 may limit the maximum extent of cross-linking relative to a pure copolymer system. In a systematic study of cross linking within worm-like micelles formed from a blend of cross-linkable and noncross-linkable copolymers, percolation behavior (indicating complete cross-linking) is observed for mol fractions of cross-linkable copolymers>0.15. Because only 1.7% (mol fraction approx.) of pluronic F-127 is used in the formation of polymersomes, rigid nanotubes after cross-linking are observed.

Nanotube Formation Using a Micropipette

Polymer nanotubes are pulled using micropipettes, by cross-linking them. The flow is generated by expelling buffer containing the appropriate cross-linking reagents from a micropipette tip into a chamber containing a fairly high density of polymersomes in buffer. This flow causes the polymersomes to move past the micropipette tip, and one of the polymersome membranes sticks to the end of the tip. The flow continues to push the polymersome downstream pulling a nanotube out of the membrane, with the other end attached to the micropipette tip.

Simultaneously, the cross-linking reagent flowing from the tip causes the resulting structure to become rigid over time. The process continues repeatedly as the polymersomes flow past the micropipette tip, generating many rigid and straight polymer nanotubes dangling from the end of the micropipette tip. As a final step, additional cross-linking reagents are added to the chamber to further stabilize all of the structures produced.

Nanotube Morphology and Diameter by Transmission Electron Microscopy (TEM) Imaging

To obtain a better measurement of the diameters and morphologies of polymer nanotubes, image them by TEM. Fig. 7.40 contains TEM images of a sample of cross-linked, polymer nanotubes. Fig. 7.40(a) shows a relatively uniform nanotube and one that is corrugated. A number (≈80%) of cross-linked nanotubes imaged in this sample appeared corrugated. The corrugation appears only after cross-linking and depends on whether the nanotube is under some tension during the cross-linking. It suggests that the corrugation is due to the cross-linking and not due to polymer vesicles fusing together linearly like a string of beads. Corrugation of tubular lipid vesicles is a Rayleigh-type instability arising during front propagation because of a sudden increase in membrane tension. Corrugated patterns observed in cylindrical polymer gels are due to stress associated with the spatial changes in the polymer density across the thickness of the polymer boundary, which limits the diffusion of water. Similarly, periodic, corrugated patterns are observed in stiff polymer films supported on relatively soft elastic media due to a strain-induced elastic buckling instability. Because chemicals are flown in to perform the free radical polymerization reaction, the polymers in the outer layer of the bilayer membrane structure, cross-link before the molecules in the inner layer, giving rise to stresses that result in corrugation of the nanotube.

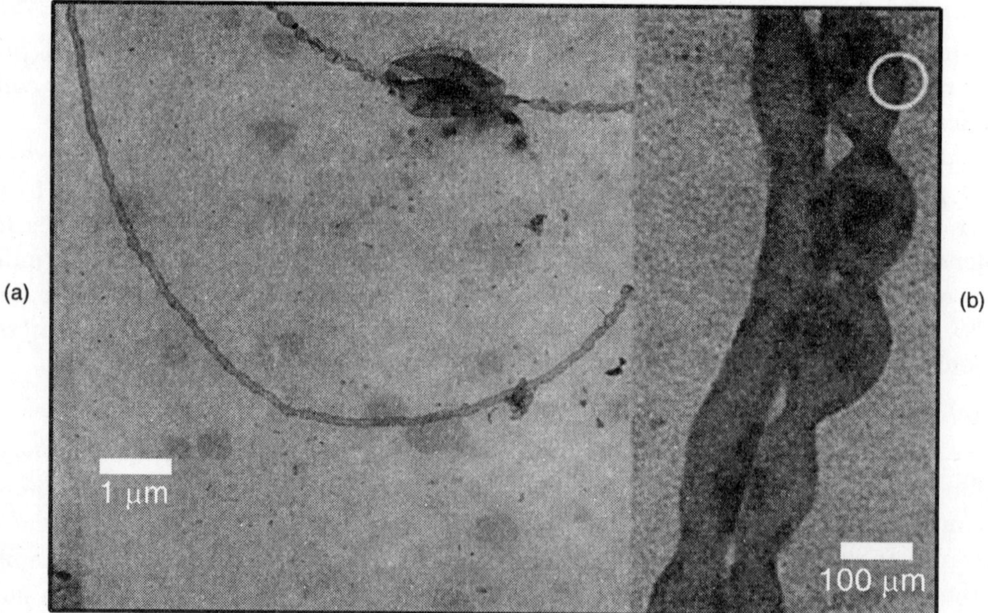

Fig. 7.40 *Low-resolution (Left) and high-resolution (Right) TEM images of the cross-linked polymer nanotubes.*

Also visible in Fig. 7.40(a) is a cross-linked polymersome that appears to have collapsed, possibly during dehydration of the sample. The rigidity of the cross-linked polymer membrane is apparent by the folds in the collapsed polymersome. Fig. 7.40(b) is a magnified view of another pair of nanotubes. Again one of the nanotubes has a relatively uniform profile, whereas the other is highly corrugated. By measuring the outer diameter of the nanotube in this image and using a (unilamellar) membrane thickness of 8 nm, an average inner diameter of ≈70 nm for the uniform nanotube ≈25 nm is obtained, at the constrictions of the corrugated nanotube. Similar diameters are obtained for nanotubes from other magnified TEM images.

Aqueous Core

For application of nanotubes as conduits for transporting biological material, it is necessary that the core of the nanotube contain water. Although our nanotubes are formed from amphiphilic copolymers, the question remains whether the core of the nanotube is aqueous or micellar, because giant worm-like micelles have been observed to form from the diblock copolymer PBd-PEO. To observe the water filled core, nanotubes from polymersomes encapsulating buffer containing the water-soluble, fluorescent dye sulforhodamine B are puled to. Because the permeability of the polymer membrane to water is >1 order of magnitude less than the corresponding permeability for lipid membranes, it is a reasonable assumption that any water in the core of the pulled polymer nanotube comes from the water enclosed by the parent polymersome. Fig. 7.41(b) is a fluorescence image, taken with a laser scanning confocal microscope, of a cross-linked, pulled polymer nanotube and the parent polymersome. The fluorescence from the excited sulforhodamine B dye evident in the parent polymersome is also seen in the pulled nanotube, indicating that buffer containing dye is in the core of the nanotube.

Polymer Nanotube-Vesicle Network

The polymer membranes can be manipulated, with either optical tweezers or micropipettes. Nanotube structures such as Y-junctions, similar to those demonstrated with lipid membranes are formed from Fig. 7.41(a) and 7.41(b) contains a composite image of a network composed of polymer nanotubes and polymersomes. As illustrated in Fig. 7.41(a) *Left*, optical tweezers are used to grab onto the polymersome membrane and stretch out a polymer nanotube. The end of the nanotube, is pulled through the membrane of another polymersome and then released. Once released, the nanotube retracts until it reaches the polymersome membrane through which it was inserted, thus forming a structure composed of two polymersomes joined by a polymer nanotube. The two Y-junctions apparent in Fig. 7.41(b) are formed by pulling an additional nanotube from a polymersome that already has a nanotube attached to it. If the additional nanotube is pulled from the membrane at a location sufficiently close to the attachment point of the first nanotube, the attachment point of the additional nanotube moves along the polymersome membrane toward the attachment point of the first nanotube and then along the length of the first nanotube, thus forming a nanotube Y-junction. (The location of the attachment point of the additional nanotube to the first nanotube is such that all nanotubes emanating from the Y-junction experience only tension and no sheer or bending force.) This process is the same mechanism of forming Y-junctions originally demonstrated with nanotubes pulled from liposome membranes using micropipettes the formation of a Y-junction by directly pulling on the wall of a nanotube to form an additional nanotube is also demonstrated.

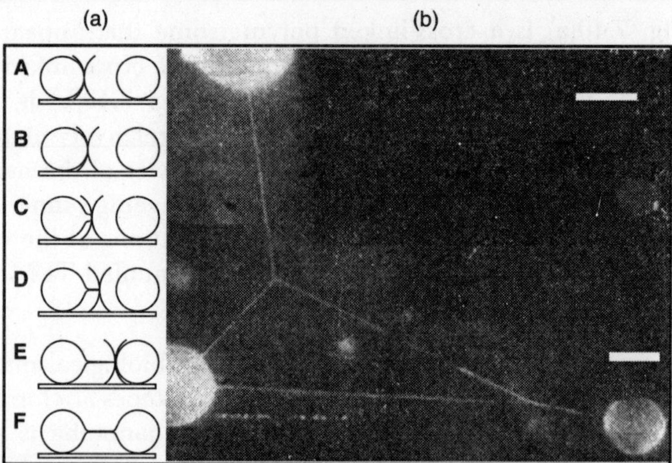

Fig. 7.41 *Creation of polymer nanotube-vesicle networks. (a) Sequence of images illustrating the pulling of a nanotube from the membrane of a polymersome using optical tweezers and the attachment of the nanotube to another polymersome. (b) Composite image from video fluorescence microscopy of a network of polymer nanotubes and polymersomes. Containing the membrane dye DiO-C16, assembled using optical tweezers. (c) (Inset) Scanning confocal microscopy image of a nanotube pulled from a polymersome encapsulating sulforhodamine B dye in buffer. Variations in the intensity are due to movement of the nanotube during the scan.*

Polymersome Preparation

Polymersome are formed by using an electroformation technique. So a mixture of Dibloc copolymer poly (butadiene-*b*-ethylene oxide) PBd-*b*-PEO. at 20 mg ml solution in chloroform, 8 wt % pluronic F-127, and 10m/M 3,3'-dihexadecyloxacarbocyanine perchlorate membrane dye is prepared. The detergent-like triblock copolymer PEOpoly (propylene)-PEO, pluronic F-127, which incorporates into the membranes is added to increase membrane fluidity. F-127 is chosen because very low concentrations (0.06 wt %) can permanently disrupt liposome membranes at room temperature .Because of the structural similarity between diblock PBd-PEO and triblock F-127, polymersome formulations incorporating <10 wt % F-127 stably form vesicles. Fifty microliters of this polymer mixture is applied to ≈1-cm length of two parallel platinum wires (1 mm diameter polymer film on the platinum wires then is hydrated in an electroformation chamber, containing ≈1 ml of 0.3 M sucrose solution, at 10 V and 10 Hz for 18 h. Five hundred microliters of the polymersome suspension is transferred to a centrifuge tube containing 10 ml of buffer (10 mM Mops/50 mM NaC1, pH 7.0) and centrifuged at 672 × g for 5 min. A sample from the bottom of the centrifuge tube is used for experiments. For some of the experiments, 1 m/M sulforhodamine B (Sigma) dissolved in the 0.3 M sucrose solution is used.

Flow Cell and Cross-Linking

The polymersomes and nanotubes are cross linked by free redical polymerization. A mixture of 90 μl of potassium persulfate (12.5 mg ml^{-1}), 8 μ of 9 mg ml^{-1} sodium metabisulfite solution, and 8 μl of 130 mg/300 μl ferrous sulfate heptahydrate solution is applied to the polymer structures by using a flow cell consists of a channel constructed using two strips of double-sided tape that was sandwiched between a microscope slide, with a 1-cm-diameter hole drilled through the center, and a glass coverslip (Fig. 7.42). The 1-cm hole, sealed off on the lower side by another glass coverslip, using silicone high-vacuum grease (Dow-Corning), serves as the sample chamber. The polymersomes, have a density greater than water because of the encapsulated sucrose, eventually (wait 60 min) reside at the bottom of a chamber and stick to the coverslip surface, after which they are readily accessible for optical manipulation and imaging on an inverted microscope. The fluid channel, created by the double-sided tape and coverslip, is open to air at one end and empties into the top portion of the chamber at the other end. Cross-linking reagents are introduced at the open end of the channel and move toward and empty into the chamber region by capillary flow.

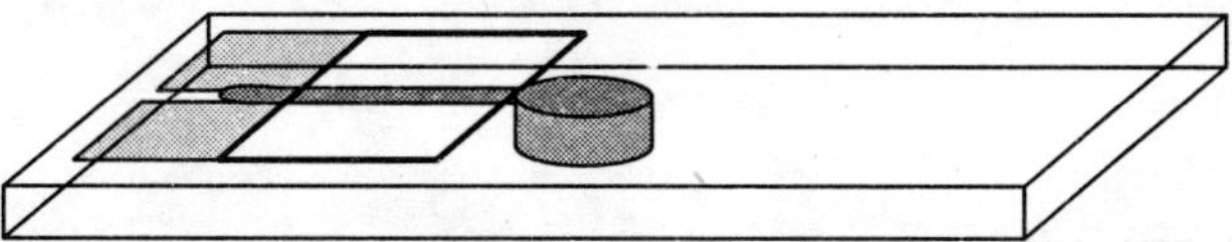

Fig. 7.42 *Illustration of the flow-cell device used for cross-linked polymer nanotubes. The sample chamber consists of a microscope slide with an ≈1 cm diameter hole drilled through the center. The bottom of the chamber is a glass coverslip (not shown) sealed to the microscope slide by silicon grease. The channel was formed by using tow pieces of double-sided tape (each piece is ≈2 cm in length with a thickness of ≈70μm) spaced by 2 mm between the upper surface of the microscope slide and an additional glass coverslip. This coverslip only covered the channel region up to the edge of the sample chamber. The sample chamber was left uncovered to allow for evaporation to pull fluid to be pulled through the channel.*

Although the cross-sectional area of the chamber is much larger than the cross-sectional area in the channel region. The low flow velocity in the chamber region allows the appropriate chemicals to reach the nanotube without destroying it.

Useful Acronyms

μCP	Microcontact printing
aa	amino acid
Ab	antibody
ADMET	Adsorption, distribution, metabolism, excretion and toxicology
AES	Auger electron spectroscopy
AFM	Atomic Force Microscope (Microscopy)
Ag	antigen
ALD	Atomic Layer Deposition
ALE	Atomic Layer Epitaxy
ASIC	Application-specific integrated circuit
ATMP	Advanced Therapy Medicinal Product
ATP	adenosine triphosphate
BBB	blood brain barrier
BCC	Body-centred Cubic (Crystal structure)
BEEM	Ballistic Electron Emission Microscopy
BET	Brunauer, Emmett and Teller method of measuring surface-area
BioNEMS	biofunctionalised nanoelectromechanical systems
BMR	Ballistic Magnetoresistance
BSF	Back surface field
BZ	Brillouin Zone
CARL	Chemically amplified resist lithography
CCD	Charge-coupled device
Cermet	Ceramic-Metal Composite
CFC	Chloroflourocabrons
CMOS	Complementary Metal-Oxide Semiconductor
CMP	Chemical Mechanical Polishing
CNS	central nervous system
CNT	Carbon Nano Tube
CPC	Condensation Particle Counter
CRD	chronic respiratory disease
CT	computer tomography
CVD	cardiovascular disease
CVD	Chemical Vapour Deposition
CW	Continuous wave
CWA	CEN Workshop Agreement
Cz	Czochralski

DBQW	Double-barrier quantum-well
DFB	Distributed feedback (QCL)
DFT	Density Function Theory
DLC	diamond-like carbon
DLTS	Deep level transient spectrocopy
DMA	Differential Mobility Analyzer
DNA	dexoyribonucleic acid
DOF	Depth of focus
DPN	dip-pen nanolithography
DRAM	Dynamic random access memory
DUV	Deep ultraviolet
EBIC	Electron beam induced current
ECL	Emitter-coupled logic
ECM	extracellular matrix
ECR	Electron cyclotron resonance (CVD, plasma etching)
EDP	Ethylene diamine/pyrocatechol
EDX	Energy Dispersive X-ray analysis
EELS	Electron Energy Loss Spectroscopy
EEPROM	Electrically erasable programmable read-only memory
EFA	Field emitter cathode array
EL	Electroluminescence
ESCA	(also called XPS) Electron Spectroscopy for Chemical Analysis
ESEM	Environmental Scanning Electron Microscope
ESR	Electron spin resonance
ESTOR	Electrostatic data storage
Et	Ethyl
EUV	Extreme ultraviolet
EUV	Extreme Ultra-Violet
EUVL	Extreme ultraviolet lithography
EXAFS	Extended X-ray Absorption Fine Structure
Fab	fragment antigen binding (of an antibody molecule)
FCC	Face Centred Cubic (Crystal structure)
FEG-SEM	Field Emission Gun Scanning Electron Microscope
FET	Field-Effect Transistor
FIB	Focused ion beam
FP	Fabry-Perot
FRET	fluorescence resonance energy transfer
FTIR	Fourier transform infrared
FWHM	Full Width, Half Maximum
GDOES	Glow Discharge Optical Emission Spectroscopy
GMR	Giant Magneto Resistance
HBT	Hetero bipolar transistor
HEL	Hot-embossing lightography
HEMT	High electron mobility transistor
HIT	Heterojunction with intrinsic thin layer
HOMO	Highest occupied molecular orbital
HREELS	High-Resolution Electron Energy Loss Spectroscopy
HREM	High-Resolution Electron Microscopy
hTEP	human tissue-engineered product
HVOF	High Velocity Oxygen Fuel
IC	Integrated circuit
ICP	Inductively coupled plasma
IMPATT	Impact ionization avalanche transit time
IPG	In plane gate
IR	Infrared

ITO	Indium-tin-oxide
ITRS	International technology roadmap for semiconductors
Laser	Light amplification by stimulated emission of radiation
LBIC	Light beam induced current
LCD	Liquid crystal display
LDD	Light-emitting diode
LED	Light-emitting diode
LED	Light Emitting Diodes
LEED	Low energy electron diffraction
LISA	Lithographically induced Self-Assembly
LMIS	Liquid metal ion source
LOC	Lab-on-a-chip
LPE	Liquid phase epitaxy
LSS	Lindhardt, Scharff, SchiØtt (Researchers)
LUMO	Lowest unoccupied molecular orbital
M	Metal
MAL	Mould-assisted lithography
MBE	Molecular Beam Epitaxy
MCT	Mercury cadmium telluride
Me	Methyl
MEMS	Micro Electro Mechanical Systems
MEMS	Microfabricated Electro Mechanical Systems
MFH	Magnetic Fluid Hypothermia
MFM	Magnetic Force Microscopy
MIS	Metal-insulator-semiconductor
MMIC	Monolithic microwave integrated circuit
MNT	(Forum)-Micro and nano technologies
MNT	Molecular Nanotechnology
MOCVD	Metallo-organic chemical vapor deposition
MODFET	Modulation-doped field-effect transistor
MOLCAO	Molecular orbitals as linear combinations of atomic orbitals
MOS	Metal-oxide-semiconductor
MOSFET	Metal-oxide-semiconductor field effect transistor
MPU	Microprocessor unit
MQW	Multi wall nanotubes
MRI	Magnetic Force Microscopy
MST	Micro Systems Technology
MWCNT	Multi-Walled Carbon NanoTube
MW	Molecular weight
MWNT	Multi wall nanotubes
NA	Numerical aperture
NAND	Not and
Nd:YAG	Neodymium yttrium aluminum garnet (laser)
NDR	Negative differential resistance
NEMS	Nano Electro Mechanical Systems
NEMS	Nanofabricated ElectroMechanical Systems
NIL	Nanoimprint lithography
NIL	Nanolmprint Lithography
NM	Nuclear Medicine
NMOS	n-Channel metal oxide-semiconductor (transistor)
NMR	Nuclear magnetic resonance
NMR	Nuclear Magnetic Resonance
NOR	Not or
NOx	Nitrogen Oxides
NQR	Nuclear Quadrupole Resonance

NSOM	Near-field Scanning Optical Microscopy
OCL	Quantum cascade laser
OLEDs	Organic Light Emitting Diodes
PADOX	Pattern-dependent oxidation
PCR	Polymerase Chain Reaction
PDMS	Polydimethylsiloxane
PE	Plasma etching
PECVD	Plasma-enhanced chemical vapor deposition
PEEM	PhotoEmission Electron Microscopy
PEG	Polyethylene glycol
PET	Polyethyleneterphthalate
PET	Position Emission Tomography
PL	Photoluminescence
PLAD	Plasma doped
PMMA	Polymethylmethacrylate
PREVAIL	Projection reduction exposure with variable axis immersion lenses
PTFE	Polytetrafluorethylene (Teflon®)
PVD	Physical vapor deposition
QDNs	Quantum Dot Nanotcrystals
QSE	Quantum size effect
QWIP	Quantum well infrared photodetector
RAM	Random access memory
RBS	Rutherford backscattering spectrometry
RCA	Radio Corporation of America (Company)
RF	Radio frequency
RGD	arginine, glycine and aspartic acid motif
RHEED	Reflection high-energy electron diffraction
RIE	Reactive ion etching
RITD	Resonant interband tunneling diode
RLS	Resonance Light Scattering
RNA	Ribonucleic acid
RTA	Rapid thermal annealing
RTBT	Resonant tunneling bipolar transistor
RTD	Resonant tunneling diode
SAMMS	Self Assembled Monolayer on Mesoporous Supports
SAM	Self-Assembled Monolayer
SCALPEL	Scattering with angular limitation projection electron beam lithography
SCM	Scanning Capacitance Microscopy
SCZ	Space charge zone
SECM	Scanning Electro Chemical Microscopy
SEM	Scanning Electron Microscope (Microscopy)
SET	Single Electron Transistor
SFIL	Step and flash imprint lithography
SFM	Scanning Force Microscope
SHT	Single hole transistor
SIA	Semiconductor Industry Association
SICM	Scanning Ion Conductance Microscopy
SIMOX	Separation by implantation of oxygen
SIMS	Secondary Ion Mass Spectrometry
SL	Superlattice
SMD	Surface-mounted device
SMPS	Scanning Mobility Particle Sizer
SNP	Single Nucleotide Polymorphisms
SOI	Silicon on insulator
SOS	Silicon on sapphire

SPM	Scanning Probe Microscope (Microscopy)
SPR	Surface Plasmon Resonance
STM (STEM)	Scanning Transmission Electron Microscope
STM	Scanning Tunnelling Microscope (Microscopy)
SWCNT	Single-Walled Carbon NanoTube
SWNT	Single-Wall NanoTube
TA	Thermal analysis
TED	Transferred electron device
TEM	Transmission Electron Microscope (Microscopy)
TEOS	Tetraethylorthosilicate
TFT	Thin film transistor
TMAH	Tetramethylammonium hydroxide
TOFSIMS	Time of Flight SIMS
TSI	Top surface imaging
TTL	Transistor-transistor logic
TUBEFET	Single carbon nanotube field-effect transistor
UHV	Ultrahigh vacuum
ULSI	Ultra large scale integration
UPS	UltraViolet Photoelectron Spectroscopy
UV	Ultraviolet
VHF	Very high frequency (30–300 MHz; 10–1 m)
VLSI	Very large scale integration
VMT	Velocity-modulated transistor
V-PADOX	Vertical pattern-dependent oxidation
VPE	Vapor phase epitaxy
XOR	Exclusive or
XPS	X-Ray Photoelectron Spectroscopy
XRD	X-ray diffraction
ZME	Zeolite modified electrode
ZVI	zerovalaent ion

Web Addresses

http://www.rpi.edu/dept/materials/COURSE/NANO/crawford
IBM Research | Almaden Research Center | Science and Technology |
ROBERT R. McCormick School of Engineering and Applied Science
http://web.mit.edui/isn/research
http://online.nanopolis.net/viewer.phb?subject.
The Royal Society & The Royal Academy of Engineering
http://www.zurich.ibm.com/images/st/atomic_manipulation/manipulation.gif
http://www.nottingham.ac.uk/physics/research/nano/instrumentation.htm
http://en.wikipedia.org/wiki/Carbon_nanotubes
Functional Nanomaterials
Edited by Kurt E. Geckeler and Edward Rosenberg
Wikipedia, the free encyclopedia
http://www.pnas.org/cgi/content/full/101
http://www.nano.gov/html/facts/whatIsNano.html
Advanced Semiconductor and Organic Nano-Techniques, Parts I, II and (Part III) H. Morkoç (Ed.)
K.B. Blodgett, I. Langmuir, Phys. Rev. 51 : 964–92, 1937.
B. Moore, C.M. Knobler J., Phy. Chem. 94 : 4588–4595, 1990.
http://www.scientific americans.com.
Lucent Technologies (2002)_http://wwwbell_labs.com/project/SCALPEL/mask format.html.
http:/www.research.ibm.com/journal/rd/411/seeger.html.
For further reference please refer to website.
R.R.H. Coombs and D.W. Robinson (Eds), *Nanotechnology in Medicine and Biosciences* (Gordon and Breach, New York, 1996)
A.S. Edelstein and R.C. Cammarate (Eds), *Nanomaterials: Synthesis, Properties and Applications* (Institute of Physics, 1998)

Bibliography

Nanotechnology and Nano Science Nanowork.
The Nanotechnology site.
JPK Instruments AG-Scanning Probe technology nanoHUB.
Physics Education.
Timeline of Carbon nanotubes–Wikipedia, the free encyclopedia.
Nanotechnology–Seeing Nano–Wikibooks.
Stable and Robust polymer nanotubes streched from polymer some–Reiner et al.
Instrumentation and tools for nanotechnology.
Nanostructured materials–Yahoo search.
www.nano engineer–1.com–Nano Engineer–1–
Gallery–SRG III speed reducer gear.
blog–nano Nanoscale materials.

Index